ADSORE

ADSORBENTS: FUNDAMENTALS AND APPLICATIONS

Ralph T. Yang
Dwight F. Benton Professor of Chemical Engineering
University of Michigan

WILEY-
INTERSCIENCE

A JOHN WILEY & SONS, INC., PUBLICATION

Published by John Wiley & Sons, Inc., Hoboken, New Jersey.
Published simultaneously in Canada.

Library of Congress Cataloging-in-Publication Data:

Yang, R. T.
Adsorbents : fundamentals and applications / Ralph T. Yang.
p. cm.
ISBN 0-471-29741-0 (cloth : acid-free paper)
1. Adsorption. I. Title.
TP156.A35Y36 2003
660′.284235—dc21

2003004715

Printed in the United States of America

10 9 8 7 6 5 4 3 2 1

CONTENTS

PREFACE

Since the invention of synthetic zeolites in 1959, innovations in sorbent development and adsorption process cycles have made adsorption a key separations tool in the chemical, petrochemical and pharmaceutical industries. In all future energy and environmental technologies, adsorption will likely play either a key or a limiting role. Some examples are hydrogen storage and CO removal (from hydrogen, to <1 ppm) for fuel cell technology, desulfurization of transportation fuels, and technologies for meeting higher standards on air and water pollutants. These needs cannot be fulfilled by current commercial sorbents.

The past two decades have shown an explosion in the development of new nanoporous materials: mesoporous molecular sieves, zeolites, pillared clays, sol-gel-derived metal oxides, and new carbon materials (carbon molecular sieves, super-activated carbon, activated carbon fibers, carbon nanotubes, and graphite nanofibers). The adsorption properties for most of these new materials remain largely unexplored.

This book provides a single and comprehensive source of knowledge for all commercial and new sorbent materials. It presents the fundamental principles for their syntheses and their adsorption properties as well as their present and potential applications for separation and purification.

Chapter 2 provides a simple formula for calculating the basic forces or potentials for adsorption. Thus, one can compare the adsorption potentials of two different molecules on the same site, or that of the same molecule on two different sites. The calculation of pore size distribution from a single adsorption isotherm is shown in Chapter 4. The effects of pore size and shape on adsorption are discussed in both Chapters 2 and 4. Chapter 3 aims to provide rules for sorbent selection. Sorbent selection is a complex problem because it also depends on the adsorption cycle and the form of sorbent (e.g., granules, powder, or monolith) that are to be used. The attributes sought in a sorbent are capacity, selectivity, regenerability, kinetics, and cost. Hence, Chapter 3 also includes a summary of equilibrium isotherms, diffusion steps, and cyclic processes. Simple sorbent selection criteria are also presented.

The fundamental principles for syntheses/preparation, adsorption properties, and applications of the commercially available sorbents are covered in Chapters 5–7. Mesoporous molecular sieves are discussed, along with zeolites, in Chapter 7.

The sorbent that forms a π-complexation bond with molecules of a targeted component in a mixture is named π-complexation sorbent. The π-complexation bond is a type of weak and reversible chemical bond, the same type that binds oxygen to hemoglobin in our blood. This type of sorbent has been developed in the past decade, largely in the author's laboratory. Because they have shown a tremendous potential for a number of important applications in separation and purification, they are discussed separately in Chapter 8. This chapter also presents their applications for olefin/paraffin separations, olefin purification (by removal of dienes to <1 ppm, separation of CO, as well as aromatics from aliphatics. The particularly promising application of π-complexation sorbents for sulfur removal from transportation fuels (gasoline, diesel, and jet fuels) is discussed in Chapter 10.

Chapter 9 covers carbon nanotubes, pillared clays, and polymeric resins. Polymeric resins are in widespread use for ion exchange, water treatment, and analytical chromatography.

In Chapter 10, sorbents for specific applications in separation and purification are discussed in detail. These include both well-established applications, such as air separation, and potential applications, such as gasoline desulfurization and energy storage (of hydrogen or methane).

In my research on new sorbents and in organizing my thoughts for this book, I have benefited greatly from discussions with a number of researchers in the field, particularly my former students who are now key researchers in industry, as well as my colleagues at SUNY at Buffalo and the University of Michigan.

Thanks are also due to my past and present students and associates, with whom I have had so much pleasure in learning. Finally, I would like to thank Ruby Sowards for her skillful help in the art work and the staff at Wiley for their highly professional editing and publication.

RALPH T. YANG
Ann Arbor, Michigan

1

INTRODUCTORY REMARKS

Separation may be defined as a process that transforms a mixture of substances into two or more products that differ from each other in composition (King, 1980). The process is difficult to achieve because it is the opposite of mixing, a process favored by the second law of thermodynamics. Consequently, the separation steps often account for the major production costs in chemical, petrochemical, and pharmaceutical industries. For many separation processes, the separation is caused by a mass separating agent (King, 1980). The mass separating agent for adsorption is adsorbent, or sorbent. Consequently, the performance of any adsorptive separation or purification process is directly determined by the quality of the sorbent.

Due to the progress made in sorbent and cyclic process developments, adsorption has already become a key separations tool that is used pervasively in industry. Adsorption is usually performed in columns packed with sorbent particles, or fixed-bed adsorbers. The high separating power of chromatography that is achieved in a column is a unique advantage of adsorption as compared with other separation processes. The high separating power is caused by the continuous contact and equilibration between the fluid and sorbent phases. Under conditions free of diffusion limitation, each contact is equivalent to an equilibrium stage or theoretical plate. Usually several hundred to several thousand such equilibrium stages can be achieved within a short column. Thus, adsorption is ideally suited for purification applications as well as difficult separations. Partly because of this unique advantage, adsorption is well-positioned to play a key role in the development of many future energy and environmental technologies. The simulated moving-bed technology is a good example of using adsorption to perform difficult separations, where satisfactory separations are achieved by using sorbents with separation factors as low as 2.

There are only a handful of generic sorbents that are commercially available. These are the sorbents being used in the current adsorption processes.

Adsorbents: Fundamentals and Applications, Edited By Ralph T. Yang
ISBN 0-471-29741-0

Future applications of adsorption are limited by the availability of new and better sorbents. Ideally, the sorbent should be tailored with specific attributes to meet the needs of each specific application. Development of better sorbents can also improve the performance of the current commercial processes. A good example is the invention of the LiX (Si/Al = 1) zeolite (Chao, 1989). Air separation has been performed by pressure swing adsorption, and the generic sorbents 13X (i.e., NaX) and 5A (i.e., CaA) zeolites were used prior to this invention. By switching from NaX to LiX (Si/Al = 1), the productivity of oxygen increased instantly by 1.4–2.7 times and the power consumption reduced by 21–27% depending on the operating conditions used (Leavitt, 1995).

The past two decades have shown an explosion in the developments of new nanoporous materials. Tremendous advances have been made in our capabilities to tailor the porosity and surface chemistry of oxide molecular sieves and new forms of carbon (carbon molecular sieves, super-activated carbon, activated carbon fibers, carbon nanotubes, and graphite nanofibers). However, the potential use of the adsorption properties of these new materials remains largely unexplored.

1.1. EQUILIBRIUM SEPARATION AND KINETIC SEPARATION

The adsorptive separation is achieved by one of three mechanisms: steric, kinetic, or equilibrium effect. The steric effect derives from the molecular sieving properties of zeolites and molecular sieves. In this case only small and properly shaped molecules can diffuse into the adsorbent, whereas other molecules are totally excluded. Kinetic separation is achieved by virtue of the differences in diffusion rates of different molecules. A large majority of processes operate through the equilibrium adsorption of mixture and hence are called equilibrium separation processes.

Steric separation is unique with zeolites and molecular sieves because of the uniform aperture size in the crystalline structure. The two largest applications of steric separation are drying with 3A zeolite and the separation of normal paraffins from iso-paraffins and cyclic hydrocarbons by using 5A zeolite (Yang, 1987). This type of separation is generally treated as equilibrium separation.

Although kinetic separation has had only limited applications, it holds high potentials for many more. It is an option to consider when equilibrium separation is not feasible. Air separation is a good example for which kinetic separation can complement equilibrium separation. Air separation by PSA (i.e., pressure-swing adsorption) using zeolite is based on the preferential adsorption of N_2 over O_2. It is hence used for the production of O_2 from air. N_2 constitutes about 78% of air. If an O_2-selective sorbent is used, air separation can be accomplished with about 1/4 of the work that is needed for the same separation by using zeolite. This is particularly the case with nitrogen production form air. Oxygen diffuses about 30 times faster than nitrogen in carbon molecular sieve. Although the adsorption capacity of carbon molecular sieve is only a fraction of that of zeolite, it is more economical to use carbon molecular sieve for the production of nitrogen from air. Separation of methane from CO_2 has also been performed by kinetic separation with carbon molecular sieve. The feasibility for propane/propylene separations by using $AlPO_4$-14 has been demonstrated (see Chapter 10). The upgrading of natural gas by removal of nitrogen from methane is a large potential application for kinetic separation. This subject will also be discussed in Chapter 10.

For equilibrium separation, the starting point for sorbent design/selection is to examine the fundamental properties of the targeted molecule that is to be adsorbed (compared with the other molecules in the mixture): polarizability, magnetic susceptibility, permanent dipole moment, and quadrupole moment. If the targeted molecule has high polarizability and magnetic susceptibility, but no polarity, carbon with a high surface area would be a good candidate. Sorbents with highly polar surfaces (e.g., activated alumina, silica gel, and zeolites) would be desirable for a targeted molecule that has a high dipole moment (and high polarizability). If the targeted molecule has a high quadrupole moment, sorbents with surfaces that have high electric field gradients are needed. Zeolites are the only such sorbents, as the cations are dispersed above the negatively charged oxides on their surfaces. Cations with high valences (i.e., charges) and small ionic radii would result in strong interactions. The methodology for calculating these interactions is given in Chapter 2 (for all sorbents) and Chapter 7 (for zeolites). The above discussion applies only to the bonding between the targeted molecule and the adsorption site. The targeted molecule also interacts with other atoms on the surfaces of the pore. These interactions are secondary but are also important. Monte Carlo simulation includes pairwise additivity and integrates the interactions over all sites. Sorbent design/selection is a complex problem, because the process for which the sorbent is used needs to be considered at the same time. For purification, particularly ultrapurification, strong adsorption bonds are needed. Strong bond yields high Henry's constant, which leads to ultrahigh product purity. Sorbents that form weak chemical bonds with the targeted molecule can be particularly useful. For this type of sorbents, molecular orbital theory is the most powerful tool for sorbent design, and is discussed in Chapter 8.

For kinetic separation, the pore size needs to be precisely tailored to lie between the kinetic diameters of the two molecules that are to be separated. Many microporous molecular sieves with various pore dimensions have been synthesized (Hartman and Kevan, 1999), which have yet to be used as sorbents.

1.2. COMMERCIAL SORBENTS AND APPLICATIONS

Only four types of generic sorbents have dominated the commercial use of adsorption: activated carbon, zeolites, silica gel, and activated alumina. Estimates of worldwide sales of these sorbents are (Humphrey and Keller, 1997)

Activated carbon	\$1 billion
Zeolites	\$100 million
Silica gel	\$27 million
Activated alumina	\$26 million

Some other reported figures are (according to 2001 demand) zeolites (\$1,070 million), silica gel (\$71 million), activated alumina (\$63 million), and clays (\$16 million) (*Chemical Engineering*, February 2000, p. 59).

Activated carbon has been used as an all-purpose sorbent. It is "hydrophobic." Its precedent, charcoal, was first used in the sugar industry in England in 1794 to decolorize sugar syrup. The major development of activated carbon took place during World War I, for use in filters to remove chemical agents from air. The commercial activated carbon has taken its present form since the 1930's (Jankowska et al., 1991). Silica gel and activated alumina are used mainly as desiccants, although many modified forms are available for special purification applications. Synthetic zeolites, the youngest type among the four, were invented by Milton in 1959 (Milton, 1959). The zeolites that are in commercial use today are mainly the types in Milton's invention, i.e., types A, X, and Y. It is remarkable that most of the \$100 million annual sales of zeolites and the businesses associated with the zeolites are generated by a single invention. Zeolites are used for their special adsorption properties due to their unique surface chemistries and crystalline pore structures. It should be noted, however, that a sizable portion of the commercial zeolites is used for ion exchange and as catalysts.

Polymeric resins are used increasing use in potable water purification, because for some organics they can remove to lower concentration levels than activated carbon does. Acid-treated clays and pillared clays are used for treatments of edible and mineral oils.

Table 1.1 shows examples of commercial applications of these sorbents. Both bulk separation and purification processes are given. Here bulk separation is defined (by Keller, 1983) as having the concentration of the adsorbed component above 10 wt % in the feed. For purification, the concentration of the adsorbed component is generally <2 wt % in the feed. The liquid-phase bulk separations that use the zeolites listed in Table 1.1 are accomplished with the simulated moving bed process. Not included in Table 1.1 are many liquid-phase bioseparations

Table 1.1. Examples of commercial adsorption processes and sorbents used

Separation	Adsorbent
Gas Bulk Separations	
Normal paraffins/isoparaffins, aromatics	Zeolite
N_2/O_2	Zeolite
O_2/N_2	Carbon molecular sieve
CO, CH_4, CO_2, N_2, Ar, NH_3/H_2	Activated carbon followed by zeolite (in layered beds)
Hydrocarbons/vent streams	Activated carbon
H_2O/ethanol	Zeolite (3A)
Chromatographic analytical separations	Wide range of inorganic and polymer resin agents

Table 1.1. (*continued*)

Separation	Adsorbent
Gas Purification	
H_2O/olefin-containing cracked gas, natural gas, air, synthesis gas, etc.	Silica, alumina, zeolite (3A)
CO_2/C_2H_4, natural gas, etc.	Zeolite, carbon molecular sieve
Hydrocarbons, halogenated organics, solvents/vent streams	Activated carbon, silicalite, others
Sulfur compounds/natural gas, hydrogen, liquefied petroleum gas (LPG), etc.	Zeolite, activated alumina
SO_2/vent streams	Zeolite, activated carbon
Odors/air	Silicalite, others
Indoor air pollutants — VOCs	Activated carbon, silicalite, resins
Tank-vent emissions/air or nitrogen	Activated carbon, silicalite
Hg/chlor-alkali cell gas effluent	Zeolite
Liquid Bulk Separations	
Normal paraffins/isoparaffins, aromatics	Zeolite
p-xylene/*o*-xylene, *m*-xylene	Zeolite
Detergent-range olefins/paraffins	Zeolite
p-Diethyl benzene/isomer mixture	Zeolite
Fructose/glucose	Zeolite
Chromatographic analytical separations	Wide range of inorganic, polymer, and affinity agents
Liquid Purifications	
H_2/organics, oxygenated organics, halogenated organics, etc., dehydration	Silica, alumina, zeolite, corn grits
Organics, halogenated organics, oxygenated organics, etc./H_2O — water purification	Activated carbon, silicalite, resins
Inorganics (As, Cd, Cr, Cu, Se, Pb, F, Cl, radionuclides, etc.)/H_2O — water purification	Activated carbon
Odor and taste bodies/H_2O	Activated carbon
Sulfur compounds/organics	Zeolite, alumina, others
Decolorizing petroleum fractions, syrups, vegetable oils, etc.	Activated carbon
Various fermentation products/fermentor effluent	Activated carbon, affinity agents
Drug detoxification in the body	Activated carbon

The components that are to be adsorbed are listed first (from Humphrey and Keller, 1997, with permission, and with minor modification).

and purifications accomplished by chromatography in the pharmaceutical and food industries.

1.3. NEW SORBENTS AND FUTURE APPLICATIONS

In the development of new energy technologies, such as fuel cells, adsorption can play a key enabling role. A breakthrough in sorbent development is needed to solve the critical problem of hydrogen storage for hydrogen fuel cells. The best fuel for fuel cells is gasoline (because of its high-energy density, ready availability, and safety in handling). However, to avoid poisoning of the Pt catalyst in the fuel cell, the sulfur content in gasoline needs to be reduced from the present level of $\sim$350 ppm to $<$1 ppm. These challenges cannot be met with the sorbents that are currently available.

Future needs for a clean environment will lead to increasingly higher standards for air and water pollutants. These challenges require better sorbents that are not commercially available. Traditionally, sorbents were developed based on empiricism. To meet the new challenges, tailored sorbents need to be developed based on fundamental principles. Theoretical tools, such as *ab initio* molecular orbital theory and Monte Carlo simulations can be used to speed up the sorbent design. It is one of the goals of this book to help put sorbent design on a more rational basis.

Some of the most challenging problems in separation and purification that require new sorbents are given in Table 1.2. New sorbents that can solve these problems are also given. Details of these new sorbents are discussed in Chapter 10. Further innovations are needed for meeting these and many more future challenges.

Table 1.2. Some future separation and purification applications by new sorbents

Application	Sorbent and Notes
CH_4 storage for on-board vehicular storage	Super-activated carbon and activated carbon fibers Near or meeting DOE target storage capacity
H_2 storage for on-board vehicular storage	Carbon nanotubes Possible candidate (?)
N_2/CH_4 separation for natural gas upgrading	Clinoptilolite, Sr-ETS-4 by kinetic separation
Sulfur removal from transportation fuels (gasoline, diesel and jet fuels)	π-complexation sorbents such as Cu(I)Y, AgY
CO removal from H_2 to $<$1 ppm for fuel cell applications	π-complexation sorbents such as $CuCl/\gamma$-Al_2O_3, CuY, and AgY
NO_x removal	Fe-Mn-Ti oxides, Fe-Mn-Zr oxides, Cu-Mn oxides
Removal of dienes from olefins (to $<$1 ppm)	π-complexation sorbents such as Cu(I)Y, AgY

Table 1.2. ***(continued)***

Application	Sorbent and Notes
C_3H_6/C_3H_8 (+hydrocarbons) separation	π-complexation sorbents such as $CuCl/\gamma$-Al_2O_3, $AgNO_3/SiO_2$, $AgNO_3$/clays
C_2H_4/C_2H_6 (+hydrocarbons) separation	π-complexation sorbents such as $CuCl/\gamma$-Al_2O_3, $AgNO_3/SiO_2$, $AgNO_3$/clays

Details are given in Chapters 8, 9, and 10.

REFERENCES

Chao, C. C. U.S. Patent 4,859,217 (1989).

Hartman, M. and Kevan, L. (1999) *Chem. Rev.* 99, 935.

Humphrey, J. L. and Keller, G. E., II. (1997) *Separation Process Technology*. McGraw-Hill, New York, NY.

Jankowska, H., Swiatkowski, A., and Choma, J. (1991) *Active Carbon*. Ellis Harwood, New York, NY.

Keller, G. E., II. (1983) *Industrial Gas Separations*. (T. E. Whyte, Jr., C. M. Yon, and E. H. Wagener, eds.). ACS Symp. Ser. No. 223. American Chemical Society, Washington, D.C., p. 145.

King, C. J. (1980) *Separation Processes*, 2nd Ed. McGraw-Hill, New York, NY.

Leavitt, F. W. European Patent EP 461,478 (1995).

Milton, R. M. U.S. Patents 2,882,243 and 2,882,244 (1959).

Yang, R. T. (1987) *Gas Separation by Adsorption Processes*. Butterworth, Boston, MA.

2

FUNDAMENTAL FACTORS FOR DESIGNING ADSORBENT

Selection or synthesis of adsorbents for a target adsorbate molecule is based on the adsorption isotherm. With the availability of high-speed computing, it is now possible to calculate the adsorption isotherms based on: (1) interaction potentials and (2) structure/geometry of the adsorbent. Let us begin with a review of the basic forces between the adsorbent and adsorbate, paying particular attention to adsorbent design.

2.1. POTENTIAL ENERGIES FOR ADSORPTION

Adsorption occurs when the interaction potential energy ϕ is equal to the work done to bring a gas molecule to the adsorbed state. As a first approximation, the adsorbed state is assumed to be at the saturated vapor pressure.

$$-\phi = -\Delta G = \int_{P}^{P_0} V\,dP = RT \ln \frac{P_0}{P} \tag{2.1}$$

where ΔG is the free energy change and P_0 is the saturated vapor pressure. Hence P is the pressure when adsorption occurs for the given ϕ (ϕ is actually the sorbate–sorbate interaction energy on the liquid surface).

The total potential between the adsorbate molecules and the adsorbent is the sum of the total adsorbate–adsorbate and the adsorbate–adsorbent interaction potentials:

$$\phi_{\text{total}} = \phi_{\text{adsorbate-adsorbate}} + \phi_{\text{adsorbate-adsorbent}} \tag{2.2}$$

The adsorbent has only a secondary effect on the adsorbate–adsorbate interaction. For this reason, we will focus our attention on the second term, adsorbate–adsorbent potential, and refer to this term as ϕ.

Adsorbents: Fundamentals and Applications, Edited By Ralph T. Yang
ISBN 0-471-29741-0

The three basic types of contributions to the adsorbate–adsorbent interactions are dispersion, electrostatic, and chemical bond. The latter, chemical bond, has been explored for adsorption only recently. Weak chemical bonds, particularly the broad type of bonds involving π electrons or π-complexation, offer promising possibilities for designing new and highly selective sorbents. The subject of π-complexation sorbents will be discussed separately, in Chapter 8. For physical adsorption, the adsorbate–adsorbent potential is

$$\phi = \phi_D + \phi_R + \phi_{Ind} + \phi_{F\mu} + \phi_{\dot{F}Q} \tag{2.3}$$

where ϕ_D = dispersion energy, ϕ_R = close-range repulsion energy, ϕ_{Ind} = induction energy (interaction between electric field and an induced dipole), $\phi_{F\mu}$ = interaction between electric field (F) and a permanent dipole (μ), $\phi_{\dot{F}Q}$ = interaction between field gradient ($\dot{F}$) and a quadrupole (with quadrupole moment Q).

The first two contributions ($\phi_D + \phi_R$) are "nonspecific" (Barrer, 1978), which are operative in all sorbate–sorbent systems. The last three contributions arise from charges (which create electric fields) on the solid surface. (This is a simplified view, because an adsorbate molecule with a permanent dipole can also induce a dipole in the sorbent if the sorbent is a conductor [Masel, 1996]). For activated carbon, the nonspecific interactions dominate. For metal oxides, zeolites, and ionic solids, the electrostatic interactions often dominate, depending on the adsorbate. For adsorbate with a quadrupole, the net interaction between a uniform field and the quadrupole is zero. However, the quadrupole interacts strongly with the field gradient, thus the term $\phi_{\dot{F}Q}$.

The individual contributions to the total potential have been reviewed and discussed in detail in the literature (Barrer, 1978; Masel, 1996; Razmus and Hall, 1991; Gregg and Sing, 1982; Steele, 1974; Adamson and Gast, 1997; Rigby et al., 1986; Israelachvili, 1992; Young and Crowell, 1962; Ross and Olivier, 1964). Their functional forms are summarized below. All interactions are given between an atom (or a charge) on the surface and the adsorbate molecule.

Dispersion:

$$\phi_D = -\frac{A}{r^6} \tag{2.4}$$

Repulsion:

$$\phi_R = +\frac{B}{r^{12}} \tag{2.5}$$

Field (of an ion) and induced point dipole:

$$\phi_{Ind} = -\frac{1}{2}\alpha F^2 = -\frac{\alpha q^2}{2r^4(4\pi\ \epsilon_0)^2} \tag{2.6}$$

Field (of an ion) and point dipole:

$$\phi_{F\mu} = -F\mu\cos\theta = -\frac{q\mu\cos\theta}{r^2(4\pi\,\epsilon_0)} \tag{2.7}$$

Field gradient $(\dot{F})$ and linear point quadrupole:

$$\phi_{\dot{F}Q} = \frac{1}{2}Q\dot{F} = -\frac{Qq(3\cos^2\theta - 1)}{4r^3(4\pi\,\epsilon_0)} \tag{2.8}$$

where A and B are constants, $\alpha =$ polarizability, $F =$ electric field, $q =$ electronic charge of ion on surface, $\epsilon_0 =$ permittivity of a vacuum, $\mu =$ permanent dipole moment, $\theta =$ angle between the direction of the field or field gradient and the axis of the dipole or linear quadrupole, $Q =$ linear quadrupole moment (+ or −). The important parameter, r, is the distance between the centers of the interacting pair. It can be shown that the field-quadrupole interaction is always zero for all θ.

The dispersion and repulsion interactions form the Lennard–Jones potential (Barrer, 1978; Masel, 1996; Razmus and Hall, 1991; Gregg and Sing, 1982; Steele, 1974; Adamson and Gast, 1997; Rigby, et al., 1986), with an equilibrium distance (r_0) at which point $\phi_D + \phi_R = 0$. This distance is taken as the mean of the van der Waals radii of the interacting pair. Once the attractive, dispersion constant, A, is known, B is readily obtained by setting $d\phi/dr = 0$ at r_0. Hence, $B = A{r_0}^6/2$. Interestingly, at r_0, $\phi_D = -2\phi_R$. The most commonly used expression for calculating A is the Kirkwood–Müller formula:

$$A = \frac{6mc^2\alpha_i\alpha_j}{(\alpha_i/\chi_i) + (\alpha_j/\chi_j)} \tag{2.9}$$

where m is the mass of electron, c is the speed of light, χ is the magnetic susceptibility, and i and j refer to the two interacting atoms or molecules. For ϕ_{Fu} and $\phi_{\dot{F}Q}$, the maximum potentials are obtained when the dipole or quadrupole is arranged linearly with the charge on the surface.

The dispersion potential, Eq. 2.4, was derived by F. London in 1930, starting from Eq. 2.6, and summarized by Adamson and Gast, 1997. The repulsion term, Eq. 2.5, was not rigorously derived. Equation 2.6 can be obtained from $\mu = \alpha F$, where μ is the induced dipole moment and α is, by definition, the polarizability. The derivation of Eqs. 2.7 and 2.8 is straightforward.

2.2. HEAT OF ADSORPTION

In 2.1, we summarized the different contributions to the potential energy for the interactions between an adsorbate molecule (or atom) and an atom on the solid surface. Pairwise additivity is generally assumed when calculating the interaction energy between the adsorbate molecule and all atoms on the surface. The task is then to add the interactions, pairwise, with all atoms on the surface, by integration.

It can be shown (Barrer, 1978; Ross and Olivier, 1964) that the isosteric heat of adsorption (ΔH) at low coverage is related to the sorbate–sorbent interaction potential (ϕ) by

$$\Delta H = \phi - RT + F(T) \tag{2.10}$$

where $F(T)$ arises due to the vibrational and translational energies of the adsorbate molecule, and for monatomic classical oscillators, $F(T) = 3RT/2$ (Barrer, 1978). For ambient temperature, $\Delta H \approx \phi$.

2.3. EFFECTS OF ADSORBATE PROPERTIES ON ADSORPTION: POLARIZABILITY (α), DIPOLE MOMENT (μ), AND QUADRUPOLE MOMENT (Q)

For a given sorbent, the sorbate–sorbent interaction potential depends on the properties of the sorbate. Among the five different types of interactions, the nonspecific interactions, ϕ_D and ϕ_R, are nonelectrostatic. The most important property that determines these interactions (and also ϕ_{Ind}) is the polarizability, α. On a surface without charges, such as graphite, $\phi_{Ind} = 0$. The value of α generally increases with the molecular weight because more electrons are available for polarization. From the expressions for ϕ_D, ϕ_R, and ϕ_{Ind}, it is seen that these energies are nearly proportional to α. The dispersion energy also increases with the magnetic susceptibility, χ, but not as strongly as α.

Table 2.1 summarizes interaction energies for a number of sorbate–sorbent pairs. Here, groupings are made for the theoretical nonelectrostatic $(\phi_D + \phi_R + \phi_{Ind})$ and the electrostatic $(\phi_{F\mu} + \phi_{\dot{F}Q})$ energies.

The nonelectrostatic energies depend directly on the polarizability of the sorbate molecule; χ makes a contribution to the dispersion energy, and χ also increases with molecular weight.

Two types of sorbents are included in Table 2.1, one without electric charges on the surface (graphitized carbon) and one with charges (three zeolites). On carbon, dispersion energy dominates. On zeolites, the permanent dipole and quadrupole can make significant contributions toward, and indeed can dominate, the total energy. N_2 has a moderately strong quadrupole but no permanent dipole, hence $\phi_{F\mu} = 0$. From Table 2.1, it is seen that $\Phi_{\dot{F}Q}$ accounts for about 1/3 of the energies on chabazite and Na-Mordenite. Na-X zeolite contains more Na^+ ions because its Si/Al ratio is lower than the other two zeolites. Consequently $\phi_{\dot{F}Q}$ contributes about 1/2 of the interaction energies for N_2 on Na-X. The other sorbate molecules included in Table 2.1 both have strong dipoles and quadrupoles (except H_2O, which has a strong dipole only). For adsorption of these molecules on zeolites, the $(\phi_{F\mu} + \phi_{\dot{F}Q})$ interactions clearly dominate.

A comparison of N_2 and O_2 holds particular interest for the application of air separation. Both molecules are nonpolar and have very similar polarizabilities and magnetic susceptibilities. However, their quadrupole moments differ by nearly

Table 2.1. Contributions (theoretical) to initial (near zero loading) heat of adsorption

Sorbent	Sorbate*	$\alpha \times 10^{24}$ cm^3/molec.	$-\Delta H$	$-(\phi_D + \phi_R + \phi_{Ind})$**	$-(\phi_{F\mu} + \phi_{\dot{F}Q})$
Graphitized Carbon	Ne	0.396	0.74	0.73	0
	Ar	1.63	2.12	1.84	0
	Kr	2.48	2.8	2.48	0
	Xe	4.04	3.7	3.1	0
Chabazite	N_2	1.74	8.98	6.45	2.55
	N_2O	3.03	15.3	9.07	6.18
	NH_3	2.2	31.5	7.5	23.8
Na-Mordenite	N_2	1.74	7.0	4.5	2.50
	CO_2	2.91	15.7	6.73	8.93
Na-X	N_2	1.74	6.5	3.10	3.4
	CO_2	2.91	12.2	4.20	7.98
	NH_3	2.2	17.9	3.75	14.2
	H_2O	1.45	≈33.9	2.65	≈31.3

*Permanent dipole moments (μ, debye): $N_2O = 0.161$, $NH_3 = 1.47$, $H_2O = 1.84$, all others = 0. Quadrupole moments (Q, erg$^{1/2}$ cm$^{5/2}$ $\times 10^{26}$): $N_2 = -1.5$, $N_2O = -3.0$, $NH_3 = -1.0$, $CO_2 = -4.3$, all others ≈0.

**For graphitized carbon, $\phi_{Ind} = 0$.

Experimental, $-\Delta H$, kcal/mol (Barrer, 1978; Ross and Olivier, 1964).

a factor of 4 ($Q = -0.4$ esu for O_2 and -1.5 esu for N_2). As a result, the adsorption isotherms of N_2 and O_2 on carbon are similar, whereas the isotherm of N_2 is much higher than that of O_2 on zeolites. The contribution of interaction between the field gradient and the quadrupole moment of N_2 accounts for about 1/2 of the total energy for N_2 adsorption on Na-X zeolite, as mentioned above. The $\phi_{\dot{F}\mu}$ energy for O_2 is approximately 1/4 of that for N_2 (see Eq. 2.8).

2.4. BASIC CONSIDERATIONS FOR SORBENT DESIGN

2.4.1. Polarizability (α), Electronic Charge (q), and van der Waals Radius (r)

For van der Waals (dispersion) interactions, the polarizabilities of the sorbate molecule and the atoms on the sorbent surface are both important (see Eq. 2.9). In electrostatic interactions, for a given sorbate molecule, the charges and van der Waals radii of the surface atoms are important. The roles of these parameters are discussed separately.

For a given sorbate molecule, its dispersion interaction potential with a surface atom increases with the polarizability of that surface atom. The polarizability increases with atomic weight for elements in the same family, and decreases with increasing atomic weight for elements in the same row of the periodic table as the outer-shell orbitals are being increasingly filled. The polarizabilities of

Table 2.2. Polarizabilities (α) of ground state atoms and ions (in 10^{-24} cm^3)

Atom	α	Atom	α	Atom	α
C	1.76	K	43.4	Co	7.5
N	1.10	Rb	47.3	Ni	6.8
O	0.802	Cs	59.6		
F	0.557	Mg	10.6	Li^+	0.029
S	2.90	Ca	22.8	Na^+	0.180
Cl	2.18	Sr	27.6	K^+	0.840
Br	3.05	Ba	39.7	Ca^{2+}	0.471
I	5.35	Al	6.8	Sr^{2+}	0.863
		Si	5.38	Ba^{2+}	1.560
Li	24.3	Fe	8.4		
Na	24.08				

selected atoms are given in Table 2.2. It can be seen that the alkali and alkaline earth metal atoms have very high polarizabilities. Hence these elements, when present on the surface, can cause high dispersion potentials. When these elements are present as cations, however, the polarizabilities are drastically reduced. The polarizabilities of selected cations are also included in Table 2.2 for comparison.

For electrostatic interactions, the charges (q) and the van der Waals radii of the surface atoms (or ions) are most important. For ionic solids with point charges distributed on the surface, the positive and negative fields can partially offset when spaced closely. However, anions are normally bigger than cations. Consequently, the surface has a negative electric field. All electrostatic interaction potentials are proportional to q ($\Phi_{F\mu}$ and $\Phi_{\dot{F}Q}$) or q^2 (Φ_{Ind}) and are inversely proportional to r^n (where $n = 2$–4, see Eqs. 2–8). Here, r is the distance between the centers of the interacting pair, which should be the sum of the van der Waals radii of the two interacting atoms. Hence, the van der Waals radii of the ions on the surface are important. The strong effects of charge (q) and ionic radius of the cation on the adsorption properties of ion-exchanged zeolites will be discussed in Chapter 7.

Because the ionic radius determines the distance r, it has a strong effect on the electrostatic interactions. The ionic radii of selected cations are given in Table 2.3. The ionic radius is a crucially important factor when considering ion-exchanged zeolites and molecular sieves as sorbents.

2.4.2. Pore Size and Geometry

The potentials discussed above are those between two molecules/atoms. The interactions between a molecule and a flat solid surface are greater because the molecule interacts with all adjacent atoms on the surface, and these interactions are assumed pairwise additive. When a molecule is placed between two flat surfaces, i.e., in a slit-shaped pore, it interacts with both surfaces, and the potentials on the two surfaces overlap. The extent of the overlap depends on

Table 2.3. Ionic Radii, r_i (Å)

Ion	r_i	Ion	r_i
Li^+	0.68	Al^{3+}	0.51
Na^+	0.97	Ce^{3+}	1.03
K^+	1.33	Cu^{+1}	0.96
Rb^+	1.47	Cu^{2+}	0.72
Cs^+	1.67	Ag^+	1.26
Mg^{2+}	0.66	Ag^{2+}	0.89
Ca^{2+}	0.99	Au^{+1}	1.37
Sr^{2+}	1.12	Ni^{2+}	0.69
Ba^{2+}	1.34	Ni^{3+}	0.62

Table 2.4. Theoretical threshold pressure for adsorption in different pore sizes and shapes

Pore Size (Å)	P/P_0 for Slit-Shaped	P/P_0 for Cylindrical Shape	P/P_0 for Spherical Shape
4	6.3×10^{-7}	1.3×10^{-12}	3.2×10^{-51}
5	9.1×10^{-6}	2.9×10^{-10}	1.1×10^{-42}
6	3.5×10^{-5}	8.3×10^{-9}	2.5×10^{-36}
7	1.2×10^{-4}	6.5×10^{-8}	6.2×10^{-32}
9	6.1×10^{-4}	3.5×10^{-6}	3.1×10^{-24}
12	2.6×10^{-3}	2.3×10^{-5}	1.2×10^{-20}
15	6.1×10^{-3}	3.2×10^{-4}	1.7×10^{-16}
20	1.4×10^{-2}	1.2×10^{-3}	6.1×10^{-13}

N_2 on carbon at 77 K. $P_0 = 1$ atm.

the pore size. For cylindrical and spherical pores, the potentials are still greater because more surface atoms interact with the adsorbate molecule.

The effects of the pore size and pore geometry are best illustrated by Table 2.4. Table 2.4 lists the threshold pressure for adsorption in different pore sizes and geometries for N_2 on carbon. The calculation was based on the Horvath–Kawazoe (HK) model (Horvath and Kawazoe, 1983), using the corrected version by Rege and Yang (2000). The corrected HK model has been shown to give pore dimensions from N_2 isotherms that agreed well with the actual pore dimension for a number of materials, including carbon and zeolites (Rege and Yang, 2000). The model is based on equating the work done for adsorption (Eq. 2.1) to the total sorbate–sorbent and sorbate–sorbate interactions. The sorbate–sorbent interactions are the sum over all sorbent surface atoms using the Lennard–Jones potentials. A detailed discussion of the HK models, as well as other models, are given in Chapter 4 for calculating pore size distribution from a single isotherm. The results in Table 2.4 exhibit the remarkable attraction forces acting on the adsorbate molecule due to the overlapping potentials from the surrounding walls. The same carbon atom density on the surface was assumed for all geometries, i.e. 3.7×10^{15} 1/cm^2. The experimental data on two molecular sieve carbons agreed with predictions for slit-shaped pores. Scarce or no experimental data are available for cylindrical pores and spherical pores of carbon. Data on these shapes may become available with the availability of carbon nanotubes and fullerenes (if an opening to the fullerene can be made).

As expected, the total interaction energies depend strongly on the van der Waals radii (of both sorbate and sorbent atoms) and the surface atom densities. This is true for both HK type models (Saito and Foley, 1991; Cheng and Yang, 1994) and more detailed statistical thermodynamics (or molecular simulation) approaches (such as Monte Carlo and Density Functional Theory). By knowing the interaction potential, molecular simulation techniques enable the calculation of adsorption isotherms (see for example, Razmus and Hall, 1991; Cracknell et al., 1995; Barton et al. 1999).

NOTATION

A	constant in the 6–12 potential
B	constant in the 6–12 potential; dispersion constant
c	speed of light
C	dispersion constant; average number of sorbate molecules per cage in zeolite
E	interaction energy
F	electric field strength
G	Gibbs free energy
h	Planck constant
H	enthalpy
k	Boltzmann constant
m	mass of electron
P	total pressure
P_0	saturation vapor pressure
q	electronic charge of ion or surface
Q	heat of adsorption; linear quadrupole moment
r	distance between centers of pair; pore radius
r_i	ionic radius
R	gas constant
T	temperature
V	molar volume
α	polarizability
γ	activity coefficient
ε	potential energy field over surface; emittivity

θ angle between field and dipole; fractional surface coverage;
μ permanent dipole moment; chemical potential
ρ density
ϕ potential energy function
χ magnetic susceptibility
ω lateral interaction energy

REFERENCES

Adamson, A. W. and Gast, A. P. (1997) *Physical Chemistry of Surfaces*, 6th Ed. Wiley, New York, NY.

Barrer, R. M. (1978) *Zeolites and Clay Minerals*. Academic Press, New York, NY.

Barton, T. J., Bull, L. M., Klemperer, W. G., Loy, D. A., McEnaney, B., Misono, M., Monson, P. A., Pez, G., Sherer, G. W., Vartuli, J. A., and Yaghi, O. M. (1999) *Chem. Mater.* 11, 2633.

Cheng, L. S. and Yang, R. T. (1994) *Chem. Eng. Sci.* 49, 2599.

Cracknell, R. E., Gubbins, K. E., Maddox, M., and Nicholson, D. (1995) *Acc. Chem. Res.* 28, 281.

Gregg, S. J. and Sing, K. S. W. (1982) *Adsorption, Surface Area and Porosity*, 2nd Ed. Academic Press, New York, NY.

Horvath, G. and Kawazoe, K. (1983) *J. Chem. Eng. Japan* 16, 470.

Israelachvili, J. (1992) *Intermolecular and Surface Forces*, 2nd Ed. Academic Press, San Diego, CA.

London, F. (1930) *Z. Phys. Chem.* B11, 222.

Masel, R. I. (1996) *Principles of Adsorption and Reaction on Solid Surfaces*. Wiley, New York, NY.

Razmus, D. M. and Hall, C. K. (1991) *AIChE J.* 37, 769.

Rege, S. U. and Yang, R. T. (2000) *AIChE J.* 46, 734.

Rigby, M., Smith, E. B., Wakeham, W. A., and Maitland, G. C. (1986) *The Forces Between Molecules*. Oxford University Press, New York, NY.

Ross, S. and Olivier, J. R. (1964) *On Physical Adsorption*. Wiley, New York, NY.

Saito, A. and Foley, H. C. (1991) *AIChE J.* 37, 429.

Steele, W. A. (1974) *The Interaction of Gases with Solid Surfaces*. Pergamon Press, New York, NY.

Young, D. M. and Crowell, A. D. (1962) *Physical Adsorption of Gases*. Butterworth, London.

3

SORBENT SELECTION: EQUILIBRIUM ISOTHERMS, DIFFUSION, CYCLIC PROCESSES, AND SORBENT SELECTION CRITERIA

The selection of a proper sorbent for a given separation is a complex problem. The predominant scientific basis for sorbent selection is the equilibrium isotherm. Diffusion rate is generally secondary in importance. The equilibrium isotherms of all constituents in the gas mixture, in the pressure and temperature range of operation, must be considered. As a first and oversimplified approximation, the pure-gas isotherms may be considered additive to yield the adsorption from a mixture. Models and theories for calculating mixed gas adsorption (Yang, 1987) should be used to provide better estimates for equilibrium adsorption. Based on the isotherms, the following factors that are important to the design of the separation process can be estimated:

1. Capacity of the sorbent, in the operating temperature and pressure range.
2. The method of sorbent regeneration — for example, temperature or pressure swing — and the magnitude of the required swing.
3. The length of the unusable (or unused) bed (LUB).
4. The product purities.

The LUB is approximately one-half the span of the concentration wavefront, or the mass transfer zone. The LUB is primarily determined by the equilibrium isotherm (Yang, 1987). A sharp concentration front, or a short LUB, is desired because it results in a high sorbent productivity as well as a high product purity.

Adsorbents: Fundamentals and Applications, Edited By Ralph T. Yang
ISBN 0-471-29741-0

Consideration should also be given to other factors. Activated carbon is the only major commercial sorbent used for wet gas stream processing. (A pre-dryer is required for other sorbents.) Sorbent deactivation, primarily by coke deposition, is an important consideration in the processing of hydrocarbon containing gases. Coke is formed catalytically, and zeolites are excellent catalysts for these reactions due to their acidities. Pore-size distribution can play a role in the LUB, but not as important as the equilibrium isotherm, since the commercial sorbent pellets are designed to minimize the pore-diffusion resistance. Kinetic separation, that is, separation based on the difference between pore diffusivities of two gases, has found only one major application: the production of nitrogen from air by molecular sieve carbon. Dehydration of cracked gases with 3A zeolite and the separation of normal and iso-paraffins with 5A zeolite are based on selective molecular exclusion. All other commercial processes are based on the equilibrium isotherms. Temperature for activation and regeneration of the sorbent should also be considered. A high temperature (in the 300 °C range) is required for zeolites, whereas activated carbon usually requires the lowest temperature for regeneration.

The total void space in the bed, which varies with the sorbents, is also an important factor. A low void space is desired for high product recoveries because the gas mixture remaining in the void space of the saturated bed is usually not recovered as a useful product. Silica gel and activated alumina have the lowest void fractions, usually slightly below 70%. Activated carbon has the highest void fraction, at nearly 80%.

Bulk separation refers to separation of a mixture that contains over approximately 10% in concentration for the component to be adsorbed. Sorbent selection depends on the nature of the separation (i.e., bulk separation vs. purification) as well as the process by which the separation will be accomplished (i.e., pressure swing vs. temperature swing). As mentioned, the most important basis for sorbent selection is the equilibrium isotherm, while diffusivity is a secondary factor for consideration. A brief summary of the equilibrium isotherms and diffusivities will be given first. Because extensive reviews on isotherms and diffusion are available elsewhere (Yang, 1987; Do, 1998), the summary here will only cover those which are directly relevant to the discussion that follows.

3.1. EQUILIBRIUM ISOTHERMS AND DIFFUSION

3.1.1. Langmuir Isotherms for Single and Mixed Gases

The Langmuir-type isotherms remain to be the most widely used for practical applications. The Langmuir isotherm for pure component adsorption is

$$\theta\left(=\frac{q}{q_m}\right)=\frac{BP}{1+BP} \tag{3.1}$$

$$B=\frac{\alpha}{\beta(2\pi\ mkT)^{1/2}}e^{Q/RT} \tag{3.1a}$$

The parameter α is the sticking probability or accommodation coefficient for adsorption (upon a collision on the surface), and β is the rate constant for desorption. The constant B is called Langmuir constant. At low pressures, the Langmuir isotherm reduces to a linear form, or Henry's law form:

$$q = KP \text{ and } K = Bq_m \tag{3.1b}$$

where K is Henry's constant.

All isotherms should reduce to the Henry's law form at extreme dilution. The Henry's constant is the most important factor for purification. From Eq. 3.1a, B is exponentially proportional to Q, the heat of adsorption. Here $Q(Q = -\Delta H)$ is positive by definition, because ΔH, the enthalpy change for physical adsorption, must be negative (Yang, 1987). For physical adsorption, Q is equal to the bond energy between the adsorbate molecule and the sorbent. Therefore, the bond energy is critical for purification. Strong bonds are needed for ultrapurification.

Zeldowitsch (1934), assuming an exponentially decaying function of site density with respect to Q, obtained the classical empirical isotherm:

$$q = KP^{1/n} \tag{3.2}$$

which is known as Freundlich isotherm. To avoid indefinite increase in adsorption with pressure, the following isotherm is consequently proposed (Sipps, 1948; 1950; Yang, 1987):

$$\theta\left(\frac{q}{q_m}\right) = \frac{BP^{1/n}}{1 + BP^{1/n}} \tag{3.3}$$

which is the Langmuir–Freundlich isotherm. This isotherm can be derived from the Langmuir isotherm by assuming each sorbate molecule occupies n sites (Yang, 1987). It can also be considered as the Langmuir isotherm on nonuniform surfaces.

The Langmuir isotherm for pure-component adsorption can readily be extended to an n-component mixture, known as the extended Langmuir isotherm (Yang, 1987):

$$q_i = \frac{q_{mi} B_i P_i}{1 + \sum_{j=1}^{n} B_j P_j} \tag{3.4}$$

In Eq. 3.4, the monolayer amount for component i in the mixed adsorbed phase remains the same as that in pure component adsorption. It is assumed that each species maintains its own molecular area (the area covered by one molecule that is not influenced by the presence of other species on the surface). This does not meet the requirement of thermodynamic consistency (which requires all monolayer amounts be equal). An empirical mixing rule is available for dealing with this problem (Yang, 1987). Nonetheless, Eq. 3.4 remains useful for practical design purposes.

As in the extended Langmuir equation for mixtures, the hybrid Langmuir–Freundlich equation (Eq. 3.3) can also be extended to an n-component mixture (Yon and Turnock, 1971):

$$q_i = \frac{q_{mi} B_i P_i^{1/n_i}}{1 + \sum_{j=1}^{n} B_j P_j^{1/n_j}} \tag{3.5}$$

This equation is referred to as loading ratio correlation (LRC), and has been very useful for practical design and process simulation.

3.1.2. Potential Theory Isotherms for Single and Mixed Gases

The isotherms derived from the potential theory have found utility in interpreting adsorption by capillary condensation, or pore filling. Thus they are especially useful for adsorption on microporous materials such as activated carbon. However, because the characteristic curve, to be described later, is assumed to be independent of temperature, which applies to adsorption by the temperature-independent dispersion forces, the resulting isotherms are applicable only to relatively nonpolar surfaces. The theory, nonetheless, is general in that it encompasses multilayer adsorption on energetically nonuniform surfaces.

The potential theory is empirical. It assumes, by Polanyi in 1914 (Yang, 1987), that the cumulated volume of the adsorbed space, V, is a function of the potential, ε:

$$\varepsilon = \Delta F = \int_{P}^{P_0} v dP = RT \ln \frac{P_0}{P} \tag{3.6}$$

where ΔF is the free energy change upon adsorption and P_0 is the saturated vapor pressure. The volume in the adsorbed space is

$$V = nV_m \tag{3.7}$$

where n = number of moles adsorbed per unit mass of sorbent and V_m = molar volume of adsorbate.

Dubinin (1960) assumed the following empirical form for the adsorbed amount:

$$V = V_0 \exp\left(-k\frac{\varepsilon^2}{\beta^2}\right) \tag{3.8}$$

where V_0 is the limiting volume of the adsorbed space, which equals micropore volume, and β is the affinity coefficient characterizing the polarizability of the adsorbate. Eq. 3.8 is referred to as the Dubinin–Radushkevich (or D–R) equation. The D–R equation can be recast into:

$$V = V_0 \exp\left[-\left(C \ln \frac{P_0}{P}\right)^2\right] \tag{3.9}$$

where

$$C = \frac{RT}{\beta E_0} \tag{3.10}$$

where E_0 is the characteristic energy of adsorption of a standard vapor (usually benzene). A theoretical basis has been given for the D–R equation (Chen and Yang, 1994), by a simple statistical mechanical derivation assuming a mean field that is related to the characteristic energy and some simplifying manipulations. This mean field was later related to the pore dimension and other properties (Chen and Yang, 1996; Hutson and Yang, 1997).

The exponent "2" in the D–R equation can be replaced by n, which is called the Dubinin–Astakhov equation (or D–A equation). The value of n empirically ranges from below 1 to about 14 (Kapoor and Yang, 1988; Kapoor et al., 1989a). The parameter n can be related to heterogeneity (Jaroniec and Madey, 1988; Rudzinski and Everett, 1992). The theoretical basis given by Chen and Yang (1994) is also valid for the D–A equation.

The potential theory isotherm can be extended to adsorption of mixed gases, as done by Bering et al. (1963 and 1965), and reviewed in Yang (1987). The model by Grant and Manes (1966) has been discussed in detail by Yang (1987). A simple and explicit model has been proposed by Doong and Yang (1988), which is given below. Doong and Yang (1988) extended the D–R equation to mixed-gas adsorption in a simple way by using the concept of maximum available pore volume without any additional equations such as the Lewis relationship (see Yang, 1987). For binary mixtures:

$$V_1 = (V_{01} - V_2)\exp\left[-\left(C_1 \ln \frac{P_{01}}{P_1}\right)^2\right] \tag{3.11}$$

$$V_2 = (V_{02} - V_1)\exp\left[-\left(C_2 \ln \frac{P_{02}}{P_2}\right)^2\right] \tag{3.12}$$

In Eq. 3.11, V_{01} is the limiting pore volume for component 1 and V_2 is the actual adsorbed amount for component 2. All parameters that characterize the gas-sorbent system for the single gases remain unchanged for the mixture. The two equations can be solved easily. This model can be readily extended to multicomponent mixtures. This model has been applied favorably for fitting experimental data (Doong and Yang, 1988). It has been used recently for the adsorption of mixtures of CO_2/H_2O on NaX zeolite and γ-alumina (Rege and Yang, 2001a), and it compared favorably against the ideal adsorbed solution theory. The D–A equation can be extended in the same manner as the D–R equation, by retaining the individual exponent (n_i) for each component (Takahashi et al., 2001).

Wood (2002) has recently made an extensive comparison of different models for predicting mixture adsorption from single-component D–R isotherms. The data of a total of 93 binary mixtures of organic vapors on activated carbon were compared. Despite the simplicity of the model (Eqs. 3.11 and 3.12), predictions were among the best.

3.1.3. Ideal Adsorbed Solution Theory for Mixture and Similarities with Langmuir and Potential Theories

The ideal adsorbed solution (IAS) theory of Myers and Prausnitz (1965) was the first major theory for predicting mixed gas adsorption from pure component isotherms, and it remains the most widely accepted. There have been approximately a dozen other theories that have been discussed in Yang (1987); however, they are not repeated here.

The adsorbed mixture is treated as a two-dimensional phase. From the Gibbs isotherm, one can calculate a spreading pressure for each component based on its pure component isotherm. The basic assumption of the IAS theory is that the spreading pressures are equal for all components at equilibrium.

The spreading pressure, π, is given by:

$$\frac{\pi A}{RT} = \int_0^P \frac{q}{P} dP \tag{3.13}$$

where q is related to P by any pure component isotherm. Assuming the activity coefficients are unity for all components in the adsorbed mixture, the IAS model consists of the following set of equations, for a two-component mixture:

$$\int_0^{P_1^0} \frac{q_1}{P} dP = \int_0^{P_2^0} \frac{q_2}{P} dP \tag{3.14}$$

The Raoult's law, or Eq. 3.15, is written for both species:

$$PY_1 = P_1^0 X_1 \tag{3.15}$$

$$PY_2 = P_2^0 X_2 = P_2^0 (1 - X_1) \tag{3.16}$$

where P_i^0 is the equilibrium "vapor pressure" for pure i adsorption at the same spreading pressure, π, and the same T as the adsorbed mixture. These three equations (Eqs. 3.14, 3.15, and 3.16) define the adsorbed mixture. For example, if P and Y_1 (and Y_2) are given (T is already given), the three equations are solved for P_1^0, P_2^0, and X_1. Equation 3.14 can be integrated to yield an algebraic equation if the isotherms have certain forms like Langmuir and Freundlich equations. Otherwise, Eq. 3.14 is solved numerically. With the values of P_1^0, P_2^0, and X_1 determined, the following equations are used to calculate q_1, q_2, and q_t:

$$\frac{1}{q_t} = \frac{X_1}{q_1(P_1^0)} + \frac{X_2}{q_2(P_2^0)} \tag{3.17}$$

$$q_1 = q_t X_1 \text{ and } q_2 = q_t X_2 \tag{3.18}$$

For systems following the Langmuir isotherm, the IAS theory is identical to the extended Langmuir equation for mixtures, if the saturated amounts are equal for all components. This can be shown by substituting the Langmuir isotherm into Eq. 3.14:

$$q_{m1}\int_0^{P_1^0}\frac{B_1}{1+B_1P}dP = q_{m2}\int_0^{P_2^0}\frac{B_2}{1+B_2P}dP \tag{3.19}$$

If $q_{m1} = q_{m2}$, then $B_1P_1^0 = B_2P_2^0$. Substituting this result into Eq. 3.15, the following is obtained:

$$X_1 = \frac{B_1P_1}{B_1P_1 + B_2P_2} \tag{3.20}$$

This equation can be readily obtained from the extended Langmuir model, Eq. 3.4, for the binary system. Thus, when the saturated amounts for the pure gases are the same, the IAS is identical to the extended Langmuir model.

It is interesting to note that similarity also exists between the IAS theory and the potential model of Grant and Manes (1966). The similarity has been shown by Belfort (1981) and summarized in Yang (1987).

3.1.4. Diffusion in Micropores: Concentration Dependence and Predicting Mixed Diffusivities

For diffusion in gases and colloidal systems, concentration is the origin (or "driving force") for diffusion. Einstein first showed in 1905 that from the concentration gradient, the diffusivity is (Kauzmann, 1966)

$$D = \frac{\delta^2}{2\Delta t} \tag{3.21}$$

where δ is the mean distance between collisions, or Brownian motion in time Δt in the x direction, and D is the diffusivity that relates the flux with the concentration gradient by Fick's law:

$$j = -D\frac{dc}{dx} \tag{3.22}$$

For diffusion in liquid mixtures, it has been argued that chemical potential is the "driving force." (Haase and Siry, 1968).

The mechanisms of diffusion in these two systems (gas and liquid) are different and unrelated; diffusion in gases is the result of the collision process, whereas that in liquids is an activated process (Bird et al., 1960). Diffusion in microporous materials is neither gaseous nor liquid diffusion. The closest case for such diffusion is surface diffusion, where molecules "hop" within the surface force field (see review by Kapoor et al., 1989b). Fick's law is used for both application (in modeling of adsorption processes) and experimental measurement of diffusion. Extensive reviews are available on diffusion in microporous materials and zeolites (Karger and Ruthven, 1992; Do, 1998). A lucid discussion on the nonlinear, and in some cases peculiar, phenomena in zeolite diffusion was given by Wei (1994). Wei's analysis included concentration dependence of single-file diffusion as well as some unsolved problems of zeolite diffusion. Here we will only briefly discuss the concentration dependence of diffusivity and prediction of multicomponent diffusivities from pure-component diffusivities.

The Fickian diffusivities for diffusion in zeolites and microporous materials are generally concentration dependent. Although a variety of concentration dependence has been reported (Yang, 1987), an increase in the diffusivity with concentration is generally the rule. The observed concentration dependence is similar to that seen for surface diffusion, that is,

$$\frac{D_{s,\theta}}{D_{s,\theta=0}} = \frac{1}{1-\theta} \tag{3.23}$$

where θ is the fractional surface coverage and D_s is the surface diffusivity. This dependence can be explained by the HIO model (Higashi et al., 1963), based on the random walk (or hop) of molecules. It assumes that the transit time between sites is negligibly relative to the residence time, Δt, at each site, given by:

$$\frac{1}{\Delta t} = ve^{-\Delta E/RT} \tag{3.24}$$

where v is the vibrational frequency of the bond that holds the molecule to the site, and ΔE is the effective bond energy, that is, the difference in energy between the states corresponding to adsorption at the ground vibrational level of the bond and the free mobility on the surface. The surface diffusivity is thus obtained by the Einstein equation, Eq. 3.21. It is further assumed that when a molecule encounters a site already occupied by another molecule, it immediately bounces off and continues without stopping until it finds an unoccupied site at which to rest. The average number of jumps a molecule takes to find an empty site at surface coverage θ is

$$\eta_\theta = \sum_{k=1}^{\infty} k(1-\theta)\theta^{k-1} = \frac{1}{1-\theta} \tag{3.25}$$

which takes the same length of time, Δt. Thus the relation in Eq. 3.23 is obtained.

The HIO model predicts values in agreement with experimental data reasonably well up to $\theta = 0.6$ or 0.7. At higher values of θ, the values predicted become greater than those measured. When $\theta = 1$, the model would give a value of infinity. This discrepancy has been circumvented by a modified model in which multilayer adsorption is allowed and a finite residence time is assigned to the second and higher-number layers (Yang et al., 1973). To account for the second-layer adsorption, the result from the modified model gives:

$$\frac{D_s}{D_{s,\theta=0}} = \frac{1}{1-\theta+\theta(v_1/v_2)\exp[-(\Delta E_1 - \Delta E_2)/RT]} \tag{3.26}$$

where the subscripts 1 and 2 denote first and second layers, respectively. The modified model improves data correlation from various sources and, more importantly, predicts a finite value at $\theta = 1$. It is interesting to note that, under certain assumptions (ideal gas behavior and Langmuir adsorption; Yang, 1987), the Darken relationship (Darken, 1948; Shewmon, 1963) could also lead to the same concentration dependence as Eq. 3.23. The Darken relationship was derived from a flux equation based on the chemical potential gradient for atomic diffusion in metal alloys, which is a highly activated process.

Because v_1/v_2 is approximately 1, Eq. 3.26 can be recast into:

$$\frac{D_s}{D_{so}} = \frac{1}{1-(1-\lambda)\theta} \tag{3.27}$$

The parameter λ reflects the difference between the effective bond energy of molecule A and vacant site (ε_{AV}) and that of molecule A on a site already occupied by $A(\varepsilon_{AA})$ (Yang et al., 1973):

$$\lambda_A = \frac{\text{sticking probability on adsorbed } A}{\text{sticking probability on vacant site}} = e^{-(\varepsilon_{AV}-\varepsilon_{AA})/RT} \tag{3.28}$$

A simple kinetic-theory derivation was made by Chen and Yang (1992) for predicting binary and multicomponent diffusivities from pure component diffusivities. The multicomponent flux equation is given by:

$$\overline{J} = -[D]\nabla\overline{q} \tag{3.29}$$

where $\overline{J}$ and $\overline{q}$ are two vectors of n components, respectively, for flux and adsorbate concentration. The multicomponent diffusivity matrix $[D]$ is given by (Chen and Yang, 1992; Chen et al., 1994):

$$D_{ii} = D_{io}\frac{\left[1-\sum_{j=1}^{n}(1-\lambda_{ij})\theta_j\right]_{j\neq i}}{1-\sum_{j=1}^{n}(1-\lambda_{ij})\theta_j} \tag{3.30}$$

$$D_{ij} = D_{io}\frac{[(1-\lambda_{ij})\theta_i]_{j\neq 1}}{1-\sum_{j=1}^{n}(1-\lambda_{ij})\theta_j} \tag{3.31}$$

These expressions were derived from the classical transition state theory (Chen and Yang, 1992). Here an adsorbed molecule A has to be activated to form A^*, which can hop to an empty site or a site that is occupied by either A or B.

Equations 3.30 and 3.31 can be obtained by adding all the elementary rate steps and setting the rate of change of the concentration of A^* to zero.

For binary diffusion of components A and B, the fluxes are

$$J_A = -D_{AA}\frac{\partial q_A}{\partial x} - D_{AB}\frac{\partial q_B}{\partial x} \tag{3.32}$$

$$J_B = -D_{BA}\frac{\partial q_A}{\partial x} - D_{BB}\frac{\partial q_B}{\partial x} \tag{3.33}$$

and the concentration-dependent Fickian diffusivities are

$$D_{AA} = D_{AO}\left[\frac{1-(1-\lambda_{AB})\theta_B}{1-(1-\lambda_A)\theta_A-(1-\lambda_{AB})\theta_B}\right] \tag{3.34}$$

$$D_{AB} = D_{AO}\left[\frac{(1-\lambda_{AB})\theta_A}{1-(1-\lambda_A)\theta_A-(1-\lambda_{AB})\theta_B}\right] \tag{3.35}$$

$$D_{BA} = D_{BO}\left[\frac{(1-\lambda_{BA})\theta_B}{1-(1-\lambda_B)\theta_B-(1-\lambda_{BA})\theta_A}\right] \tag{3.36}$$

$$D_{BB} = D_{BO}\left[\frac{1-(1-\lambda_{BA})\theta_A}{1-(1-\lambda_B)\theta_B-(1-\lambda_{BA})\theta_A}\right] \tag{3.37}$$

where D_{AO} and D_{BO} are pure-component diffusivities at zero coverage, given by (for D_{AO}):

$$\frac{D_A}{D_{AO}} = \frac{1}{1-(1-\lambda_A)\theta_A} \tag{3.38}$$

Because the values of q_s for the two components are not very different, one may adopt the adsorbed-phase averaging scheme of Innes et al. (Chapter 3, Yang, 1987) for calculating q_s:

$$\frac{1}{q_s} = \frac{X_A}{q_{sA}} + \frac{X_B}{q_{sB}} \tag{3.39}$$

and

$$\theta_A = \frac{q_A}{q_s};\theta_B = \frac{q_B}{q_s};\theta_A+\theta_B \le 1 \tag{3.40}$$

where X is the adsorbed-phase mole fraction at equilibrium.

The parameter λ is given by Eq. 3.28:

$$\lambda_A = e^{-(\varepsilon_{AV}-\varepsilon_{AA})/RT} \tag{3.41}$$

Similarly,

$$\lambda_B = e^{-(\varepsilon_{BV}-\varepsilon_{BB})/RT} \tag{3.42}$$

$$\lambda_{AB} = e^{-(\varepsilon_{AV}-\varepsilon_{AB})/RT} \tag{3.43}$$

$$\lambda_{BA} = e^{-(\varepsilon_{BV}-\varepsilon_{BA})/RT} \tag{3.44}$$

For surface diffusion, (ε_{AV}) and (ε_{BV}) may be assumed to be the heats of adsorption (Chen and Yang, 1992). For diffusion in zeolite, the derivation is also valid, provided that the activation energies for diffusion are used for (ε_{AV}) and (ε_{BV}). In contrast to surface diffusion, in which the activation energy is always lower than the heat of adsorption, the activation energy for diffusion in zeolite can be higher than heat of adsorption because the repulsive forces between the diffusing molecule and the aperture or pore surfaces can influence the diffusion process.

From the experimental values of λ_A and ε_{AV} the value for ε_{AA} can be obtained from Eq. 3.41. Likewise, the value of ε_{BV} can be obtained form λ_B. Finally, the interaction energies between unlike molecules, ε_{AB} or ε_{BA}, can be calculated by assuming the mixing rule:

$$\varepsilon_{AB} = \varepsilon_{BA} = (\varepsilon_{AA}\varepsilon_{BB})^{1/2} \tag{3.45}$$

Equations 3.34–3.37 give the main-term and cross-term diffusivities. The only needed data are the pure component diffusivities and their concentration dependence (which yield the interaction energies). This model has been used successfully for predicting the mixed nitrogen/oxygen diffusivities in carbon molecular sieve (Chen and Yang, 1992; Chen et al., 1994). A prediction comparison with experimental data for diffusion of nitrogen/oxygen mixture in molecular sieve carbon will be given in Chapter 5 (under molecular sieve carbon) (Chen et al., 1994).

A molecular dynamics approach can also be used to predict mixed gas diffusivities in microporous materials, at the expense of computation cost (e.g., Qureshi and Wei, 1990; Chitra and Yashonath, 1995; Trout et al., 1997; Snurr and Karger, 1997). The empirical correlation of Vignes (1966) for binary diffusivities in liquid solutions and also metallic alloys has been used extensively for calculating binary diffusivities, using the Maxwell–Stefan formalism for flux equations (e.g., Krishna, 1990).

3.2. TEMPERATURE SWING ADSORPTION AND PRESSURE SWING ADSORPTION

Because the sorbent needs to be regenerated for most commercial applications, adsorption processes are necessarily cyclic. A number of cyclic adsorption processes are available, depending on the way the sorbent is regenerated. These processes have been discussed extensively elsewhere (e.g., Yang, 1987; Humphrey and Keller, 1997).

Adsorptive gas separation processes can be divided into two types: bulk separation and purification. The former involves adsorption of a significant fraction, 10% by weight or more from a gas stream according to Keller's definition (Keller, 1983). Whereas in purification, $<$10% (usually $<$2%) by weight of a gas stream is adsorbed. Such a differentiation is useful because, in general, unique process cycles are used for different types of separation.

For purification, temperature swing adsorption (TSA) is generally the process of choice. For bulk separation, pressure swing adsorption (PSA) is more suitable.

However, PSA is also used for drying of air and industrial gases. More applications continue to be discovered for PSA, including air prepurification (i.e., purification of air prior to cryogenic distillation by removal of carbon dioxide, water, and hydrocarbons; see Rege et al., 2000). For air prepurification, PSA has been used only after ca. 1990. TSA and PSA are both currently being used.

Besides TSA and PSA, inert purge cycles have also been used for desorbing weakly adsorbed compounds. This cycle is similar to the temperature swing cycle except that preheating of the purge gas is not required. Usually a fraction of the light product (raffinate) is used as the inert purge gas. The inert purge cycles are similar to TSA and will not be discussed separately.

Simulated moving bed (SMB) or the UOP Sorbex processes are used only for liquid-phase separations. No gas phase separation has been performed commercially by SMB due to the axial dispersion problem in the fixed beds.

3.2.1. Temperature Swing Adsorption

In this process cycle, the bed is regenerated by raising the temperature. The most convenient way for raising the temperature is by purging the bed with a preheated gas. This is the oldest and most completely developed adsorption cycle. Because heating is a slow and often a rate-limiting step, the length of each cycle usually ranges from several hours to over a day. In order to make the time length of the adsorption step comparable with that of regeneration, the cycle is used only for purification purposes.

Dual-bed systems are most commonly used for TSA. A number of the dual bed arrangements can be made in which no external source of regenerant (e.g., air, steam, or other available gases) is used. A three-bed design has also been used for systems with a long length of unused bed (LUB, which is approximately one-half the mass transfer zone or the span of the concentration front). In the three-bed system, a guard bed is located between the adsorber and that being regenerated (Chi and Cummings, 1978). When the concentration of the adsorber effluent reaches nearly that of the feed stream, the beds are switched. The loaded primary adsorber goes to regeneration, the former guard bed assumes the primary position, and the freshly activated bed becomes the guard. By this rotation, the LUB section is always contained in the guard bed, and the primary adsorber is always fully loaded to its capacity when regeneration begins. Thus both a high-purity product and an economic regeneration are achieved. The cost of the additional bed, however, may outweigh its benefits for many common purification processes.

The extensive literature on fixed-bed adsorber modeling and analysis has been discussed elsewhere (Wankat, 1986; Yang, 1987; Suzuki, 1990; Tien, 1994; Basmadjian, 1997; Crittenden and Thomas, 1998). The theoretical analyses primarily resort to equilibrium theory, that is, mass and heat transfer rates are assumed to be instantaneous. Because the adsorption and desorption steps in TSA are operated slowly, each spanning for hours, the equilibrium theories are indeed good. Often, quantitative agreements are obtained between theory and experiment (see

Chapter 5, Yang, 1987). Mass and heat transfer are considered dispersive forces, which have a dispersive or smearing effect on the concentration and temperature fronts. The detailed bed profiles and breakthrough curves can be calculated only by the numerical solution of the mass and heat balance equations (Yang, 1987), coupled with the equations for equilibrium adsorption from mixture.

Experimental and theoretical results on the TSA process focused on the first cycle of adsorption, followed by regeneration by using hot gas purge (Pan and Basmadjian, 1970; Basmadjian et al., 1975a and 1975b; Carter, 1975). Because the bed regenerated by hot purge is usually very clean, the cyclic steady state is approached in the first cycle. This fact is implied but not clearly stated in several early studies where the cyclic experiments were performed but only the first-cycle results were reported (Carter, 1975; Chi and Cummings, 1978). This finding is in contrast with the inert (cold) purge cycles, where several cycles are needed to approach a cyclic steady state (Bunke and Gelbin, 1978). The difference is due to the fact that cold purge is relatively ineffective and that the TSA cycle requires the times for adsorption and regeneration to be equal in length. Using hot-air purge for *n*-hexane-activated carbon, Davis and LeVan (1989) showed that cyclic steady state was reached after 2 or 3 cycles.

Regeneration by steam is more practical, but more difficult to understand (Scamehorn, 1979; Wankat and Partin, 1980; Schork and Fair, 1988; Schweiger and LeVan, 1993). For steam regeneration, one needs to consider the condensing flow as well as the difficult problem of adsorption equilibrium for hydrocarbon–water mixtures. Schork and Fair (1988) observed a high-temperature middle zone in the bed due to the heat of adsorption of steam. The system of *n*-hexane-carbon-steam has been studied in detail by LeVan and co-workers. Three plateau zones were observed in the regenerating bed. The middle zone, at an intermediate temperature between the feed steam and the front zone containing the initial gas in the bed, is where the regeneration actions occur. Schweiger and LeVan (1993) described such actions as "steam distillation" of the water-immiscible hydrocarbon. In this zone, steam adsorbs and condenses, causing the fluid velocity to reduce to a near stop beyond this zone. They have also given a complete model that could fully describe their experimental observations in steam regeneration.

The regeneration step in the TSA cycle requires time to heat, desorb, and cool the bed. It is often the time-limiting step in the TSA cycle and is also the most complex and least understood one. The following discussion on TSA focuses on the regeneration step and presents some simple rules for design.

Minimum Purge Temperature. The minimum purge temperature was derived based on the equilibrium theory by Basmadjian et al. (1975a and 1975b), which has been discussed in detail elsewhere (Yang 1987; Basmadjian, 1997).

Efficient desorption is accomplished at temperatures above the "characteristic temperature," T_0. The characteristic temperature is equal to the temperature at which the slope of the adsorption isotherm at the origin is equal to C_{ps}/C_{pb}, the ratio of the heat capacities of the solid phase and the inert carrier gas. For a

Langmuir isotherm:

$$q = \frac{K(T)P}{1 + B(T)P} \tag{3.46}$$

and T_0 is determined by:

$$K(T_0) = \frac{C_{ps}}{C_{pb}} \tag{3.47}$$

where q is the amount adsorbed; P is pressure; and C_{ps} and C_{pb} are, respectively, the heat capacities for the solid-phase (sorbent plus adsorbate) and inert carrier gas. The values for some systems are given in Table 3.1. As the temperature is increased beyond the characteristic temperature, however, the energy cost also increases without a significant gain in desorption. Thus the characteristic temperature is actually the optimal temperature for desorption.

Materials problems must also be considered in selecting the regeneration temperature. In the presence of hydrocarbons, coke deposition occurs on zeolites as a result of the catalytic decomposition at temperatures as low as 100 °C, depending on the hydrocarbon compound, its partial pressure, and other factors. Severe oxidation occurs in activated carbon at slightly above 100 °C in the presence of trace amounts of oxygen.

3.2.2. Pressure Swing Adsorption

A multitude of sophisticated PSA processes and designs have been implemented for various separations. However, there are not many basic ideas on which these designs are made. These ideas can be combined and synthesized, with minor modifications, to accomplish the desired separations.

To facilitate the discussions that follow, some terminologies will first be defined. The performance of the separation process is measured by three parameters: (1) product purity, (2) product recovery, and (3) adsorbent productivity. If more than one of the constituents are to be recovered, (1) and (2) apply to all products. The product purity is a volume-averaged quantity, as the effluent

Table 3.1. Characteristic temperatures, T_0, for various systems at 1 atm

System	T_0, °F
CO_2/CH_4/5A zeolite	≈230
H_2O/air/5A zeolite	>600
H_2O/air/silica gel	≈250
H_2S/CH_4/5A zeolite	≈400
Acetone/air/activated carbon	≈300

Source: Basmadjian et al. (1975b). Reprinted with permission. Values are accurate within ±10%, because of uncertainties in K, C_{ps}, and C_{pb}.

concentration and flow rate from a PSA process, without the use of surge or mixing tanks, vary with time. Product recovery measures the amount of component that is contained in the product stream divided by the amount of the same component in the feed mixture. Adsorbent productivity is measured by the amount of product or feed mixture processed per unit amount of sorbent per unit time. For a given separation, the product purity is predetermined, the energy requirement is usually proportional to the recovery, and the size of the sorbent bed is inversely proportional to the sorbent productivity. It is important to bear in mind that these three parameters are interrelated for any given PSA process, and the interrelationship can be determined only through a model for the process. In addition to these parameters, the term "strong adsorptive" will be used to refer to the strongly adsorbable gas, whereas "weak adsorptive" will be used for the weakly adsorbable gas. The product that mostly contains the weak adsorptive may also be called "loosely light product" or "raffinate", the latter is in analogy to liquid–liquid extraction.

The basic PSA cycle was described by Skarstrom in 1960 (Skarstrom, 1960; 1972). A similar cycle was the Guerin–Domine Cycle, invented at about the same time (Guerin–Domine, 1964; see a detailed account of these inventions in Yang, 1987). The latter was the basis for the modern vacuum swing cycle. The major additions to these cycles are co-current depressurization (CD) and pressure equalization (PE) steps. The CD step was added to increase product recovery, whereas the pressure equalization step was added to conserve the mechanical energy.

Skarstrom Cycle. The Skarstrom cycle uses a two-bed apparatus as shown in Figure 3.1. There are two main cycle steps: high-pressure adsorption and low-pressure purge. In between these two steps are two pressure-changing steps. These cycle steps are illustrated in Figure 3.2. After the high-pressure adsorption step, the first bed is depressurized to atmospheric (or a low) pressure. At the same time the compressed feed mixture is switched to the second bed to repressurize it. It then starts the adsorption step at the feed pressure. A fraction of the purified effluent from the second bed is passed through the first bed, countercurrent to the feed direction, to purge the bed at atmospheric (or low) pressure. After the purge, the unit is ready for the next cycle. Thus each bed undergoes two half-cycles, the times of which are equal. For a purification process, steady and continuous flows of both feed and purified product are achieved.

Two useful intuitive ideas were also described by Skarstrom's invention. First, short cycles and low throughput per cycle should be maintained in order to "conserve the heat of adsorption." This idea was derived from the observation that long cycles and high throughputs would result in hot beds during adsorption and cold beds during regeneration, both detrimental to separation. The second idea was that the volume of both purge and feed, at their respective pressures, should be at least equal to ensure complete purge, or displacement, during regeneration. These two basic ideas have been determined to be very useful in later developments.

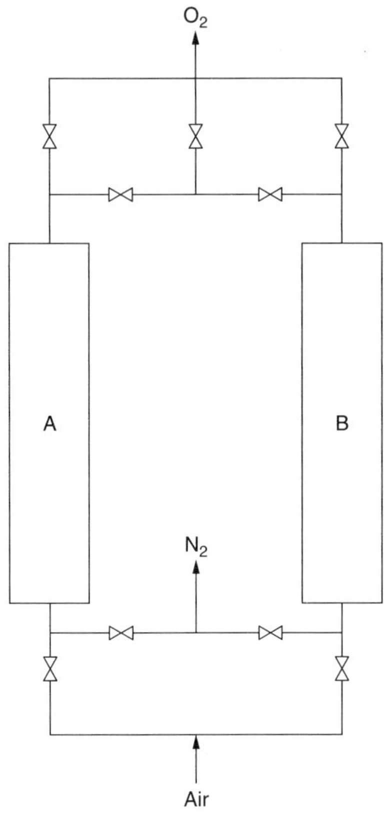

Figure 3.1. The Skarstrom cycle. Two steps are involved in the cycle for each bed: (1) repressurization followed by feed in bed A, while O_2 is withdrawn as the product, and (2) blowdown and purge with O_2 (both downward) in bed A while N_2 is withdrawn. The functions of A and B are reversed in the next cycle. Note that this is not used commercially for air separation, but is used for drying.

The Skarstrom cycle has been widely used for air-drying. For oxygen production from air and other bulk separations, however, the product recovery was too low, and hence the energy requirement too high, to be economical.

Co-Current Depressurization. The first major process improvement after the invention of the Skarstrom and Guerin–Domine cycles was the introduction of the co-current depressurization step. To incorporate this step into the Skarstrom cycle, the adsorption step is cut short well before the breakthrough point, that is, the concentration front is far from reaching the outlet of the bed. The adsorption step is immediately followed by co-current depressurization before the bed is desorbed by further blowdown and purge, as required in the Skarstrom cycle.

The large void space in the adsorber plays a limiting role in the recovery of the strong adsorptive. The void fraction ranges from approximately 67% for silica and

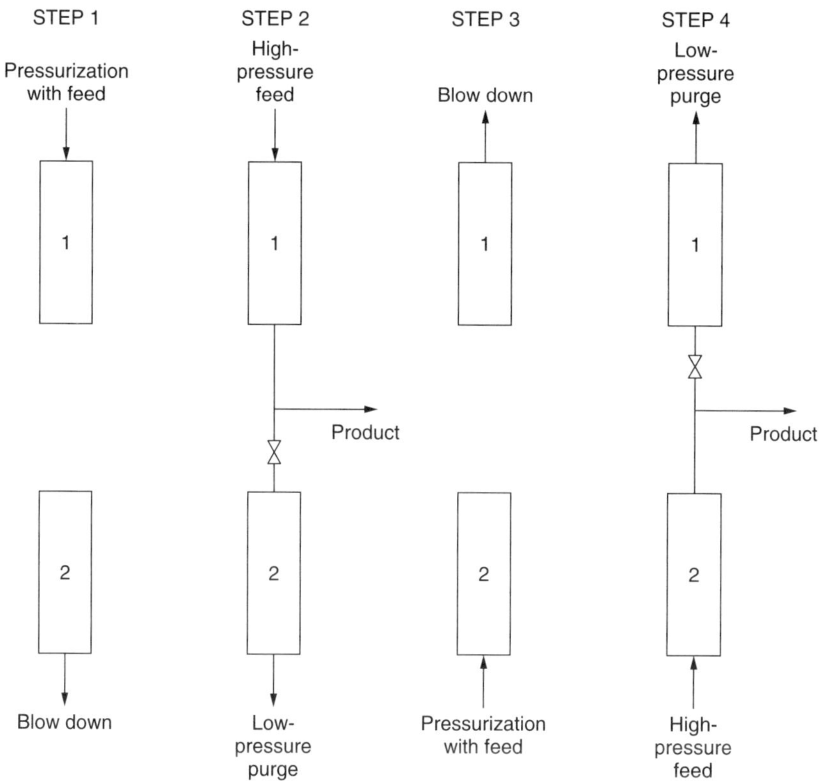

Figure 3.2. Cycle steps in the two-bed Skarstrom Cycle.

alumina to 74% for zeolites and 78% for activated carbon. The composition of the gas contained in the voids of the saturated bed is the same as the feed mixture. At the feed pressure, the amount of gas in the voids can be equal to or greater than the amount adsorbed on the sorbent surface. The total concentration of the strong adsorptive in the bed (in the voids plus the adsorbate) is not high. This is the gas that will be desorbed under the best regeneration conditions in the Skarstrom and Guerin–Domine cycles. The major function of the co-current depressurization step is to increase the concentration of the strong adsorptive in the bed. This is done by lowering the pressure in the voids, which enhances the concentration of the strong adsorptive in both phases. When the bed is depressurized, the strong component desorbs more than the light component does, thus increasing the gas-phase concentration of the strong component. Consequently, a lighter component will be desorbed in the subsequent steps. This point has been illustrated by Cen and Yang by using a bed-loading analysis (Cen and Yang, 1986).

The net result of incorporating the CD step in the Skarstrom cycle is the product purity enhancement of the strong adsorptive, which in turn increases the product recovery of the weak adsorptive.

The CD step was first used by Union Carbide in the IsoSiv process in a Texas PSA plant in 1961 (Avery and Lee, 1962; Keller, 1983; Cassidy and Holmes, 1984). The process separated 1,000 barrels per day of "natural gasoline" feed into n-paraffins and branched/cyclic hydrocarbons. The feed contained 54.4% n-paraffins and 45.6% branched/cyclic hydrocarbons (Symoniak, 1980). The n-paraffin (strong adsorptive) product purity was 95–98%, whereas the purity of the "isomer" was 98–99% (Symoniak, 1980). Additional and larger IsoSiv plants have been built since 1961. The CD step was mentioned, as part of more complex PSA processes, in several patents that were all filed later than its first commercial practice in 1961 (Kiyonaga, 1965; Wagner, 1969).

Pressure Equalization. The term "pressure equalization" (PE) refers to the action by which the pressures in two interconnected beds are equalized. The main purpose of the PE step is to conserve the mechanical energy that is contained in the gas of the high-pressure bed. With the PE steps, the pressure in a regenerated bed is increased in a sequence of steps by the gases admitted from other beds, which are in various stages of depressurization. The energy reduction, as well as other improvements resulting from the PE steps (to be discussed shortly in this section), made large-scale PSA separations economically feasible.

The idea of pressure equalization was first suggested in a patent granted to Marsh et al. in 1964 (Marsh et al., 1964). The process described in the patent required an empty tank in addition to the two beds used in the Skarstrom cycle. The tank was used to store a portion of the compressed gas from a saturated bed, and the gas was used to later purge the same bed. The primary objective was to recover the components contained in the compressed gas. The pressure equalization step, as it is currently used in commercial processes, was disclosed in the patents to Berlin (1966) and Wagner (1969). Four- to five-bed arrangements were given by Wagner. No empty tanks are required in this process.

Besides energy conservation and increased product recovery, the flow of the strong adsorptive product is "smoothed" by the pressure equalization steps. The four-bed process has been subsequently developed into the Polybed process, consisting of 9 to 10 beds, which is successfully used for large-scale production of high-purity hydrogen.

Pressure equalization is accomplished by connecting the ends of two beds. It is natural to use the CD step of one bed to re-pressurize another bed that has been purged. The problem with the EQ step is that the concentration of the unwanted component in the effluent from the CD step is increasing with time. The "cleanest" portion enters the bed that is to be re-pressurized first. Consequently the concentration profile in the re-pressurized bed is in the wrong direction, i.e., with the feed end the cleanest and the discharge end the dirtiest. This prevents the production of a clean product during the ensuing cycle. To alleviate this problem, multibeds are used to accommodate more than one PE step (Yang, 1987). In the multibed PSA (or "Polybed" PSA of Union Carbide), typically three PE steps are used. The effluent from the CD step is divided into three portions, with the first portion fed to the bed that is in its third stage of repressurization (Fuderer and

Rudelstorfer, 1976; Yang, 1987). In the Polybed PSA, typically 9 to 10 beds are used. This has been practiced for hydrogen purification, where product recovery is important. For air separation, however, oxygen recovery is not as important (versus increased costs from more beds), and no polybed PSA is used. In some cases, the EQ step is accomplished by connecting the CD effluent to the mid-point of another bed that is to be re-pressurized. This can also alleviate the problem with the "reversed" concentration profile.

Still another clever method to alleviate the "wrong" concentration profile problem of the EQ step is the use of "parallel equalization," (PEQ) by Hirose and co-workers (Yoshida et al., 1998; 2000). This was done for the enrichment of Xe from air, in the second-stage PSA, where five parallel connections were made between the two beds. By doing so, the Xe is more enriched in the discharge end of the bed and consequently a more enriched Xe can be produced.

Pretreatment Beds. Water vapor and carbon dioxide, both very strongly adsorbed on zeolites, created a major problem in the separations of air as well as other mixtures when using zeolites. These gases are not easily desorbed and hence tend to accumulate in the bed, leading to a stoppage in operation. The problem can be solved by using a separate pretreatment bed outside the PSA system, which must be regenerated separately from the PSA. For process simplicity, it is desirable to include the pretreatment bed(s) within the PSA system so the integrated system can be regenerated as a whole.

The idea of integrating the pretreatment beds into the PSA system was first suggested by Heinze (1962), and improved by Tamura (1967), mainly by allowing a flow in the adsorption step. The sorbent in the pretreatment bed should be different from that in the main bed since a relatively inefficient sorbent is used in the former, unlike the efficient sorbent used in the main bed. Although both of the aforementioned processes were commercialized for oxygen production (Lee and Stahl, 1973), a more efficient process was later developed by Sircar and Zondlo (1977). In the latter process, a desiccant such as silica gel or alumina is placed in the pretreatment bed, while the main bed contains a zeolite. By this arrangement, water vapor and carbon dioxide are adsorbed in the pretreatment beds, which are easily regenerated along with the main beds by pressure swing. An additional advantage of using small pretreatment beds is that the temperature fluctuations in the smaller beds are less severe.

The idea of integrating the pretreatment beds in PSA has also been used in hydrogen purification, in which an inefficient sorbent in the pretreatment beds adsorbs pentane and heavier compounds, whereas the main beds adsorb methane, ethane, and propane (Alexis, 1967). This idea should be useful in many multicomponent separations.

The pretreatment bed can be combined with the main bed; that is, two sorbents are packed in the same bed with the desiccant (alumina or silica gel) placed near the feed end followed by the zeolite bed. This arrangement is widely used in the production of oxygen from air (Armond, 1979), as well as other separations.

Purge by Strong Adsorptive. As discussed above, under *Co-Current Purge*, the feed gas stored in the void space of the saturated bed severely limits the separation performance, especially for the product purity of the strong adsorptive—hence the product recovery for the weak adsorptive. The CD step can improve the separation. An illustrative example was given for the separation of a 50/50 hydrogen/methane mixture on activated carbon (Yang and Doong, 1985). Using the Skarstrom cycle, the highest product purities were approximately 99% for hydrogen and 80% for methane. With the addition of the CD step in the cycle, the methane product purity was increased to approximately 90%. A more drastic modification of the cycle must be made in order to further increase the methane product purity.

A more effective method is to purge out the void spaces, after adsorption, by using the strong adsorptive gas. The idea was first suggested in a patent to Tamura (1974). In this cycle, a high-pressure purge step, in which the strong adsorptive gas is used, follows the high-pressure adsorption step in the Skarstrom cycle. The purge is conducted co-current to the feed direction.

There are two concentration wavefronts during co-current purge by the strong adsorptive: (1) a zone filled with the pure strong adsorptive near the feed end; (2) a displaced zone, which was the zone saturated with the feed mixture prior to purge. Owing to the favored isotherm of the strong adsorptive, the two wavefronts are both the self-sharpening type and result in good separation.

A disadvantage of the PSA cycle is that compression is required for the strong adsorptive gas, which is available from the PSA process at a low pressure. There is, however, room for optimization between the CD and strong adsorptive purge steps. An optimal operation should exist in which the purge is performed after some CD. During CD, the weak adsorptive is eluted from the saturated zone (Yang and Doong, 1985), and the bed is enriched in the strong adsorptive. More importantly, with the CD step, the bed pressure is reduced and hence lowers the energy requirement for compressing the purge gas. A combination of these two steps apparently has not been used.

Recent/Current Developments and Future Prospects for PSA

Simultaneous Purification and Sorbate Enrichment. The use of PSA as a tool for enrichment gas been studied only recently (Ritter and Yang, 1991; Kikkinides and Yang, 1991; Kikkinides and Yang, 1993; Kikkinides et al., 1993; Chue et al., 1995; Chue et al., 2000). These were studies for air purification and concentration in a variety of sorbates including CO_2, SO_2, and gasoline vapors. The basic PSA cycles were typically: (1) feed pressurization; (2) high-pressure feed; (3) countercurrent evacuation; and (4) countercurrent vacuum purge. Evacuation during desorption is critically important for enrichment, while the high pressure is not important. An optimal purge/feed ratio is needed for the highest enrichment. This type of process has been developed to the pilot plant stage at Korea Institute of Energy Research by Cho and co-workers (Cho, 2001).

A recent study for enrichment was performed by Hirose et al. (Yoshida, 1998; 2000) for enriched Xe from air. A two-stage PSA employing two ideas was used to achieve an enrichment ratio of 4000 (from 0.1 ppm in the feed air to 400 ppm in the desorption product). The two basic ideas were reflux (or rinse) and parallel pressure equalization (as discussed above). The first PSA was super-atmospheric, and the second one was sub-atmospheric, both containing 13X zeolite.

As the environmental applications are becoming increasingly more important for PSA, more studies are needed for the enrichment and recovery of sorbate. As already seen, the fundamental as well as operational issues for enrichment are different from those involved in purification and bulk separation and have not been well understood.

Radial-Bed PSA. The development of radial-bed PSA is aimed at oxygen production from air (Rota and Wankat, 1991; Poteau and Eteve, 1993; Chiang and Hong, 1995a and 1995b; Smolarek et al., 1998). Due to the relatively linear isotherms of nitrogen and oxygen on zeolites, O_2-PSA has adopted the vacuum swing mode since the 1980's, that is, adsorption at approximately 1 atm with desorption at 0.2–0.4 atm, depending on the sorbent (Rege and Yang, 1997). With the invention/improvements of the sorbent (notably the invention of LiLSX zeolite by Chao, 1989), the trend for O_2-PSA has been to use shorter cycle times and higher flow rates. Eventually the bed pressure drops and, to a lesser degree, the pellet diffusion rates have become the limitation. As a consequence, shorter beds are used. The development of annular beds, or radial flow beds, is a further improvement of the shallow beds. In these beds, the sorbent is contained in an annular layer. The feed air flows radially from the outside toward the inner part. The reason for the feed from outside in the radial bed is to reduce the velocity variation. As the nitrogen is adsorbed, the gas velocity reduces and the cross sectional area in the radial bed is also reduced. In addition, it is possible that, due to the increased interstitial flow velocity toward the center, the concentration wavefront is sharpened, as the isotherm is of the favorable type (i.e., concave shape) (Yang, 1987). This further helps the separation. Chiang's invention was unique in that very small particles (3 μm crystals) were used in thin annular beds where pressure drop was exploited for the separation (Chiang and Hong, 1995a and 1995b).

The radial-bed development will further improve O_2-PSA. Significant (and likely larger) improvements can also be achieved from the invention of new sorbents. With the availability of better sorbents, nitrogen production from air with zeolites could be competitive with molecular sieve carbon.

PSA Reactor. The idea of a PSA reactor was first suggested by Vaporciyan and Kadlec (1987; 1989). The basic idea is to combine sorption and catalytic reaction in order to shift the thermodynamic equilibrium of the reaction. The sorbent selectively adsorbs one of the products and is regenerated during the low pressure half-cycle. By doing so, the conversion is increased and simultaneous separation is also accomplished. Vaporciyan and Kadlec (1989) demonstrated the idea for CO oxidation reaction by mixing a 5A zeolite sorbent and a Pt/alumina catalyst in the adsorber. This idea was pursued by Sircar and co-workers (Carvill et al., 1996; Hufton et al., 1999; 2001) for the feasibility of hydrogen production from the water gas shift reaction. Their sorbent was hydrotalcite for the adsorption of CO_2, and a noble metal as the catalyst at a bed temperature near 490 °C. The enhancement of catalytic reactions by combining with PSA has been analyzed and discussed in detail by Alpay and co-workers (1993; Chatsiriwech et al., 1994; Cheng et al., 1998; Yongsunthon and Alpay, 1998, 2000). They also studied the PSA reactors for steam reforming of methane by using hydrotalcite CO_2 sorbent and supported Ni catalyst (Ding and Alpay, 2000) and l-butene dehydrogenation by using CrO_2-alumina catalyst and a zeolite K-Y adsorbent (Sheikh et al., 2001). Chou et al. (1998) performed a simulation of the PSA reactors for CO oxidation (i.e., the system studied by Vaporciyan and Kadlec, 1989) and cyclohexane dehydrogenation (Chou et al., 1998). The series reaction $A \rightarrow B \rightarrow C$ was studied with the PSA reactor by removing A by Kodde et al. (2000).

The PSA reactor clearly holds potential for equilibrium-limited reactions, such as water–gas shift and dehydrogenation reactions. However, the limiting factor remains to be the sorbent. The PSA reactor must be operated at a relatively high temperature in order for the catalyzed reaction to proceed at a realistic rate. High temperature is not favorable for sorption (or sorption reaction). The challenge for this problem is to find the proper sorbent that could react and sorb one of the products at a high rate and with a high capacity. At the same time, the sorbent must be readily regenerable. For H_2 production by water gas shift, a better sorbent than hydrotalcite is needed for CO_2 (forming carbonate), and the reaction needs to be run at a temperature higher than 500 °C for reasonable rates. Other alkaline earth oxides and perovskites may hold promise for this application. Sorbent integrity upon cyclic sorption/desorption (or decomposition) can also be a severe limitation.

Analytical Solution for Skarstrom Cycle. It is desirable to have a simple, analytical solution for PSA that can be used for estimation purposes, e.g., for sizing the beds and comparing different sorbents. As Knaebel quoted Chilton, "The simpler things become in a piece of research or development, the closer one has come to the truth." (Ruthven et al., 1994). The following is the solution by Chan et al. (1981) which has proved to be quite useful. The PSA cycle is a simple two-step Skarstrom cycle with linear isotherms.

The assumptions for the model are

1. Linear isotherms are followed by both components A and B. The strong adsorptive, A, is at a trace level. The two isotherms are non-interfering.
2. The cycle is isothermal.
3. The interstitial flow velocity, u, is constant during the adsorption and purge steps.
4. Heat and mass transfer are instantaneous.
5. Plug flow is assumed, with no axial or radial dispersion.

6. Ideal gas law is obeyed.
7. Pressure gradient across the bed is negligible.

The mass balance yields, for A:

$$\varepsilon\left[\frac{\partial C_A}{\partial t}+\frac{\partial(uC_A)}{\partial z}\right]+(1-\varepsilon)\frac{\partial q_A}{\partial t}=0 \tag{3.48}$$

and for B:

$$\varepsilon\left[\frac{\partial C_B}{\partial t}+\frac{\partial(uC_B)}{\partial z}\right]+(1-\varepsilon)\frac{\partial q_B}{\partial t}=0 \tag{3.49}$$

where C is gas-phase concentration, q is the adsorbed amount per volume of pellet, t is time, z is distance, ε is the interstitial (or interpellet) void fraction and u is the interstitial velocity.

Both components adsorb independently:

$$q_A = B_A C_A; \quad q_A = B_A C_A \tag{3.50}$$

where B is Henry's constant and $B_A > B_B$. The Henry's constant may also be expressed as:

$$\beta_A = \frac{\varepsilon}{\varepsilon+(1-\varepsilon)B_A}; \quad \beta_B = \frac{\varepsilon}{\varepsilon+(1-\varepsilon)B_B} \tag{3.51}$$

where β indicates the ratio of gas-phase capacity to the total capacity of the sorbent.

The penetration distances Z during the high- and low-pressure steps are given by:

$$Z_H = B_A u_H \Delta t \tag{3.52}$$

$$Z_L = B_A u_L \Delta t \tag{3.53}$$

where subscripts L and H denote the low-(purge) and high-(adsorption) pressure steps, with equal time length, Δt.

By using the method of characteristics, the propagation velocity of the concentration front can be expressed. The two pressure-changing steps can be accounted for using ideal gas law and assuming zero axial pressure drop.

For complete purification, that is, complete removal of A in the high-pressure product stream at steady-state operation, the following two conditions must be satisfied:

1. Breakthrough of feed into the high-pressure product stream does not occur.
2. Purge/feed ratio must be greater than a critical value such that the net displacement of a concentration wavefront during a complete cycle is positive.

Starting from saturated beds as the initial condition, Chan et al. (1981) derived solutions for product purity and recovery:

Product Purity:

The mole fraction in the nth cycle is

$$Y_n = Y_f (P_H/P_L)^{2n(\beta-1)} \tag{3.54}$$

where Y_f is the mole fraction of A in the feed, and β is given by:

$$\beta = \frac{\beta_A}{\beta_B} < 1 \tag{3.55}$$

Product Recovery (ρ):

$$\rho = \frac{(P_H/P_L)^{1-\beta} - 1}{(P_H/P_L)^{1-\beta}(\beta L/Z_L)(P_H - P_L)/P_L} \tag{3.56}$$

where L is the length of the beds and Z_L is given by Eq. 3.53.

The bed size is given by the bed length and hence the sorbent productivity can be calculated from the feed velocity.

3.3. SIMPLE CRITERIA FOR SORBENT SELECTION

Detailed mathematical models are needed for accurately assessing a given sorbent for either PSA or TSA. Many models are available in the literature (Yang, 1987; Ritter and Al-Muhtaseb, 2001). A number of commercial computing packages are already available, such as ADSIM (by Aspen) and SiMoBe (by Prosim).

During the process development stage, several adsorbents may be identified as suitable materials to effect a viable separation for a particular application. The preliminary viability can be judged by a visual comparison of the isotherms and/or the uptake rates of the pure components of the gas mixture. However, a cursory visual inspection can be misleading and does not quantify the merit or the efficiency of the sorbent for the separation under consideration. For a detailed comparison of the adsorbent performance, it is necessary to perform lab scale PSA experiments by using a suitable cycle, or computer simulations with a proven mathematical model, as mentioned. It is obvious that these steps are time-consuming. A quick and easily calculable parameter would be highly desirable in selecting the optimum sorbent for a particular gas-separation application. A simple selection parameter for sorbent selection for PSA has been given by Rege and Yang (2001a and b), and will be given here.

A simple method for determining sorbent selectivity was proposed by Knaebel (1995), which involved taking the ratio of Henry law constants. This rough estimate for selectivity can serve as a crude sorbent selection parameter. A more precise estimate can be obtained by comparing the isothermal binary working selectivity of the sorbents as defined by Gaffney et al. (1993). An adiabatic separation factor, computed by taking the ratio of working capacities determined under non-isothermal, multicomponent conditions, was proposed by Ackley (1998). By taking the product of working capacity and adiabatic separation factor, Ackley et al. (2000) compared the relative merits of different sorbents for a particular gas separation application. Lastly, an example of a sorbent parameter appears in a patent by Notaro et al. (1998), which mainly concerns itself with the separation of N_2 from air. This parameter, referred to as "Adsorption Figure of Merit (AFM)," was defined as follows:

$$\text{AFM} = \Delta N_2 \cdot \frac{\alpha_{ads}^2}{\alpha_{des}} \tag{3.57}$$

where ΔN_2 is the delta N_2 loading (or the difference between N_2 adsorption amounts at high pressure and low pressure), and α_{ads} and α_{des} are the adsorption selectivities (for N_2/O_2) under adsorption and desorption conditions. The AFM was used to assign the zones in which different adsorbents would be used in a layered bed configuration. However, the reasoning behind the AFM was not explained by Notaro et al. (1998), and it is obvious that this parameter is an empirical rule of thumb. Also, it should be noted that this parameter is not dimensionless and thus restricts its generalization.

Building on the various methods described above, Rege and Yang (2001a and b) proposed a simple parameter when comparing two adsorbents for a particular binary gas separation on the basis of their equilibrium adsorption capacities. The parameter is composed of two factors. One is related to the equilibrium selectivity of the sorbent, which is an inherent property of the sorbent. The other factor is related to the working capacity of the sorbent, which is defined as the difference in the adsorbed amount at the adsorption and the desorption pressures. Thus, the latter factor is related to the particular PSA cycle used for performing the separation. The applicability of the parameter has been demonstrated by using two examples from air separation.

Theoretical Basis for Simple Parameter: Sorbent Selection Parameter (S). A parameter for the selection of an adsorbent should preferably have the following attributes: It should be readily estimable without complicated calculations, and it should incorporate the essential nature of the isotherms under the PSA operating conditions, which will dictate the selectivity and hence the performance of the adsorbents. The PSA parameter discussed here applies only to sorbents with negligible mass transfer limitations or at most, having close diffusivity values for both components. It is further assumed that the adsorption isotherms follow Langmuir behavior, although this parameter can be easily extended to other isotherm models as well. By Langmuir behavior it is meant that the sorbents exhibit a limiting coverage as the saturation pressure is approached and that they have an energetically homogeneous surface. The Langmuir equation,

Eq. 3.1, is followed by both components. For a binary mixture, the extended Langmuir equation, Eq. 3.4., is

$$q_i = \frac{q_{m_i} b_i P_i}{1 + b_1 P_1 + b_2 P_2} \tag{3.58}$$

The first factor in the PSA parameter is the equilibrium selectivity ($\alpha_{1,2}$) and is defined as follows:

$$\alpha_{1,2} = \frac{x_1}{x_2} \cdot \frac{y_2}{y_1} \tag{3.59}$$

where x_1, x_2 are the mole fractions of the two components on the adsorbent surface, whereas y_1, y_2 are the corresponding mole fractions in the gas phase. It will be assumed that component (1) is the more strongly adsorbed species. Using the extended Langmuir equation in conjunction with the definition given above (Eq. 3.59), it can be shown that the adsorbent selectivity for component (1) is a constant for the entire range of partial pressures as follows:

$$\alpha_{1,2} = \frac{q_{m_1} b_1}{q_{m_2} b_2} \left(= \frac{K_1}{K_2} \right) \tag{3.60}$$

The product ($q_{mi} b_i$) corresponds to the initial slope of the isotherm, or Henry's constant (K), for component i. Hence, the adsorbent selectivity is equivalent to the ratio of the initial slopes of the isotherms of the two components, or K_1/K_2. It should be noted that the selectivity has resulted in a constant value simply because of the nature of the Langmuir isotherm. If, however, a different model such as the loading ratio correlation (Eq. 3.5) is used, the selectivity is likely to be dependent on the operating pressures of the PSA cycle.

Another important factor for PSA separations is the change in the adsorbed amounts of the two components upon cycling the pressure. The working capacity of a sorbent typically refers to the strongly adsorbed species and is defined as the difference between the adsorbed amounts at the adsorption (high) pressure and the desorption (low) pressure. Strictly speaking, the working capacity should be defined with respect to the adsorbed amounts under the mixture conditions (that is, using a binary component isotherm). However, using a pure component isotherm can suffice to make the calculation of the parameter facile. A ratio of the working capacities of the two components would give an idea about the adsorption performance for a particular pressure swing cycle. Hence, the second factor to the parameter is the working capacity selectivity ratio and is defined as:

$$W = \frac{\Delta q_1}{\Delta q_2} \tag{3.61}$$

Having defined the two contributing factors to the parameter, the PSA sorbent selection parameter (S) can be written as follows:

$$S = W \cdot a_{1,2} \tag{3.62}$$

or in its expanded form (using the Langmuir isotherm):

$$S = \frac{\Delta q_1}{\Delta q_2} \cdot \frac{q_{m_1} b_1}{q_{m_2} b_2} \tag{3.63}$$

The above parameter can thus be used to compare the performance of two sorbents. The better sorbent will have a correspondingly higher value for S.

Returning to the AFM discussed before (Eq. 3.57), its similarity to the parameter S (Eq. 3.63) now becomes obvious. If the equilibrium adsorption data seem to give a good fit to the Langmuir equation (as is usually the case for commonly occurring gas separation applications), then the selectivity $\alpha_{1,2}$ would not differ significantly under adsorption and desorption conditions. As a result, $\alpha_{ads}^2/\alpha_{des}$ would reduce to simply $\alpha_{1,2}$, which is indeed the case with the parameter S defined above. Also the ΔN_2 (or Δq_1) factor is common to both the parameters, the only difference is the addition of the parameter Δq_2 in the denominator. The latter parameter serves to increase the sensitivity of the parameter and, more importantly, makes it a non-dimensional value.

Test of Sorbent Selection Parameter (S). To demonstrate the usefulness of the S parameter in comparing adsorbent performances, examples from published PSA simulation results were used. Both of these examples deal with the production of O_2 by separating it from N_2 present in atmospheric air. The two pairs of sorbents to be considered for comparison are as follows: (1) LiX or LiLSX (Si/Al = 1.0, 100% Li^+ – exchange) and NaX (or 13X), and (2) LiX (Si/Al = 1.0, 100% Li^+ – exchange) and LiAgX (Si/Al = 1.0, 1.1 Ag^+ per unit cell and ~94 Li^+ per unit cell). These two pairs have been previously analyzed for their performance for air separation by PSA using computer simulations of a proven model by Rege and Yang (1997) and Hutson et al. (1999).

A five-step PSA cycle (i.e., a commercially used cycle) was used for both groups of sorbents in this test. The steps involved in each cycle are as follows: (1) pressurization with the feed gas, namely 22% O_2 (mixture of O_2 with Ar impurity included) and 78% N_2; (2) high-pressure adsorption with the feed gas or feed step; (3) co-current depressurization; (4) countercurrent blowdown; (5) countercurrent low-pressure purge with the product of the feed step (oxygen). All the steps above involved equal duration (30 s). Thus the time required for the completion of each PSA cycle was 2.5 min. The model assumed only two adsorbable components, namely O_2 and N_2. The less-strongly adsorbed species, like Ar, were grouped with O_2 with the assumption that all contaminants in air like CO_2 and water vapor were removed completely prior to feeding by using pretreatment beds. The product of each cycle comprised a volumetric mixture of the output stream of the feed step and the co-current depressurization step. This product stream was partly used to purge the bed countercurrently in step (5). The gas was fed to the PSA beds at 298 K.

The model used assumed an adiabatic bed, negligible pressure drop in bed and axial dispersion. Further details about the simulation model used as well as the numerical method are discussed in detail elsewhere (Rege and Yang, 1997; Hutson et al., 1999). The merit of the sorbent pairs was judged by subjecting them to similar PSA cycles and by studying the performance parameters, such as product purity, recovery, and throughput. The cycle parameters were so adjusted that two of these performance parameters were nearly constant and the third was compared to determine the better sorbent.

The first group of sorbents considered includes LiX (Al/Si = 1.0, 100% Li^+ – exchange) and NaX (13X) zeolites. The Langmuir equation parameters for the N_2 and O_2 isotherms at 298 K for this group, as well as LiAgX, are summarized in Table 3.2. The results of the PSA simulation runs with corresponding process parameters previously given by Rege and Yang (1997) are shown in Table 3.3. The performance of the adsorbents for air separation to give about 95% pure O_2 product (with Ar impurity included in the O_2 product) was determined by keeping the product throughput fixed at approximately 0.03 kg O_2/h/kg sorbent at constant product purity. A wide range of pressure ratios (P_H/P_L) from 2 to 10 were considered. It has been shown that LiX is a superior sorbent compared with NaX in the entire range of pressure ratios (Rege and Yang, 1997). The sorbent parameter S was calculated for each of these cases and was plotted against the O_2 product recovery as shown in Figure 3.3. The figure clearly shows an increasing product recovery with the value of the sorbent parameter. More importantly, the value of S for LiX sorbent is much greater than that for NaX, which shows that it is a valid parameter for comparing sorbents.

The second group of sorbents consisted of LiLSX and LiAgX sorbents. The N_2 and O_2 isotherms for these sorbents, as well as a comparison of their performance as sorbents for air separation, appear in Hutson et al. (1999). The parameters of the PSA cycle, as well as the corresponding PSA sorbent selection parameters, are shown in Table 3.3. Two different cycles were used with different adsorption and desorption pressures. In this case the product purity and recovery in each run were kept roughly the same for both sorbents by manipulating the PSA parameters, and the product throughputs were compared. As was shown by Hutson et al. (1999) the LiAgX sorbent (with 1 Ag^+-ion per unit cell) showed a 12% higher product throughput compared with LiLSX sorbent. This is the result of a slightly higher N_2 loading on LiAgX compared with that on LiX. The product throughput for the two runs was plotted against the parameter S and is shown in Figure 3.4.

Table 3.2. Langmuir parameters for N_2 and O_2 isotherms at 298 K on the adsorbents used

Adsorbent	N_2		O_2	
	q_s (mmol/g)	b (atm^{-1})	q_s (mmol/g)	b (atm^{-1})
LiX (Si/Al = 1)	2.653	0.946	2.544	0.086
NaX	0.982	0.901	0.276	0.624
LiAgX	2.635	1.170	1.863	0.131

From Rege and Yang, 2001.

Table 3.3. PSA simulation operating conditions, performance results, and parameter S for the comparison of LiX (Al/Si = 1, 100% Li^+-exchange) vs. NaX adsorbents for air separation (P_H = 1.0 atm, average product purity = 95.5%, average product throughput = 2.7 x 10^{-2} kg O_2/h/kg sorbent)

(A) LiX or LiLSX (Si/Al = 1, 100% Li^+) adsorbent

Pressure Ratio P_H/P_L	*Purge-to-Feed Ratio P/F*	P_{CD} *(atm)*	O_2 *Product Recovery (%)*	*Sorbent Selection Parameter, S*
2	0.44	0.80	53.5	201.7
3	0.23	0.70	61.9	220.9
4	0.13	0.70	63.3	232.0
5	0.09	0.70	64.1	239.3
6	0.06	0.65	65.3	244.4
7	0.04	0.65	65.3	248.1
8	0.03	0.65	64.5	251.0
10	0.02	0.65	64.7	255.2

(B) NaX (13X) adsorbent

Pressure Ratio P_H/P_L	*Purge-to-Feed Ratio P/F*	P_{CD} *(atm)*	O_2 *Product Recovery (%)*	*Sorbent Selection Parameter, S*
2	0.80	0.85	20.0	110.40
3	0.49	0.80	42.3	118.29
4	0.46	0.70	53.3	122.82
5	0.35	0.70	53.4	125.77
6	0.27	0.70	53.2	127.83
7	0.21	0.70	52.8	129.36
8	0.18	0.67	52.8	130.54
10	0.14	0.67	53.1	132.24

From Rege and Yang, 2001.

Although the two sorbents have a subtle difference in performance, the better one involving the LiAgX sorbent is reflected accurately by the parameter. In both cases S was higher for LiAgX compared with LiLSX under the said operating pressures. Thus the sorbent selection parameter is shown to be sensitive to even slight differences in the adsorption isotherms.

The S parameter has been defined for Langmuir systems, that is, the isotherms for both components are Langmuirian. For non-Langmuirian systems, α_{12} probably should be replaced by $\alpha_{ads}^2/\alpha_{des}$, as given by the AFM, because this ratio was empirically determined. It seems that the α during the adsorption step is more important than during the desorption step in determining the PSA performance.

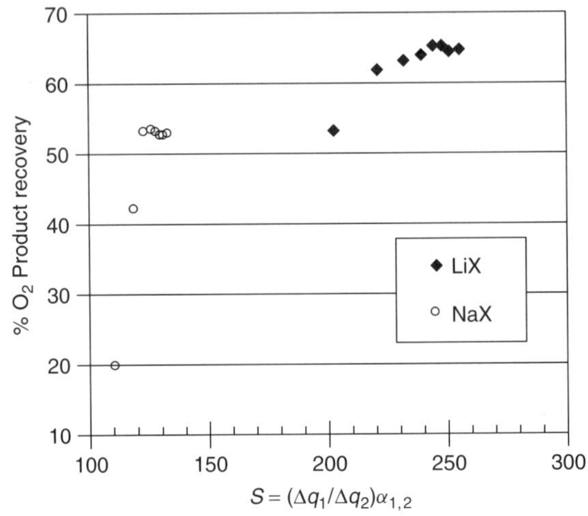

Figure 3.3. O_2 product recovery (%) vs. sorbent selection parameter S for the comparison of LiX (LiLSX) and NaX adsorbents for air separation at different pressure ratios. Refer to Table 3.3 for additional details (from Rege and Yang, 2001).

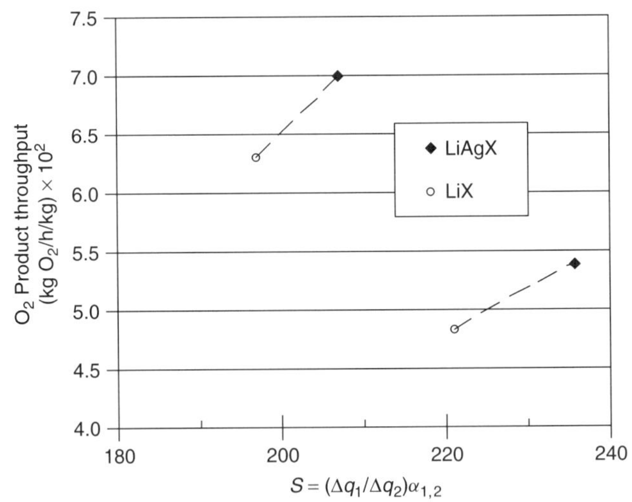

Figure 3.4. O_2 product throughput (kg O_2/h/kg adsorbent) × 10^2 vs. sorbent selection parameter S for the comparison of LiX (LiLSX) and LiAgX adsorbents for air separation under different cycle conditions. Refer to Table 3.4 for additional details (from Rege and Yang, 2001).

Sorbent Selection Parameters for Other Processes. The above analysis pertains to bulk separation. For purification processes, that is, PSA, and TSA, as well as liquid phase applications, no study has been reported for establishing a sorbent selection parameter. In this case, obviously $\alpha_{1,2}$ for the adsorption step

Table 3.4. PSA simulation operating conditions, performance results, and sorbent selection parameter S for the comparison of LiAgX (1 Ag^+-ion and 95 Li^+ per unit cell) vs. LiX (Si/Al = 1, 100% Li^+-exchange) adsorbents for air separation ($P_H = 1.0$ atm)

Run 1

Sorbent	P_H *(atm)*	P_L *(atm)*	P_{CD} *(atm)*	O_2 *Product Purity (%)*	O_2 *Product Recovery (%)*	*Product Throughput* (kg O_2/H/kg *Adsorbent)* $\times 10^2$	*Parameter S*
LiLSX	1.0	0.33	0.70	96.11	62.03	4.84	220.9
LiAgX	1.0	0.33	0.69	96.42	62.74	5.40	235.6

Run 2

Sorbent	P_H *(atm)*	P_L *(atm)*	P_{CD} *(atm)*	O_2 *Product Purity (%)*	O_2 *Product Recovery (%)*	*Product Throughput* (kg O_2/H/kg *Adsorbent)* $\times 10^2$	*Parameter S*
LiLSX	1.2	0.4	0.70	90.68	78.02	6.31	196.9
LiAgX	1.2	0.4	0.71	90.83	78.48	7.01	207.0

From Rege and Yang, 2001.

will be the dominating factor, and the factor $\Delta q_1/\Delta q_2$ should not be weighted equally as $\alpha_{1,2}$.

For kinetic separations by PSA, a simple parameter could be defined as the ratio of the amounts of uptake for the two competing components during the adsorption step. Assuming a step change in the gas phase concentration, clean beds initially, and linear isotherms, the amounts can be expressed by the short-time solution of the diffusion equation (Carslaw and Jaeger, 1959):

$$\frac{q_t}{q_\infty} = \frac{6}{r_c}\sqrt{\frac{Dt}{\pi}} \tag{3.64}$$

and for linear isotherms:

$$q_t = KP\frac{6}{r_c}\sqrt{\frac{Dt}{\pi}} \tag{3.65}$$

Habgood (1958) expressed the separation factor for kinetic separation as the product of the equilibrium separation factor and the square root of the diffusivity ratio. This definition provides a reasonable estimate of the true separation factor in the short time regime. Following this idea, the kinetic separation factor is given by (Ruthven et al., 1994):

$$\alpha_k = \frac{q_1/P_1}{q_2/P_2} = \frac{K_1}{K_2}\sqrt{\frac{D_1}{D_2}} \tag{3.66}$$

This parameter is incomplete because it does not include the actual amount adsorbed during the adsorption step. The sorbent productivity is directly dependent upon the amount adsorbed. Thus a complete parameter should include this amount, as

$$S_k = q_t \frac{K_1}{K_2}\sqrt{\frac{D_1}{D_2}} = \frac{6K_1^2 P_1}{\pi^2 K_2}\sqrt{\frac{D_1}{D_2}}\sqrt{\frac{D_1 t}{r_c^2}} \tag{3.67}$$

Here S is the sorbent selection parameter, component "1" is the fast diffusing component, t is the step time for adsorption, and P is the partial pressure in the feed mixture. For a given feed mixture and step time t, the parameter may be simplified as follows when comparing different sorbents:

$$S_k = \frac{K_1^2}{K_2}\frac{D_1}{\sqrt{D_2}} \tag{3.68}$$

where D may be replaced by D/r_c^2.

Neither of these parameters has been tested. The main commercial kinetic separation is nitrogen production from air by using molecular sieve carbon. The cycle used is a simple Skarstrom cycle. This parameter may be tested for this system.

As discussed, for the full assessment of a sorbent, PSA simulation is needed. Kinetic PSA simulations with various degrees of simplifications are available. For instance, Schork et al. (1993) reported a kinetic PSA model with the assumption that the solid-phase concentration profiles along the bed are linear, and the computation can be substantially reduced. Their simplified model was used to evaluate carbon molecular sieves for N_2 production from air.

NOTATION

a	partial molar area
A	surface area of sorbent
b	Langmuir constant
B	Langmuir constant
c	concentration
C	dispersion constant; average number of sorbate molecules per cage in zeolite
D	diffusivity
D_s	surface diffusivity
D_{so}	surface diffusivity at zero coverage
E	interaction energy
f	fugacity; partition function
$f,$	fugacity at saturation
F	field strength
G	Gibbs free energy
h	Planck constant

H	enthalpy
k	Boltzmann constant; Henry's constant
K	Henry's constant
M	mass
n	amount adsorbed in moles/g
N,	Avogadro number
P	total pressure
P_0	saturation vapor pressure
q	amount adsorbed, usually in mmol/g
q_t	total amount adsorbed
q_m	monolayer or saturated amount adsorbed
Q	heat of adsorption; quadrupole moment
r	pore radius; distance of separation
R	gas constant
S	sorbent selectivity parameter
T	temperature
v	molar volume
V	amount adsorbed in volume (STP)/g
V_0	monolayer amount adsorbed
V_m	molar volume of adsorbate
V_O	limiting volume of adsorbed space (= micropore volume)
W	working capacity ratio
X	mole fraction in adsorbed phase
Y	mole fraction in gas phase
α	polarizability; adsorption constant (= sticking probability); selectivity ratio
β	desorption constant; affinity coefficient; or parameter defined by Eqs. 3.51 and 3.55
δ	mean distance of each movement
ε	potential energy field over surface; effective bond energy
ε_{AB}	pairwise interaction parameter
λ	ratio of sticking probabilities
μ	chemical potential; permanent dipole
ν	vibrational frequency of the bond
π	spreading pressure
θ	fractional surface coverage
ρ	density

REFERENCES

Ackley, M. W. Multilayer Adsorbent Beds for PSA Gas Separation. EP 0875279A2, (1998), to Praxair.

Ackley, M. W., Stewart, A. B., Henzler, G. W., Leavitt, F. W., Notaro, F., and Kane, M. S. PSA apparatus and process using adsorbent mixtures. U.S. Patent 6,027,548 (2000), to Praxair.

Alexis, R. W. (1967) *Chem. Eng. Prog.* 63(5), 69.

Alpay, E., Kenny, C. N., and Scott, D. M. (1993) *Chem. Eng. Sci.* 48, 3173.

Armond, J. W. (1979) *Properties and Applications on Zeolites*. (R. P. Townsend, ed.), Special Publ. No. 33. Chemical Society, London, p. 92.

Avery, W. F. and Lee, M. N. Y. (1962) *Oil and Gas J.* June, 121.

Baksh, M. S. A. and Notaro, F. Optimal pressure swing adsorption refluxing, U.S. Patent 5,565,018 (1996), to Praxair, Inc.

Basmadjian, D. (1997) *The Little Adsorption Book*. CRC Press, Boca Raton, FL.

Basmadjian, D., Ha, K. D., and Pan, C. Y. (1975a) *Ind. Eng. Chem. Proc. Des. Dev.* 14, 328.

Basmadjian, D., Ha, K. D., and Proulx, D. P. (1975b) *Ind. Eng. Chem. Proe. Des. Dev.* 14, 340.

Belfort, G. (1981) *AIChE J.* 27, 1021.

Bering, B. P., Serpinsky, V. V., and Surinova, S. I. (1963) *Dokl. Akad. Nauk. SSSR* 153, 129.

Bering, B. P., Serpinsky, V. V., and Surinova, S. I. (1965) *Izv. Skad. Nauk, SSSR, Otd. Khim. Nauk*, 769.

Berlin, N. H. U.S. Patent 3,280,536 (1966), to Esso Research and Engineering Company.

Bird, R. B., Stewart, W. E., and Lightfoot, E. N. (1960) *Transport Phenomena*. Wiley, New York.

Bunke, G. and Gelbin, D. (1978) *Chem. Eng. Sci.* 33, 101.

Carslaw, H. S. and Jaeger, J. C. (1959) *Conduction of Heat in Solids*, 2nd Ed. Oxford University Press, Oxford, UK.

Carter, J. W. (1975) *AIChE J.* 21, 380.

Carvill, B. T., Hufton, J. R., Anand, M., and Sircar, S. (1996) *AIChE J.* 42, 2765.

Cassidy, R. T. and Holmes, E. S. (1984) *AIChE Symp. Ser.* 80 (No. 233), 68.

Cen, P. L. and Yang, R. T. (1986) *Ind. Eng. Chem. Fundam.* 25, 758.

Chan, Y. N. I., Hill, F. B. and Wong, Y. W. (1981) *Chem. Eng. Sci.*, 36, 243.

Chao, C. C. Process for separating nitrogen from mixtures thereof with less polar substances. U.S. Patent, 4,859,217 (1989), to UOP.

Chatsiriwech, D., Alpay, E., and Kershenbaum, L. S. (1994) *Catal. Today.* 20, 351.

Chen, S. G. and Yang, R. T. (1994) *Langmuir* 11, 4244.

Chen, S. G. and Yang, R. T. (1996) *J. Coll. Interf. Sci.* 177, 298.

Chen, Y. D. and Yang, R. T. (1992) *Chem. Eng. Sci.*, 47, 3895.

Chen, Y. D., Yang, R. T. and Uawithya (1994) *AIChE J.*, 40, 577.

Cheng, Y. S., Alpay, E., and Kershenbaum, L. S. (1998) *Comput. Chem. Eng.* 22, S45.

Chi, C. W. (1978) *AIChE Symp. Ser.* 74 (No. 179), 42.

Chi, C. W. and Cummings, W. P. (1978) Adsorptive separation processes: gases. In *Kirk Othmer Encyclopedia of Chemical Technology*, 3rd Ed., Vol. I. Wiley Interscience, New York.

Chiang, A. S. T. and Hong, M. C. Radial flow rapid pressure swing adsorption, Patent 239,082, Republic of China, January 21, (1995a).

Chiang, A. S. T. and Hong, M. C. (1995b) *Adsorption* 1, 153.

Chitra, R. and Yashonath, S. (1995) *Chem. Phys. Lett.* 234, 16.

Crittenden, B. and Thomas, W. J. (1998) *Adsorption Technology and Design*. Butterworth-Heinemann, Oxford, UK.

Cho, S. H. (2001) Korea Institute of Energy Research, Taejon, Korea, personal communication.

Chou, C. T., Peng, F. S., and Hsia, L. H. (1998) *Fundamentals of Adsorption* (F. Meunier, ed.). Elsevier, Amsterdam, p. 781.

Chue, K. T., Kim, J. N., Yoo, Y. J., Cho, S. H., and Yang, R. T. (1995) *Ind. Eng. Chem. Res.* 34, 591.

Chue, K. T., Bum, H. T., Kim, D. S., and Cho, S. H. (2000) *Adsorption Science and Technology*. (D. D. Do, ed.). World Scientific Publishers, River Ridge, NJ, p. 159.

Darken, L. (1948) *Trans. AIME* 174, 184.

Davis, M. M. and LeVan, M. D. (1989) *Ind. Eng. Chem. Res.* 28, 778.

Ding, Y. and Alapy, E. (2000) *Chem. Eng. Sci.* 55, 3461.

Do, D. D. (1998) *Adsorption Analysis: Equilibrium and Kinetics*. Imperial College Press, London.

Doong, S. J. and Yang, R. T. (1988) *Ind. Eng. Chem. Res.* 27, 630.

Dubinin, M. M. (1960) *Chem. Rev.* 60, 235.

Fuderer, A. and Rudelstorfer, E. U.S. Patent 3,986,849 (1976), to Union Carbide Corporation.

Gaffney, T. R., Kirner, J. F., Kumar, R., Maliszewskyj, R. J., and Schmidt, W. P. O_2 VSA process with low O_2 capacity adsorbents. U.S. Patent 5,266,102 (1993), to Air Products and Chemicals, Inc.

Grant, R. J. and Manes, M. (1966) *Ind. Eng. Chem. Fundam.* 5, 490.

Guerin de Montgareuil, P. and Domine, D. U.S. Patent 3,155,468 (1964), to Soci&~ L'Air Liquide, Paris.

Haase, R. and Siry, M. (1968) *Z. Phys. Chem. (Frankfurt)* 57, 56.

Habgood, H. W. (1958) *Can. J. Chem.* 36, 1384.

Heinze, G. Belgian Patent 613,267 (1962), to Farbenfabriken Bayer A.G.

Higashi, K., Ito, H., and Oishi, J. (1963) *J. Atomic Energy Soc. Japan* 5, 846.

Hufton, J. R., Mayorga, S., and Sircar, S. (1999) *AIChE J.* 45, 248.

Hufton, J. R., Mayorga, S., and Sircar, S. (1999) *AIChE J.* 45, 248.

Humphrey, J. L. and Keller, II, G. E. (1997) *Separation Process Technology*. McGraw Hill, New York.

Hutson, N. D., Rege, S. U., and Yang, R. T. (1999) *AIChE J.* 45, 724.

Hutson, N. D. and Yang, R. T. (1997) *Adsorption* 3, 189.

Jaroniec, M. and Madey, R. (1988) *Physical Adsorption on Heterogeneous Solids*. Elsevier, Amsterdam.

Kapoor, A., Ritter, J. A., and Yang, R. T. (1989a) *Langmuir* 5, 1118.

Kapoor, A. and Yang, R. T. (1988) *Gas Separ. Purification* 3, 187.

Kapoor, A., Yang, R. T., and Wong, C. (1989b) Surface Diffusion, *Catal. Rev.- Sci. Eng.* 31, 129.

Karger, G. and Ruthven, D. M. (1992) *Diffusion in Zeolites and Other Microporous Materials*. Wiley, New York.

Kauzmann, W. (1966) *Kinetic Theory of Gases*. Benjamin, New York.

Keller 11, G. E. (1983) Gas-adsorption processes: state of the art. In *Industrial Gas Separations*. (T. E. Whyte, Jr., C. M. Yon, and E. H. Wagener, eds.), ACS Symp. Ser. 223. American Chemical Society, Washington, D.C.

Kikkinides, E. S. and Yang, R. T. (1991) *Ind. Eng. Chem. Res.* 30, 1981.

Kikkinides, E. S. and Yang, R. T. (1993) *Ind. Eng. Chem. Res.* 32, 2365.

Kikkinides, E. S., Yang, R. T., and Cho, S. H. (1993) *Ind. Eng. Chem. Res.* 32, 2714.

Kiyonaga, K. U.S. Patent 3,176,444 (1965), to Union Carbide Corporation.

Knaebel, K. S. (1995) *Chem. Eng.* 102(11), 97.

Kodde, A. J., Fokma, Y. S. and Bliek, A. (2000) *AIChE J.* 46, 2295.

Krishna, R. (1990) *Chem. Eng. Sci.* 45, 1779.

Lee, H. and Stahl, D. E. (1973) *AIChE Symp. Ser.* 69 (No. 134), 1.

Marsh, W. D., Pramuk, F. S., Hoke, R. C., and Skarstrom, C. W. U.S. Patent 3,142,547 (1964), to Esso Research and Engineering Company.

Myers, A. L. and Prausnitz, J. M. (1965) *AIChE J.* 11, 121.

Notaro, F., Mulhaupt, J. T., Leavitt, F. W., and Ackley, M. W. Adsorption process and system using multilayer adsorbent beds. U.S. Patent 5,810,909 (1998), to Praxair.

Pan, C. Y. and Basmadjian, D. (1970) *Chem. Eng. Sci.* 25, 1653.

Poteau, M. and Eteve, S. Adsorber comprising annular superposed beds of adsorbent materials, U.S. Patent 5.232,479 (1993), to L'Air Liquide.

Qureshi, W. R. and Wei, J. (1990) *J. Catal.* 126, 126.

Rege, S. U. and Yang, R. T. (1997) *Ind. Eng. Chem. Res.* 36, 5358.

Rege, S. U. and Yang, R. T. (2001a) *Chem. Eng. Sci.* 56, 3781.

Rege, S. U. and Yang, R. T. (2001b) *Separ. Sci. Tech.* 36, 3355.

Rege, S. U., Yang, R. T., and Buzanowski, M. A. (2000) *Chem. Eng. Sci.* 55, 4827.

Ritter, J. A. and Al-Muhtaseb, S. A. Paper presented at AIChE Annual Meeting, Reno, (November, 2001).

Ritter, J. A. and Yang, R. T. (1991) *Ind. Eng. Chem. Res.* 30, 1023.

Rota, R. and Wankat, P. C. (1991) *Proc. Adsorption Processes for Gas Separation*. (F. Meunier and M. D. LeVan, eds.). GFGP, Nancy, France, pp. 143–148.

Rudzinski, W. and Everett, D. H. (1992) *Adsorption of Gases on Heterogeneous Surfaces*. Academic Press, San Diego.

Ruthven, D. M., Farooq, S., and Knaebel, K. S. (1994) *Pressure Swing Adsorption*. VCH, New York.

Scamehorn, J. F. (1979) *Ind. Eng. Chem. Proc. Des. Dev.* 18, 210.

Schork, J. M. and Fair, J. R. (1988) *Ind. Eng. Chem. Res.* 27, 1547.

Schork, J. M., Srinivasan, R., and Auvil, S. R. (1993) *Ind. Eng. Chem. Res.* 32, 2226.

Schweiger, T. A. J. and LeVan, M. D. (1993) *Ind. Eng. Chem. Res.* 32, 2418.

Sheikh, J., Kershenbaum, L. S., and Alpay, E. (2001) *Chem. Eng. Sci.* 56, 1511.

Shewmon, P. G. (1963) *Diffusion in Solids*, Chap. 4. McGraw-Hill, New York.

Skarstrom, C. W. U.S. Patent 2,944,627 (1960), to Esso Research and Engineering Company.

Skarstrom, C. W. (1972) Heatless fractionation of gases over solid sorbents. In *Recent Developments in Separation Science*. (N. N. Li, ed.), Vol. 2. CRC Press, Cleveland.

Sips, R. (1948) *J. Chem. Phys.* 16, 490.

Sips, R. (1950) *J. Chem. Phys.* 18, 1024.

Sircar, S. and Zondlo, J. W. U.S. Patent 4,013,429 (1977), to Air Products and Chemicals, Inc.

Smolarek, J., Leavitt, F. W., Nowobilski, J. J., Bergsten, V. E., and Fassbaugh, J. H. Radial bed vacuum/pressure swing adsorber vessel, U.S. Patent, 5,759,242 (1998), to Praxair Technology, Inc.

Snurr, R. Q. and Karger, J. (1997) *J. Phys. Chem. B.* 101, 6469.

Suzuki, M. (1990) *Adsorption Engineering*. Elsevier, Amsterdam.

Symoniak, M. F. (1980) *Hydrocarbon Process.* 110, May.

Takahashi, A., Yang, R. T., Munson, C. L., and Chinn, D. (2001) *Ind. Eng. Chem. Res.* 40, 3979.

Tamura, T. French Patent 1,502,458 (1967); Ger. Offen. 1,817,004 (1969); U.S. Patent 3,533,221 (1970), to T. Tamura of Tokyo, Japan.

Tamura, T. U.S. Patent 3,797,201 (1974), to T. Tamura, Tokyo, Japan.

Tien, C. (1994) *Adsorption Calculations and Modeling*. Butterworth-Heinemann, Wonurn, MA.

Trout, B. L., Chakraborty, A. K., and Bell, A. T. (1997) *Chem. Eng. Sci.* 52, 2265.

Vaporciyan, G. G. and Kadlec, R. H. (1987) *AIChE J.* 33, 1334.

Vaporciyan, G. G. and Kadlec, R. H. (1989) *AIChE J.* 35, 831.

Vignes, A. (1966) *Ind. Eng. Chem. Fundam.* 5, 189.

Wagner, J. L. U.S. Patent 3,430,418 (1969), to Union Carbide Corporation.

Waldron, W. E., Hufton, J. R., and Sircar, S. (2001) *AIChE J.* 47, 1477.

Wankat, P. C. (1986) *Large-Scale Adsorption and Chromatography*, Volumes 1 and II. CRC Press, Boca Raton, FL.

Wankat, P. C. and Partin, L. R. (1980) *Ind. Eng. Chem. Proc. Des. Dev.* 19, 446.

Wei, J. (1994) *Ind. Eng. Chem. Res.* 33, 2467.

Wood, G. O. (2002) *Carbon* 40, 231.

Yang, R. T. (1997) *Gas Separation by Adsorption Processes*. Butterworth, 1987, reprinted by Imperial College Press, London.

Yang, R. T., Fenn, J. B., and Haller, G. L. (1973) *AIChE J.* 19, 1052.

Yang, R. T. and Doong, S. J. (1985) *AIChE J.* 11, 1829.

Yon, C. M. and Turnock, P. H. (1971) *AIChE Symp. Ser.*, No. 117, 67, 3.

Yongsunthon, I. and Alpay, E. (1998) *Chem. Eng. Sci.* 53, 691.

Yongsunthon, I. and Alpay, E. (2000) *Chem. Eng. Sci.* 55, 205.

Yoshida, M., Kodama, A., Goto, M., and Hirose, T. (1998) *Fundamentals of Adsorption*. (F. Meunier, ed.). Elsevier, Amsterdam, p. 817.

Yoshida, M., Kodama, A., Goto, M., Hirose, T., and Ritter, J. A., (2000) *Adsorption Science and Technology*. (D. D. Do, ed.). World Scientific Publishers, River Ridge, NJ, p. 254.

Zeldowitsch, J. (1934) *Acta Physicochim. U.R.S.S.* 1, 961.

4

PORE SIZE DISTRIBUTION

According to the IUPAC classification of pores, the size ranges are micropores (<2 nm), mesopores (2–50 nm), and macropores (>50 nm) (IUPAC, 1972). All useful sorbents have micropores. The quantitative estimation of pore size distribution (PSD), particularly for the micropores, is a crucial problem in the characterization of sorbents. Numerous methods exist, of which three main methods will be described: Kelvin equation (and the BJH method), Horváth–Kawazoe approach, and the integral equation approach.

4.1. THE KELVIN EQUATION

The vapor pressure (P) of a liquid over a meniscus that stretches across a cylindrical pore is given by (Adamson and Gast, 1997; Gregg and Sing, 1982; Rouquerol et al., 1999):

$$\ln \frac{P}{P_0} = -\frac{2\gamma V}{rRT} \tag{4.1}$$

Here P_0 is the saturated vapor pressure at absolute temperature T, γ is the surface tension, V is the molar volume of the liquid, r is the radius of the cylindrical pore, and R is the gas constant.

The Kelvin equation is used to interpret type IV isotherms, which are the general isotherms for adsorption on mesoporous materials (Gregg and Sing, 1982; Yang, 1987). This equation applies to the range of the isotherm that corresponds to capillary condensation, that is, the section of the isotherm above the first "knee." Using the Kelvin equation, one can calculate the value of the pore radius for any point on the isotherm. That is, for any set of (q, P/P_0) along the isotherm, there is a corresponding pore radius. According to the capillary condensation

Adsorbents: Fundamentals and Applications, Edited By Ralph T. Yang
ISBN 0-471-29741-0

hypothesis, all the pores with radii smaller than that size become filled at that pressure. Consequently one can obtain the cumulative pore volume as a function of pressure. The cumulative pore volume is obtained by assuming that the pores are filled by liquid, or q/ρ, where ρ is the liquid density. The pore size distribution is then obtained by taking the derivative of the cumulative pore volume as a function of pore radius.

N_2 isotherms at 77 K are used for practical reasons (e.g., simultaneous determination of the BET surface area). The use of the Kelvin equation was a popular approach for estimating the pore size distribution. Many procedures were proposed for calculating the pore size distribution from the N_2 isotherms over the period between 1945 and 1970 (Rouquerol et al., 1999). The method proposed by Barrett, Joyner, and Halenda (1951), known as the BJH method, continues to be used today. In the BJH method, the desorption branch of the isotherm is used, which is the desorption branch of the usual hysteresis loop of the isotherm for the mesoporous sorbent. The underlying assumptions for this method are

(a) All pores are non-intersecting, cylindrical pores.
(b) Hemispherical meniscus with zero contact angle, or complete wetting.
(c) The simple Kelvin equation is applied.
(d) Validity of the correction for multilayer adsorption.

Many questions have been raised concerning the validity of the Kelvin equation, in particular, the lower limit of the pore size that one could use with this approach. Clearly, this method does not apply when the pore size approaches molecular dimensions, or a few molecular sizes. Molecular simulations by Jessop et al. (1991) showed that the Kelvin equation fails to account for the effects of fluid-wall interactions. The density functional theory study of Lastoskie et al. (1993), as well as other work, indicated that the Kelvin equation would underestimate the pore size and should not be extended below a pore size of ~7.5 nm.

4.2. HORVÁTH–KAWAZOE APPROACH

A simple and popular method for evaluating PSD of microporous materials was proposed by Horváth and Kawazoe (1983). This technique has been successfully used for the determination of pore size distribution in microporous sorbents such as activated carbons and zeolites (e.g., Seifert and Emig, 1987; Venero and Chiou, 1988; Davis et al., 1988, 1989; Beck et al., 1992; Horváth et al., 1998, Kane et al., 1998). This technique utilizes adsorption isotherm data at a temperature below or equal to the critical temperature of the adsorbate, which is typically nitrogen (at 77 K), argon (at 87 K), or an organic vapor (e.g., CH_3Cl) at ambient temperature (Mariwala and Foley, 1994). By assuming micropore-filling and equating the free-energy change upon adsorption to the average interaction energy of the adsorbing molecules, the "step" in the isotherm data is translated into a pore-size distribution. The model assumes that entropic effects are negligible for small adsorbate molecules and that networking effects are insignificant if the analysis adsorbate is smaller than the adsorbent channels. The mathematical expression relating relative pressure of the adsorbing gas to the pore size of the adsorbent has come to be known as the Horváth-Kawazoe (HK) model.

Over the past two decades, different types of HK models have been developed depending upon the pore geometry. The original HK model discussed slit-shaped pores (Horváth and Kawazoe, 1983), whereas models for cylindrical pores (Saito and Foley, 1991), and spherical pores (Cheng and Yang, 1994) have also been proposed. The basic framework for all the different HK models is the same:

$$RT \ln\left(\frac{P}{P_0}\right) = U_0 + P_a \tag{4.2}$$

where U_0 and P_a denote the sorbate–sorbent and sorbate–sorbate interaction energies, respectively. Thus the R.H.S. is a function of the pore geometry and dimension, which is related to the relative pressure of the adsorbate. The calculation of the PSD is now relatively simple. By using different values of a pore dimension (pore width in case of a slit-pore, pore diameter in case of pores with curvature), the threshold sorption relative pressure (P/P_0) at which the pore filling will occur (or "filling" pressure) can be obtained over the expected pore size range. From the adsorption measurements using a suitable sorbate, the fractional adsorbed amount ($q = w/w_\infty$) is obtained as a function P/P_0. From a combination of the above two functional relationships, the adsorbed amount w/w_∞ can be plotted as a one-to-one function of pore dimension L, thus giving the cumulative PSD. A differential PSD can be further obtained by calculating the derivative $d(w/w_\infty)/dL$ as a function of L.

The original HK equation included an implicit assumption that the adsorption isotherm follows Henry's law. In order to correct for the deviation of isotherm data from Henry's law at higher relative pressures, Cheng and Yang (1994) proposed a correction factor, which incorporated a Langmuir isotherm fit to the data. The HK model with the Cheng–Yang correction (henceforth referred to as the HK–CY equation) is given as:

$$RT \ln\left(\frac{P}{P_0}\right) + \left[RT - \frac{RT}{\theta} \ln \frac{1}{1-\theta}\right] = U_0 + P_a \tag{4.3}$$

Despite the immense utility of the HK model, there exist certain conceptual defects in the original model. The derivation of the original HK model proceeds by calculating the energy potential of a single adsorbate molecule with a layer of sorbent molecules along the pore periphery of a particular geometry by using the Lennard–Jones 6–12 potential. This potential is calculated by incorporating the adsorbate–adsorbent dispersion term as calculated by the Kirkwood–Müller formalism in the potential energy minimum (Horváth and Kawazoe, 1983). In order to include the adsorbate–adsorbate interactions, the adsorbate–adsorbate dispersion constant is appended to the adsorbate–adsorbent dispersion term in the potential energy minimum, as will be shown in detail later. However, no clear explanation is provided for doing so. Moreover, the energy potential due to adsorbate–adsorbate interactions in a filled pore can be interpreted as the energy profile due to the adsorbate molecules placed, rather unrealistically, at the same position as the adsorbent molecules. This is because, according to the original model, the distance z of a sorbate molecule from another sorbate molecule is the same as that between a sorbate molecule and a sorbent molecule. Furthermore, the average interaction energy is calculated by integration of this energy profile over the entire pore characteristic length (pore width for slit-pores or radius in case of curved geometries). This implies that in a filled pore there is a continuous distribution of molecules along the characteristic length. However, micropores have dimensions comparable with molecular widths, and hence the molecules in a filled pore can occupy only discrete positions relative to each other, if minor thermal vibrations are neglected. The original HK model has also been criticized for not taking the distribution of energy into account while calculating the average interaction energy (Webb and Orr, 1997). As a result of these and other shortcomings, it has been observed that pore sizes estimated by the original HK method for materials in the higher micropore size range (8–20 Å) are unrealistically low in some cases (e.g., Carrott et al., 1998).

The original HK model will be given first, followed by the corrected model by Rege and Yang (2000).

4.2.1. The Original HK Slit-Shaped Pore Model

The basis for obtaining the energy profile in a pore is the Lennard–Jones 6–12 potential:

$$\varepsilon_{12}(z) = 4\varepsilon_{12}^{*}\left[\left(\frac{\sigma}{z}\right)^{12} - \left(\frac{\sigma}{z}\right)^{6}\right] \tag{4.4}$$

Halsey and co-workers gave the interaction energy of one adsorbate molecule with a single infinite-layer plane of adsorbent molecules to be (Sams et al., 1960)

$$\varepsilon(z) = \frac{N_S A_S}{2\sigma^4}\left[-\left(\frac{\sigma}{z}\right)^{4} + \left(\frac{\sigma}{z}\right)^{10}\right] \tag{4.5}$$

Everett and Powl (1976) extended the above result to two parallel infinite lattice planes whose nuclei are spaced at distance L apart:

$$\varepsilon(z) = \frac{N_S A_S}{2\sigma^4}\left[-\left(\frac{\sigma}{z}\right)^{4} + \left(\frac{\sigma}{z}\right)^{10} - \left(\frac{\sigma}{L-z}\right)^{4} + \left(\frac{\sigma}{L-z}\right)^{10}\right] \tag{4.6}$$

where the internuclear distance at zero-interaction energy (in Angstrom): $\sigma = (2/5)^{1/6} d_0$; d_0 is the average of the adsorbate and adsorbent molecule diameters, i.e., $(d_0 + d_A)/2$, N_S is the number of sorbent molecules per unit area, and z is the internuclear distance between the adsorbate and adsorbent molecules (in Angstrom). The dispersion constants A_S and A_A are calculated by the Kirkwood–Müller formulae (erg $\times$ cm^6) as follows:

$$A_S = \frac{6mc^2\alpha_S\alpha_A}{\dfrac{\alpha_S}{\chi_S} + \dfrac{\alpha_A}{\chi_A}} \tag{4.7}$$

$$A_A = \frac{3}{2}mc^2\alpha_S\alpha_A \tag{4.8}$$

Further, Horváth and Kawazoe (1983) proposed that the potential is increased by the interaction of adsorbate molecules within the pore. They included this additional interaction by adding an adsorbate dispersion term ($N_A A_A$) in the numerator of the depth of potential energy minimum ($N_S/2\sigma^4$) in Eq. 4.6 as follows:

$$\varepsilon(z) = \frac{N_S A_S + N_A A_A}{2\sigma_4}\left[-\left(\frac{\sigma}{z}\right)^4 + \left(\frac{\sigma}{z}\right)^{10} - \left(\frac{\sigma}{L-z}\right)^4 + \left(\frac{\sigma}{L-z}\right)^{10}\right] \tag{4.9}$$

However, no clear justification could be found in the literature for incorporating the adsorbate–adsorbate–adsorbent interaction in this manner. Further, $\varepsilon(z)$ in Eq. 4.9 may be split as follows:

$$\begin{aligned}\varepsilon(z) &= \varepsilon_{A-S}(z) + \varepsilon_{A-A}(z)\\ &= \frac{N_S A_S}{2\sigma^4}\left[-\left(\frac{\sigma}{z}\right)^4 + \left(\frac{\sigma}{z}\right)^{10} - \left(\frac{\sigma}{L-z}\right)^4 + \left(\frac{\sigma}{L-z}\right)^{10}\right]\\ &\quad + \frac{N_A A_A}{2\sigma^4}\left[-\left(\frac{\sigma}{z}\right)^4 + \left(\frac{\sigma}{z}\right)^{10} - \left(\frac{\sigma}{L-z}\right)^4 + \left(\frac{\sigma}{L-z}\right)^{10}\right]\end{aligned} \tag{4.10}$$

The first term $\varepsilon_{A-S}(z)$ gives the adsorbate-surface interaction and the second term $\varepsilon_{A-A}(z)$ gives the adsorbate–adsorbate interaction. Because the distance z of the gas molecule is the same in both $\varepsilon_{A-S}(z)$ as well as $\varepsilon_{A-A}(z)$, Eq. 4.10 gives an erroneous interpretation that the adsorbate–adsorbate interaction is caused between a gas molecule and two parallel infinite sheets of gas molecules, impractically placed at the *same* position as the sorbent molecules. Besides, the internuclear distance at zero interaction energy (σ) used in $\varepsilon_{A-A}(z)$ is expected to differ from that used in the $\varepsilon_{A-S}(z)$ expression; however, the original model does not take this fact into account.

The next step in the derivation involved obtaining the average interaction energy by integrating the above profile over free space in the slit-pore:

$$\bar{\varepsilon}(z) = \frac{\displaystyle\int_{d_0}^{L-d_0} \varepsilon(z)\,dz}{\displaystyle\int_{d_0}^{L-d_0} dz} \tag{4.11}$$

This implies that a gas molecule has a continuous distribution of positions in the filled micropore, which it is free to occupy. In reality, however, the gas molecule has only fixed positions at which it can exist when the pore is filled to capacity, especially when the pore width is only a few multiples of the adsorbate molecule diameter. Besides, the Boltzmann law of distribution of energy requires that the above average be instead computed by weighing the energy terms with the probability of a molecule to possess that energy, namely $e^{-\varepsilon/kT}$. In other words the molecule is expected to occupy a position at which the energy potential is minimum. This aspect was not considered in the original work. In the present work, a discretized average, which attempts to incorporate the above described features in the model, is proposed.

Finally, the average potential energy that is calculated is related to the free energy change upon adsorption, $RT\ln(P/P_0)$. The resulting slit-pore HK model is as follows:

$$RT\ln\left(\frac{P}{P_0}\right) = N_{Av}\cdot\overline{\varepsilon}(z) = N_{Av}\frac{N_S A_S + N_A A_A}{\sigma^4(L-2d_0)} \times\left[\frac{\sigma^4}{3(L-d_0)^3} - \frac{\sigma^{10}}{9(L-d_0)^9} - \frac{\sigma^4}{3d_0^3} + \frac{\sigma^{10}}{9d_0^9}\right] \tag{4.12}$$

As can be seen from Eq. 4.12, the interaction energy of a molecule is *indeterminate* when $L = 2d_0$, that is, when the free space of the pore $(L - d_S)$ equals the diameter of an adsorbate molecule (d_A). However, a finite value for the potential exists for a molecule of this size, as is obvious from Eq. 4.9. This anomaly is overcome in the modified model.

An implicit assumption in the original derivation by Horváth and Kawazoe (1983) was that the isotherm followed Henry's law. It is well known that most isotherm data show a considerable deviation from Henry's law at pressures near the saturation pressure of the adsorbate, and the equilibrium loading usually follows a Langmuir-type behavior. Cheng and Yang (1994) introduced a correction term to incorporate the Langmuir isotherm in the model and the corrected model (referred to as the HK–CY model) was given as:

$$RT\ln\left(\frac{P}{P_0}\right) + RT\left[1+\frac{1}{\theta}\ln(1-\theta)\right] = N_{Av}\cdot\overline{\varepsilon}(z) = N_{Av}\frac{N_S A_S + N_A A_A}{\sigma^4(L-2d_0)}\left[\frac{\sigma^4}{3(L-d_0)^3} - \frac{\sigma^{10}}{9(L-d_0)^9} - \frac{\sigma^4}{3d_0^3} + \frac{\sigma^{10}}{9d_0^9}\right] \tag{4.13}$$

The Langmuir isotherm is typically represented as follows:

$$\theta = \frac{K\left(\frac{P}{P_0}\right)}{1+K\left(\frac{P}{P_0}\right)} \tag{4.14}$$

Taking a limit of Eq. 4.14 as P approaches the saturation pressure P_0, we have $\theta = K/(1+K)$. Now, applying the above limit of P/P_0 approaching 1 to Eq. 4.13, we obtain the following:

$$\lim_{P \to P_0} N_{Av} \cdot \overline{\varepsilon}(z) = RT \left[1 - \frac{K+1}{K} \ln(K+1) \right] \tag{4.15}$$

Thus, at high pressures, the average interaction energy should approach the value given by Eq. 4.15. Near the saturation pressure, the size of the largest pore, which can be filled with adsorbate, tends towards infinity. As can be seen from Eq. 4.12, as $L \to \infty$ the average interaction energy approaches zero. Intuitively, it is apparent that there will always be some adsorbate–adsorbate interaction and that the limiting interaction energy cannot go to zero even at large pore size. Thus, the original model for average interaction energy in the pore is inconsistent with the HK equation by using the Cheng–Yang correction. The proposed model is found to have a better agreement in this regard.

4.2.2. Modified HK Model for Slit-Shaped Pores

The shortcomings in the original HK model provide the motivation for developing a new model based on an improved energy profile for a micropore filled with adsorbate molecules (Rege and Yang, 2000). The new model developed by Rege and Yang (2000) proposes that molecules occupy discrete positions in the adsorbate-filled micropore. Furthermore, the interaction energy of a molecule is calculated by using an intermolecular spacing corresponding to a minimum energy potential between that molecule and the immediate neighboring molecules. The average interaction energy is obtained by a population-weighted average of energy potentials rather than by integration. New models for PSD, based on the Horváth-Kawazoe principle, have been developed for slit-, cylindrical-, as well as spherical-shaped micropores (Rege and Yang, 2000). Comparison of the average interaction energy versus pore-size profiles for the original and modified models showed a good agreement for small pore widths, but a considerable deviation was observed at pore widths measuring more than two adsorbate molecule diameters. Pore-size distribution using the modified energy profile has been obtained by using isotherm data in the literature for all three pore geometries, and the results are comparable with that used in the original model. Clear and significant improvements have been obtained by using the modified models (Rege and Yang, 2000).

The general strategy for obtaining the PSD for the three pore geometries is basically the same. The modified model for calculating the average interaction energy in a filled micropore proceeds by first estimating the number of gas adsorbate molecule layers within the filled micropore. Each gas molecule is assumed to rest preferentially at a position at which its energy potential within the pore would be minimum, in accordance with the Boltzmann law of energy distribution. Furthermore, a gas molecule is assumed to interact most effectively only with its laterally immediate molecular layer. This fact is in agreement with experimental observations that show that the energy of adsorption of a monolayer of adsorbate molecules on a clean sorbent surface is the highest and that on subsequently adsorbed layers of gas molecules is a nearly constant lower value. Any interaction with molecules not in the immediate proximity of the molecule or with those lying above or below the molecule in the same layer, is assumed to be negligible. The average interaction energy is then calculated by averaging the energy potentials of the individual gas molecule layers weighted by an approximate molecular population of each layer.

Consider a slit pore with the nuclei of the sorbent molecules in the lattice planes forming the pore-wall spaced at a distance L apart. If the diameter of the sorbent molecules is denoted as d_S and that of the adsorbate molecules as d_A, then the number of molecule layers M that can be accommodated laterally as the pore gets filled up can be estimated as:

$$M = \frac{L - d_S}{d_A} \tag{4.16}$$

It should be noted that M can be a whole number or it may also be a fraction. The physical interpretation of a fractional number of layers is that the molecules in the layers are not arranged with their centers oriented in the same straight line but are slightly skewed with respect to each other so as to afford a greater packing of molecules. Such a packing arrangement will of course result in a decrease in molecular density in the neighboring layers, but the resulting decrease in interaction energy will be accounted for when taking the average of the energy potentials.

When $1 \leq M < 2$, only one layer of molecules can be accommodated within the slit pore as shown in Figure 4.1. In this case, each adsorbate molecule will interact with the two lattice planes forming the pore wall of the sorbent. The interaction energy of this monolayer, denoted as $\varepsilon_1(z)$, is given by an expression similar to Eq. 4.7. An examination of Eq. 4.7 reveals that the minimum energy potential exists at a distance of d_0 from either sorbent lattice planes. Because the Boltzmann law of energy distribution law suggests that the molecule will most probably rest at a position at which the energy potential is the minimum, each molecule is assumed to exist at a distance of d_0 from one of the two sorbent lattice planes. The interaction energy would then be

$$\varepsilon_1(z) = \frac{N_S A_S}{2\sigma_S^4}\left[-\left(\frac{\sigma_S}{d_0}\right)^4 + \left(\frac{\sigma_S}{d_0}\right)^{10} - \left(\frac{\sigma_S}{L - d_0}\right)^4 + \left(\frac{\sigma_S}{L - d_0}\right)^{10}\right] \tag{4.17}$$

Note that the internuclear distance at zero interaction energy for an adsorbate–sorbent system is now denoted as σ_S to differentiate it from σ_A, which is the zero interaction energy distance for an adsorbate–adsorbate system. The expressions for σ_S and σ_A are as follows:

$$\sigma_S = (2/5)^{1/6} d_0 \tag{4.18}$$

$$\sigma_A = (2/5)^{1/6} d_A \tag{4.19}$$

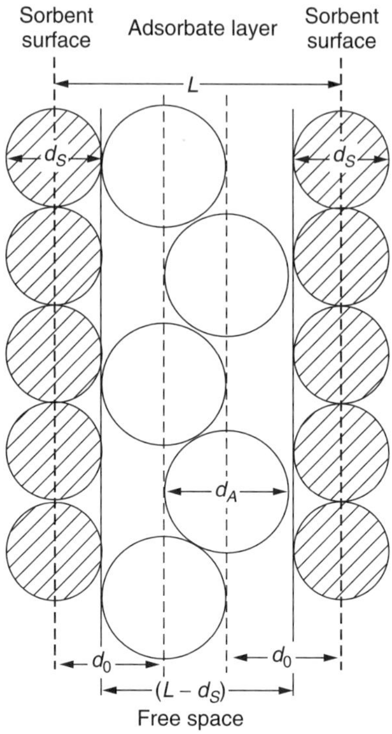

Figure 4.1. Single layer of adsorbate molecules in a slit-shaped adsorbent pore of width L (Rege and Yang, 2000, with permission).

The average energy of a slit pore having $1 < M < 2$ is the same as $\varepsilon_1(z)$, since, irrespective of M, the molecules are assumed to rest at a distance of d_0 from either pore walls. The situation differs slightly when $M \geq 2$. Three cases for this situation are depicted in Figures 4.2–4.4, wherein two different types of layers can be visualized: the first type, consisting of the two layers in the immediate vicinity of the two pore walls and at a distance of d_0 from them, and the second type of molecules present in the interior of the pore sandwiched between two adjoining adsorbate layers. The first type of molecules (marked A in the figure) will interact with the sorbent molecules on one side and with adsorbate molecules on the other. The distance between the two adsorbate molecules would be approximately d_A, using the same argument as discussed above. The interaction if energy of this layer of molecules, denoted as $\varepsilon_2(z)$, is given by:

$$\varepsilon_2(z) = \frac{N_S A_S}{2\sigma_S^4}\left[-\left(\frac{\sigma_S}{d_0}\right)^4 + \left(\frac{\sigma_S}{d_0}\right)^{10}\right] + \frac{N_A A_A}{2\sigma_A^4}\left[-\left(\frac{\sigma_A}{d_A}\right)^4 + \left(\frac{\sigma_A}{d_A}\right)^{10}\right] \tag{4.20}$$

The second type of molecules (marked B in the figure), comprising $(n - 2)$ number of layers, will interact effectively only with two parallel planes of gas

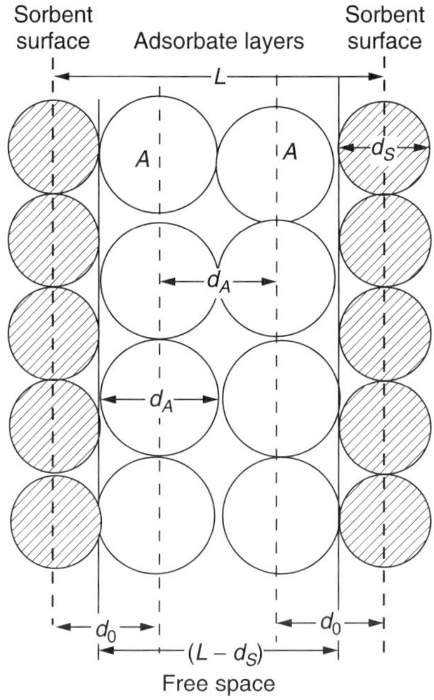

Figure 4.2. Two layers of adsorbate molecules in a slit-shaped adsorbent pore of width L (Rege and Yang, 2000, with permission).

molecules at a distance of d_A on either side of the molecule; its interaction energy $\varepsilon_3(z)$ will be given by:

$$\varepsilon_3(z) = 2 \cdot \frac{N_A A_A}{2\sigma_A^4}\left[-\left(\frac{\sigma_A}{d_A}\right)^4 + \left(\frac{\sigma_A}{d_A}\right)^{10}\right] \tag{4.21}$$

The average interaction energy for $M > 2$ can be calculated by weighing the energy potentials of the molecules by the actual number of layers of molecules possessing the corresponding energy:

$$\overline{\varepsilon} = \frac{2\varepsilon_2 + (M - 2)\varepsilon_3}{M} \tag{4.22}$$

Here M, the number of layers accommodated within the pore, is given by Eq. 4.14. The above method of averaging eliminates integration as done in the original model, which assumes that a gas molecule in a filled micropore is free to assume a continuous distribution of positions within the slit-pore free space. When M is not a whole number, the picture will be something between the two whole numbers enclosing M. For example, Figure 4.2 shows the molecule layers for $M = 2$ and Figure 4.3 shows the case when $M = 3$. For $2 < M < 3$, Figure 4.4 shows a likely configuration. Since the centers of the sorbate molecules

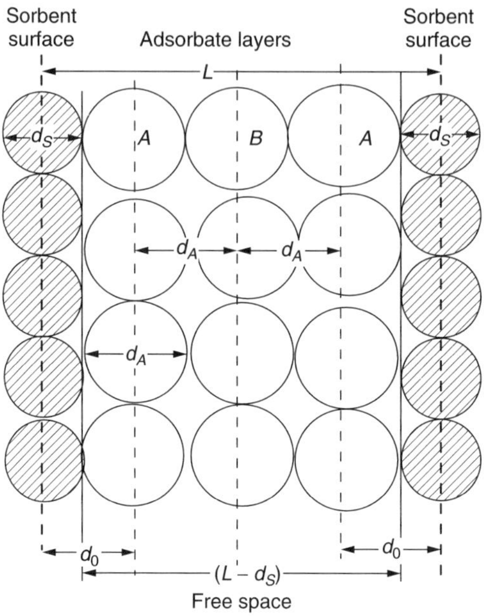

Figure 4.3. Three layers of adsorbate molecules in a slit-shaped adsorbent pore of width L (Rege and Yang, 2000, with permission).

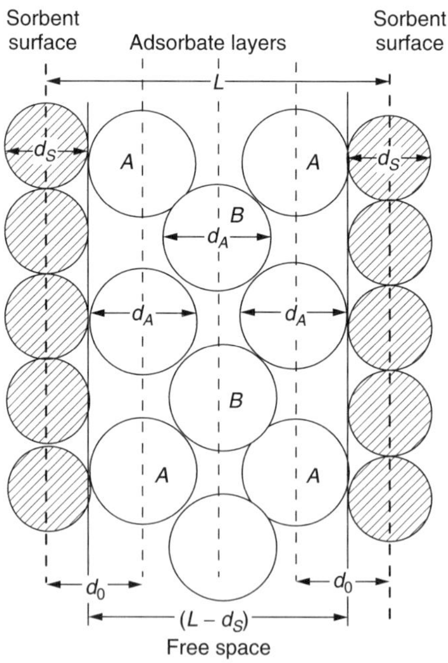

Figure 4.4. Possible configuration of adsorbate molecules in a slit-shaped adsorbent pore measuring between 2 to 3 adsorbate molecular diameters in width (Rege and Yang, 2000, with permission).

will no longer be collinear, the density of the molecules adjacent to the middle layer will be decreased and its interaction energy will be lower. This factor is taken into account by using the interpolating Eq. 4.22, which will consider only a fraction of the adsorbate–adsorbate potential $\varepsilon_3(z)$ for this molecule in the final average. Equation 4.22 will thus be exact when M is a whole number and will be an interpolated approximation when M lies between two whole numbers.

The PSD can be now determined following the methodology outlined by Horváth and Kawazoe (1983). As was explained in the previous section regarding the original HK model, the free energy change upon adsorption, $RT\ln(P/P_0)$, is equated to the molar average interaction energy, $N_{Av}\cdot\bar{\varepsilon}$. Thus the modified-HK equation can be written as:

$$RT\ln\left(\frac{P}{P_0}\right) = N_{Av}\cdot\bar{\varepsilon} \tag{4.23}$$

On applying the correction for isotherm non-linearity given by Cheng and Yang (1994), the modified HK–CY equation is obtained as:

$$RT\ln\left(\frac{P}{P_0}\right) + RT\left[1+\frac{1}{\theta}\ln(1-\theta)\right] = N_{Av}\cdot\bar{\varepsilon} \tag{4.24}$$

In Eqs. 4.23 and 4.24 above:

$$\begin{aligned} \bar{\varepsilon} &= \varepsilon_1 && \text{when} \quad 1 \le M < 2 \\ &= \frac{2\varepsilon_2 + (M-2)\varepsilon_3}{M} && \text{when} \quad M \ge 2 \end{aligned} \tag{4.25}$$

The above model is simple and can be easily implemented by using a programming code or a spreadsheet. A slight discontinuity in $\bar{\varepsilon}$ may occur at $M = 2$ as a result of the above two interpolating functions, but the difference in the values was found to be negligible for the examples studied in this work. The algorithm for determining pore-size, by using the corrected HK equation is summarized in Table 4.1.

We can now compare predictions of PSD by using different models from isotherm data from the literature. The physical parameters used in the calculation are given in Table 4.2 (Rege and Yang, 2000). The PSD of HGS-638 molecular sieve carbon predicted by the original HK Eq. 4.12, original HK–CY Eq. 4.13, modified-HK Eq. 4.23 and the modified HK–CY Eq. 4.24 are shown in Figure 4.5. It can be seen that there is virtually no shift in the peak pore size when the interaction energy model was changed. This condition is so because the peak pore size of HGS-638 lies in the 0.4–0.6 nm range, which corresponds to roughly twice the adsorbate molecule (N_2) diameter. As shown by Rege and Yang (2000), the interaction energy predicted by the two models is nearly the same until the slit width is about $2d_A$; hence the predicted pore widths are also similar in this size range. Also, it was observed that the points on the figure corresponding to the higher extreme of the pore width in the original model appeared

Table 4.1. Algorithm (pseudo-code) for obtaining effective pore size of slit-shaped pores (from Rege and Yang, 2000)

(1) Guess slit width L corresponding to adsorbate relative pressure P/P_0.

(2) Calculate number of adsorbate molecular layers M (Eq. 4.16):

$$M = \frac{L - d_S}{d_A}$$

(3) Calculate ε_1, ε_2, ε_3 and (Eqs. 4.17, 4.20, 4.21):

$$\varepsilon_1(z) = \frac{N_S A_S}{2\sigma_S^4}\left[-\left(\frac{\sigma_S}{d_0}\right)^4 + \left(\frac{\sigma_S}{d_0}\right)^{10} - \left(\frac{\sigma_S}{L - d_0}\right)^4 + \left(\frac{\sigma_S}{L - d_0}\right)^{10}\right]$$

$$\varepsilon_2(z) = \frac{N_S A_S}{2\sigma_S^4}\left[-\left(\frac{\sigma_S}{d_0}\right)^4 + \left(\frac{\sigma_S}{d_0}\right)^{10}\right] + \frac{N_A A_A}{2\sigma_A^4}\left[-\left(\frac{\sigma_A}{d_A}\right)^4 + \left(\frac{\sigma_A}{d_A}\right)^{10}\right]$$

$$\varepsilon_3(z) = 2 \cdot \frac{N_A A_A}{2\sigma_A^4}\left[-\left(\frac{\sigma_A}{d_A}\right)^4 + \left(\frac{\sigma_A}{d_A}\right)^{10}\right]$$

(4) If $M < 2$, then

$$\overline{\varepsilon} = \varepsilon_1$$

Else $(M \geq 2)$

$$\overline{\varepsilon} = \frac{2\varepsilon_2 + (M - 2)\varepsilon_3}{M}$$

(5) If $[RT \ln(P/P_0) = N_{Av}\overline{\varepsilon}]$, then
Effective pore width $= (L - d_S)$
Else
Guess new L and iterate from (1) again.

---

at a still higher pore size in the new model. This again can be explained from the fact that for the same interaction energy the new model predicts a higher pore size compared with that predicted by the original model.

The adsorption isotherm of argon on faujasite zeolite at 87 K has been given by Borghard et al. (1991). This type of zeolite actually has a spherical cavity, and the results of using the modified HK model for spherical pores will be subsequently shown. The slit-pore model was tried in the case of faujasite, particularly because it has a comparatively larger pore structure than molecular sieve carbons. The differences between the model predictions will become obvious. It has been shown previously (Cheng and Yang, 1994) that the application of the original HK model gives a highly underestimated pore-size distribution. Figure 4.6 shows the predictions of the two original and two modified models. It can be seen that while the original models predicted a pore size of 7 Å, the modified models gave a peak at much higher pore size of 13 Å. It is interesting to note that the

Table 4.2. Physical parameters used in micropore size distribution calculations

Parameter	Adsorbent			Adsorbate	
	Carbon	*Aluminosilicate Oxide Ion*	*Aluminophosphate Oxide Ion*	*Nitrogen*	*Argon*
diameter (Å)	3.40[1]	2.76[2]	2.60[2]	3.00[1] 3.72[10,9]	3.82[4,5,6,9] 2.95[3,8]
polarizability, a (cm^3)	1.02×10^{-24} [1]	2.5×10^{-24} [2]	2.5×10^{-24} [2]	1.46×10^{-24} [1]	1.63×10^{-24} [2]
magnetic susceptibility, c (cm^3)	1.35×10^{-28} [1]	1.3×10^{-29} [2]	1.3×10^{-29} [2]	2.0×10^{-29} [1]	3.24×10^{-29} [2]
molecular surface density (molecule/cm^2)	3.85×10^{15} [1]	3.75×10^{15} [3,8] 1.00×10^{15} [4,7] 8.48×10^{14} [5,2] 8.73×10^{14} [6,2] 3.27×10^{14} [10,7]	1.0×10^{15} [7]	6.7×10^{14} [1]	8.52×10^{14} [2]

(1) Horváth and Kawazoe (1983).
(2) Cheng and Yang (1994).
(3) Faujasite, slit-pore model.
(4) ZSM-5, cylindrical-pore model.
(5) Faujasite, spherical-pore model.
(6) 5A zeolite, spherical-pore model.
(7) Estimated by Rege and Yang (2000).
(8) ASAP 2010 Manual, Micromeritics Inc. (1995).
(9) Razmus and Hall (1991).
(10) MCM-41, cylindrical-pore model.

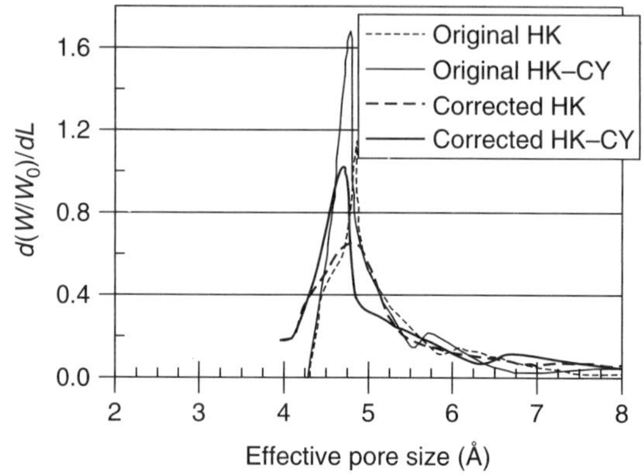

Figure 4.5. Pore size distribution of HGS-638 molecular sieve carbon as predicted by original HK models and the corrected HK models for slit-shaped pores (Rege and Yang, 2000, with permission).

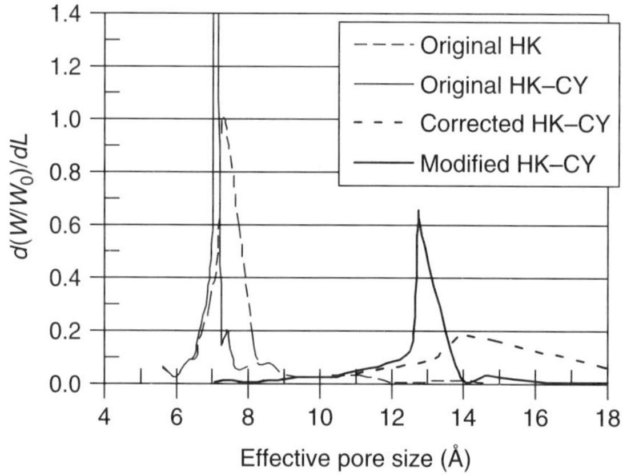

Figure 4.6. Pore size distribution of faujasite zeolite as predicted by original HK models and the corrected HK models for slit-shaped pores (Rege and Yang, 2000, with permission).

crystallographic channel size of faujasite is 7.4 Å, while the actual cavity size of faujasite is 13.7 Å. Thus the modified model is regarded as capable of determining the PSD in the higher range of pore size compared with the original model. This is a result of a more realistic picture for calculation of adsorbate–adsorbate interaction and an improved method of averaging interaction energy.

4.2.3. Modified Model for Cylindrical Pores

The derivation for the modified HK model for pores with cylindrical geometry (Rege and Yang, 2000) follows the same procedure as that for the slit-pore model. The total number of concentric layers of adsorbate molecules that can be accommodated in the cylinder of radius L is first determined. The energy potential of the adsorbate molecules in each layer is then calculated by using the appropriate equation corresponding to that given by Everett and Powl (1976). The number of sorbate molecules in each sorbate layer is estimated by using the sorbate molecule diameter, and the diameter of the corresponding concentric layers. The average interaction energy is then determined by a population-weighted average of the individual layer potentials. One of the assumptions made by the model is that because adsorption will proceed from the extreme periphery of the pore towards the center, the energy potential of each layer would correspond to only the field induced by the immediately surrounding outer layer of molecules. The interaction energy induced by the inside layer of molecules, enclosed by the molecules for which the energy potential is being calculated, is considered to be negligible.

Figure 4.7 show a cross-sectional view of a cylindrical pore. The total number of concentric molecular layers, M, that can be accommodated within a free space diameter of $(2L - d_S)$ is given by:

$$M = \text{int}\left[\left(\frac{2L - d_S}{d_A} - 1\right)\Big/2\right] + 1 \tag{4.26}$$

Note that *int* is a mathematical function that truncates the decimal part of a value and retains only the integer part.

The first layer of molecules will be enclosed by a layer of molecules belonging to the sorbent material (e.g., oxide ions). As it is assumed that molecules in a layer

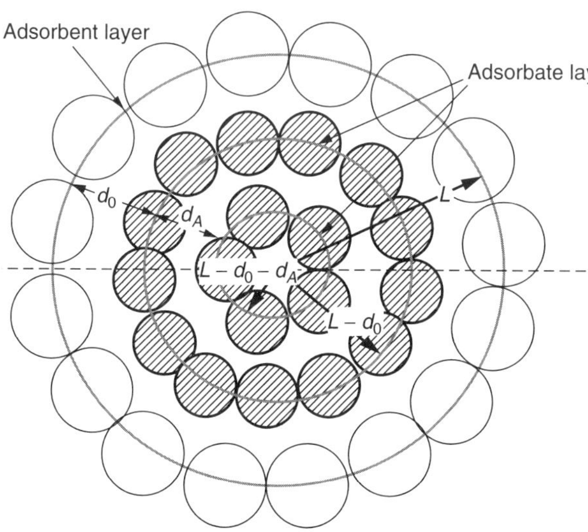

Figure 4.7. Cross-sectional view of the configuration of adsorbate molecules in a cylindrical pore of radius L (Rege and Yang, 2000, with permission).

will preferentially occupy positions at which the energy potential is minimum, the first layer of molecules is supposed to exist at a radius of $(L - d_0)$ from the center. Thus making the required substitution, the energy potential of the first layer (e_1) is obtained as:

$$\varepsilon_1 = \frac{3}{4}\pi \frac{N_S A_S}{d_0^4} \left[\frac{21}{32} a_1^{10} \sum_{k=0}^{\infty} \alpha_k (b_1^{2k}) - a_1^4 \sum_{k=0}^{\infty} \beta_k (b_1^{2k}) \right] \tag{4.27}$$

where a_1 and b_1 are given by

$$a_1 = d_0/L \text{ and } b_1 = (L - d_0)/L \tag{4.28}$$

As can be seen from Eq. 4.27, the constant a_1 (and in general, a_i) corresponds to a ratio of internuclear spacing d_0 to the radius of the enclosing layer of molecules, whereas b_1 (and in general, b_i) is the ratio of distance of the layer from the central axis of the pore to the radius of the enclosing layer of molecules. Note that in the following discussion the layers are numbered in increasing order starting from the outermost layer to the innermost.

One can write the equation for the energy potential for the ith layer $(i > 1)$ as follows:

$$\varepsilon_i = \frac{3}{4}\pi \frac{N_A A_A}{d_A^4} \left[\frac{21}{32} a_i^{10} \sum_{k=0}^{\infty} \alpha_k (b_i^{2k}) - a_i^4 \sum_{k=0}^{\infty} \beta_k (b_i^{2k}) \right] \tag{4.29}$$

and the constants a_i and b_i are given by:

$$a_i = \frac{d_A}{L - d_0 - (i-2)d_A}$$
$$b_i = \frac{L - d_0 - (i-1)d_A}{L - d_0 - (i-2)d_A} \tag{4.30}$$

The maximum number of molecules of diameter d that can be accommodated with their centers along the circumference of a circle of diameter D is given by $N = int[p/sin^{-1}(d/D)]$ for $d \leq D$ and $N = 1$ for $d > D$. For the ith layer of adsorbate molecules within the pore, the molecules lie along a circle of diameter $2[L - d_0 - (i-1)d_A]$, and hence the maximum number of molecules of diameter d_A that are present in a horizontal cross section of the ith layer (when $d_A \leq 2[L - d_0 - (i-1)d_A]$ can be written as:

$$N_i = \frac{\pi}{\sin^{-1}\left[\dfrac{d_A}{2(L - d_0 - (i-1)d_A)}\right]} \tag{4.31}$$

However, if $d_A > 2[L - d_0 - (i-1)d_A]$, $N_i = 1$.

Thus the average interaction energy $\overline{\varepsilon}$ can now be written down by a population-weighted average as follows:

$$\overline{\varepsilon} = \frac{\sum_{i=1}^{M} N_i \varepsilon_i}{\sum_{i=1}^{M} N_i} \tag{4.32}$$

The PSD for the cylindrical pore model can now be easily determined by using either the HK Eq. 4.23, or the HK–CY Eq. 4.24. The algorithm for determining pore size by using the corrected HK equation for cylindrical pores was also given by Rege and Yang (2000).

The modified HK model for spherical pores has also been derived by Rege and Yang (2000). The HK–CY version for the spherical pore was given earlier by Cheng and Yang (1994).

Rege and Yang (2000) have compared the PSD predicted from the four HK type models to assess a number of molecular sieves with well-defined shapes and sizes. In all cases, the modified or corrected HK–CY models gave significantly better results than all other models. Some of the comparisons are shown here.

The ZSM-5 zeolite is an aluminosilicate with three-dimensional channels measuring 5.6 × 5.3 Å size. High-resolution isotherm of Ar on ZSM-5 at 87 K has been measured by Venero and Chiou (1988). The pores are approximately cylindrical. The PSD of ZSM-5 predicted by the four models is shown in Figure 4.8. It is seen that for the parameters chosen (Table 4.2), no prominent peak was

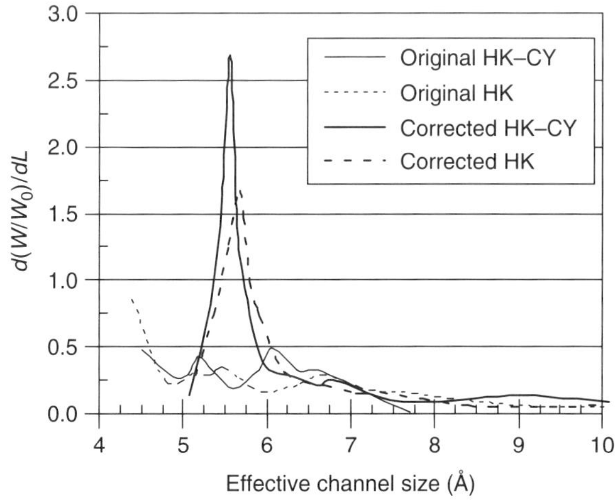

Figure 4.8. Pore-size distribution of ZSM-5 zeolite as predicted by original HK models and the corrected HK models for cylindrical pores (Rege and Yang, 2000, with permission).

visible for both the original HK as well as the HK–CY model, as no solution for the pore size existed for the first few points of the isotherm. However, a peak was obtained at 5.7 Å by using the modified HK model. On using the modified HK–CY model, a sharper peak was observed at 5.6 Å, which agrees well with the crystallographic dimension of the ZSM-5 channel.

MCM-41 materials belong to the family of large pore (mesoporous) molecular sieves and are increasingly becoming popular as catalyst supports. These synthetic materials possess a regular array of hexagonal, uniform unidimensional mesopores, which can be systematically varied from 16 to 100 Å. Due to their narrow pore size distribution, minimal network effects, and well-known surface chemistry, they are ideal candidates for testing pore-size models. Numerous studies have been conducted for studying the PSD of MCM-41 materials by using various techniques such as the BJH method (Ravikovitch et al., 1997; Sayari et al., 1999), X-ray diffraction (XRD) (Ravikovitch et al., 1995), H-NMR (Schmidt et al., 1995), and non-local density functional theory (NLDFT) (Ravikovitch et al., 1995, 1997).

In order to test the corrected HK model, N_2 adsorption data at 77.4 K on a MCM-41 material denoted "Sample C" in a previously published work (Ravikovitch et al., 1995) was used. The XRD results and the NLDFT simulation studies reported in the latter work show the pore size of the material as 45 Å. Figure 4.9 shows the PSD for the MCM-41 material predicted from the N_2 adsorption data by using the corrected and original HK cylindrical pore models. Because the isotherm does not follow Langmuirian behavior, the Cheng–Yang correction was computed by fitting the data to a Langmuir isotherm as much as possible. The figure shows that the corrected HK–CY equation accurately

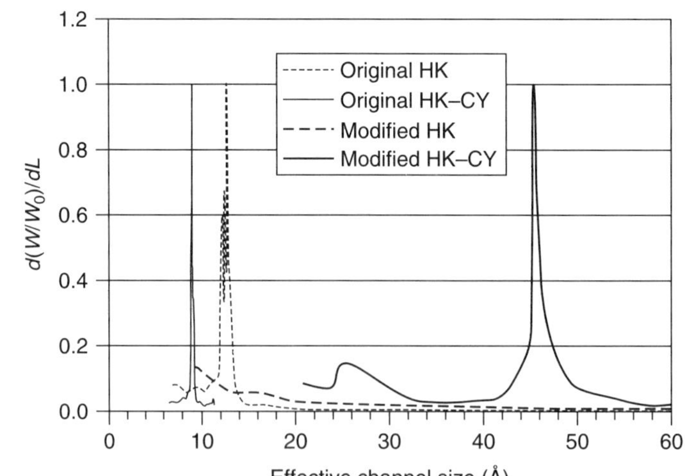

Figure 4.9. Pore size distribution of MCM-41 material as predicted by original HK models and the corrected HK models for cylindrical pores (Rege and Yang, 2000, with permission).

estimates the pore size of the material as 45 Å. By using the same model parameters (shown in Table 4.2), the original HK and HK–CY models give a grossly underestimated pore size of 12 Å and 9 Å, respectively. It must be noted that the corrected HK equation without the Cheng–Yang correction gives a highly overestimated pore size value (beyond the limits of Fig. 4.9). This is because the inflexion in the curve occurs at a high P/P_S value (~0.4). There is a severe deviation from Henry's law at this value. Further, Ravikovitch et al. (1997) have reported that the BJH method systematically underestimated the pore size of the MCM-41 materials studies by ~10 Å. Thus, the utility of the corrected HK–CY model in predicting PSD for materials in the mesoporous (>20 Å) size range is demonstrated.

Like zeolitic materials, aluminophosphates also possess well-defined structure and narrow PSD. Most aluminophosphates such as $AlPO_4$-11 and VPI-5 have a one-dimensional channel structure. Adsorption isotherms of argon on these materials have been reported by Davis et al. (1988).

Figure 4.10 shows the computed PSD of VPI-5 aluminophosphate whose isotherm is seen to have two sigmoidal bends, suggesting that it may have a bimodal PSD. The currently accepted pore size of VPI-5 is 12.1 Å (Davis et al., 1988). It was shown previously by Cheng and Yang (1994) that VPI-5 gives two peaks by use of the original HK–CY equation: a major peak at 8.6 Å and a minor one at 11.7 Å. In the work of Rege and Yang (2000) different parameters were used for the kinetic diameter of argon and the oxide ion density in the aluminophosphate framework. As a result the new peaks obtained by using the original model were situated at 6.8 Å and at 8.7 Å. With the modified HK–CY model the major peak was observed at 11.8 Å and a minor peak was observed

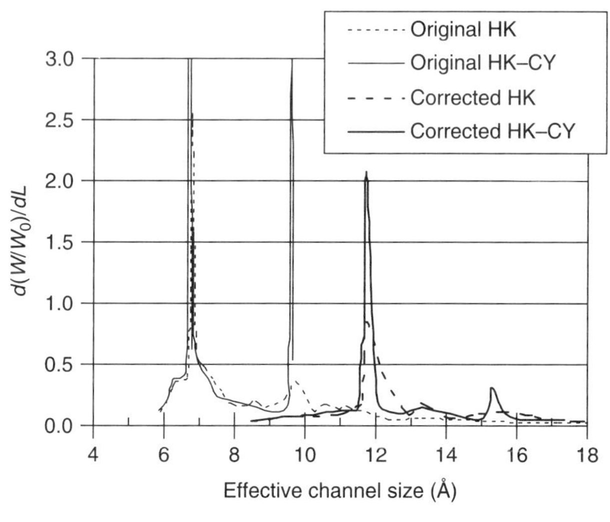

Figure 4.10. Pore size distribution of VPI-5 as predicted by original HK models and the corrected HK models for cylindrical pores (Rege and Yang, 2000, with permission).

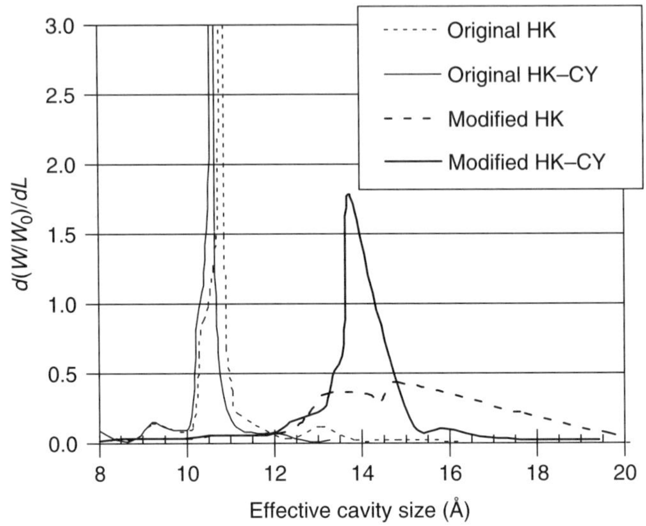

Figure 4.11. Pore-size distribution of faujasite zeolite as predicted by original HK models and the corrected HK models for spherical pores (Rege and Yang, 2000, with permission).

at 15.2 Å. Thus the modified cylindrical HK–CY model has a better agreement with the crystallographic data than the original one.

The PSD for faujasite zeolite by using the slit-pore model is shown in Figure 4.6. This zeolite is, however, known to have nearly spherical-shaped cavities measuring 13.7 Å in diameter. Hence it is apparent that the spherical pore model would be more appropriate for obtaining the PSD for this material. The spherical-shaped pore model is next used to deduce the PSD of Faujasite from the argon isotherm data of Borghard et al. (1991) (Baksh and Yang, 1991; Rege and Yang, 2000). Figure 4.11 shows the distribution of pores as calculated by the four models. For the parameters chosen, the original HK–CY equation provided a peak at about 10.5 Å, whereas the modified HK–CY model calculated one at the expected value of 13.7 Å. The modified HK model without the correction for isotherm non-linearity, however, provides a highly diminished peak at about 14.8 Å.

4.3. THE INTEGRAL EQUATION APPROACH

The amount adsorbed in a sorbent with a distribution of pore sizes is given by:

$$q(P) = \int_0^\infty f(x, P)G(x)\,dx \tag{4.33}$$

where $q(P)$ is the measured or overall isotherm, $f(x, P)$ is the local or individual pore isotherm for the pore with size x, and $G(x)$ is the pore size distribution function. The limits for the integration can be replaced by the upper and lower bounds for the pore sizes.

Equation 4.33 is the *Fredholm integral equation of the first kind*. Except for a few special cases, no solution for $G(x)$ exists (Tricomi, 1985). Numerical solutions can be used. To solve for $G(x)$ from a given $q(P)$, one needs to have an individual pore isotherm, which must be related to the pore size. Moreover, the integral equation is ill-defined, that is, the solution for $G(x)$ is not unique, unless a functional form for $G(x)$ is assumed. It is clear then that there are as many solutions for the PSD as the number of assumed functional forms.

Earlier work focused on analytical solutions for the PSD function (Stoeckli, 1977; Jaroniec and Madey, 1988; Stoeckli, 1990; Rudzinski and Everett, 1992). They remain useful because of their simplicity. An outline of the solution by Jaroniec and Choma (Jaroniec et al., 1991) is given next.

For a homogeneous or nearly homogeneous microporous material, the Dubinin-Radushkevich isotherm (Eq. 3.9) (Dubinin and Radushkevich, 1947) is applicable:

$$V = V_0 \exp\left[-B\left(\frac{RT}{\beta}\ln\frac{P_0}{P}\right)^2\right] = f(B, P) \tag{4.34}$$

where V is the amount adsorbed at relative pressure P/P_0, β is the affinity coefficient, and B is a structural parameter, which increases with an increase in pore size. A semi-empirical relationship was found between B and the half-width x of the slit-like micropores (Dubinin and Stoeckli, 1980; Jaroniec and Madey, 1988; Stoeckli, 1990; Chen and Yang, 1996):

$$B = kx^2 \tag{4.35}$$

where the constant k is characteristic of the sorbent material. The value of k can be obtained by using a limiting probe molecule (Baksh et al., 1992). A theoretical basis has been given for this relationship by Chen and Yang (1996). Jaroniec and co-workers used a Gamma-type pore size distribution function:

$$G(x) = \frac{2(ku)^{n+1}}{\Gamma(n+1)} x^{2n+1} \exp(-kux^2) \tag{4.36}$$

where u is a constant. Upon substitution of this distribution function into Eq. 4.33, Jaroniec obtained the following solution, referred to as Jaroniec–Choma isotherm:

$$\frac{V}{V_0} = \left[\frac{u}{u + \left(\dfrac{RT}{\beta}\ln\dfrac{P_0}{P}\right)^2}\right]^{n+1} \tag{4.37}$$

The constants β, k and u can be determined by using data with a probe molecule (Baksh et al., 1992). The PSD function, $G(x)$, can then be calculated from Eq. 4.36 by using N_2 isotherm at 77 K.

However, despite its experimental and theoretical simplicity, the above procedure was undermined by the empiricism and assumptions involved. Recently new statistical mechanics approaches by using either Monte Carlo techniques or the density functional theory have been proposed and are steadily gaining acceptance. In this approach, the local or individual pore isotherm, $f(x, P)$, is calculated by using any of the statistical mechanics techniques. These techniques enable the calculation of the isotherm (dependent on pore size and geometry) in conjunction with PSD functions with assumed mathematical forms. A large amount of literature has been published regarding the application of this theory for studying gas adsorption and, in particular, evaluation of PSD of microporous materials (e.g., Seaton et al., 1989; Lastoskie et al., 1993; Olivier, 1995; Webb and Orr, 1997; Ravikovitch et al., 1998). The advantage of the technique is that it is a unified method applicable to both microporous as well as mesoporous types of materials. However, the empirical parameters used in these models are difficult to obtain from the literature except for a few well-studied systems. This is a common problem with the HK approach. The results are highly sensitive to the empirically parameters that are used. Moreover, the calculations involved in the statistical mechanics approach are highly intensive and require an elaborate computer code.

A common drawback for all three approaches (i.e., the Kelvin equation, the Horváth–Kawazoe approach, and the integral equation approach) is that non-intersecting pores are assumed. This problem of intersecting pores remains unsolved. There exists an interesting conceptual difference between the statistical mechanics approach and the two former approaches. In the Kelvin equation and the HK approaches, a pore is either filled or empty; whereas in the statistical mechanics approach, all pores are partially and gradually filled. The experimental data in the literature on high-resolution isotherms cannot distinguish, unambiguously, which is the correct picture. Obviously the actual picture of the pore filling also depends on the homogeneity of the sites on the pore walls.

REFERENCES

Adamson, A. W. and Gast, A. P. (1997) *Physical Chemistry of Surfaces*, 6th Ed. Wiley, New York, NY.

ASAP 2010 Operator's Manual, V. 2.00. (1995) Micromeritics Instrument Inc., Norcross, GA.

Baksh, M. S. A. and Yang, R. T. (1991) *A.I.Ch.E. J.* 37, 923.

Baksh, M. S. A. and Yang, R. T. (1992) *A.I.Ch.E. J.* 38, 1357.

Barrett, E. P., Joyner, L. G., and Halenda, P. H. (1951) *J. Am. Chem. Soc.* 73, 373.

Beck, J. S., Vartuli, J. C., Roth, W. J., Leonowicz, M. E., Kresge, C. T., Schmitt, K. D., Chu, C. T.-W., Olson, D. H., Sheppard, E. W., McCullen, S. B., Higgins, J. B., and Schlenker, J. L. (1992) *J. Am. Chem. Soc.* 114, 10835.

Borghard, W. S., Sheppard, E. W., and Schoennagel, H. J. (1991) *Rev. Sci. Instrum.* 62, 2801.

Carrott, P. J. M., Carrott, M. M. L. R., and Mays, T. J. (1998) Comparison of methods for estimating micropore sizes in active carbons from adsorption isotherms. In *Proceedings Sixth International Conference Fundamental of Adsorption*. (F. Meunier, ed.). Elsevier, Paris, 677.

Chen, S. G. and Yang, R. T. (1996) *J. Coll. Interf. Sci.* 177, 298.

Cheng, L. S. and Yang, R. T. (1994) *Chem. Eng. Sci.* 49, 2599.

Davis, M. E., Montes, C., Hathaway, P. E., Arhancet, J. P., Hasha, D. L., and Graces, J. E. (1989) *J. Am. Chem. Soc.* 111, 3919.

Davis, M. E., Saldarriaga, C., Montes, C., Garces, J., and Crowder, C. (1988) *Zeolites*, 8, 362.

Dubinin, M. M. and Radushkevich, L. V. (1947) *Doklady Akad. Nauk. S.S.S.R.* 55, 327.

Dubinin, M. M. and Stoeckli, H. F. (1980) *J. Coll. Interf. Sci.* 75, 34.

Dyer, A. (1988) *An Introduction to Zeolite Molecular Sieve*. Wiley, New York, NY.

Everett, D. H. and Powl, J. C. (1976) *J. Chem. Soc. Faraday Trans.* 72, 619.

Gregg, S. J. and Sing, K. S. W. (1982) *Adsorption, Surface Area and Porosity*, 2nd Ed. Academic Press, New York, NY.

Horváth, G., Halász-Laky, V., Dékány, I., and Berger, F. (1998) Pore characterization of different adsorbents in micropore domain based on CO_2 adsorption. In *Proceeding Sixth International Conference Fundamental of Adsorption*. (F. Meunier, ed.). Elsevier, Paris, France 611.

Horváth, G. and Kawazoe, K. (1983) *J. Chem. Eng. Japan*, 16, 470.

IUPAC Manual of Symbols and Terminology, Appendix 2, Pt. 1 (1972) *Pure Appl. Chem.* 31, 578.

Jaroniec, M., Choma, J., and Lu X. (1991) *Chem. Eng. Sci.* 46, 3299.

Jaroniec, M. and Madey, R. (1988) *Physical Adsorption on Heterogeneous Solid*. Elsevier, Amsterdam, The Netherlands.

Jessop, C. A., Riddiford, S. M., Seaton, N. A., Waltomn, J. P. R. B., and Quirke, N. (1991) In *Characterization of Porous Solids II*. (F. Rodriguez-Reinoso, J. Rouquerol, K. S. W. Sing, and K. K. Unger, eds.). Elsevier, Amsterdam, The Netherlands p. 123.

Kane, M. S., Bushong, J. H., Foley, H. C., and Brendley, W. H. (1998) *Ind. Eng. Chem. Res.* 37, 2416.

Lastoskie, C., Gubbins, K. E., and Quirke, N. (1993) *J. Phys. Chem.* 97, 4786.

Mariwala, R. K. and Foley, H. C. (1994) *Ind. Eng. Chem. Res.* 33, 2314.

Olivier, J. P. (1995) *J. Porous Materials* 2, 9.

Ravikovitch, P. I., Domhnaill, S. C. Ó., Neimark, A. V., Schüth, F., and Unger, K. K. (1995) *Langmuir* 11, 4765.

Ravikovitch, P. I., Haller, G. L., and Neimark, A. V. (1998) Adsorption characterization of nanoporous materials of M41S type. In *Proceeding Sixth International Conference on Fundamental of Adsorption*. (F. Meunier, ed.). Elsevier, Paris, 545.

Ravikovitch, P. I., Wei, D., Haller, G. L., and Neimark, A. V. (1997) *J. Phys. Chem. B* 101, 3671.

Razmus, D. M. and Hall, C. K. (1991) *A.I.Ch.E. J.* 37, 769.

Rege, S. U. and Yang, R. T. (2000) *A.I.Ch.E. J.* 46, 734.

Rouquerol, F., Rouquerol, J., and Sing, K. (1999) *Adsorption by Powders and Porous Solids*. Academic Press, San Diego, CA.

Rudzinski, W. and Everett, D. H. (1992) *Adsorption of Gases on Heterogeneous Surfaces*. Academic Press, San Diego, CA.

Saito, A. and Foley, H. C. (1991) *A.I.Ch.E. J.* 37, 429.

Sams, J. R., Jr., Constabaris, G., and Halsey, G. D. (1960) *J. Phys. Chem.* 64, 1689.

Schmidt, R., Hansen, E. W., Stöcker, M., Akporiaye, D., and Ellestad, O. H. (1995) *J. Am. Chem. Soc.* 117, 4049.

Seaton, N. A., Walton, J. P. R. B., and Quirke, N. (1989) *Carbon*, 27, 853.

Sayari, A., Yong, Y., Kruk, M., and Jaroniec, M. (1999) *J. Phys. Chem. B* 103, 3651.

Seifert, J. and Emig, G. (1987) *Chem. Eng. Technol.* 59, 475.

Stoeckli, H. F. (1977) *J. Colloid Interface Sci.* 59, 184.

Stoeckli, H. F. (1990) *Carbon* 28, 1.

Tricomi, F. G. (1985) *Integral Equations*. Dover Publications, New York, NY.

Venero, A. F. and Chiou, J. N. (1988) *MRS Symp. Proc.* 111, 235.

Webb, P. A. and Orr, C. (1997) *Analytical Methods in Fine Particle Technology*, Chap. 3. Micromeritics Inc. Norcross, GA.

Yang, R. T. (1997) *Gas Separation by Adsorption Process*. Butterworth, Boston, MA, Imperial College Press, River Edge, NJ.

Yang, R. T. and Baksh, M. S. A. (1991) *A.I.Ch.E. J.* 37, 679.

5

ACTIVATED CARBON

Activated carbon is the most widely used sorbent. Its manufacture and use date back to the 19th century (Jankowska et al., 1991). Its usefulness derives mainly from its large micropore and mesopore volumes and the resulting high surface area. The pore size distribution of a typical activated carbon is given in Figure 5.1, compared with several other sorbents. Comprehensive reviews on activated carbon are available elsewhere (Mattson and Mark, 1971; Cheremisinoff and Ellerbusch, 1980; Bansal et al., 1988; Jankowska et al., 1991; Rouquerol et al., 1999).

5.1. FORMATION AND MANUFACTURE OF ACTIVATED CARBON

The modern manufacturing processes basically involve the following steps: raw material preparation, low-temperature carbonization, and activation. The conditions are carefully controlled to achieve the desired pore structure and mechanical strength. The raw materials for activated carbon are carbonaceous matters such as wood, peat, coals, petroleum coke, bones, coconut shell, and fruit nuts. Anthracite and bituminous coals have been the major sources. Starting with the initial pores present in the raw material, more pores, with desired size distributions, are created by the so-called activation process. Two standard activation methods are used: gas and chemical. After initial treatment and pelletizing, gas activation first involves carbonization at 400–500 °C to eliminate the bulk of the volatile matter, and then partial gasification at 800–1000 °C to develop the porosity and surface area. A mild oxidizing gas such as CO_2 and steam, or flue gas, is used in the gasification step because the intrinsic surface reaction rate is much slower than the pore diffusion rate, thereby ensuring the uniform development of pores throughout the pellet. The activation process is usually performed in fixed beds, but in recent years fluidized beds have also been used. The activated carbon created by the activation

Adsorbents: Fundamentals and Applications, Edited By Ralph T. Yang
ISBN 0-471-29741-0

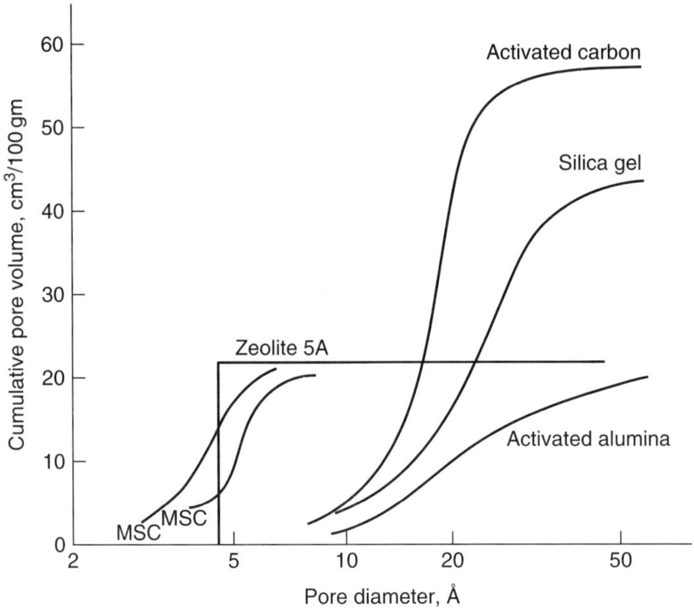

Figure 5.1. Pore-size distributions for activated carbon, silica gel, activated alumina, two molecular-sieve carbons (MSC), and zeolite 5A (Yang, 1997, with permission).

process is used primarily for gas and vapor adsorption processes. The other activation process that is also used commercially depends on the action of inorganic additives to degrade and dehydrate the cellulosic materials and, simultaneously, to prevent shrinkage during carbonization. This is called chemical activation, usually used for lignin-based precursors such as sawdust. No pre-carbonization step is required. Lignin, usually the raw material that is blended with activators such as phosphoric acid, zinc chloride, potassium sulfide, or potassium thiocyanate, is carbonized at temperatures between 500 to 900 °C. The remaining chemicals are removed from the carbon by acid or water wash. The product, usually in powder form, is used for aqueous or gas phase purposes. The inorganic material contained in activated carbon is measured as ash content, generally in the range between 2 and 10%. Beside granulated activated carbon (GAC), powdered activated carbon (PAC) is also used commercially (for example, for dioxin removal from incinerators by injection of PAC; Yang et al., 1999).

Although the process for manufacturing activated carbon has been largely empirical, a considerable understanding of the fundamental processes is now available (e.g., Barton et al., 1999). During carbonization of the precursor raw material, condensation of polynuclear aromatic compounds and breakage of side-chain groups occur, resulting in a carbon residue. In the meantime, cross-linking reactions occur that inhibit the development of a pre-graphite structure. Hence activated carbon is referred to as non-graphitizing, as opposed to graphitizing. The graphitizing type applies to pitches derived from petroleum, coal tar, and aromatic hydrocarbons, which form an intermediate liquid crystal phase called mesophase (Marsh and Walker, 1980; Walker, 1990; Andresen et al., 1998). Carbonization alone is not adequate to develop large porosity, hence activation by gasification is needed. The resulting activated carbon can be described by various models (e.g., Barton et al., 1999; Bandosz et al., 2001). The most representative model is a twisted network of defective hexagonal carbon layer planes, cross-linked by aliphatic bridging groups. The layer planes consist of single layers or layers of two, three, or four with variable interlayer spacings that typically range from 0.34 to 0.8 nm. The size of the layer planes varies but is typically about 5 nm wide. Simple functional groups (e.g., C−OH, C=O), as well as heteroatoms (mainly oxygen and hydrogen), are incorporated in the network.

By judicious choice of the precursor and also careful control of both carbonization and activation steps, it is possible to tailor the pore structure for particular applications (e.g., Barton et al., 1999). Mesoporosity (near or larger than 30 Å) is desirable for liquid-phase applications, whereas smaller pore sizes (10 to 25 Å) are required for gas-phase applications (Yang, 1997). The following table (Table 1) shows the pore size distribution as a function of the rank of the precursor coal (Wilson, 1981).

The activation step is crucial for tailoring the pore size distribution. As mentioned, this step is basically controlled gasification, by using steam, CO_2 or O_2, as well as added catalysts. The gasification reaction needs to be performed in the kinetic controlled regime, that is, under conditions where the intrinsic reaction (rather than internal pore diffusion or external gas film diffusion) is the rate-controlling step (Walker et al., 1959). The relative rates of gasification of graphite by these gases have been compared at 800 °C and 0.1 atm (Walker et al., 1959), shown below in Table 5.2.

Table 5.1. Pore volumes (cm^3/g) and PSD for coal-based activated carbons

Coal Precursor	Micropores	Mesopores	Macropores
Anthracite	0.51	0.07	0.11
Bituminous	0.43	0.17	0.26
Lignite	0.22	0.58	0.33

Table 5.2. Relative intrinsic rates of carbon gasification reactions (at 800 °C and 0.1 atm)

	Approx. Relative Rates
$C + O_2 \Rightarrow CO_2$	1×10^5
$C + H_2O \Rightarrow CO + H_2$	3
$C + CO_2 \Rightarrow 2CO$	1

With steam or CO_2 at atmospheric pressure, the kinetically controlled regime is below approximately 900 °C; hence this step is usually performed at 750–900 °C. The mechanisms of these gasification reactions are different and have been studied extensively (Walker et al., 1959; Yang, 1984). A main function of gasification is to widen the pores, creating large mesoporosity. These reactions are catalyzed further by minerals that catalyze the reactions by undergoing a variety of actions such as tunneling, channeling, and pitting (Yang, 1984; Baker, 1986), hence helping create and propagate fine pores. Another major function by chemical activation is the removal of oxygen and hydrogen. With the added catalysts, oxygen and hydrogen are selectively and nearly completely removed at below 700 °C (Jankowska et al., 1991). The main catalysts are zinc chloride, phosphoric acid, and potassium sulfide. The mechanism of activation by phosphoric acid has been studied (Jagtoyen et al., 1992; 1993; Solum et al., 1995). Studies by use of ^{13}C NMR have shown that phosphoric acid promotes cross-linking reactions and dehydration at low temperatures, thereby bonding otherwise volatile compounds and hence increasing the carbon yield. Different chemical agents also yield different pore structures; for example, carbons activated by zinc chloride were more mesoporous, whereas those activated by KOH were more microporous, although both yielded high surface areas around 1000 m^2/g, with that from $ZnCl_2$ slightly higher (Ahmadpour et al., 1998).

A breakthrough was made by Wennerberg and O'Grady (1978) by using a novel activation method, after a series of patents by Wennerberg beginning in 1971 (Wennerberg, 1971). The activation process is accomplished in molten KOH. The method was initially aimed at petroleum coke (a rather pure form of coke generated as the tail-end product in petroleum refining, used as anodes in aluminum smelting). It works also with other precursors for activated carbon, such as coals and nutshells. In this method, KOH and coke are mixed at a ratio of about 3/1 KOH/coke, heated to ~700–800 °C (the melting point of KOH is 360 °C) in an inert atmosphere (or in a closed system) for about 2 h. A small amount of water is used for pasting. A large microporosity is formed during the activation, with "cage-like" pores, mainly <2 nm. The BET surface area is typically 2800 m^2/g and reaches as high as 4057 m^2/g, depending on the activation condition. The mechanism of activation is not understood. The 3/1 KOH/coke ratio is equivalent to ~0.6 KOH/C. This amount of KOH is adequate to provide oxygen for gasification of carbon to create the micropores. Activation with a 1/1 mole mixture of KOH and NaOH is also effective (Audley and Holder, 1988). These carbons have been proven to be the best for methane storage, as will be discussed in Chapter 10 (10.4). Further developments and the commercial status of these carbons are also discussed in Chapter 10.4.

5.2. PORE STRUCTURE AND STANDARD TESTS FOR ACTIVATED CARBON

Activated carbons are characterized by a large surface area between 300 and ~4000 m^2/g, as measured by the BET method, and are the largest among all sorbents. Commercial grades of activated carbon are designated for either gas phase or liquid phase, depending on its application. A majority of the pore volume is from pores near or larger than 30 Å in diameter for liquid-phase carbons, whereas the pores of gas-phase carbons are mostly in the range from 10 to 25 Å in diameter. The need for larger pores in liquid-phase carbons is due to the large size of many dissolved adsorbates and the slower diffusion in liquid than in gas for molecules of the same size.

A polymodal pore-size distribution is generally found in activated carbon. The pore structure may be pictured as having many small pores branching off from larger ones, which are open through the entire particle. The larger pores are called feeder or transport pores; the smaller ones, which may be dead-end, are called adsorption pores. The cumulative pore-volume distribution of the fine pores for a typical gas-phase activated carbon is shown, along with four other sorbents, in Figure 5.1. The larger pores are mostly submicrometer in size, and their total volume amounts to a fraction of that found in the fine pores. The micropore structures of activated carbons can be determined by a variety of adsorption methods (Rodriquez–Reinoso and Linares–Solano, 1986).

Testing of activated carbons involves measurements of bulk density, absolute density (or helium density), particle density, particle size distribution, pressure drop across packed beds, and mechanical strength. Standardized ASTM tests are available for these measurements. Details of these tests are available from ASTM. The mechanical strength is defined by the abrasion number, which is a measure of the change of the mean particle size, expressed as a percentage, of a sample after 3 min of vigorous agitation with smooth steel balls. In addition, numerous adsorption tests are used for the characterization of the adsorbent properties of commercial carbons. The effectiveness of activated carbon is usually specified by the amount of a certain test chemical it can adsorb under standardized conditions. For gas-phase applications, CCl_4 is commonly used, whereas for aqueous-phase or liquid-phase applications, iodine and molasses are used.

Carbon Tetrachloride Number or Butane Number: The carbon tetrachloride number, as defined by ASTM D3467-99, is the ultimate capacity for CCl_4 by adsorption of its vapor in air. The carbon tetrachloride number of a typical commercial GAC is 60, meaning 60% (wt./wt.). Due to the adverse effects of CCl_4 to the environment and human health, the butane number is adopted, which measures the capacity for *n*-butane (ASTM D5228-92). Empirically, the CCl_4 number can be obtained by multiplying the butane number by 2.55.

Iodine Number: The iodine number is defined as the amount of iodine (in milligrams) adsorbed by powdered carbon (per gram) from 0.02 N iodine aqueous solution (ASTM D4607-94). Iodine is in the form of I_2, with a very small amount of I_3^- anion. A typical iodine number for activated carbon is 900, with values >1000 for better grades of carbon. The iodine number has been roughly correlated to the surface area of pores >10 Å diameter. It is regarded approximately as the total pore volume.

Molasses Number: The molasses number is calculated from the ratio of the optical densities of the filtrate of a molasses solution treated with a standard

Table 5.3. ASTM tests of typical activated carbons from different precursors

	Bituminous	Sub-Bituminous	Lignite	Nut-Shells
Iodine number	1000–1100	800–900	600	1000
Molasses number	235	230	300	—
Abrasion number	80–90	75	60	97
Bulk density (lb/ft^3)	26–28	25–26	23	29–30

activated carbon and one treated with the activated carbon being tested. The molasses number has been correlated with the surface area of pores >28 Å diameter. This number is a very rough indication of the surface area of the larger pores because molasses is a mixture of many organic molecules.

Some typical properties of commercially activated carbons derived from different precursors are given in Table 5.3. The results from these simple tests (Table 5.3) do agree qualitatively with the results of the more elaborate measurements of pore size distributions, shown in Table 5.1.

5.3. GENERAL ADSORPTION PROPERTIES

The unique surface property of activated carbon, in contrast to the other major sorbents, is that its surface is nonpolar or only slightly polar as a result of the surface oxide groups and inorganic impurities. This unique property gives activated carbon the following advantages:

1. It is the only commercial sorbent used to perform separation and purification processes without requiring prior stringent moisture removal, such as is needed in air purification. For the same reason, it is also widely used as a sorbent for processes treating aqueous solutions.
2. Because of its large, accessible internal surface (and large pore volume, shown in Figure 5.1), it adsorbs more nonpolar and weakly polar organic molecules than other sorbents do. For example, the amount of methane adsorbed by activated carbon at 1 atmosphere (atm) and room temperature is approximately twice that adsorbed by an equal weight of molecular sieve 5A (Fig. 5.2).
3. The heat of adsorption, or bond strength, is generally lower on activated carbon than on other sorbents. This is because only non-specific, van der Waals forces are available as the main forces for adsorption (see Chapter 2). Consequently, stripping of the adsorbed molecules is relatively easier and results in relatively lower energy requirements for regeneration of the sorbent.

It is not correct, however, to regard activated carbon as hydrophobic. The equilibrium sorption of water vapor on an anthracite-derived activated carbon

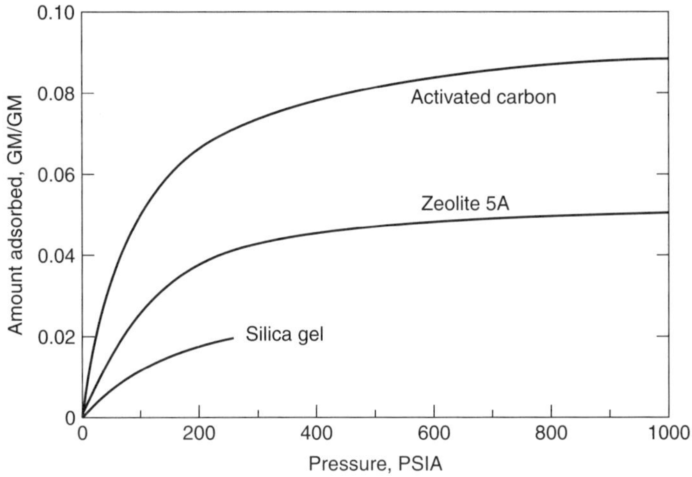

Figure 5.2. Equilibrium adsorption of methane at 25 °C on silica gel, zeolite 5A, and activated carbon. *Source*: Saunders (1982); Yeh (1989).

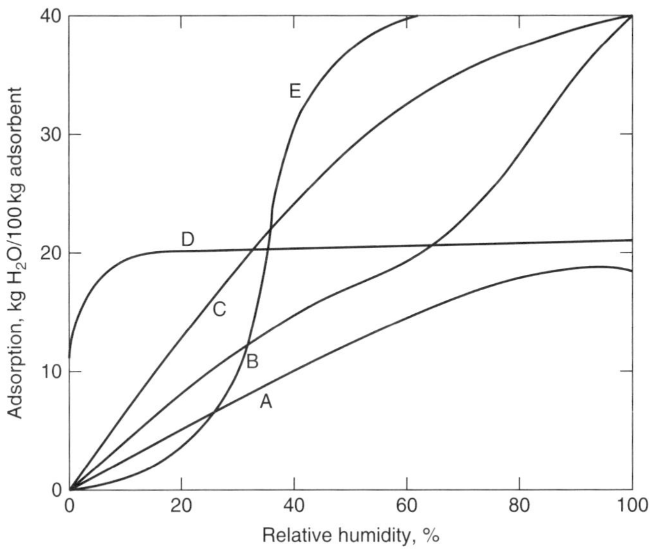

Figure 5.3. Equilibrium sorption of water vapor from atmospheric air at 25 °C on (A) alumina (granular), (B) alumina (spherical), (C) silica gel, (D) 5A zeolite, (E) activated carbon. The vapor pressure at 100% R.H. is 23.6 Torr. *Source*: (Yang, 1997; with permission).

is compared with that of other sorbents in Figure 5.3. The sorption of water vapor on activated carbon follows a Type V isotherm with a sigmoidal or S-shaped curve (according to the BDDT classification, Yang, 1997). The reason for the low adsorption at low vapor pressures is due to the low van der Waals interactions. Once a few water molecules are adsorbed, adsorbate–adsorbate interaction commences, leading to cluster formation and eventually pore filling or capillary condensation in the micropores. Activated carbon is used, nonetheless, in processes dealing with humid gas mixtures and water solutions because the organic and nonpolar or weakly polar compounds adsorb more strongly, and hence preferentially, on its surface than water does.

Adsorption of water vapor on activated carbon has been studied extensively because of its scientific as well as practical importance. Chemical modification can significantly alter the adsorption behavior. It has long been known that oxidation and reduction affect the "hydrophobicity" of carbon. As mentioned, the water isotherm generally follows an S-shaped curve, with little or no adsorption at P/P_0 below 0.3 or 0.4. In this region, water molecules are bonded to certain oxygen complexes, likely by hydrogen bonding and electrostatic forces (the nonspecific interactions by Lennard–Jones 6–12 potential are insignificant). At higher P/P_0, clusters and eventually pore filling occur through hydrogen bonding. Pore structure comes into play only in the latter stage. Oxidation of the surface increases the oxygen complexes, hence shifting the threshold P/P_0 for water adsorption. The extensive literature on this subject has been discussed elsewhere (Jankowska et al., 1991; Rouquerol et al., 1999; Leon y Leon and Radovic, 1994; Rodriquez–Reinoso et al., 1992; Carrasco-Marin et al., 1997; Salame and Bandosz, 1999; and others to be discussed specifically later).

5.4. SURFACE CHEMISTRY AND ITS EFFECTS ON ADSORPTION

As described above, activated carbon can be represented by a model of a twisted network of defective hexagonal carbon layer planes (typically 5 nm wide), which are cross-linked by aliphatic bridging groups. Heteroatoms are incorporated into the network and are also bound to the periphery of the planes. The heteroatoms bound to the surfaces assume the character of the functional groups typically found in aromatic compounds, and react in similar ways with many reagents. These surface groups play a key role in the surface chemistry of activated carbon. They are particularly important for adsorption in aqueous solutions and the catalytic properties of carbon. The surface chemistry of activated carbon has been a subject of long-standing scientific interest, and many reviews of the subject are available (Mattson and Mark, 1971; Puri, 1970; Cookson, 1978; Bansal et al., 1988; Zawadzki, 1989; Jankowska et al., 1991; Leon y Leon et al., 1994; Radovic and Rodriquez–Reinoso, 1996; Boehm, 2002). Chemical modifications, particularly oxidation, have been used to effectively tailor the adsorption and catalytic properties. Some typical surface groups are given in Figure 5.4 (Puri, 1970; Zawadzki, 1989; Radovic and Rodriquez–Reinoso, 1996; Boehm, 2002).

The surface groups shown in Figure 5.4 are acidic groups. Concentration of these groups can be created or increased by oxidation with oxygen at elevated temperatures (or by aging at mild temperatures) or with liquid oxidants, typically nitric acid (Noh and Schwarz, 1990). The acidic surface shows cation exchange properties in aqueous solutions. If the carbon is de-gassed at a high temperature,

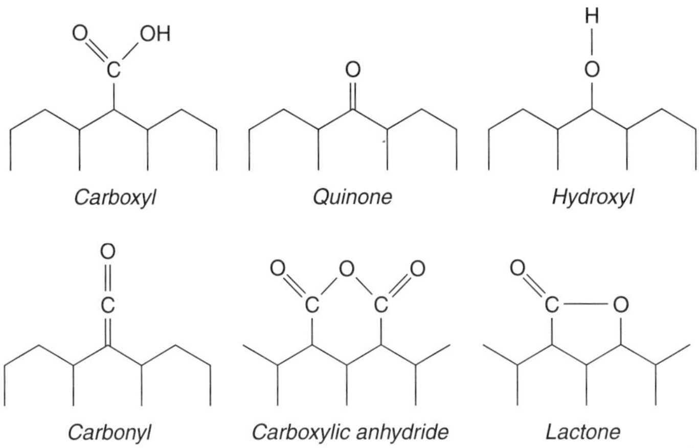

Figure 5.4. Simplified schematic of some acidic surface groups on activated carbon. These groups are bonded to aromatic rings.

for example, 950 °C, in vacuo or under an inert atmosphere, and subsequently oxidized at room temperature after cooling, it exhibits basic character and hence acquires an anion exchange capacity. Although the nature of the acidic surface sites is quite well understood, the origin of the basic properties is still a subject for debate (Boehm, 2002).

Numerous experimental techniques have been used to determine the acidity or basicity of activated carbon (Boehm, 2002). Titration has been the most useful technique. Others include: 1) infrared spectroscopy (Zawadzki, 1989; Sellitti et al., 1990; Meldrum and Rochester, 1990; Fanning and Vannice, 1993; Zhang et al., 1994; Mawhiney et al. 1999; Moreno–Castillo et al., 2000); 2) X-ray photoelectron spectroscopy or ESCA (Papirer et al., 1994; Biniak at al., 1997; Nanse et al., 1997; Moreno–Castillo et al., 2000); 3) temperature-programmed desorption or TPD (Otake and Jenkins, 1993; Figureido et al., 1998; Moreno–Castillo et al., 1998; Haydar et al, 2000); and 4) electrokinetic measurements (Boehm and Voll, 1970; Menendez et al., 1995).

The acidic properties are caused by the oxygen-containing groups shown in Figure 5.4. These groups exhibit different acid strength following the order (Boehm, 2002):

$$\textit{Hydroxyls} > \textit{Lactones and Lactols} > \textit{Carboxyl (and their anhydrides)}$$

The amount as well as the strength of acidity can be determined by titrating with basic solutions of different alkalinity (Boehm, et al., 1964; Voll and Boehm, 1970; Bandosz, et al., 1993; Boehm, 1994; Contescu et al., 1997; Biniak et al., 1997; Boehm, 2001; Boehm, 2002). The three types of acid groups above can be neutralized by 0.05 N solutions of, respectively, NaOH, Na_2CO_3, and $NaHCO_3$ (known as Boehm titration). The acidity of any functional group is influenced by its local chemical environment, that is, the size and shape of the polyaromatic layers, other neighboring groups, and their charges. However, such influence is minor, so the different types of groups can be differentiated by the simple titration technique. By careful continuous titration, the distribution of different acid groups on the same carbon can be delineated by different peaks at different acidity constants (pK) (e.g., Contescu et al., 1997).

Continuous titration can also show the existence of basic sites of different pK or basicity strengths on the surface of activated carbon (Contescu et al., 1998). Two types of basic groups that have been proposed are shown in Figure 5.5.

The pyrone-type group (first suggested by Boehm and Voll, 1970; Voll and Boehm, 1971) contains two non-neighboring oxygen atoms to constitute one basic site. Upon contact with a proton in the aqueous medium, the semiquinone oxygen is protonated to a hydroxyl, leaving a positive charge on the other oxygen. The two oxygen atoms are located at two different graphite rings so as to favor resonance stabilization of the positive charge. The net positive charge gives rise to its anion exchange capacity and basicity. Evidence for the pyrone-type site as the basic site has been cumulating, including acid titration and TPD (Temperature programmed desorption) (Leon y Leon et al., 1992) as well as theoretical calculations (Suarez et al., 1999; 2000; Boehm, 2002). The chromene model as a basic site was suggested by Garten and Weiss (1957). Upon reaction with a proton and O_2, a positive charge is introduced to the aromatic ring, hence the basicity. Beside the pyrone and chromene types, it has long been thought that the basic behavior of carbon surfaces may derive from the π basicity (or Lewis basicity) of the aromatic rings (Boehm and Voll, 1970; Leon y Leon et al., 1992; Montes-Moran et al., 1998). However, the basicity from the aromatic rings is weak; and the main basicity is still attributed to oxygen-containing groups.

Much has been learned from thermal decomposition or TPD of surface oxides. When the TPD results are coupled with titration and chemical analysis, one can determine the specific forms of surface oxides that are being desorbed (e.g., Cookson, 1978). During TPD, water, carbon dioxide, carbon monoxide, and small amounts of hydrogen are the evolved gases. Generally, the evolution of CO_2 and H_2O commences in the 200–300 °C range, ending in the 400–500 °C range for H_2O and the 700–800 °C range for CO_2. The evolution of CO commences in

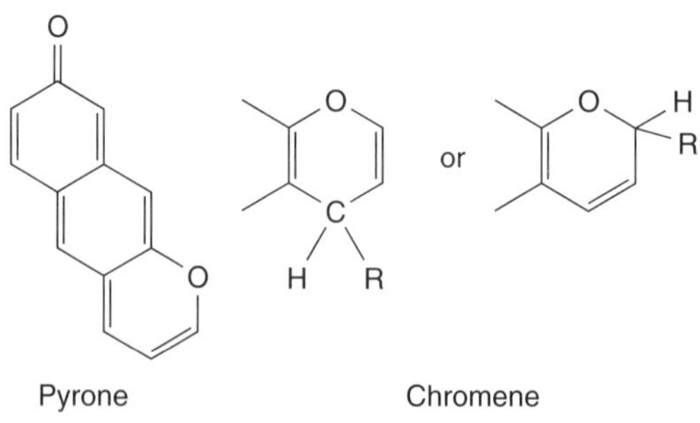

Figure 5.5. Schematic of possible basic groups on activated carbon.

the 500–700 °C range and ends around 1000 °C. Evolution of small quantities of hydrogen occurs at the 600–1000 °C range, and some hydrogen is retained even after heating at 1200 °C. This is generally because the C–H bond is stronger than the C–C bond. It is generally believed that carboxylic groups and their derivatives, such as lactones, decompose to give CO_2, whereas quinone or semiquinone groups give CO and hydroquinones and phenols give CO and H_2O.

Acidic surface oxides can be generated by oxidation with oxygen at elevated temperatures or with liquid oxidants. Aging can also generate such oxides. However, aging in air or water (at temperatures <200 °C; Puri, 1970) after the carbon is degassed at 700–950 °C would create surface oxide that is basic in nature (Puri, 1970; Boehm, 2002). Obviously, little is understood about the aging process. The surface oxides can also be removed, as seen in TPD, at high temperatures in vacuo or in an inert atmosphere. Adsorption on these cleaned carbons has also been studied.

Basicity of carbon surfaces can also be introduced by "nitriding," that is, reacting the carbon with anhydrous ammonia at elevated temperatures (Radovic et al., 1996). Unlike oxygen, only very small amounts of nitrogen can be incorporated as functionalities, which, however, can yield basicity on the surface.

5.4.1. Effects of Surface Functionalities on Gas Adsorption

Adsorption of gas molecules on activated carbon is dominated by the van der Waals forces (i.e., dispersion and repulsion forces). The electrical charges on activated carbon (or modified carbons) are too weak (compared with other sorbents) or the positive and negative charges are too close to each other to exert any significant electric field or field gradient on the surface. In addition, the small pore sizes and the large surface area of activated carbon play the major role in gas adsorption. Consequently, the surface groups have significant effects only for adsorption of polar gases. The most extensively studied gas adsorbate molecule is water. Other polar molecules that have been studied are alcohols, amines, ammonia, SO_2 (Jankowska et al., 1991), NO, and other "supercritical" gases (Kaneko, 1998).

With polar molecules that have strong permanent dipole moments, the field-dipole interaction term (see Chapter 2) becomes significant. Hence the adsorption of these molecules can be significantly increased by introducing oxide groups on the surface. An example for water adsorption is shown in Figure 5.6. For the adsorption of water, hydrogen bonding between water and the surface oxide has been proposed (Puri, 1966). The amount of increase in adsorption seems to be correlated to the amount of the oxygen-containing groups that release CO_2 upon thermal desorption, and there appears to be one water molecule per such group (Puri, 1966).

Figure 5.6 shows the significant increase in adsorption of water molecules at low vapor pressures by oxidation (Barton et al, 1984). This increase is caused by the introduction of surface oxygen functionalities. Changes in pore volume and surface area also accompany oxidation. The ultimate capacity (at P/P_0 near 1)

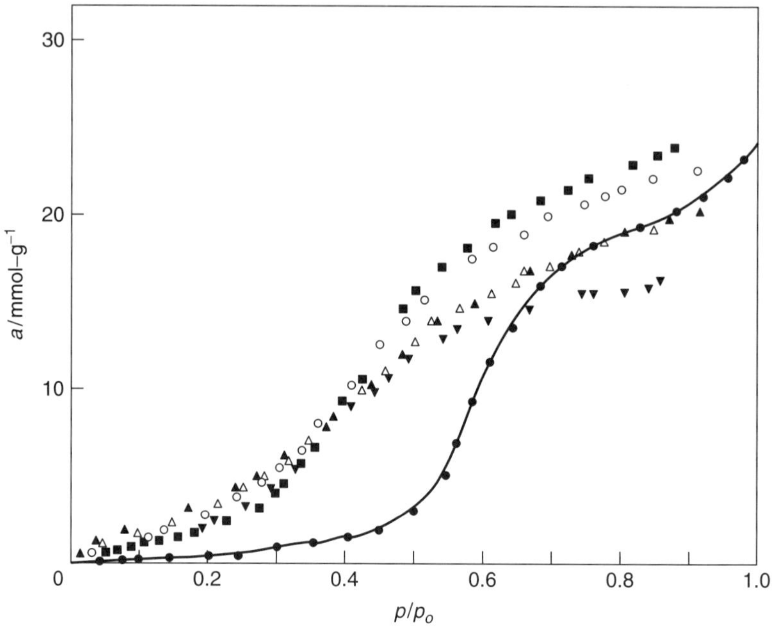

Figure 5.6. Adsorption isotherms of water vapor on BPL carbon (bituminous coal-based, gas-phase carbon, from Calgon Corp.) at 25 °C. Untreated BPL (•) and BPL oxidized by 6 M HNO_3 at time lengths in hours: HNO_3 – 0.5 (■); HNO_3 – 1(△); HNO_3 – 2(o); HNO_3 – 4(■); HNO_3 – 7(▼). Data taken from Barton et al., 1984.

remains unchanged by oxidation. Gubbins and co-workers have studied adsorption of water on activated carbon with surface functional groups by molecular simulation and were able to obtain agreement with experimental data (Muller et al., 1996; MacCallum et al., 1999). A detailed mechanism of water adsorption in confined micropores of carbon is not available. However, rather convincing evidence has been shown for the formation of water clusters in the micropores of carbon (Kaneko, 2000). Kaneko and co-workers have made a series of studies with two samples of activated carbon fiber, with narrow pore size distributions around 0.75 and 1.1 nm. Very different adsorption/desorption and hysteresis behaviors have been seen with these two samples (Hanzawa and Kaneko, 1997; Kaneko et al., 1999; Kaneko, 2000). Much less water could be adsorbed at saturation on the carbon with smaller pores (0.75 nm), that is, ~0.2 g/g versus 0.8 g/g for the carbon with 1.1 nm pores. From these studies, the formation of pentamers of water, $(H_2O)_5$, was proposed. The cluster formation proposal was further supported by results from small-angle X-ray scattering (Kaneko, 2000), X-ray diffraction (Iijima et al., 1997), and differential scanning calorimeter (Kaneko et al., 1998) of the water adsorbed in these two carbon samples (Kaneko, 2000). Based on the formation of pentamers, an isotherm was derived for water in carbons by Do (2000), which fit experimental data well. Functional groups are obviously the sites for anchoring the clusters.

By more intense oxidation of the carbon, the amount of water vapor adsorbed at low relative pressures (<4 Torr at 25 °C) can be drastically increased. For example, Walker and co-workers showed a 100-fold increase in water vapor adsorption by activated carbon after strong surface oxidation by HNO_3 (Mahajan et al., 1982). Exchange of the surface H-ions by cations (Li, Na, K, Ca) on the oxidized carbon further increased the moisture capacity at low vapor pressures to amounts comparable with that on zeolites. The ion-exchanged carbon was fully regenerated at 140 °C, in contrast to temperatures >350 °C that are required for zeolite regeneration (Mahajan et al., 1982).

As mentioned above, the surface groups can also be removed by thermal desorption. As expected, less water and polar molecules would be adsorbed on the cleaned carbon samples. An example of the decreased water adsorption on a partially cleaned activated carbon (desorbed at 500 °C) is shown in Figure 5.7. It was further found (Puri, 1966; Jankowska et al., 1991) that the amount of adsorbed water was proportional to the oxygen-containing groups that yielded CO_2 upon thermal desorption (not to the total amount of oxygen functionalities), that is, carboxylic groups and their derivative, such as lactones. Hydrogen bonding was proposed to have been formed between water and these groups. The bonds were sufficiently strong so that irreversible adsorption, or hysteresis, was reported (Puri, 1966).

Although the effects of oxidation on water adsorption are expected, the effects of incorporation with other atoms do not yield expected results. Mild chlorination of the surface of activated carbon results in slightly more hydrophobicity (Hall and Holmes, 1993). Mild fluorination has shown drastically increased

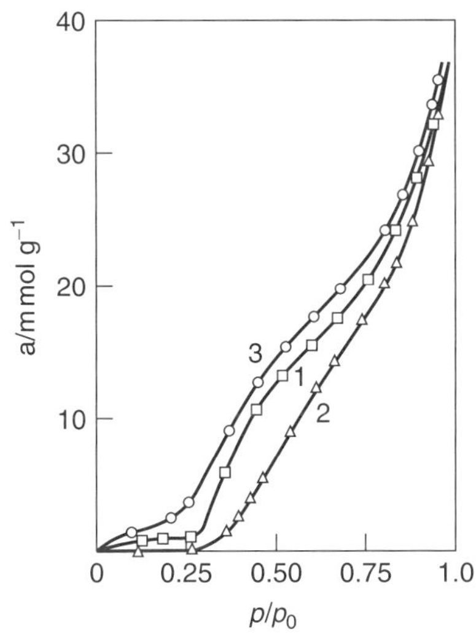

Figure 5.7. Adsorption isotherms of water vapor at 20 °C on activated carbon CWN-2. Sample 1: untreated. Sample 2: heat treated under vacuum at 500 °C. Sample 3: oxidized with air at 400 °C (from Jankowska et al., 1991, with permission).

hydrophobicity, at even $P/P_0 \approx 1$ (Kaneko et al., 1995; Kaneko et al., 2000). The work on fluorination by Kaneko and co-workers was performed on active carbon fibers (ACFs), but their observations are expected on activated carbon as well. Incorporation of nitrogen atoms on ACFs, on the contrary, decreased the hydrophobicity (Kaneko et al., 2000). A further understanding of the different effects by different chemical modifications is clearly needed.

5.5. ADSORPTION FROM SOLUTION AND EFFECTS OF SURFACE FUNCTIONALITIES

Adsorption from liquid solution is more complicated than that from the gas phase. In a two-component liquid solution, both the solvent and the solute will be adsorbed to different extents. Usually the adsorption of the solute is of interest. The experimental procedure and expression for the amount adsorbed both differ from those used for gas adsorption. The basis for expressing the amount adsorbed from liquids is the concept of surface excess given by Gibbs in 1878. Gibbs surface excess is the difference in the amount of a given component in the surface layer (per unit surface area) over that in the bulk liquid.

The actual experiment involves measurements of the initial and final concentrations of the solution. Let n^o be the total number of moles in the initial solution, and $n^o = n_1^o + n_2^o$, where subscripts 1 and 2 denote the solute and solvent, respectively. Let x^o and x denote the initial and final mole fractions, respectively. Then the Gibbs adsorption or the apparent adsorption is given, for 1 g of sorbent, by:

$$n_1^{\sigma(n)} = n_o(x_1^o - x_1) = n^o \Delta x_1 = n_1^s - (n_1^s + n_2^s)x_1 \tag{5.1}$$

where n_1^s and n_2^s are the number of moles of component 1 and 2, respectively, in the surface layer. For totally miscible solutions, two general types of adsorption isotherms of Gibbs adsorption, also referred to as the composite adsorption isotherm, are shown in Figure 5.8.

For each composite isotherm, there are two real adsorption isotherms, $n_1^s(x_1)$ and $n_2^s(x_2)$, which can also be measured and are always positive. These two general types have been further classified into five types by Nagy and Schay (1960). The composite isotherm shown in Figure 5.8 clearly indicates the relative strengths of adsorption: n-butyl amine > benzene > methyl acetate.

The composite isotherm can be changed by introduction of surface functionalities such as oxygen. Oxidation will increase the preference for the more polar component of the solution. This is shown by the example of adsorption of the ethanol–benzene solution, in Figure 5.9. The concentrations of the four groups (phenolic−OH, carboxylic, lactone, and carbonyl) were measured and were increased significantly by oxidation with nitric acid (Jankowska et al., 1983). The total amount of oxygen-containing groups was increased from 2.05 to 6.1%. The preference for adsorption of ethanol over benzene is evident in Figure 5.9. Degassing of the surface reduced the surface oxygen to 0.35%, and resulted in a more preferential adsorption of benzene.

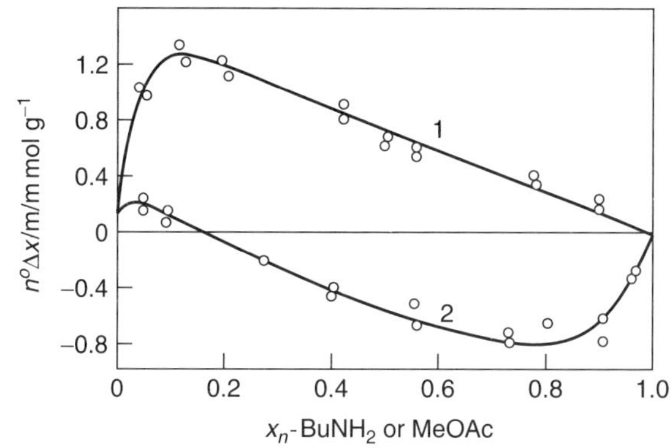

Figure 5.8. Isotherm of excess adsorption on activated carbon of (1) n-butylamine and (2) methyl acetate, from their respective solutions in benzene (Blackburn et al., 1957), where x is mole fraction in solution and the excess adsorption is given by Eq. 5.1.

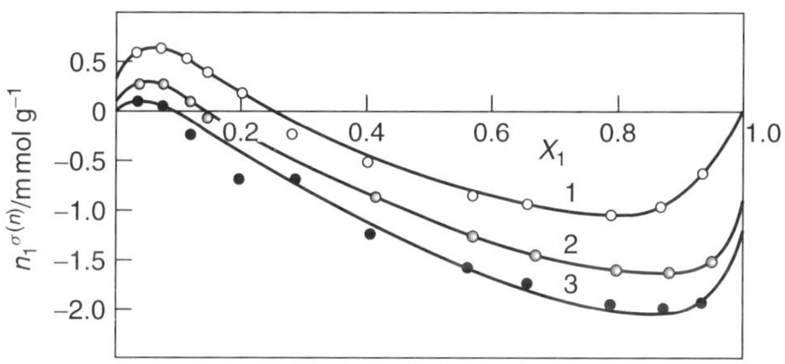

Figure 5.9. Isotherm of excess adsorption of ethanol from ethanol-benzene solution at 20 °C on CWZ-3 activated carbon: (1) oxidized with nitric acid; (2) untreated; and (3) heat treated in nitrogen at 1100 °C (from Jankowska et al., 1983, with permission). X_1 is the mole fraction of ethanol in solution, and its excess adsorption is given by Eq. 5.1. The total amounts of oxygen functionalities = 6.1% (1), 2.05% (2), and 0.35 % (3).

5.5.1. Adsorption from Dilute Solution (Particularly Phenols)

Activated carbon is widely used for potable water and wastewater treatments. A large compilation of the equilibrium adsorption data, expressed in the form of Freundlich isotherm, is available in Faust and Aly (1987) for organic compounds in dilute aqueous solutions. These compounds include the Environmental Protection Agency organic priority pollutants, such as phenolic compounds, aromatic and chlorinated aromatic compounds, chloroethylenes and other volatile organic compounds (VOC), carbon tetrachloride, and organic pesticides. It is also a good source of reference data for adsorption of inorganic compounds such as those of As, Ba, Cd, Cr, Cu, Pb, Se, Hg, F, and Cl.

Adsorption from liquid solution is complicated by the presence of the solvent. Interactions between solute–surface, solvent–surface as well as solute–solvent are all involved. The effects of solubility on adsorption have long been known.

Generally, low solubility favors adsorption. The dramatic effects of the solute–solvent interaction on adsorption may be illustrated by adsorption of aqueous solutions of thiolane and its derivative, sulfolane. Sulfolane is of interest because it has been used for more than 40 years for solvent extraction in a number of industrial processes (e.g., the Sulfinol process for CO_2 removal, the UOP Sulfolane process for recovery of aromatics from hydrocarbons, processes for extraction of fatty acids, etc.). It has now become a serious pollutant for ground water, however. The structures of sulfolane and thiolane are shown in Figure 5.10. The adsorption isotherms of their aqueous solutions on a clay sorbent are given in Figure 5.11 (Kim et al., 1999). The clay had a cation exchange capacity of 0.3 meq/g, and its surface was relatively polar (more polar than carbon but less polar than silica gel and activated alumina).

Comparison of thiolane and sulfolane shows 1) sulfolane is polar and thiolane is nonpolar, and 2) sulfolane has a significantly larger polarizability than thiolane. Hence the interaction between sulfolane and the clay surface should be much

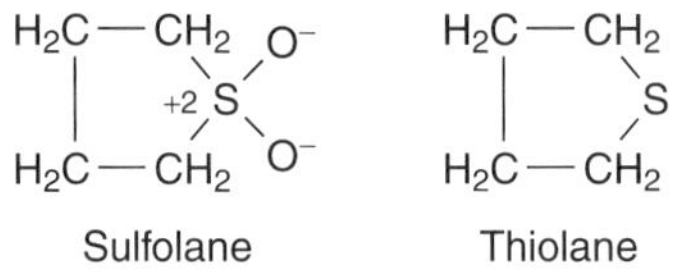

Figure 5.10. Structures of sulfolane and thiolane.

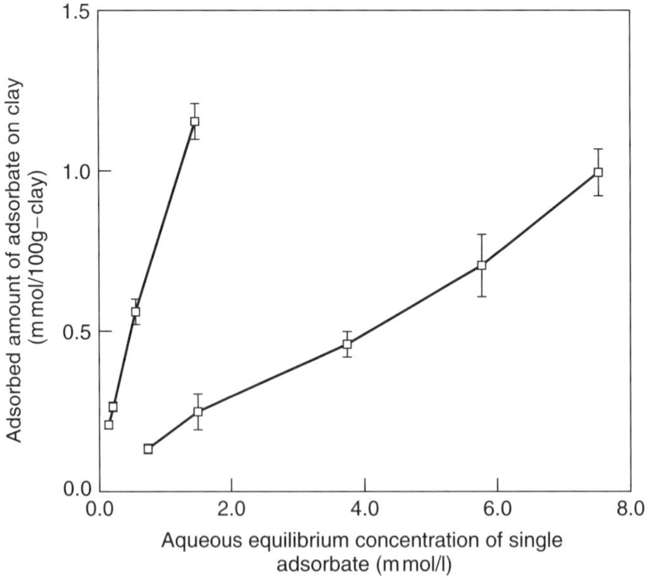

Figure 5.11. Adsorption isotherms of thiolane (upper curve) and sulfolane (lower curve) on clay from their respective aqueous solutions at 18 °C (from Kim et al., 1999 with permission).

stronger than that between thiolane and clay. For adsorption from their aqueous solutions, on the contrary, Figure 5.11 shows that the adsorption isotherm of sulfolane is much lower than that of thiolane. The low adsorption of sulfolane is due to the thermodynamic stability of the sulfolane–water interactions that are comparable with or stronger than the interactions between sulfolane and the negatively charged sites of the clay. The opposite holds for thiolane. This example illustrates the importance of the solute–solvent interaction and also the complexity of adsorption from liquid solution. The effects of solvent have been studied for adsorption of "Sudan III" and "Butter Yellow" from different solvents, and the amounts adsorbed varied over an order of magnitude (Manes and Hofer, 1969). Another good example for the strong solvent effect is seen in the adsorption of glycols and sugars with multiple −OH groups from their aqueous solutions on activated carbon (Chinn and King, 1999). Three adsorbates are compared: propylene glycol (M.W. = 76, 2 hydroxyls), glycerol (M.W. = 92, 3 hydroxyls), and glucose (M.W. = 180, 5 hydroxyls). The solubility follows the order of the number of hydroxyl groups, that is, glucose is the most soluble. Without the solvent, the sorbate–sorbent bond strength follows the order of the molecular weight. From the aqueous solution, however, the order of the heat of adsorption is completely reversed (20–30 kJ/mol for propylene glycol and ∼10 kJ/mol for the other two).

Phenol is the most extensively studied adsorbate for the adsorption of aqueous solutions on carbon, for practical (in waste water treatment) as well as scientific reasons. It is also a good model compound for organic pollutants in wastewater. Numerous factors are known to have significant effects on the adsorption of phenol: pH of solution, type of carbon, carbon surface functionalities, oxygen availability ("oxic" vs. "anoxic" condition), mineral contents of carbon, and addition of electrolytes. Typical isotherms, as a function of solution pH, are shown in Figure 5.12. The amount adsorbed is decreased at both high and low pH values.

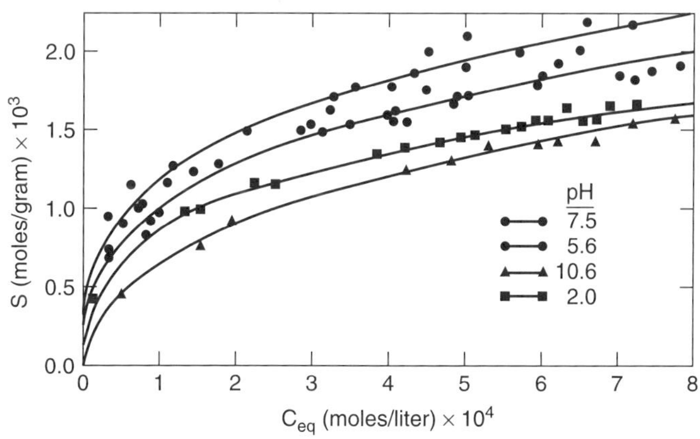

Figure 5.12. Isotherms for phenol on Columbia carbon (from National Carbon Co.) at 25 °C and different pH (Snoeyink et al., 1969, with permission). Data were taken after >10 days.

The interpretation for the pH effects is based on the dissociation of phenol into phenolate anion and proton (Snoeyink et al., 1969). The pKa value for phenol is 9.89, hence the principal adsorbing species above this pH is mostly anionic. Because phenolate anion has more affinity for the aqueous phase and the surface is acidic, repulsion between the surface layer and the anionic phenolate results in reduced adsorption. The low pH value was obtained by using an acid solution, which introduced additional protons in the solution. The infrared study of Mattson et al. (1969) indicated that surface carbonyl was the adsorption site that bonded phenol and nitrophenol by forming an acceptor-donor-type charge-transfer complex. The added protons in the solution would compete for the carbonyl sites, hence the reduced adsorption at low pH (Snoeyink et al., 1969). The presence of an adsorption maximum with pH could be explained by a model given by Muller et al. (1980). The model was based on the surface charge in equilibrium with the solution pH (as characterized by equilibrium constant, K) and the electrostatic interaction potential of the ionized solute with the charged surface (which yields the value of K).

It has long been known that some of the phenol and its derivatives adsorb on carbon irreversibly, that is, the irreversibly adsorbed phenol cannot be desorbed in water or by heating (to temperatures as high as 800 °C) (Coughlin and Ezra, 1968; Mattson et al., 1969; Snoeyink et al., 1969; Pahl et al., 1973; Suzuki et al., 1978; Cooney, 1983; Sutikno and Himmelstein, 1983; Mange and Walker, 1986; Martin and Ng, 1987; Grant and King, 1990; Leng and Pinto, 1996). The nature and mechanism of the irreversible adsorption of phenol have been clarified by a definitive study by Grant and King (1990). Because the adsorption equilibrium is reached slowly at room temperature, >5 days equilibration time was typically used in all studies. The irreversible amounts could be measured by extraction with acetone or as the difference between the total amount and the reversibly adsorbed amount (Grant and King, 1990). The reversibly and irreversibly adsorbed amounts are shown in Figure 5.13.

The effect of pH on the total amount adsorbed, by adding the reversibly and irreversibly adsorbed amounts in Figure 5.13, agree well with that shown in Figure 5.12. The irreversibly adsorbed phenol was extracted with acetone, and the extracted solutions were analyzed by mass spectrometry. The mass spectra are shown in Figure 5.14. Their results showed clearly that polymerization of phenol occurred on the surface of activated carbon. The abundance of the polymer decreased with molecular weight. No polymers were detected in the control solution. This product distribution is consistent with that from oxidative coupling reactions. Grant and King further showed that the oxidative coupling reaction was not caused by any minerals in the carbon and did not require oxygen or water (although oxygen increased the amount of the irreversibly adsorbed phenol). Numerous substituted phenols have also been studied by Grant and King (1990), and all showed irreversible adsorption. The ordering of the irreversible adsorption follows: *p*-methoxyphenol > 2,4-dimethylphenol = *p*-chorophenol > phenol > aniline > *p*-nitrophenol = *p*-hydroxybenzaldehyde. Reactivities of these

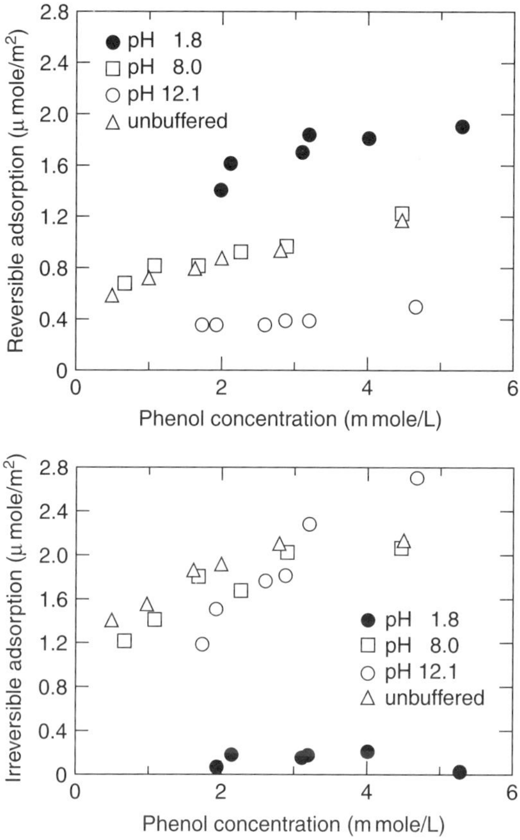

Figure 5.13. Reversibly and irreversibly adsorbed amounts of phenol from aqueous solutions with various values of pH, on Columbia carbon (from Witco, BET area = 900 m^2/g), at 80 °C after 5 days (Grant and King, 1990, with permission).

compounds in oxidative coupling reactions corresponded well with the ordering above. The reactivities were correlated with the "critical oxidation potential" (Grant and King, 1990). Apparently oxidative coupling reactions occur on the carbon surface with all these compounds.

Polymerization by oxidative coupling of phenol on carbon has been further studied and confirmed by Vidic et al. (1993; 1994). As shown by Grant and King (1990), the presence of oxygen increased the irreversibly adsorbed phenol.

The active sites for oxidative coupling are not understood. By using several polymeric resins as the sorbents, Grant and King (1990) showed that carboxylic acid and ester groups on these resins do not promote irreversible adsorption. However, evidence indicates that the phenolic groups do. It is known that activated carbon can catalyze numerous reactions at near ambient temperatures, particularly oxidation reactions (e.g., reviews by Leon y Leon et al. 1994; Radovic and Rodriquez–Reinoso, 1996). Surface oxygen functionalities are generally needed

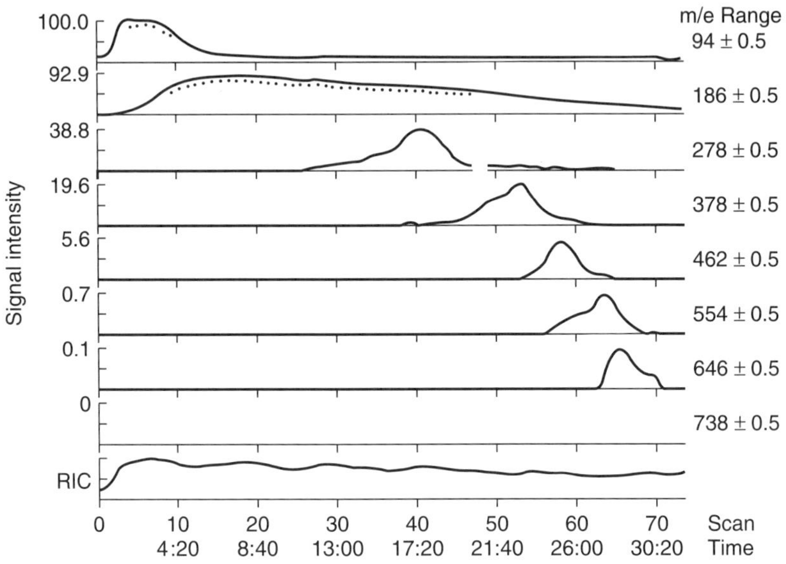

Figure 5.14. Mass spectra for the irreversibly adsorbed phenol on Columbia carbon (extracted by acetone). Not shown here are the spectra of the control solution of phenol in acetone, which shows only the species of m/e = 94. The mass number 94 corresponds to phenol, and those of multiples of 94 (minus H from −OH) correspond to polymers of phenol ($C_{6n}H_{4n+2}O_n$, where $n = 2$–7) (from Grant and King, 1990, with permission).

for these reactions. The catalytic activities of oxidized carbons for the oxidation of SO_2 (to SO_3) and NO to (NO_2) at ambient temperature in both gas phase and aqueous phase are well known and have been studied extensively (Komiyama and Smith, 1975; Py et al., 1995; Govindarao and Gopalakrishna, 1995; Rubel and Stencel, 1997; Mochida et al., 1994; Mochida, 1997; Lisovskii et al., 1997; Muniz et al., 1998; Mochida et al., 2000).

The effects of surface functionalities on adsorption of phenols have been studied extensively. However, in light of the findings of Grant and King (1990), such effects need to be re-examined, that is, reversible and irreversible adsorption should be treated separately.

A large body of experimental data is available in the literature on equilibrium adsorption isotherms of organic compounds from aqueous solutions. Because much of the data is for dilute solutions where no plateau is seen in the isotherm, the data can be best fitted with the Freundlich isotherm:

$$q = KP^{1/n} \tag{5.2}$$

The Freundlich isotherm constants for a few selected organic compounds are given in Table 5.4. The Freundlich isotherm constants for adsorption of some 600 organic compounds and pollutants on activated carbon have been compiled by Faust and Aly (1987).

Table 5.4. Freundlich isotherm constants for adsorption on activated carbon from aqueous solution at 25 °C

Compound	$K[(mg/g)(L/mg)^{1/n}]$	$1/n$
Benzene	1.0	1.60
Toluene	100	0.45
p-Xylene	200	0.42
Ethylbenzene	175	0.53
o-Xylene	174	0.47
Chlorobenzene	28	0.40
1,2-Dichlorobenzene	54	0.41
1,2,4-Trichlorobenzene	129	0.44
Nitrobenzene	68	0.43
Phenol	21	0.54
2-Nitrophenol	99	0.34
4-Nitrophenol	76	0.25
Chloroform	2.6	0.73
Benzoic acid	0.76	1.80
Styrene	327	0.48
PCB	14100	1.03

From Dobbs and Cohen, 1980; Weber and DiGiano, 1996.

The interactions outlined in Chapter 2 are operative in the adsorption of organics. In addition to these interactions, hydrogen bonds as well as chemical bonds may also form. The solubility is also a factor, that is, the solute–solvent interactions, as discussed above. In general, adsorption depends on molecular weight, functional groups, polarizability (which scales roughly with molecular weight), polarity, and hydrophobicity (or lack of solubility). Numerous studies have been made in attempting to correlate adsorption with these molecular properties (Al-Bahrani and Martin, 1976; Abe et al., 1980; Jankowska et al., 1991).

5.5.2. Effects of Surface Functionalities on Adsorption

Aqueous solution of phenol is the most studied solution on the effects of functionalities on adsorption. As mentioned above, in light of the results of Grant and King (1990), the effects of functionalities on adsorption of phenols should be studied separately for reversible and irreversible adsorption. The results summarized below are for the total adsorption at equilibrium.

Dramatic effects of oxidation and reduction of the carbon surface on the adsorption of phenol and nitrobenzene were first shown by Coughlin and Ezra (1968). Surface oxygen functionalities were increased by oxidation with ammonium thiosulfate solution, and reduction was performed with zinc amalgam in HCl solution. The isotherms of phenol on three carbon samples are shown in

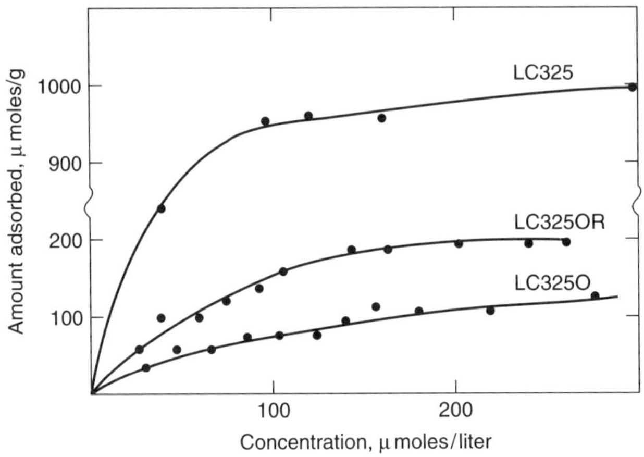

Figure 5.15. Isotherms of phenol at 30 °C on Columbia carbon. LC325: untreated. LC325O: oxidized. LC325OR: oxidation followed by reduction. From Coughlin and Ezra, 1968, with permission.

Figure 5.15. The sample labeled LC325 was a Columbia carbon (from Union Carbide, coconut shell-based, as were other Columbia carbons), which was washed with dilute HCl to remove residual alkalies. This sample contained the least amount of oxygen groups. The sample labeled LC 325O was oxidized, whereas that labeled LC 325OR indicated reduction followed by oxidation. The latter sample actually contained more oxygen groups than the LC 325 sample.

The concentrations of different oxygen-containing functional groups on these carbons were also analyzed. The groups on the oxidized carbon (LC325O) were mainly carboxyl and hydroxyl and a very small quantity of carbonyl groups. It was concluded that the carboxyl and hydroxyl groups hindered the adsorption of the aromatic compounds. The explanation for this effect by Coughlin and Ezra (1968), which is generally accepted today, is that the adsorption of aromatics is governed by the $\pi-\pi$ dispersion interaction between the basal plane of carbon and the aromatic ring of the adsorbate. Oxygen bonded to the edges of the graphite can localize electrons and thereby remove them from the π electron system of the basal plane. Consequently the $\pi-\pi$ interaction is weakened. Indeed, other studies with aromatic compounds showed that the carboxyl and hydroxyl oxygen groups on activated carbon decreased their adsorption, for example, benzenesulfonate (Coughlin et al., 1968), *p*-hydroxybenzaldehyde (Ishizaki and Cookson, 1974; Cookson, 1978), and benzoic acid (as benzoate anion, Radovic et al., 1996).

Contrary to the effects of surface carboxyl and hydroxyl groups, the surface quinone (or carbonyl) groups actually increased the adsorption of aromatics. Epstein et al. (1971) observed these effects with the adsorption of *p*-nitrophenol. Their explanation was that the carbonyl groups aid the adsorption of aromatics by involving the formation of an electron donor–acceptor complex of the aromatic ring with the surface carbonyl groups, as proposed earlier by Mattson et al. (1969).

Simple experimental procedures are known for generating the two different types of acid oxygen groups on carbon (Cookson, 1978). Surface oxides developed by chemical treatment and dry oxidation at temperatures <400 °C are mainly the carboxyl and hydroxyl types. At temperatures >400 °C, the dry oxidation treatment yields mainly carbonyl groups (in the form of quinone and hydroquinone).

The effects of oxygen functionalities on the adsorption of aliphatic compounds from their aqueous solutions have also been studied (Cookson, 1978; Jankowska et al., 1991; Radovic, 1996). The adsorption capacities of butyl disulfide and decane were both decreased by surface oxides (Cookson, 1978). Hence it was concluded that surface oxides hindered adsorption of nonpolar aliphatic compounds.

Radovic et al. (1996) investigated the effects of "nitriding" the surface on adsorption from solution. Reacting with ammonia at elevated temperatures introduced pyridine functional groups on carbon. Reaction at 200 °C forms amides, imides, imines, amines, and nitriles; while reaction at 250 °C results in bonding of ammonia to the carbon double bonds (Vinke et al., 1994). The effects of nitriding (at 250 °C) were similar to that of oxidation. Nitriding also hindered the adsorption of benzoate and aliphatic anions, oxalate, and fumarate.

The effects of surface functionalities on adsorption of organic electrolytes, including weak electrolytes such as phenols, are significantly more complicated to assess. One needs to consider the surface charge of the carbon as well as the extent of ionization of the solute. The surface charge of carbon is a function of pH of the solution. The surface charge of a typical commercial activated carbon is shown in Figure 5.16. The pH at which the surface charge is zero is called the point of zero charge (PZC), also referred to in the literature as zero charge point (ZCP) and zero point charge (ZPC). The surface is positively charged at pH below ZCP and is negatively charged at pH above ZCP. The isoelectric point (IEP), that is, the pH of zero ζ-potential, is usually near ZCP, but is lower than ZCP for activated carbon (Boehm, 2002).

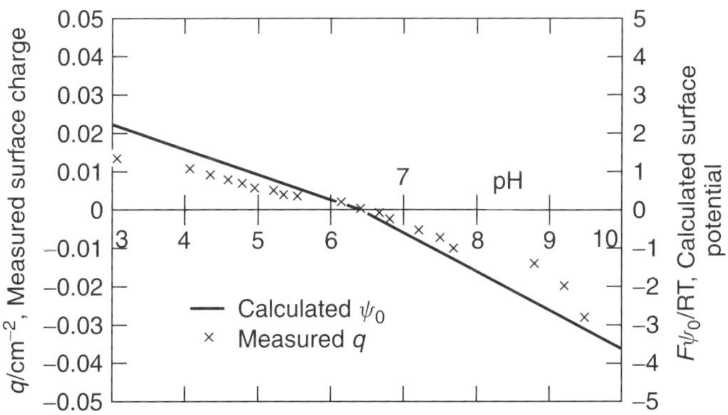

Figure 5.16. Surface charge of an untreated activated carbon as a function of pH (Muller et al., 1980, with permission).

It is important to consider the charge of the surface because it determines the capacity of the carbon for ion exchange. In the aqueous solution of an electrolyte, an electrical double layer (or diffuse cloud) of dissociated H^+ and OH^- is formed on a charged surface. Hydroxide ions (OH^-) compose the inner layer of the electrical double layer on a positively charged surface, whereas protons (H^+) form the inner layer on a negatively charged surface. Anion exchange occurs on the positively charged carbon surface via:

$$C^+ \cdots OH^- + H^+ + A^- \Rightarrow C^+ \cdots A^- + H_2O \qquad (5.3)$$

Cation exchange occurs on the negatively charged surface by:

$$C^- \cdots H^+ + K^+ + A^- \Rightarrow C^- \cdots K^+ + H^+ + A^- \qquad (5.4)$$

and it is accompanied by acidification of the solution.

The ZCP can be altered easily by oxidation or other surface treatments. A good example is shown by Noh and Schwarz (1990) by use of nitric acid treatment. Table 5.5 shows the ZCP values of a commercial activated carbon that has been oxidized with nitric acid at various concentrations (0.2, 0.4, 1, and 2 M) at room temperature.

It is seen that the ZCP is strongly dependent on acid treatment. The results of titration by using Boehm's method are also given in Table 5.5. The untreated sample of this particular activated carbon (North American carbon, low "ash", <2%) is basic, with ZCP = 10. Carboxyl is the most abundant group on all acid treated carbons. The approximate distribution of different acid groups is seen from the titration results.

The extent of ionization in the aqueous solution depends on the *pH* and is given by the dissociation constant, *K*. An acid HA and a base B (Bronsted acid and base) take part in the following equilibria in water:

$$HA + H_2O \Rightarrow A^- + H_3O^+ \qquad K_a = [H^+][A^-]/[HA] \qquad (5.5)$$

$$B + H_2O \Rightarrow BH^+ + OH^- \qquad K_b = [BH^+][OH^-]/[B] \qquad (5.6)$$

Table 5.5. Zero charge point (ZCP) of unoxidized and oxidized activated carbon

HNO_3 Treatment	0 M	0.2 M	0.4 M	1 M	2 M
ZCP	10	7.8	6	5.5	3.5
Neutralization (meq/g)					
HCl	0.74	0.60	0.31	0.16	0.14
NaOH	0.04	0.27	0.48	0.89	1.71
Na_2CO_3	—	—	0.13	0.35	0.50
$NaHCO_3$	—	—	—	0.33	0.12

Note: Neutralization yields the following surface groups: carboxyl ($NaHCO_3$), carboxyl+lactone (Na_2CO_3), carboxyl+lactone+phenolic OH(NaOH).

Table 5.6. *pK* of dissociation constants at 25 °C in aqueous solutions

	pK_a		pK_b
Acetic acid	4.756	Aniline	9.42
Butyric acid	4.820	Ethylamine	3.25
Lactic acid	3.860	Diethylamine	3.02
Succinic acid	4.207	Triethylamine	3.24
Phenol	9.895	Glycine	11.65
Ammonium ion	9.245	Hydrazine	5.52
Methylammonium ion	10.624	Ammonium hydroxide	4.75
Phosphoric acid	2.148		

Atkins, 1982.

Table 5.7. Dependence of adsorption of ions on surface reduction and oxidation of activated carbon

	Conc. of Functional Groups, mmol/g					
Pretreatment	−COOH	−COO	−OH	=CO	q(Br^- or Cl^-)	q(Na^+)
Reduction in vacuo at 500 °C For 24 h	Total = 0.063				0.0023(Br^-)	0.0014
Untreated	0.04	0.03	0.11	0.03	0.189 (Br^-) 0.0508 (Cl^-)	0.0858 0.0205
Air oxidized at 400 °C for 6 h	0.90	0.29	0.78	0.03	0.361 (Br^-) 0.119 (Cl^-)	0.392 0.153

Jankowska, 1991.

Like pH, pK is defined as:

$$pK_a = -\log_{10} K_a, \qquad pK_b = -\log_{10} K_b \tag{5.7}$$

A list of pK values are given in Table 5.6.

The effects of carbon oxidation and reduction on adsorption of strong electrolytes have been studied with NaCl and NaBr (Jankowska, 1991). Their results are summarized in Table 5.7. The strong acidic groups introduced by oxidation undergo cation exchange with the sodium ion, hence acidifies the solution. The acidified solution significantly increases the potential that favors the adsorption of anions. This explains the simultaneous increases in the adsorption of both anions and the cation.

5.6. ACTIVATED CARBON FIBERS

Carbon fibers have been produced commercially since ca. 1960. The precursors for carbon fibers were polymeric fibers (polyacrylonitrile, or PAN; phenolic formaldehyde, or Novalak; polyvinylidene chloride, or Saran), celluose (and rayon derived from cellulose), and pitches (coal tar pitch and petroleum pitch). These carbon fibers have high tensile strength and high elasticity and are considerably more graphitic than activated carbon because a mesophase is usually formed during the carbonization process of the fibers. Gas activation (usually with carbon dioxide) of these carbon fibers results in ACFs. Consequently there are as many types of ACFs as there are precursors. Comprehensive reviews on ACFs have been made by Suzuki (1994) and Mays (1999), including a history of its development and the present commercial production. The pioneering development work on ACF (Abbott, 1962; Rodgers et al., 1965; Doying, 1966; Peters, 1966; Boucher et al., 1970; Economy and Lin, 1973; Bailey and Maggs, 1972) has also been reviewed.

Activated carbon fibers have high surface areas, with BET surface areas ranging from around 1000 m^2/g to well over 2000 m^2/g. Beside their fibrous form, they have the following unique properties (compared with GAC and PAC):

(1) Narrow and uniform pore size distribution (hence stronger interactions with sorbates).
(2) Small and uniform fiber diameter (hence faster uptake and desorption).
(3) Graphitic (hence more conductive and more heat resistant).
(4) High strength and elasticity (allowing greater flexibility in the shapes and forms of the sorbent such as ACF clothes and papers)

As a result of these unique properties, there are many advantages for using them as sorbents. Despite these properties, their present application is limited by its high cost. The total annual worldwide production of carbon fibers in 1993 was 7300 tons, of which less than 2% was ACF (Mays, 1999). Five companies in Japan (Suzuki, 1994) provide the main supply of ACFs. However, as production processes improve further and new demands arise for environmental applications, as well as consumer products applications, the use of ACF will undoubtedly increase.

As a consequence of the four unique properties above, the adsorption properties of ACF are discussed accordingly.

(1) Narrow and uniform pore size distribution: The pore sizes in ACF are generally <20 Å and mostly in the 8–10 Å. range. Table 5.8 shows the pore size ranges for three commercial ACFs (Kaneko et al. 1994; Suzuki, 1994).

From their N_2 adsorption data, Jaroniec et al. (1991) determined the pore sizes of PAN-based and cellulose-based ACFs, and found that 85% of the pore volumes were composed of uniform micropores with a diameter ~10 Å. In fact, the pore sizes of ACF in all reported literature are ~10 Å.

There are two likely reasons for the fine pore sizes and uniformity (or narrowness) of their size distribution. First, the ACFs have essentially zero "ash"

Table 5.8. Pore sizes of ACF determined by N_2 and He adsorption

		Cellulose-Based	Pitch-Based	PAN-Based
N_2	BET Surface Area	1147	795	743
	Micropore Vol. (cm^3/g)	0.58	0.37	0.35
	Pore Size (Å)	10.1	9.4	9.5
He	Narrow Pore Vol. (cm^3/g)	0.54	0.42	0.32
	Wide Pore Vol. (cm^3/g)	0.04	0	0.01
	Pore Size (Å)	9.3	7.6	8.9

Kaneko et al. 1994.

content because their precursors (polymers and pitches) are ash-free. As discussed above, any minerals would act as catalyst particles during activation, and these particles are known to undergo channeling, tunneling, and edge recession actions during the carbon gasification process (Baker, 1986; Yang 1984; Goethel and Yang, 1989). These catalytic actions create large pores. During gas activation of the carbon fibers, the pores are elongated but not widened. The basic reason is the graphitic structure of the carbon fiber. The carbon atoms on the basal plane of graphite are not active and are not gasified, whereas the edge atoms are the active sites for gasification (Yang, 1984). Moreover, the reactivities of the edge atoms are anisotropic, that is, the zigzag edges are more active than the armchair edge sites (Yang, 1984; Yang and Duan, 1985). The real edges of the graphite planes are a combination of different crystalline edge sites, with different reactivities. The inter-plane spacing in graphite is 3.35 Å. Once gasification is initiated on an edge site, the gasification will continue on neighboring edge sites along the same graphite (or graphene) sheet. Hence the pore is elongated within two graphite sheets, with a pore dimension of approximately 7 Å. If gasification is initiated on edge atoms of two adjacent sheets, the pores will be confined to within two graphite sheets ~10 Å apart. Consequently most of the pores in ACFs are 10 Å in size. To produce larger pores and a distribution of pore sizes, catalysts must be added before gas activation. This was shown is a series of publications by Freeman et al. (1989). A schematic of pore development during gas activation is shown in Figure 5.17.

Stronger interactions with adsorbate molecules is a direct consequence of the small and uniform pore sizes. Smaller pores give risc to stronger overlap of

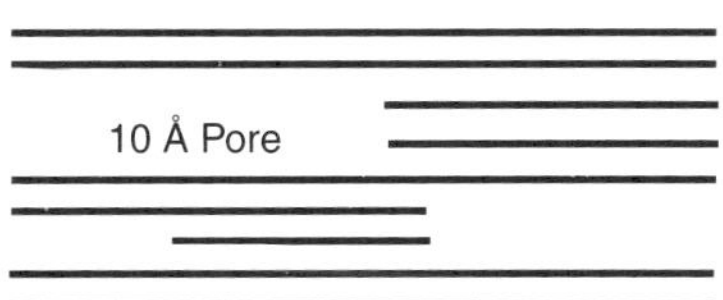

Figure 5.17. Schematic showing pore development during activation of graphitic fibers by gasification, initiated from active sites on edges. The pores are approximately either 7 or 10 Å in size.

potentials from surrounding surfaces, as discussed in Chapters 2 and 4. This point is further discussed in 5.6.1 (Adsorption Isotherms).

(2) Small and uniform fiber diameter: The precursors of ACF are carbon fibers. Because the fibers are made from a variety of spinning techniques (e.g., Edie and McHugh, 1999), they are available in remarkably uniform diameters.

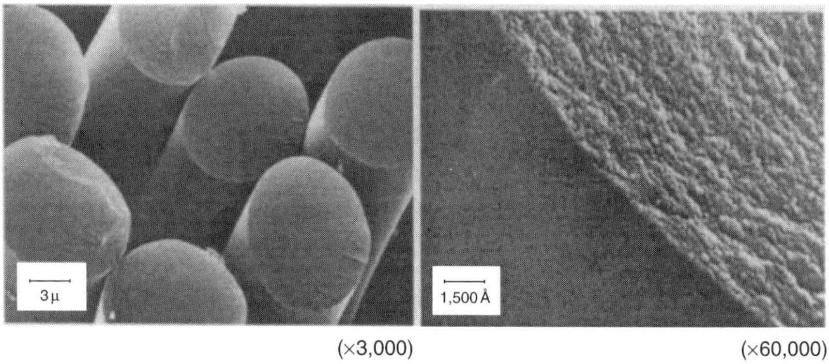

Figure 5.18. Scanning electron microscope images of activated carbon fibers. "Kynol" ACF, activated from phenolic resin fibers by Kuraray Chemical Co., Ltd., Japan, with permission.

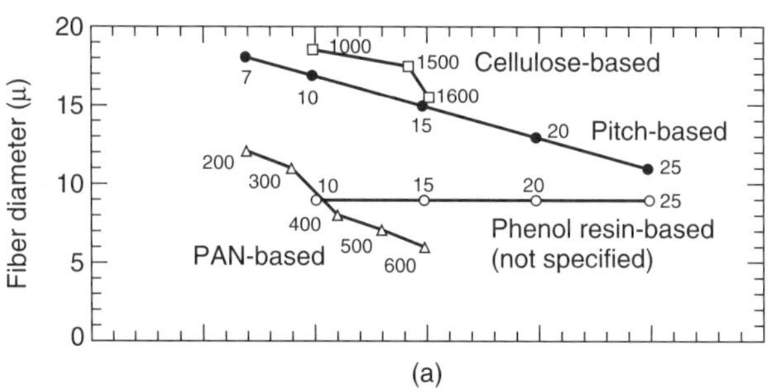

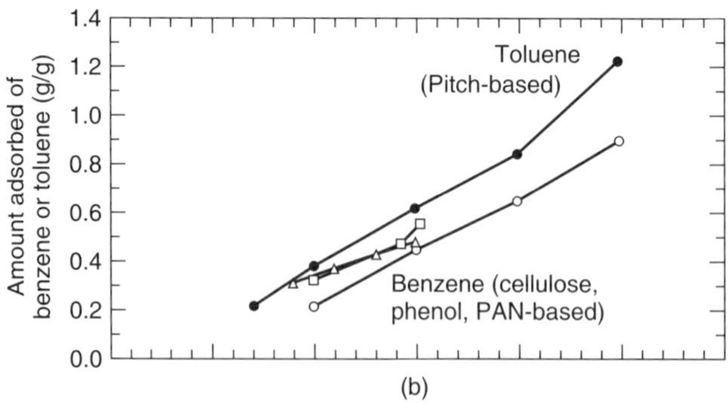

Figure 5.19. Fiber diameters and adsorption capacities (at 20 °C) of various ACF's (from Suzuki, 1994, with permission). The corresponding capacities for GAC are 0.33 (benzene), 0.35 (toluene), and 0.30 g/g (phenol).

The diameters are retained after activation. Most commercial activated carbon fibers have a diameter of nearly 10 μm, although other sizes in the range of 8–20 μm are also available. Figure 5.18 shows a SEM image of a commercial ACF. The fiber diameters of ACFs derived from various precursors are shown in Figure 5.19.

The small and uniform fiber diameter has a direct and important implication on the mass-transfer rates for both adsorption and desorption. The uptake and desorption rates are directly related to the diffusion time constant, D/R^2 (where D is diffusivity and R is radius). The value of R^2/D is approximately the time for accomplishing 99% diffusion upon a step change for a spherical particle with clean initial condition. The value decreases rapidly with decreasing R. ACF has only micropores of a diffusion length less than R (i.e., fiber radius), whereas GAC has both micropores and meso/macropores. Hence two diffusion time constants are involved. The interplay of these two diffusion time constants has been delineated from the analysis of diffusion in a bidisperse porous structure by Ruckenstein et al. (1971). In all commercial sorbents, the resistance by the macropore (in pellet) diffusion is as important as the micropore diffusion because of the much larger diffusion distance in the macropore.

Numerous studies have been reported on the breakthrough curves of VOCs in packed beds of ACF and were compared with that from beds of GAC (Lin and Chen, 1995; Schmidt et al., 1997; Lordgooei et al., 1998). Schmidt et al. (1997) showed that the adsorption of methylene blue from aqueous solution in a rayon-based ACF was 2 orders of magnitude faster than in a granulated activated carbon, and 1 order of magnitude faster than in a powdered activated carbon. The work of Suzuki (1990) best illustrates the comparison between ACF and GAC, and is shown in Figure 5.20. The breakthrough behavior of trichloroethylene (TCE) from a contaminated ground water in a packed bed of ACF is compared directly with that in a GAC bed. Not only the amount of TCE adsorbed in the GAC bed is less, but the breakthrough curve in the ACF bed is much sharper. The sharper breakthrough curve is due to the higher mass-transfer rates, rather than less axial dispersion (Suzuki, 1994). The role of mass transfer resistance in spreading the wavefront is well understood (Yang, 1987).

Axial dispersion in the beds of ACF was studied by Suzuki. The axial dispersion coefficient was proportional to flow velocity, and the proportionality constants for different beds could be correlated (increased) with the bed densities (in g/cc) (Suzuki, 1994). Using the dispersion coefficient and a Freundlich isotherm, Suzuki could predict the breakthrough curves in ACF beds (Suzuki, 1994).

(3) Graphitic structure, high conductivity, and high strength: It has been pointed out early in the development of ACF that it has the advantages of greater rates as well as possible in situ electrical regeneration of the sorbent due to its high electrical conductivity (Lin and Economy, 1973; Economy and Lin, 1976). The high conductivity of ACF (or low electrical resistivity) derives from its graphitic structure. The electrical resistivity (as defined by Ohm's law) of isotropic ACFs are in the range of 4–6 mΩ-cm (manufacturers' data quoted from Burchell et al., 1997). These values are only about 3–5 times higher than that

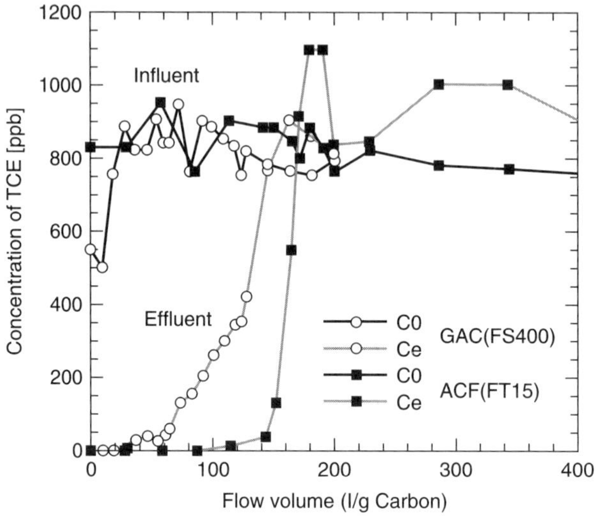

Figure 5.20. Breakthrough curves of trichloroethylene in contaminated ground water in packed beds of GAC (Filtrasorb 400) and ACF (phenolic resin-based, FT 15) (from Suzuki, 1990, with permission).

of graphite (1.38 mΩ-cm). The electrical resistivities of ACF clothes and fabrics are, however, substantially higher. Subrenat et al. (2001) measured the electrical resistivities of fabrics of a number of rayon-based ACFs and reported values as high as 600 mΩ-cm. The electrical resistivity of a ACF composite monolith was 130 mΩ-cm. Comparing different ACF-clothes, the electrical resistivity increased with surface area due to higher void fraction (Subrenat et al., 2001).

The high conductivity enables sorbent regeneration by resistance heating of the beds, which has been referred to as electrothermal desorption. Electrothermal desorption has been studied by a number of groups (Petrovska et al., 1991; Burchell et al., 1997; Lordgooei et al., 1998; Sullivan et al., 2001; Subrenat et al., 2001). This approach is particularly promising for purification such as removal of VOCs, as demonstrated by Lordgooei et al. (1998) by using a fiber-cloth adsorber.

The graphitic nature also makes ACFs heat resistant, so they are inflammable up to 1000 °C. Lack of dust/fines from the ACF has also been claimed as an advantage.

The high tensile strength of the activated carbon fibers gives ACF the versatility to be formed into a large variety of different types of cloths, fabrics, papers, felts, and composites. Combined with high adsorption capacities and rates, these different materials are excellent candidates for fabricating low-volume, high-throughput adsorption devices. For example, a rotor-type adsorber (with counter-current thermal regeneration) made from corrugated ACF has been demonstrated for solvent removal from air (Suzuki, 1994). This type of adsorber, or monolithic wheel, has been discussed in some detail by Humphrey and Keller (1997).

Monolithic composites of activated carbon fibers with phenolic resin as the binder have been prepared for a variety of possible applications, including gas separation (Burchell, 1999; Kimber et al., 1996). The low-density composites with densities <0.25 g/cm^3 are particularly promising for gas separation as well as energy storage (CH_4 and H_2) applications. A detailed discussion on these types of materials has been given by Burchell (1999). Methane storage will be discussed in Chapter 10.

5.6.1. Adsorption Isotherms

A large body of experimental data on the adsorption isotherms exists in the literature (both journal publications and commercial brochures). With the exception of methane storage, most of the data are related to purification, particularly for environmental applications. Comparing different ACFs, the amounts adsorbed are, in general, directly dependent on the BET surface area and the micropore volume of the ACF. However, the heats of adsorption on ACF are higher than those on the granulated activated carbon. This is shown in Table 5.9 for hydrocarbons and CO_2. The data in Table 5.9 are taken from Kuro-Oka et al. (1984), Meredith and Plank (1967), Laukhuf and Plank (1969), and Reich et al., (1980).

Adsorption of SO_2, NO, and various VOCs on ACF has been studied. Figure 5.19 shows the adsorption capacities of benzene, toluene, and phenol for a number of ACFs. The corresponding capacities for commercial activated carbons (GACs) are 0.33 g/g (benzene), 0.35 g/g (toluene), and 0.30 g/g (phenol). These are near the lowest values of the various ACFs. The adsorption of NO by ACFs has been studied extensively by Kaneko (1998). Comparing a number of GACs and ACFs, the amounts adsorbed of NO were higher on the ACFs. For example, for two samples with the same BET surface area (860 m^2/g), the amounts were 17 mg/g (at 13 kPa and 303 K) for the GAC compared with 65 mg/g for a PAN-based ACF; and at 80 kPa and 303 K, these values were 47 mg/g for GAC and 115 mg/g for the ACF. The adsorption of SO_2 by ACF was studied extensively by Mochida and co-workers (e.g., Mochida et al., 1997a; 1997b). However, the focus of their studies was on the catalytic activity for oxidation of SO_2 to SO_3 in the presence of both O_2 and liquid water (to form sulfuric acid). Moderate activities at room temperature were reported. The mechanism of the reaction is not understood, although molecular orbital studies have indicated the importance of surface oxides on carbon for the catalytic activity (Yang and Yang, 2002).

Enhanced adsorption capacities by the small pores of ACF have also been observed for adsorption from aqueous solutions. Sakoda et al. (1987) reported the adsorption isotherms of trichloroethylene and tetrachloroethylene on a number of ACFs derived from phenolic resin, as shown in Figure 5.21. These data are compared with that on GAC, and the enhancement is clearly seen.

5.7. CARBON MOLECULAR SIEVES

Because they are less hydrophilic than zeolites but have molecular sieving properties, carbon molecular sieves (CMS's or MSC's) can be used more efficiently

Table 5.9. Comparison of adsorption (q, mmol/g) and heat of adsorption ($-\Delta H$, kJ/mol) between activated carbon (BPL) and activated carbon fiber (KF-1500)

	CO_2(1 atm)		CH_4(1 atm)		C_2H_6(1 atm)		C_3H_6(0.5 atm)		C_4H_{10}(0.5 atm)*	
	q	$-\Delta H$	q	$-\Delta H$	q	$-\Delta H$	q	$-\Delta H$	q	$-\Delta H$
GAC (BPL)	1.5 @ 303 K	14	0.65 @ 303 K	21	3.3 @ 303 K	31	4.0 @ 298 K	23	3.8 @ 293 K	37
ACF (KF-1500)	2.5 @ 298 K	45	1.12 @ 298 K	24	4.4 @ 298 K	33	5.4 @ 298 K	37	5.2 @ 298 K	—

*For butane, the data were for Nuxit carbon (Szepesy and Illes, 1963) and a phenolic-based ACF with approximately the same BET surface area of the GAC (Foster et al., 1992).

Data taken from sources cited by Valenzuela and Myers, 1989.

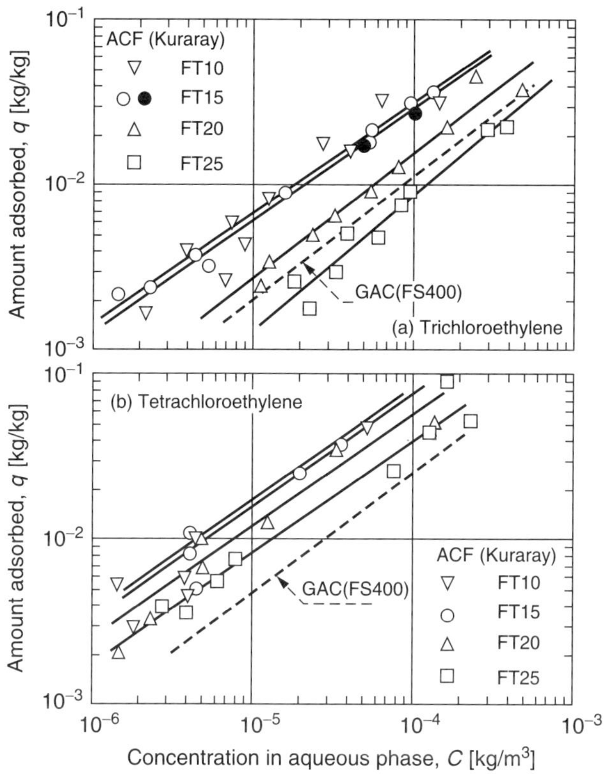

Figure 5.21. Adsorption isotherms of trichloroethylene and tetrachloroethylene from aqueous solution on a number of ACF's at 20 °C (Sakoda et al., 1987, with permission).

in separation processes involving wet-gas streams. Their molecular sieving properties derive from their unique pore structure (see Figure 5.1 for pore size distribution). As a result of the foregoing and a number of other promising features (Walker et al., 1966; Juntgen, 1977; Walker, 1990; Cabrera et al., 1993; Jones and Koros, 1994; Koros, 1995; Acharya and Foley, 2000), CMS's have attracted considerable interest. Since the 1970's, CMS's have been produced commercially on the basis of proprietary processes. Although the details of the processes vary, they are similar and are based on the pioneering work of Walker. CMS's have been used worldwide for the production of nitrogen from air (Notaro et al., 1999).

The early research focused on the preparation of CMS's and their sieving properties. The following three approaches were taken for their preparation:

1. Carbonization of polymers, such as poly(vinylidene chloride) (PVDC) (Walker et al., 1966; Dacey and Thomas, 1954; Dubinin et al., 1964; Lamond et al., 1965); Saran (90/10 mixture of vinylidene chloride and vinyl chloride); and cellulose, sugar, and coconut shell.

2. Slightly carbonizing coals, especially anthracite (Walker et al., 1966; Metcalfe et al., 1963; Mason and Eberly, 1965; Patel et al., 1972).
3. Coating of the pore mouths of the commercial activated carbon with a carbonized or coked thermosetting polymer (Walker et al., 1966).

A variety of interesting molecular sieving properties were found for these samples. For example, PVDC carbonized at 700 °C was thought to have slit-shaped pores because it adsorbed flat molecules such as benzene and naphthalene but not spherical ones such as neopentane. Carbonized Saran had a pore entrance of ~6 Å and thus showed a striking separation between isobutane (with a kinetic diameter of 5.0 Å, which was admitted) and neopentane (with a kinetic diameter of 6.2 Å, which was rejected) (Lamond et al., 1965). With the carbonized PVDC, molecular-sieve effects with regard to neopentane were not seen until the carbonization temperature reached 1200 °C (Walker et al., 1966). Anthracite heat-treated in hydrogen at 650 °C adsorbed n-butane at an amount about five times larger than that of isobutane (Mason and Eberly, 1965). At carbon burnoff (by oxygen at 427 °C) below 6.9%, anthracite admitted CO_2, less N_2, and almost no neopentane (Patel et al., 1972). CO_2 is a linear molecule and is thought to have the smallest minimum diameter, 3.7 Å, among the three adsorbates (Patel et al., 1972). For the coke-coating technique, Walker et al. (1966) prepared samples by forming carbon on activated carbons from furfuryl alcohol, polymerized with phosphoric acid. The pores were thought to be nearly 5 Å in diameter since the samples had a large capacity for n-butane (kinetic diameter = 4.3 Å), a small capacity for isobutane, and negligible capacity for neopentane. In a later work (Kamishita et al., 1977), carbon was deposited into the pores of a lignite char by cracking methane at 855 °C. With nearly 3% carbon deposited, the samples showed significant molecular sieving between CO_2 (admitted) and N_2 (hindered). The early development on CMS has been reviewed by Walker et al. (1966) and Spencer (1967).

The patent literature on the processes for manufacturing CMS's has been reviewed elsewhere (Cabrera et al., 1993; Armor, 1994), including the Bergbau patents (Munzner et al., 1974; Munzner et al., 1976) and Japanese patents (Eguchi et al., 1974; Ohsaki and Abe, 1984). The general procedure is the same, that is, carbonization/activation followed by carbon deposition, as described below.

At present, CMS's are produced commercially by Bergbau-Forschung GmbH in Germany and by Takeda Chemical Company in Japan, among others. The detailed procedures of the manufacturing processes, although not revealed in the open literature, are based on the carbonization of coal or nutshell followed by coating of carbon on the char by using a variety of hydrocarbon vapors. Carbonization of polymers such as PVDC is not used due to economic reasons.

The general procedure for the manufacture of CMS's used by Bergbau-Forschung is shown in Figure 5.22 (Jüntgen et al., 1981). Two types of CMS's are produced, although that designated CMS N2 has O_2/N_2 sieving properties and is used for nitrogen production from air. The raw material is a bituminous coal, ground to 90% passing 40 μm. The coal is first oxidized by air at temperatures

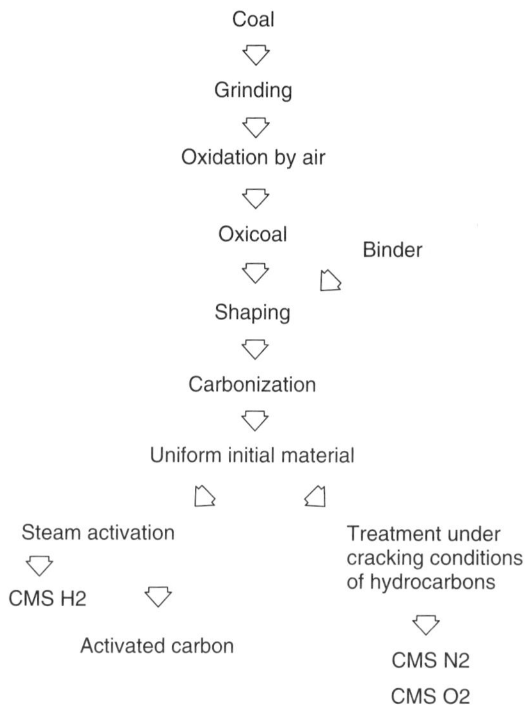

Figure 5.22. Procedures for the manufacture of molecular-sieve carbons. *Source*: Jüntgen et al. (1981). Reprinted with permission. Benzene is used for deposition of carbon at pore mouths (Cabrera et al., 1993).

below the ignition point in a fluidized bed to form "oxicoal." Oxicoal is pelletized with a binder into granules 2–3 mm in diameter and then carbonized in a rotary drum. The uniform material formed at this point is further treated to produce two types of CMS: one by creating and enlarging the pores (CMS H2) and one by partially blocking the pore mouths (CMS N2). CMS H2 is made by slight steam activation. CMS N2 is produced by cracking hydrocarbons, such as benzene, in order to deposit a thin layer of carbon at the pore mouths.

The former type is used for H_2 and He purification (by equilibrium separation) without pre-drying (which can also be done by activated carbon), whereas the latter is used for N_2 production from air. The sizes of the bottlenecks in CMS N2 are near 5 Å, as shown in Figure 5.1, which allow much faster penetration of oxygen than nitrogen into the pores. The pores of the two types of carbon are depicted in Figure 5.23. Using CMS N2 in pressure-swing adsorption, dry and CO_2-free nitrogen at 99.9% purity can be produced without the need to pre-dry the air feed. The CMS's manufactured by Takeda Chemical Company in Japan, designated CMS's 3A, 4A, and 5A, indicating the approximate main aperture sizes. CMS 5A has a micropore volume of 0.18 cm^3/g for pores with a nominal size of 5 Å, and a macropore volume of 0.38 cm^3/g for pores of

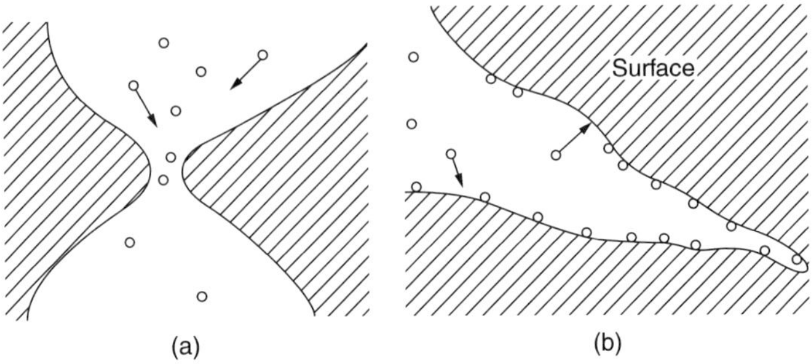

Figure 5.23. Molecular-sieve carbons made by Bergbau-Forschung: (A) Type CMS N2 with bottlenecks near 5 Å formed by coke deposition at the pore mouth; (B) Type CMS H2 formed by steam activation. *Source*: Jüntgen et al., 1981. Reprinted with permission.

2.0 μm (Kawazoe et al., 1974). Takeda 3A is O_2/N_2 selective and is used for air separation.

5.7.1. Carbon Deposition Step

The key step in making CMS's is the carbon deposition step. This step involves cracking a hydrocarbon under an inert atmosphere so that carbon is deposited at the pore mouths. For a given hydrocarbon, this step is accomplished by careful control of a combination of conditions, including concentration, temperature, and time. For example, 5–12% benzene vapor at 800 °C for 20 min was used in the Bergbau Forschung process (Munzner, 1974). Vapors of pitch, benzene, and furfural alcohol have been used as the cracking gas in Walker's early work (Walker, 1966). For air separation, the pore entrance needs to be narrowed to a dimension that lies between the kinetic diameters of O_2 (3.46 Å) and N_2 (3.64 Å). It is equally important that the pore volume is large so a large capacity for O_2 is obtained, as illustrated in Figure 5.23A. It is quite remarkable that tonnage quantities of CMS's with controlled and consistent qualities can be produced.

Deposition of carbon at the pore mouths rather than throughout the pore walls is the crucial step for producing a good sorbent for kinetic separation (Armor, 1994). This principle has been demonstrated well by the experiment of Gaffney et al. (1994). The sorption kinetics of O_2 versus N_2 were studied on a number of CMS's prepared by deposition with two different hydrocarbons: propylene and isobutylene. Takeda 4A and 5A (which are not O_2/N_2 selective due to large pores) were used as the starting materials, and identical cracking conditions were used (600 °C for 2 h). The CMS's prepared with isobutylene cracking were far superior than those with propylene for O_2/N_2 separation. The CMS's with isobutylene cracking showed high kinetic selectivity for O_2 and no significant loss in O_2 capacity. The desired properties of the CMS's prepared with isobutylene cracking were attributed to carbon deposition at the pore mouths, whereas propylene cracking resulted in deposition inside the pores.

A two-step carbon deposition process using isobutylene was developed by Cabrera and Armor (1991). Numerous optimized deposition processes have been described by Cabrera et al. (1993). In the two-step scheme, the carbon support with pore sizes of about 4.5 to 20 Å is contacted with two different concentrations of a hydrocarbon. The concentration in the first step is larger than that in the second step. In this fashion, the pore openings of the support micropores are narrowed successively in two distinct steps without excessive pore filling.

For rational design of the carbon deposition step, one needs to have a Thiele modulus for the cracking reaction to be within an optimal range for which the effectiveness factor is kept low while still so allowing reasonably high reaction rates. The range for the Thiele modules of between 10 to 100 seems to be an optimal range. For a first-order cracking reaction, the effectiveness factor is given by (Levenspiel, 1972):

$$E = \frac{\tanh(mL)}{mL} \quad \text{and} \quad mL = L\sqrt{\frac{k}{D}} \tag{5.8}$$

where E is the effectiveness factor, L is the pore length, k is the first-order rate constant for cracking, D is the diffusivity of the cracking molecule in the pore, and mL is the Thiele modulus. At effectiveness factor = 1, uniform carbon deposition throughout the pore is expected. For $mL = 10$–100, $E = 0.1$–0.01. Cracking in this range would ensure deposition at the pore mouths rather than throughout the pore walls.

5.7.2. Kinetic Separation: Isotherms and Diffusivities

Because the finishing step in producing CMS's is carbon deposition in an inert atmosphere at a moderately high temperature, the surface of CMS's is quite uniformly covered by carbon. Unlike activated carbon that has a considerable amount of surface functionality, CMS's not have detectable surface functionality (Armor, 1994). Moreover, they should have fewer exposed inorganic compounds than activated carbon and not have cations. Consequently, adsorption of gas molecules on CMS's involves only nonspecific dispersion forces (see Chapter 2). For these reasons, CMS's should also be more hydrophobic than activated carbon.

The main use for CMS's is nitrogen production from air and CH_4/CO_2 separation, both by PSA. The latter is applied for: (1) landfill gas that contains approximately 50% each of CH_4/CO_2, and (2) tertiary oil recovery where the effluent gas contains ~80% CO_2 and 20% of CH_4 plus other light hydrocarbons. The PSA separation of CH_4/CO_2 with Bergbau Forschung CMS has been discussed in detail by Kapoor and Yang (1989) and by Baron (1994), who also discussed several other possible applications.

The difference in the kinetic diameters of N_2 and O_2 is ~0.2 Å. That between CH_4 and N_2 is also ~0.2 Å. Given the importance of CH_4/N_2 separation in the field of natural gas upgrading, it is surprising that a suitable CMS has not yet been developed. Attention is certainly warranted for developing such a CMS.

In kinetic separation, both equilibrium amount and diffusion rates are important. A sorbent selection parameter (S_k) has been given in Chapter 3. It involves both diffusivity and equilibrium constant. The equilibrium isotherms and diffusion time constants for a number of important sorbates on CMS's are given below.

The equilibrium isotherms of oxygen and nitrogen on Bergbau Forschung CMS's (Chen et al., 1994) are shown in Figure 5.24. The isotherm of oxygen on 5A zeolite is also shown for comparison. The adsorption of O_2 on 5A zeolite is mainly by dispersion forces. The ion-induced dipole interaction (see Chapter 2) is likely minor. Although the surface areas of these two sorbents are similar, their O_2 isotherms are also similar, as expected. The slight curvature of the isotherm

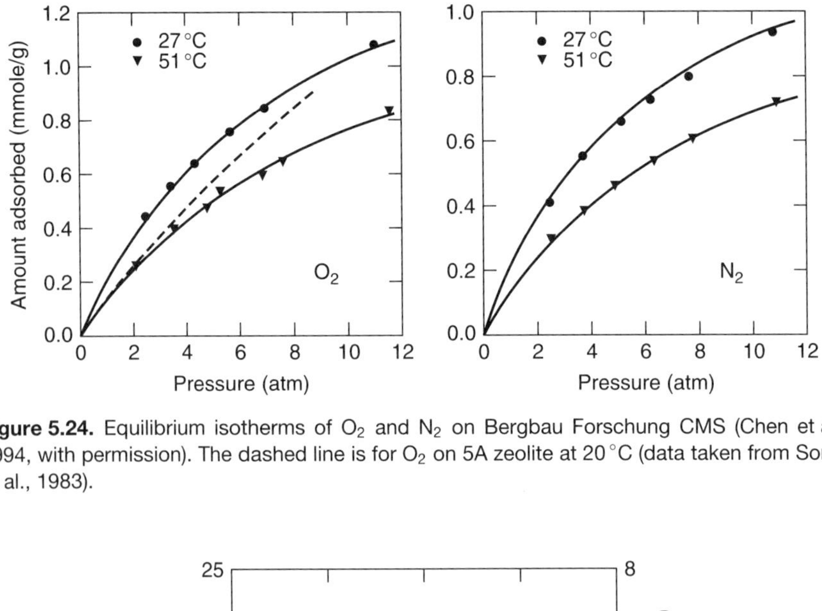

Figure 5.24. Equilibrium isotherms of O_2 and N_2 on Bergbau Forschung CMS (Chen et al., 1994, with permission). The dashed line is for O_2 on 5A zeolite at 20 °C (data taken from Sorial et al., 1983).

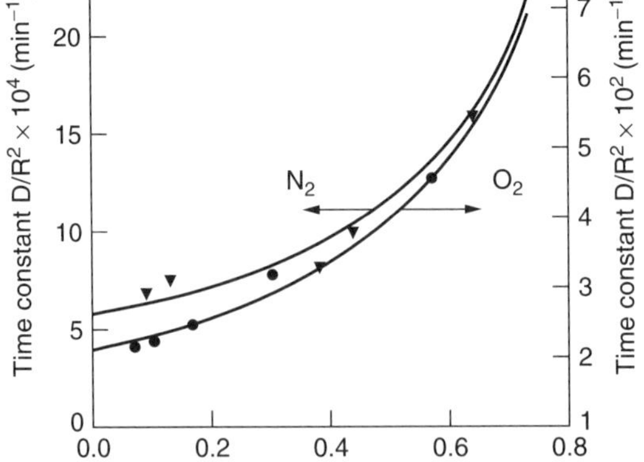

Figure 5.25. Concentration-dependent diffusion time constants for O_2 and N_2 in Bergbau Forschung CMS at 27 °C (Chen et al., 1994; Chen, 1994, with permission).

on CMS's is possibly due to higher heterogeneity of the CMS's. The diffusion time constants (Chen et al., 1994) are shown in Figure 5.25. These diffusion time constants have also been measured by a number of groups. As noted by Chen et al. (1994), those groups used activation conditions (prior to diffusion measurements) similar to those used for zeolite activation, and consequently higher diffusion time constants (by an order of magnitude) were obtained due to pore enlargement by high-temperature activation. A further discussion will be given later on sorbents for air separation.

Rates of binary O_2/N_2 diffusion in CMS's were also reported by Chen et al. (1994). It was found that the cross-term diffusivities, D_{ij}, are quite significant, hence the diffusion rates cannot be predicted by approximating the system as pure component diffusion. A simple kinetic theory is derived in Chapter 3 for predicting binary (or mixture) diffusion from pure-component diffusivities (Chen and Yang, 1992). The concentration dependence of the pure component diffusivities are needed for this prediction. Figure 5.26 compares the rates predicted by the kinetic model with experimental data. Also shown are results predicted from pure component diffusivities. It is clear that pure component diffusivities cannot be used for predicting rates for binary diffusion. A statistical theory by MacElroy et al. (1997) has also been used to predict binary diffusion in CMS's with satisfactory results.

The diffusion time constants for O_2, Ar, N_2, and CH_4 in Takeda 3A CMS's were reported by Ma et al. (1991) and are shown in Figure 5.27. This sorbent shows selectivity for N_2/CH_4, although higher selectivities are needed for this important separation.

5.7.3. Carbon Molecular Sieve Membranes

Promising developments have recently taken place in CMS membranes. CMS membranes are fabricated in the same manner as described above, that is, by

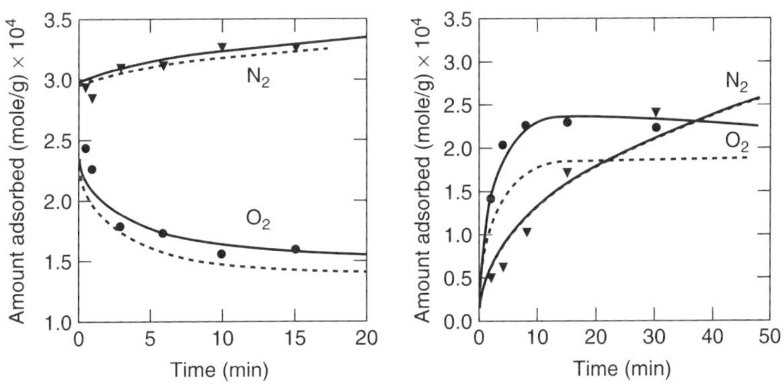

Figure 5.26. Counter-diffusion (left) and co-diffusion (right) of O_2/N_2 in Bergbau Forschung CMS at 27 °C (Chen et al., 1994; Chen, 1994, with permission). Symbols are experimental data points; solid lines are prediction by simple kinetic theory for mixture diffusion (Chen and Yang, 1992, and Chapter 3); dashed lines are prediction from pure component diffusivities.

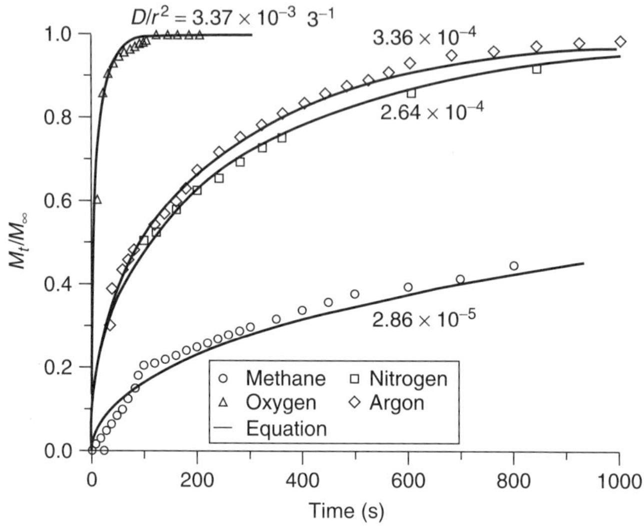

Figure 5.27. Uptake rates of (from top down) O_2, Ar, N_2, and CH_4 in Takeda 3A CMS at 303 K (Ma et al., 1991). The values of D/r^2 (in 1/s) are, respectively, 3.37×10^{-3}, 3.36×10^{-4}, 2.64×10^{-4}, and 2.86×10^{-5}.

controlled pyrolysis of thermosetting polymeric membranes. These membranes have been prepared by either directly pyrolyzing hollow-fiber polymeric materials or pyrolizing a thin layer of polymer coated on a macroporous support. Both sintered stainless steel tubing and macroporous alumina tubing have been used as the support. Compared with traditional polymeric membranes for gas separation (Paul and Yampol'skii, 1994), the CMS membranes showed both superior separation properties (i.e., permeances and selectivity) and higher stability (Jones and Koros, 1994a). The higher stability (i.e., greater resistance to solvents, chemical attack, and heat) offers the advantage of operating in environments prohibitive to polymeric membranes.

The earliest report on CMS membranes obtained from hollow fiber polymeric membranes appears to be from Koresh and Soffer (1983) and Soffer et al. (1987). By comparing CMS membranes derived from different polymeric membranes, Jones and Koros (1994a) found that the ones from aromatic polyimides yielded the best separation and mechanical properties. The polymers tested by Jones and Koros were cellulose acetate, polyaramides, and polyimides. Polyfurfural alcohol was used by Foley and co-workers (Foley, 1995; Shiflett and Foley, 1999; Strano and Foley, 2002). A comparison of the O_2 permeances and O_2/N_2 selectivities showed that the CMS membrane from polyimide was indeed much better than that from polyfurfural alcohol, as will be seen shortly.

The quality of the CMS membrane is determined not only by the type of the polymer precursor, but also by the pyrolysis protocols (Jones and Koros, 1994a) as well as the manner that the polymer film is deposited on the support (Shiflett and Foley, 1999). The pyrolysis protocol includes temperature program of heating rate and the final temperature and time, as well as the inert gas used versus vacuum. The ideal membrane has pore sizes in the range between 3 to 7 Å for the separation of the permanent gas molecules (e.g., H_2, O_2, N_2, Ar, CH_4, and CO_2). A narrow range in the pore size is needed for a specific separation. The ideal membrane also needs to have a uniform thickness of about 10–30 μm and is mechanically strong and crack-free.

Jones and Koros (1994a) prepared CMS membranes by temperature-programmed pyrolysis of an asymmetric hollow-fiber polyimide membrane. The heating program involved decreasing heating rates until the final temperature of 500 or 550 °C was reached. Pyrolysis under vacuum was found to be better than pyrolysis with inert gas purge. The membrane thickness was 30–35 μm. They measured binary mixture separation properties with a single-fiber permeation system by using a CMS membrane with a diameter of 170 μm. The 500 °C pyrolysis protocol produced membranes with O_2/N_2 selectivities ranging from 8.5 to 11.5, with O_2 fluxes of 20 to 50 GPU (gas permeation units). The 550 °C pyrolysis protocol produced membranes with O_2/N_2 selectivities ranging from 11 to 14, with O_2 fluxes of 15 to 40 GPU. The separation factor is defined as:

$$\alpha_{AB} = (Y_{A1}/Y_{B1})/(Y_{A2}/Y_{B2}) \tag{5.9}$$

where α_{AB} is the separation factor for component A and B, Y is mole fraction, and subscripts 1 and 2 indicate downstream (1) and upstream (2). The gas permeation unit (GPU) is defined as:

$$\text{GPU} = 10^{-6}\text{cc(STP)}/[\text{s} - \text{cm}^2 - \text{cmHg}] = 3.35 \times 10^{-7}\text{gmol}/[\text{s} - \text{m}^2 - \text{kPa}] \tag{5.10}$$

The separation properties of these CMS membranes were compared with a very good asymmetric polysulfone membrane for air separation (Jones and Koros, 1994a). The polysulfone membrane has an O_2/N_2 selectivity ranging from 5.5 to 6.2, with an O_2 flux between 20 to 30 GPU. Hence the CMS membrane has better separation properties. Jones and Koros also measured the separation properties of these CMS membranes for CO_2/N_2, CO_2/CH_4, and H_2/CH_4 and showed that the CMS membranes outperformed the traditional polymeric membranes.

The effects of the pyrolysis conditions on the separation properties of the CMS membranes from polyimides were further studied by Geiszler and Koros (1996). It was shown that by raising the pyrolysis temperature from 500 to 800 °C, the effective pore size was reduced, thereby making the CMS membranes more selective but less productive (i.e., lower permeation fluxes). Vacuum pyrolysis versus that in inert gases had the same effect as raising the temperature.

Drawbacks of the CMS membranes from hollow-fiber polymeric membranes have also been reported (Jones and Koros, 1994b; Jones and Koros, 1995). These membranes were vulnerable to organic vapors as well as moisture. These molecules seemed to adsorb strongly in the pores and reduced the fluxes of the permanent gas molecules. Membrane deterioration was observed with organic contaminations at concentrations as low as 0.1 ppm. However, a promising regeneration process was reported by the use of propylene; significant recovery of the fluxes was obtained by exposure of the membranes to polypropylene (Jones and Koros, 1994b). Jones and Koros (1995) also reported a solution to the problem concerning the adverse effects of humidity by coating a thin layer of hydrophobic polymer (e.g., Teflon) on the CMS membranes. Although small losses in selectivity and productivity occur as a result of the resistance caused by adding the polymer layer, the composite membranes are still very attractive when compared with conventional polymer membranes (Jones and Koros, 1995).

An interesting method for coating the polymer layer on macroporous support was described by Shiflett an Foley (1999). The polymer solution was sprayed on the support with an ultrasonic nozzle, which generated droplets with relatively uniform sizes and also minimized penetration into the support. A sintered stainless steel tubing was used as the support, and a 25 wt % solution of polyfurfural alcohol (PFA) in acetone was sprayed through the nozzle. The ultrasonic nozzle was operated at 40 kHz (the standard frequency for laboratory ultrasound cleaner). The deposited film was subsequently pyrolyzed at a heating rate of 5 °C/min to 450 °C for 2 h in He. SEM images showed the resulting membrane surface to be uniform and defect-free. A comparison of the performance of this CMS membrane and that prepared by pyrolysis of hollow-fiber polyimide membrane is given in Table 5.10. The performance of a polysulfone hollow-fiber membrane is also included in the comparison. Polysulfones are used for air separation, and the dependence of their air separation properties on the structures of different polysulfones has been discussed in detail by Pixton and Paul (1994). Although the separation factor is higher for the PFA-derived membrane than the others, its O_2 flux is much lower.

Diffusion in CMS membranes clearly follows that of surface diffusion or diffusion of adsorbate in micropores. The simple kinetic theory of Chen and Yang (1992) described in Chapter 3 for predicting binary diffusivities from single-component diffusivities should be applicable. Indeed, the predicted binary diffusivities for CH_4/C_2H_6 in CMS prepared from pyrolysis of PFA by using the simple theory agreed well with the experimental data (Chen and Yang, 1994).

Table 5.10. Comparison of fluxes and separation factors at 3 atm feed and room temperature for three membranes

Membrane	Thickness (μm)	Fluxes (GPU)			Separ. Factor
		O_2	N_2	H_2	O_2/N_2
CMS[1]	21.3	0.557	0.018	6.05	30.4
CMS[2]	30–35	24	1.7	150	12
Polysulfone[3]		20	3.6		5.5

[1]CMS membrane by pyrolysis of polyfurfural alcohol ultrasonically sprayed on sintered stainless steel (Shiflett and Foley, 1999).
[2]CMS membrane by pyrolysis of hollow-fiber polyimide membrane (Jones and Koros, 1994a).
[3]From Jones and Koros, 1994a.

A promising development in CMS membrane separation is the "selective surface flow" carbon membranes, by Rao, Sircar, and co-workers (Rao et al., 1992; Rao and Sircar, 1993a; Rao and Sircar, 1993b; Rao et al., 1994). Three types of fluxes occur in pores: convective flux, Knudsen diffusion flux, and flux by surface diffusion. Surface diffusion has been reviewed and discussed extensively by Kapoor et al. (1989). When the surface concentration is high (e.g., for a strongly adsorbed component at a high concentration) or when the other fluxes are low (e.g., in liquid phase), surface diffusion can become the dominant flux for pore diffusion. Furthermore, when the pore dimension is about twice that of the strongly adsorbed molecule, the (back-to-back) adsorbed molecules effectively reduce the pore opening, thereby eliminating or hindering the fluxes for the other non-adsorbed or weakly adsorbed molecules. The net result is enhanced separation.

This concept of "selective surface flow" was first proposed in the 1950's when the effect one adsorbed species had on blocking the flux of another species was lively discussed in the literature (e.g., Kammermeyer and Wyrick, 1958). Kammermeyer and Wyrick (1958) studied the separation of propane and carbon dioxide by a plug of porous glass, and enrichment factors as high as 100 for propane were achieved. The separation was caused by differences in surface fluxes because the gas-phase fluxes were about equal. However, the pores in the porous glass were not small enough to achieve complete blockage by the adsorbed molecules. This was achieved subsequently by Ash et al. (1963). Ash et al. studied separations of four binary mixtures (H_2/SO_2, N_2/CO_2, Ne/CO_2, and A/N_2) by a plug of carbon black powder that was highly compacted. They used the case of H_2/SO_2 for discussion, that the flux of H_2 would be enriched if the carbon powder was not highly compacted. When it was highly compacted, the enrichment was reversed, that is, SO_2 was enriched due to surface flow. Their results are shown in Figure 5.28. The flux of H_2 was reduced to nearly 20% when the relative saturated adsorption of SO_2 reached 20% and was completely blocked at 60% SO_2 adsorption. Similar results were obtained for other mixtures. This separation principle was subsequently demonstrated and studied by Barrer, Ash and co-workers, as well as others-including Okazaki, Toei, and their co-workers (e.g., Ash et al., 1973).

Successful application of this principle for separation depends on the fabrication of membranes with controlled and uniform pore sizes. This was done by Rao, Sircar, and co-workers by controlled pyrolysis of PVDC supported on macroporous alumina tubes. A schematic of their CMS membrane and the principle of separation by the membrane is shown in Figure 5.29. The membrane has pores of uniform sizes in the range between 5–10 Å. The pore sizes can be tailored for different separations.

Golden et al. (1998) described a demonstration unit with seven CMS/alumina tubes (3.5-ft long) mounted in a shell-and-tube type housing. The feed mixture was passed through the bore of the membrane, and the permeate was withdrawn counter-current to the feed. Recovery of H_2 from H_2/hydrocarbon mixtures was first demonstrated. The H_2 product is produced at a high pressure, which

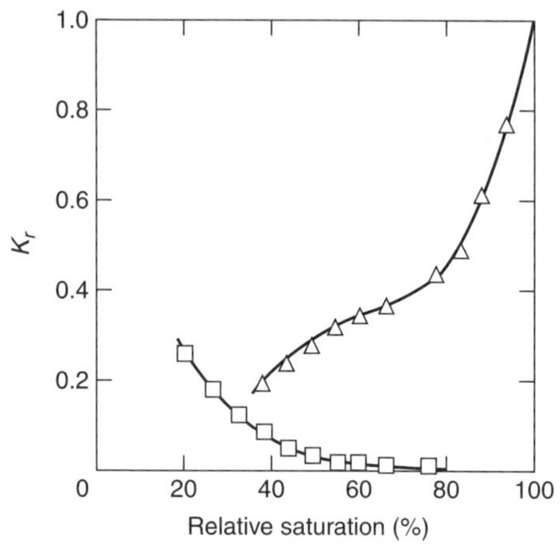

Figure 5.28. Relative permeabilities, K_r (= permeability of blocked medium/permeability of unblocked medium), as functions of degree of filling of pore space for the separation of H_2/SO_2 by a highly compacted carbon plug (Ash et al., 1963, with permission). Relative saturation refers to adsorption in the binary mixture on the feed side of the carbon plug. Squares: H_2 in H_2/SO_2; triangles: SO_2 in H_2/SO_2.

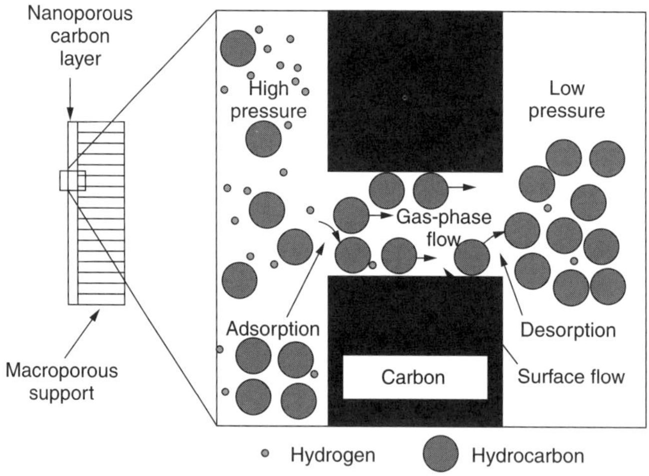

Figure 5.29. Schematic of exclusive surface diffusion (or, "selective surface flow") CMS membrane (from Golden et al., 1998, with permission). (*Note*: The size of H_2 is ~2.8 Å, whereas that of CH_4 is 3.8 Å).

eliminates the need for recompression of H_2 for further processing. Other applications include the separation of H_2S from CO_2 and H_2, and air drying by using a CMS membrane oxidized with nitric acid. The oxidation step generated surface acidic sites giving rise to increased water adsorption capacity, as discussed in Section 5.4 of this chapter. The largest unit described by Golden et al. (1998)

was a shell-and-tube assembly (containing 9-ft-long tubes) for H_2 enrichment from refinery waste gas with a design capacity of 1 MMSCF/d of feed.

REFERENCES

Abe, I., Hayashi, K., Kitagawa, M. et al. (1980) *Bull. Chem. Soc. Japan* 53, 1199.

Abbott, W. F. US Patent 3053775 (1962).

Acharya, M. and Foley, H. C. (2000) *AIChE J.* 46, 911.

Ahmadpour, A., King, B. A., and Do, D. D. (1998) *Ind. Eng. Chem. Res.* 37, 1329.

Al-Bahrani, K. S. and Martin, R. J. (1976) *Water Res.* 10, 731.

Andresen, J. M., Martin, Y., Moinelo, S., Maroto-Valer, M. M., and Snape, C. E. (1998) *Carbon* 36, 1043.

Armor, J. N. (1994) *Separation Technology*. (E. F. Vansant, ed.) Elsevier, Amsterdam, The Netherlands. p. 163.

Ash, R., Barrer, R. M., and Pope, C. G. (1963) *Proc. Royal Soc. London, Ser. A.* 271, 19.

Ash, R., Barrer, R. M., and Pope, C. G. (1973) *J. Chem. Soc., Faraday Trans.* 69, 266.

Atkins, P. W. (1982) *Physical Chemistry*, 2nd Ed. Freeman, San Francisco, CA.

Audley, G. J. and Holder, K. A. U.S. Patent 4,769,359 (1988).

Bailey, A. and Maggs, F. A. P. G.B. Patent 1301101 (1972).

Baker, R. T. K. (1986) *Carbon and Coal Gasification Science and Technology*, NATO ASI.105. Kluwer, Dordrecht, p. 231.

Bandosz, T. J., Biggs, M. J., Gubbins, K. E., Hattori, Y., Iijima, T., Kaneko, K., Pikunic, J., and Thomson, K. (2001) *Chemistry and Physics of Carbon*. (P. L. Thrower, ed.). Dekker, New York, NY.

Bandosz, T. J., Jagiello, J., Contescu, C., and Schwarz, J. A. (1993) *Carbon* 31, 1193.

Baron, G. V. (1994) Industrial gas separation using PSA. In *Separation Technology*. (E. V. Vansant, ed.). Elsevier, Amsterdam, p. 201.

Bansal, R. C., Donnet, J. B., and Stoeckli, F. (1988) *Active Carbon*. Dekker, New York, NY.

Barton, T. J., Bull, L. M., Klemperer, W. G., Loy, D. A., McEnaney, B., Misono, M., Monson, P. A., Pez, G., Sherer, G. W., Vartuli, J. A., and Yaghi, O. M. (1999) *Chem. Mater*. 11, 2633.

Barton, S. S., Evans, M. J. B., Holland, J., and Koresh, J. E. (1984) *Carbon* 22, 265.

Biniak, S., Szymanski, J., Suedlewski, J., and Sviatkowski, A. (1997) *Carbon* 35, 1799.

Blackburn, A., Kipling, J. J., and Tester, D. A. (1957) *J. Chem. Soc.* 2373.

Boehm, H. P. (1994) *Carbon* 32, 759.

Boehm, H. P. (2001) *Graphite and Precursors*. (P. Delhaes, ed.). Gordon and Breach, New York, NY, pp. 141–178.

Boehm, H. P. (2002) *Carbon* 40, 145.

Boehm, H. P. and Voll, M. (1970) *Carbon* 8, 227.

Boehm, H. P., Diehl, E., Heck, W., and Sappok, R. (1964) *Angew. Chem., Int. Ed. Engl.* 33, 669.

Boucher, E. A., Copper, R. N., and Everett, D. H. (1970) *Carbon* 8, 597.

Burchell, T. D. (1999) Porous carbon fiber-carbon composite. In *Carbon Materials for Advanced Technologies*. (T. D. Burchell, ed.). Elsevier Science, Oxford, UK, Chapter 6.

Burchell, T. D., Judkins, P. R., Rogers, M. R., and Williams, A. M. (1997) *Carbon* 35, 1279.

Cabrera, A. L. and Armor, J. N. U.S. Patent 5,071,450 (1991).

Cabrera, A. L., Zehner, J. E., Coe, C. G., Gaffney, T. R., Farris, T. S., and Armor, J. N. (1993) *Carbon* 31, 969.

Carrasco-Marin, E., Mueden, A., Centeno, I. A., Stoeckli, F., and Moreno-Castillo, C. J. (1997) *Chem. Soc. Faraday Trans*. 93, 2211.

Chen, Y. D. and Yang, R. T. (1992) *Chem. Eng. Sci.* 47, 3895.

Chen, Y. D. and Yang, R. T. (1994) *Ind. Eng. Chem. Res.* 33, 3146.

Chen, Y. D., Yang, R. T., and Uawithya, P. (1994) *AIChE J.* 40, 577.

Cheremisinoff, P. N. and Ellerbusch, F. eds., (1980) *Carbon Adsorption Handbook*. Ann Arbor Science, Ann Arbor, MI.

Chinn, D. and King, C. J. (1999) *Ind. Eng. Chem. Res.* 38, 3738.

Contescu, A., Contescu, C., Putyera, K., and Schwarz, J. A. (1997) *Carbon* 35, 83.

Contescu, A., Vass, M., Contescu, C., Putyera, K., and Schwarz, J. A. (1998) *Carbon* 36, 247.

Cookson, J. T. Jr. (1978) Adsorption mechanisms: the chemistry of organic adsorption on activated carbon. In *Carbon Adsorption Handbook*. (P. N. Cherernisinoff and F. Ellerbusch eds.). Ann Arbor Science, Ann Arbor, MI.

Cooney, D. O., Nagerl, A., and Hines, A. L. (1983) *Water Res.* 17, 403.

Coughlin, R. W. and Ezra, F. S. (1968) *Env. Sci. Tech.* 2, 291.

Coughlin, R. W., Ezra, F. S., and Tan, R. N. (1968) *J. Coll. Interf. Sci.* 28, 386.

Dacey, J. R. and Thomas, D. G. (1954) *Trans. Faraday Soc.* 50, 740.

Do, D. D. (2000) *Carbon* 38, 767.

Dobbs, R. A. and Cohen, J. M. (1980) *Carbon Adsorption Isotherms for Toxic Organics*. EPA-600/8-80-023, U.S. Environmental Protection Agency.

Doying, E. G. US Patent 3256206 (1966).

Dubinin, M. M., Kadlec, O., Botlik, I., Zaverina, E. D., Zukal, A., and Sumec, B. (1964) *Dokl. Akad. Nauk SSSR* 157, 656.

Economy, J. and Lin, R. Y. German Patent 2246572 (1973).

Economy, J. and Lin, R. Y. (1976) *Appl. Polymer Symp.* 29, 199.

Edie, D. D. and McHugh, J. J. (1999) High performance carbon fibers. In *Carbon Materials for Advanced Technologies*. (T. D. Burchell, ed.). Elsevier Science, Oxford, UK, pp. 119–138.

Eguchi, Y., Itoga, K., Sawada, T., and Nishino, H. Jpn. Patent Sho 49–37036 (1974).

Epstein, B. D., Dalle-Molle, E., and Mattson, J. S. (1971) *Carbon* 9, 609.

Fanning, P. E. and Vannice, M. A. (1993) *Carbon* 31, 721.

Figureido, J. L., Pereira, M. F. R., Freitas, M. M. A., and Orfao, J. J. M. (1998) *Carbon* 37, 1379.

Faust, S. D. and Aly, O. M. (1987) *Adsorption Processes for Water Treatment*. Butterworths, Boston, MA.

Foley, H. C. (1995) *Micropor. Mater.* 4, 407.

Foster, K. L., Fuerman, R. G., Economy, J., Larson, S. M., and Rood, M. J. (1992) *Chem. Mater.* 4, 1068.

Freeman, J. J., Gimblett, F. R. G., and Sing, K. S. W. (1989) *Carbon* 27, 85.

Gaffney, T. R., Braymer, T. A., Farris, T. S., Cabrera, A. L., Coe, C. G., and Armor, J. N. (1994) In *Separation Technology*. (E. F. Vansant, ed.). Elsevier, Amsterdam, The Netherlands, p. 317.

Garten, V. A. and Weiss, D. E. (1957) *Australian J. Chem.* 10, 309.

Geiszler, V. C. and Koros, W. J. (1996) *Ind. Eng. Chem. Res.* 35, 2999.

Goethel, P. J. and Yang, R. T. (1989) *J. Catal.* 119, 210.

Golden, C. M. A., Rao, M. B., and Sircar, S. (1998) In *Fundamentals of Adsorption*. (F. Meunier, ed.). Elsevier, Amsterdam, The Netherland, p. 1083.

Govindarao, V. M. H. and Gopalakrishna, K. V. (1995) *Ind. Eng. Chem. Res.* 34, 2258.

Grant, T. M. and King, C. J. (1990) *Ind. Eng. Chem. Res.* 29, 264.

Hall, C. R. and Holmes, R. J. (1993) *Carbon* 31, 881.

Hanzawa, Y. and Kaneko, K. (1997) *Langmuir* 13, 5802.

Haydar, S., Moreno-Castillo, C., Ferro-Garcia, M. A., Carraso-Marin, F., Rivera-Utrilla, J. et al. (2000) *Carbon* 38, 1297.

Humphrey, J. L. and Keller, G. E. II. (1997) *Separation Process Technology*. McGraw-Hill, New York, NY.

Iijima, T., Nishikawa, K., Suzuki, T., and Kaneko, K. (1997) *Chem. Phys. Lett.* 274, 152.

Ishizaki, C. and Cookson, J. T. (1974) *Chemistry of Water Supply Treatment and Distribution*. (A. J. Rubin, ed.). Ann Arbor Science, Ann Arbor, MI, Chapter 10.

Jagtoyen, M. and Derbyshire, F. J. (1993) *Carbon* 31, 1185.

Jagtoyen, M., Thwaites, M., Stencil, J., McEnaney, B., and Derbyshire, F. J. (1992) *Carbon* 30, 1089.

Jankowska, H., Swiatkowski, A., and Choma, J. (1991) *Active Carbon*. Ellis Harwood, New York, NY.

Jankowska, H., Swiatkowski, A., Oscik, J., and Kusak, R. (1983) *Carbon* 21, 117.

Jaroniec, M., Gilpin, R. K., Kaneko, K., and Choma, J. (1991) *Langmuir* 7, 2719.

Jones, C. W. and Koros, W. J. (1994a). *Carbon* 32, 1419.

Jones, C. W. and Koros, W. J. (1994b). *Carbon* 32, 1426.

Jones, C. W. and Koros, W. J. (1995) *Ind. Eng. Chem. Res.* 34, 164.

Jüntgen, H. (1977) *Carbon* 15, 273.

Jüntgen, H., Knoblauch, K., and Harder, K. (1981) *Fuel* 60, 817.

Kammermeyer, K. and Wyrick, D. D. (1958) *Ind. Eng. Chem.* 50, 1309.

Kamishita, M., Mahajan, O. P., and Walker, P. L. Jr. (1977) *Fuel* 56, 444.

Kaneko, K. (1998) *Studies in Surface Science and Catalysis*. (A. Dabrowski, ed.). Elsevier, Amsterdam, The Netherlands, Vol. 120, p. 635.

Kaneko, K. (2000) *Carbon* 38, 287.

Kaneko, K., Hanzawa, Y., Iijima, T., Kanda, T., and Suzuki, T. (1999) *Adsorption* 5, 7.

Kaneko, K., Miyawaki, J., Wanatabe, A., and Suzuki, T. (1998) *Fundamentals of Adsorption*. (F. Meunier, ed.). Elsevier, Amsterdam, The Netherlands, p. 51.

Kaneko, Y., Ohbu, K., Uekawa, N., Fugie, K., and Kaneko, K. (1995) *Langmuir* 11, 708.

Kaneko, K., Setoyama, N., and Suzuki, T. (1994) *Characterization of Porous Solids III*. (J. Rouquerol, ed.), Elsevier, Amsterdam, p. 593.

Kaneko, K., Yang, C. M., Ohkubo, T., Kimura, T., Iiyama, T., and Touhara, H. (2000) *Adsorption Science and Technology*. (D. D. Do, ed.). World Scientific Publishers, Singapore.

Kapoor, A. and Yang, R. T. (1989) *Chem. Eng. Sci.* 44, 1723.

Kapoor, A., Yang, R. T., and Wong, C. (1989) *Surface Diffusion, Catal. Rev.* 31, 129–214.

Kawazoe, K., Suzuki, M., and Chihara, K. (1974) *J. Chem. Eng. Japan* 7, 151.

Kim, C. G., Clarke, W. P., and Lockington, D. (1999) *Korean J. Chem. Eng.* 16, 215.

Kimber, G. M., Jagtoyen, M., Hei, Y. Q., and Derbyshire, F. J. (1996) *Gas Sep. Purif.* 10, 131.

Komiyama, H. and Smith, J. M. (1975) *AIChE J.* 21, 664.

Koresh, J. E. and Soffer, A. (1983) *Sep. Sci. Tech.* 18, 723.

Koros, W. J. (1995) *Chem. Eng. Progress* 91(10), 68.

Kuro-Oka, M., Suzuki, T., Nitta, T., and Katayama, T. (1984) *J. Chem. Eng. Japan* 17, 588.

Lamond, T. G., Metcalfe, J. E., III, and Walker, P. L. Jr. (1965) *Carbon* 3, 59.

Laukhuf, W. L. S. and Plank, C. A. (1969) *J. Chem. Eng. Data* 14, 48.

Leng, C. C. and Pinto, N. G. (1996) *Ind. Eng. Chem. Res.* 35, 2024.

Leon y Leon, C. A. and Radovic, L. R. (1994) Interfacial chemistry of carbon surfaces. In *Chemistry and Physics of Carbon*. (P. A Thrower, ed.). Dekker, New York, NY, Vol. 24, p. 213.

Leon y Leon, C. A., Solar, J. M., Calemma, V., and Radovic, L. R. (1992) *Carbon* 30, 797.

Levenspiel, O. (1972) *Chemical Reaction Engineering*, 2nd ed. Wiley New York, NY.

Lin, R. Y. and Economy, J. (1973) *Appl. Polymer Symp.* 21, 143.

Lin, S. H. and Chen, Y. V. (1996) *J. Environ. Sci. Health*, A31, 1279 and 1292.

Lisovskii, A., Semiat, R., and Aharoni, C. (1997) *Carbon* 10, 1639.

Lordgooei, M., Sagen, J., Rood, M. J., and Rostam-Abadi, M. (1998) *Energy & Fuels* 12, 1079.

Ma, Y. H., Sun, W., Bhandarkar, M., Wang, J., and Miller, G. W. (1991) *Separ. Tech.* 1, 90.

MacCallum, C. L., Bandosz, T. J., McGrother, S. C., Muller, E. A., and Gubbins, K. E. (1999) *Langmuir* 15, 533.

MacElroy, J. M. D., Seaton, N. A., and Friedman, S. P. (1997) Sorption rate processes in carbon molecular sieves. In *Equilibrium and Dynamics of Gas Adsorption on Heterogeneous Solid Surfaces*. (W. Rudzinski, W. A. Steele, and G. Zgrablich, ed.), Studies in Surface Sci. and Catal., Col. 104. Elsevier, Amsterdam, The Netherlands.

Mahajan, O. P., Youssef, A., and Walker, P. L. Jr. (1982) *Sep. Sci. Tech.* 17, 1019.

Manes, M. and Hofer, L. J. E. (1969) *J. Phys. Chem.* 73, 584.

Mange, P. and Walker, P. L. Jr. (1986) *Carbon* 24, 101.

Marsh, H. and Walker, P. L. Jr. (1980) The formation of graphitizable carbons via mesophase: chemical and kinetic considerations. In *Chemistry and Physics of Carbon*. (P. L. Walker, Jr., ed.). Dekker, New York, NY.

Martin, R. J. and Ng, W. J. (1987) *Water Res.* 21, 961.

Mason, R. B. and Eberly, P. E. Jr. U.S. Patent 3,222,412 (1965).

Mattson, J. S. and Mark, H. B. Jr. (1971) *Activated Carbon*. Dekker, New York, NY.

Mattson, J. S., Mark, H. B., Malbin, M. D., Weber, W. J., Jr., and Crittenden, J. C. (1969) *J. Coll. Interf. Sci.* 31, 116.

Mawhiney, D. B., Rossin, J. A., Gerhart, K., and Yates, J. T. Jr. (1999) *Langmuir* 15, 4617.

Mays, T. J. (1999) Activated carbon fibers. In *Carbon Materials for Advanced Technologies*. (T. D. Burchell, ed.). Elsevier Science, Oxford, UK, Chapter 3.

Meldrum, B. J. and Rochester, C. H. (1990) *J. Chem. Soc. Faraday Trans.* 86, 1881.

Menendez, J. A., Illan-Gomez, M. J., Leon y Leon, C. A., and Radovic, L. R. (1995) *Carbon* 33, 1655.

Meredith, J. M. and Plank, C. A. (1967) *J. Chem. Eng. Data* 12, 259.

Metcalfe, J. E., Kawahata, M., and Walker, P. L. Jr. (1963) *Fuel* 42, 233.

Mochida, I., Kawabuchi, Y., Kawano, S., Matsumura, Y., and Yoshikawa, M. (1997) *Fuel*.

Mochida, I., Kisamori, S., Hironaka, M., Kawano, S., Matsumura, Y., and Yoshikawa, M. (1994) *Energy Fuels* 8, 1341.

Mochida, I., Korai, Y., Shirahama, M., Hada, T., Seo, Y., Yoshikawa, M., and Yasutake, A. (2000) *Carbon* 38, 227.

Mochida, I., Kuroda, K., Kawano, S., Matsumura, Y., Yoshikawa, M., Grulke, E., and Andrews, R. (1997a) *Fuel* 76, 537.

Mochida, I., Kuroda, K., Miyamoto, S., Sotowa, C., Korai, Y., Kawano, S., Sakanoishi, K., Yasutaki, A., and Yoshikawa, M. (1997b) *Energy Fuels* 11, 272.

Montes-Moran, M. A., Menendez, J. A., Fuente, E., and Suarez, D. (1998) *J. Phys. Chem. B* 102, 5595.

Moreno-Castillo, C., Carraso-Marin, F., Maldonado-Hodar, F. J., and Rivera-Utrilla, J. (1998) *Carbon* 36, 145.

Moreno-Castillo, C., Lopez-Ramon, M. V., and Carrasco-Marin, F. (2000) *Carbon* 38, 1995.

Muller, E. A., Rull, L. F., Vega, L. F., and Gubbins, K. E. (1996) *J. Phys. Chem.* 100, 1189.

Muller, G., Radke, C. J., and Prausnitz, J. M. (1980) *J. Phys. Chem.* 84, 369.

Muniz, J., Herrero, J. E., and Fuertes, A. B. (1998) *Appl. Catal.* 18, 171.

Munzner, H., Heimbach, H., Korbacher, W., Peters, W., Juntgen, H., Knoblauch, K., and Zundorf, D. U.S. Patent 3,801,513 (1974).

Munzner, H., Heimbach, H., Korbacher, W., Peters, W., Juntgen, H., Knoblauch, K., and Zundorf, D. U.S. Patent 3,960,769 (1976).

Nagy, L. G. and Schay, G. (1960) *Magyar Kemp. Folyoirat* 66, 31.

Nanse, G., Papirer, E., Fioux, P., Moquet, F., and Tressaud, A. (1997) *Carbon* 35, 175.

Noh, J. S. and Schwarz, J. A. (1990) *Carbon*, 28, 675.

Notaro, F., Ackley, M. W., and Smolarek, J. (1999) *Chem. Eng.*, April, p. 104.

Ohsaki, T. and Abe, S. U.S. Patent 4,458,022 (1984).

Otake, Y. and Jenkins, R. G. (1993) *Carbon* 31, 109.

Pahl, R. H., Mayhan, K. G., and Bertrand, G. L. (1973) *Water Res.* 7, 1309.

Papirer, E., Lacroix, R., and Donnet, J. B. (1994) *Carbon* 32, 1341.

Patel, R. L., Nandi, S. P., and Walker, P. L. Jr. (1972) *Fuel* 51, 47.

Paul, D. R. and Yampol'skii, Y. P. (1994) *Polymeric Gas Separation Membranes*. CRC Press, Boca Raton, FL.

Peters, E. M. U.S. Patent 3235323 (1966).

Petrovska, M., Tondeur, D., Grevillot, G., and Granger, J. (1991) *Separ. Sci. Tech.* 26, 425.

Pixton, M. R. and Paul, D. R. (1994) *Polymeric Gas Separation Membranes*. (D. R. Paul and Y. P. Yampol'skii, ed.). CRC Press, Boca Raton, FL, Chapter 3.

Puri, B. P. (1966) *Carbon* 4, 391.

Puri, B. P. (1970) Surface complexes on carbon. In *Chemistry and Physics of Carbon*. (P. L. Walker, Jr. ed.), Vol. 6, 191. Dekker, New York, NY.

Py, X., Roizard, X., and Midoux, N. (1995) *Chem. Eng. Sci.* 50, 2069.

Radovic, L. R. and Rodriquez–Reinoso, F. (1996) Carbon Materials in Catalysis. In *Chemistry and Physics of Carbon*. (P. A. Thrower, ed.). Dekker, New York, NY.

Radovic, L. R., Ume, J. I., and Scaroni, A. W. (1996) *Fundamentals of Adsorption*. (M. D. LeVan, ed.), Kluwer, Boston, p. 749.

Rao, M. B., Golden, T. C., and Sircar, S. U.S. Patent 5,104,425 (1992).

Rao, M. B., Golden, T. C., and Sircar, S. U.S. Patent 5,431,864 (1994).

Rao, M. B. and Sircar, S. (1993a) *J. Membrane Sci.* 85, 253.

Rao, M. B. and Sircar, S. (1993b) *Gas Separ. Purif.* 4, 279.

Reich, R., Ziegler, W. T., and Rogers, K. A. (1980) *Ind. Eng. Chem. Proc. Des. Dev.* 19, 336.

Rodgers, S. J., Udavcak, R. J., and Mausteller, J. W. (1965) *International Symposium of Fission Product Release and Transport under Accident Conditions*. Oak Ridge National Laboratories, TN, 1204.

Rodriquez–Reinoso, F. and Linares–Solano, A. (1986) Microporous structure of activated carbons as revealed by adsorption methods. In *Chemistry and Physics of Carbon*. (P. A. Thrower, ed.). Dekker, New York, NY, Vol. 21.

Rodriguez–Reinoso, E., Molina-Sabio, M., and Munecas, M. A. (1992) *J Phys. Chem.* 96, 2707.

Rouquerol, F., Rouquerol, J., and Sing, K. (1999) *Adsorption by Powders and Porous Solids*. Academic Press, San Diego, CA.

Rubel, A. M. and Stencel, J. M. (1997) *Fuel* 76, 521.

Ruckenstein, E., Vaidyanathan, A. S., and Youngquist, G. R. (1971) *Chem. Eng. Sci.* 26, 1306.

Sakoda, A., Kawazoe, K., and Suzuki, M. (1987) *Water Research* 21, 717.

Salame, I. I. and Bandosz, T. J. (1999) *J. Coll. Interf Sci.* 210, 367.

Saunders, J. T. (1982) M.S. Thesis, State University of New York at Buffalo.

Schmidt, J. L., Pimenov, A. V., Lieberman, A. I., and Cheh, H. Y. (1997) *Sep. Sci. Tech.* 32, 2105.

Sellitti, C., Koenig, J. L., and Ishida, H. (1990) *Carbon* 28, 221.

Shiflett, M. B. and Foley, H. C. (1999) *Science* 285, 1902.

Snoeyink, V. L., Weber, W. J., Jr., and Mark, H. B. (1969) *Environ. Sci. Tech.* 3, 918.

Soffer, A., Koresh, J. E., and Saggy, S. U.S. Patent 4,685,940 (1987).

Solum, M. S., Pugmire, R. J., Jagtoyen, M., and Derbyshire, F. J. (1995) *Carbon* 33, 1247.

Sorial, G. A., Granville, W. H., and Daly, W. O. (1983) *Chem. Eng. Sci.* 38, 1517.

Spencer, D. H. T. (1967) *Porous Carbon Solids*. (R. L. Bond, ed.). Academic Press, New York, NY, pp. 87–154.

Strano, M. S. and Foley, H. C. (2002) *Carbon*, 40, 1029.

Suarez, D., Menendez, J. A., Fuente, E., and Montes-Moran, M. A. (1999) *Langmuir* 15, 3897.

Suarez, D., Menendez, J. A., Fuente, E., and Montes-Moran, M. A. (2000) *Angew Chem., Int. Ed. Engl.* 39, 1376.

Subrenat, A., Baleo, J. N., Le Cloirec, P., and Blanc, P. E. (2001) *Carbon* 39, 707.

Sullivan, P. D., Rood, M. J., Hay, K. J., and Qi, S. (2001) *J. Envir. Eng*., 127, 217.

Sutikno, T. and Himmelstein, K. J. (1983) *Ind. Eng. Chem. Fundam.* 22, 420.

Suzuki, M. (1990) *Water Sci. Tech.* 23, 1649.

Suzuki, M. (1994) *Carbon* 32, 577.

Suzuki, M., Misic, D. M., Koyama, O., and Kawazoe, K. (1978) *Chem. Eng. Sci.*, 33, 271.

Szepesy, L. and Illes, V. (1963) *Acta Chim. Hungary* 35, 37.

Valenzuela, D. P. and Myers, A. L. (1989) *Adsorption Equilibrium Data Handbook*. Prentice-Hall, Englewood Cliffs, NJ.

Vidic, R. D., Suidan, M. T., and Brenner, R. C. (1993) *Environ. Sci. Tech.* 27, 2079.

Vidic, R. D., Suidan, M. T., and Brenner, R. C. (1994) *Water Res.* 28, 263.

Vinke, P., van der Eijk, M., Verbree, M., Voskamp, A. F., and van Bekkum, H. (1994) *Carbon* 32, 675.

Voll, M. and Boehm, H. P. (1970) *Carbon* 8, 741.

Voll, M. and Boehm, H. P. (1971) *Carbon* 9, 481.

Walker, P. L. Jr. (1990) *Carbon* 28, 261.

Walker, P. L., Jr., Austin, L. G., and Nandi, S. P. (1966a) *Chemistry and Physics of Carbon*. (P. L. Walker, Jr., ed.), Vol. 2. Marcel Dekker, New York, NY.

Walker, P. L., Jr., Lamond, T. G., and Metcalfe, J. E. (1966) *2nd Conference Industrial Carbon Graphite*, Society of Chemistry and Industry, London, UK. p. 7.

Walker, P. L., Jr., Rusinko, F., Jr., and Austin, L. G. (1959) *Adv. Catal*. 11, 133.

Weber, W. J., Jr. and DiGiano, F. A. (1996) *Process Dynamics in Environmental Systems*, Wiley, New York, NY.

Wennerberg, A. N. U.S. Patent 4,624,004 (1971).

Wennerberg, A. N. and O'Grady, T. M. U.S. Patent 4,082,694 (1978).

Wilson, J. (1981) *Fuel* 60, 823.

Yang, R. T. (1984) Etch-decoration electron microscopy studies of gas–carbon reactions. In *Chemistry and Physics of Carbon*. (P. A. Thrower, ed.). Dekker, New York, NY, Vol. 19, 163.

Yang, R. T. (1987) *Gas Separation by Adsorption Processes*. Butterworth, Boston, MA, Imperial College Press, London, UK.

Yang, R. T. and Duan, R. Z. (1985) *Carbon* 23, 325.

Yang, R. T., Long, R., Padin, J., Takahashi, A., and Takahashi, T. (1999) *Ind. Eng. Chem. Res.* 38, 2726.

Yang, F. H. and Yang, R. T. (2002) Unpublished results, University of Michigan.

Yeh, Y. T. (1989) Ph.D. Thesis, State University of New York at Buffalo.

Zawadzki, J. (1989) Infrared spectroscopy in surface chemistry of carbons. In *Chemistry and Physics of Carbon*. (P. A. Thrower, ed.). Dekker, New York, NY, Vol. 21, pp. 147–380.

Zhang, Q. L., Kyotani, T., and Tomita, A. (1994) *Energy Fuels* 8, 714.

6

SILICA GEL, MCM, AND ACTIVATED ALUMINA

Silica gel is the most widely used desiccant because of its large capacity for water (~40% by weight) and ease in regeneration (~150 °C, compared with 350 °C for regenerating zeolites). In addition, its surface can be readily modified by reacting (or grafting) with a monomolecular layer of organic ligand, and these modified silica gels are being applied in an increasing number of applications in chromatography. All aspects of silica gel and its modification have been reviewed and discussed extensively (Iler, 1979; Unger, 1979; Vansant et al., 1995).

The MCM-type materials belong to a new family of ordered, mesoporous silicate/aluminosilicate prepared by hydrothermal formation of silica gels in the presence of surfactant templates (Beck et al., 1992). They were discovered only recently, by Beck et al. in 1992, and hold promise for a number of interesting applications. Hence they are included in this chapter.

Activated alumina is also widely used as a desiccant because of the same advantages for which silica gel is used. Unlike silica gel, which is amorphous, activated alumina is crystalline. Oxygen vacancies (defects) are easily formed on its surfaces, thus alumina has both Lewis and Brønsted acid sites. The surface chemistry, as well as the pore structure of activated alumina, can be modified, for example, by treatment with acid (HCl or HF) or alkaline (to alter the acidity) and controlled thermal treatment (to tailor the pore structure). As a result, activated alumina is more versatile than silica gel and has been applied more often as a sorbent.

6.1. SILICA GELS: PREPARATION AND GENERAL PROPERTIES

The commercial silica gel sorbents are mesoporous, that is, with pores mostly larger than 20 Å. Silica gel can be formed by two routes: (1) polymerization

Adsorbents: Fundamentals and Applications, Edited By Ralph T. Yang
ISBN 0-471-29741-0

of silicic acid, and (2) aggregation of particles of colloidal silica. Silicic acid, $Si(OH)_4$, has a strong tendency to polymerize and form a network of siloxane (Si-O-Si), leaving a minimum number of uncondensed Si-O-H groups. The pH value is important in the polymerization process (Iler, 1979). The second route involves coagulation of silica sols of rather uniform sizes. These submicrometer particles can coagulate by van der Waals forces or by cations bridging as coagulants. Commercially, silica is prepared through the first route by mixing a sodium silicate solution with a mineral acid, such as sulfuric or hydrochloric acid. The reaction produces a concentrated dispersion of finely divided particles of hydrated SiO_2, known as silica hydrosol or silicic acid:

$$Na_2SiO_3 + 2HCl + nH_2O \rightarrow 2NaCl + SiO_2{\cdot}nH_2O + H_2O \tag{6.1}$$

The hydrosol, on standing, polymerizes into a white jellylike precipitate, which is silica gel. The resulting gel is washed, dried, and activated. Various silica gels with a wide range of properties, such as surface area, pore volume, and strength, can be made by varying the silica concentration, temperature, pH, and activation temperature (Iler, 1979). Two common types of silica gel are known as regular-density and low-density silica gels, although they have the same densities (true and bulk). The regular-density gel has a surface area of 750–850 m^2/g and an average pore diameter of 22–26 Å, whereas the respective values for the low-density gel are 300–350 m^2/g and 100–150 Å.

The silica gel is amorphous. By use of high-resolution electron microscopy, it is known that its amorphous framework is composed of small globular (primary) particles having sizes of 10 to 20 Å (Rouquerol et al., 1999). Still another route for preparing silica gel involves reactions of silicon alkoxides with water, and a wide variety of materials can be produced in this manner (Jones, 1989; Brinker and Sherer, 1990). The processes based on this route are referred to as "sol-gel" processing, and they offer many promising possibilities. For silica gel, the reaction is

$$Si(OR)_4 + 2H_2O \xrightarrow{ROH} SiO_2 + 4ROH \tag{6.2}$$

Silicic acids are also formed by hydrolysis:

$$-\overset{|}{\underset{|}{Si}}-OR + H_2O = -\overset{|}{\underset{|}{Si}}-OH + ROH \tag{6.3}$$

The silicic acids thus formed can then polymerize via:

$$-\overset{|}{\underset{|}{Si}}-OH + HO-\overset{|}{\underset{|}{Si}}- = -\overset{|}{\underset{|}{Si}}-O-\overset{|}{\underset{|}{Si}}- + H_2O \tag{6.4}$$

or

$$-\overset{|}{\underset{|}{Si}}-OH + RO-\overset{|}{\underset{|}{Si}}- = -\overset{|}{\underset{|}{Si}}-O-\overset{|}{\underset{|}{Si}}- + ROH \tag{6.5}$$

The reaction products are high-molecular-weight polysilicates (a sol), which form a 3-D porous network filled with solvent molecules (a gel). The more recent development of MCM-41 (to be discussed separately) is a derivative of the sol-gel route. The pore structure, as well as the surface chemistry, can be tailored in the sol-gel route. The pH value in the initial stages (Reactions 3–5) is a main factor in controlling the pore dimensions. Low pH (e.g., by adding HCl) leads to microporosity, whereas high pH (e.g., by adding ammonium hydroxide) results in mesoporosity (Brinker and Sherer, 1990). Apparently, pH influences the size distribution of the globular, primary particles and also how these particles agglomerate, hence the final pore structure. Sol-gel processing is highly versatile. It is not limited to silica and is also applicable to many other main group and transition metal oxides.

The most-used property for silica gel is as a desiccant, that is, adsorbent for moisture. As mentioned, this quality is due to its relatively weak bonds with water as well as its large pore volume and mesoporosity. The pore-size distribution of silica gel is given in Figure 6.1, along with other major sorbents for comparison. The water isotherms on silica gel and other sorbents are given in Figure 6.2. From Figures 6.1 and 6.2, the reasons that silica are the most used desiccant are clear: It adsorbs a large amount of water at low humidity and has the highest total water capacity, in addition to its ease in regeneration.

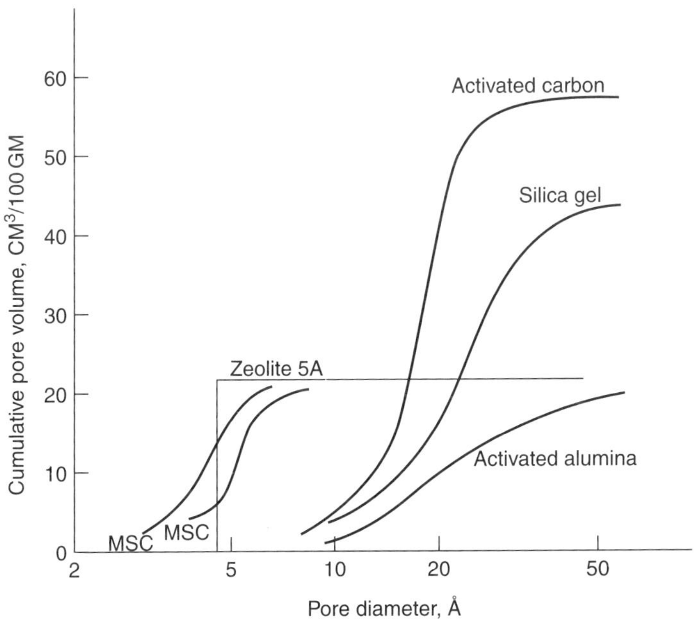

Figure 6.1. Pore-size distribution for activated carbon, silica gel, activated alumina, two molecular-sieve carbons, and zeolite 5A. *Source*: Yang, 1997, with permission.

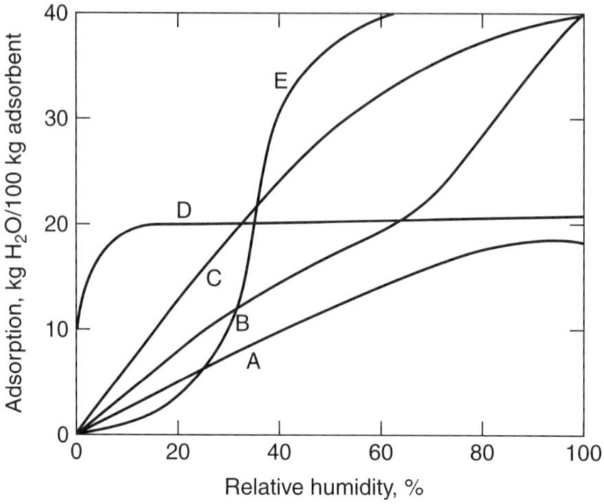

Figure 6.2. Equilibrium sorption of water vapor from atmospheric air at 25 °C on (A) alumina (granular); (B) alumina (spherical); (C) silica gel; (D) 5A zeolite; (E) activated carbon. The vapor pressure at 100% R.H. is 23.6 Torr. *Source*: Yang, 1997, with permission.

6.2. SURFACE CHEMISTRY OF SILICAS: THE SILANOL GROUPS

The surface chemistry of silicas is dominated by the surface hydroxyl groups, or silanols (Si-O-H). The silanol groups participate in adsorption (for water and other compounds, including organic compounds) as well as the chemical modification of silica surfaces. The subject of surface silanols has been studied extensively over the past seven decades and has been reviewed (e.g., Iler, 1979; Unger, 1979; Zhuravlev, 1993; Bergna, 1994; Vansant et al., 1995).

Three basic types of surface silanols are shown in Figure 6.3. Another type, vicinal hydroxyls, applies to free hydroxyls that are adjacent to each other on the surface. These four groups are formed by two main processes. First, they are formed in the course of silica gel preparation, that is, during the polymerization condensation of silicic acid, $Si(OH)_4$. Upon drying, the *hydrogel* becomes *xerogel*, leaving some of the hydroxyls on the surface. The second process is the reaction of the siloxane (Si-O-Si) surface with water, under ambient conditions, to form hydroxyl groups.

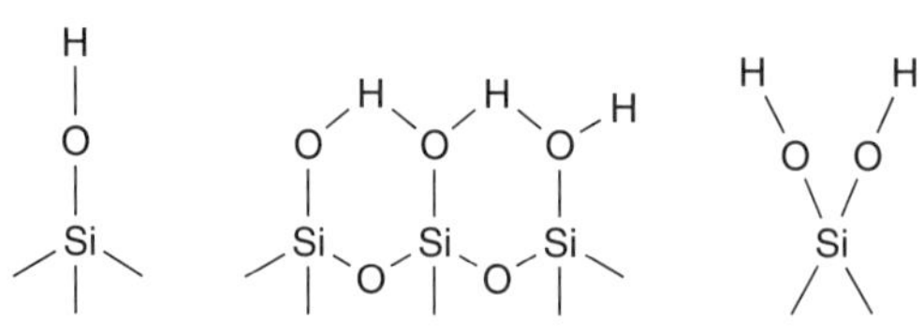

Figure 6.3. Three basic types of silanol (hydroxyl) groups on silicas.

Hydrogen bonding is the most important type of bonding for adsorption of water as well as other compounds. For water to adsorb on the first layer of silanol, it has been shown that water sits "oxygen down" on the SiOH group (Klier and Zettlemoyer, 1977). As more water molecules adsorb, hydrogen-bonded clusters form. The heats of adsorption are (Iler, 1979; Zhuravlev, 1993) the following:

First Adsorbed Layer:

$$\mathrm{Si-OH + H_2O \Rightarrow Si-OH{:}OH_2} \quad \Delta H = -6.0\ \mathrm{kcal/mol}$$

Higher Layers:

$$\mathrm{Si-OH{:}OH_2 + x\ H_2O \Rightarrow Si-OH{:}OH_2(OH_2)} \quad \Delta H = -10.5\ \mathrm{kcal/mol}$$

The value of 10.5 kcal/mol is very close to the latent heat of condensation of water (10.8 kcal/mol).

Many techniques have been used to identify and quantify the different silanol groups. The most extensively used is infrared spectroscopy (IR). IR spectroscopy of the adsorbed water and hydroxyls (as well as other compounds) has been discussed extensively by Hair (1967) and by Kiselev and Lygin (1975). The most useful absorption band is the OH stretching frequency. In addition to the surface OH groups, there also exist internal OH groups in the silica structure. The characteristic OH stretching vibrational frequencies for the different groups are (Iler, 1979; Hair, 1967; Kiselev and Lygin, 1975) as follows:

Type of OH	Approximate IR Frequency, cm^{-1}
Isolated OH	3750
Paired or H-Bonded on Surface	3540
Internal OH, H-Bonded	3650
Liquid Water	3400

The stretching frequency band for the isolated OH groups is a very sharp one. All other bands are broad. This sharp band appears only after dehydration, for example, >200–400 °C in vacuo.

6.3. THE SILANOL NUMBER (OH/NM^{-1})

The density of the OH groups on the surface is referred to as the silanol number, α_{OH}, expressed in OH groups per square nanometer. The silanol number can be determined by deuterium exchange (Kiselev and Lygin, 1975; Zhuravlev, 1993) or by various titration methods, for example, by titration with hexamethyldisilazane (Vansant et al., 1995):

$$2Si - OH + (CH_3)_3 - Si - NH - Si - (CH_3)_3 \Rightarrow 2Si - O - Si(CH_3)_3 + NH_3$$

For deuterium exchange, D in D_2O exchanges with only the H atoms of the hydroxyl groups on the surface but not with those in the interior of silica.

Many early calculations were made on the "theoretical" silanol number (Iler, 1979). The silica gels are mostly heat-stabilized amorphous silica. Cristobalite and tridymite, which are crystalline, have been used as models for amorphous silicas because their densities and refractive indices are similar. Using the (111) face of the octahedral cristobalite and assuming 1 OH per surface Si, De Boer and Vleeskens (1958) and De Boer (1958) calculated the silanol number as 4.55 OH/nm^2. A similar calculation using tridymite gave 4.6 OH/nm^2 (Iler, 1979).

Zhuravlev (1987; 1993) has shown that for 100 different samples of amorphous silicas with a wide range of surface areas (from 9.5 to 945 m^2/g), the silanol numbers fell within a narrow range of 4.3–5.8 OH/nm^2. These samples differed in the methods of synthesis and in their structural characteristics.

It seems that on the heat-stabilized, amorphous silicas that are fully hydroxylated, the silanol number is 4–5 OH/nm^2. However, this "universal" number could be misleading because it could be very different for many new silica materials. Unger et al. (2001) showed that the silanol numbers are

Silica	Silanol Number, OH/nm^2
MCM-41	2–4
MCM-48	3
MCMoidal	5–6
Silica Xerogel	8–9

Dehydroxylation commences upon heat-treatment at temperatures above 200 °C. The silanol number decreases with temperature. Typical data on the temperature dependence of the silanol number are shown in Figure 6.4. There are two regions in this figure. At temperatures <400 °C, both free OH groups and H-bonded OH groups are being dehydroxylated. At temperatures >400 °C, only free hydroxyls are dehydroxylated. By drawing a straight line through the second region (i.e., the high-temperature region) and extrapolating to 200 °C, the difference between the two lines gives the amount of the H-bonded OH groups that are being dehydroxylated.

The direct involvement of the surface hydroxyls in adsorption of a large variety of gas molecules on silica has been studied extensively by using IR spectroscopy (Hair, 1967; Kiselev and Lygin, 1975). Upon adsorption, the stretching vibration frequency of the free OH groups (at 3750 cm^{-1}) is shifted toward lower

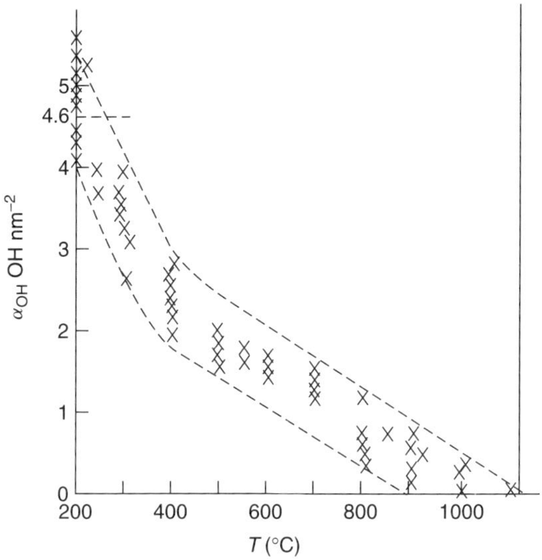

Figure 6.4. Silanol number (α_{OH}) as a function of temperature in vacuo for different silica samples (Zhuravlev, 1993, with permission); 4.6 is the theoretical number.

frequencies. Accompanying the shift is a broadening of the band, and also an increase in intensity. These findings are evidence for intermolecular hydrogen bonding formed upon adsorption. The magnitude of the shift in the OH stretching frequency is directly related to the strength of the bonding. In fact, linear correlations have been found between the frequency shift and the heat of adsorption for a large number of hydrocarbons and water on silica (Anderson, 1965; Kiselev and Lygin, 1975).

For adsorption of water on silica, that is, hydration, water molecules first attach themselves to the silanol groups by hydrogen bonding, followed by H-bonding to the already adsorbed water. Hence the IR band is gradually lowered and broadened to 3400 cm^{-1}. For the same adsorbate, the frequency shift depends on the temperature as well as the amount adsorbed. The literature data on the frequency shifts by different adsorbate molecules have been discussed by Hair (1967) and by Kiselev and Lygin (1975) and are summarized in Table 6.1. The experiments for the permanent gases (e.g., N_2 and CH_4) were performed at low temperatures ($-190\,^\circ$C), while that for all others were at room temperature. The frequency shifts are denoted by $\Delta\nu$ (cm^{-1}), lowered from 3750 cm^{-1}.

The literature data showed that the adsorbate molecules fall approximately into three classes in regard to strength of interaction with the hydroxylated surfaces (Iler, 1979):

Low: Ar, N_2, CCl_4, *n*-alkanes, cyclohexane

Medium: CH_3NO_2, C_6H_6, $CH_3COOC_2H_5$, CH_3CN, tetrahydrofuran, dioxane, acetone, $(C_2H_5)_2O$

Strong: NH_3, pyridine, aniline, $(C_2H_5)_3N$

Table 6.1. Frequency shift (lowering from 3750 cm^{-1}) by adsorption on silicas

Adsorbate	$\Delta\nu$ (Lowered, cm^{-1})	Adsorbate	$\Delta\nu$ (Lowered, cm^{-1})
Ar	8	Benzene	110
O_2	12	Toluene	130
N_2	24	Aniline	550
CH_4	32	Phenol	350
NH_3	800	Nitrobenzene	140
CO	130	*n*-Hexane	50
CH_3OH	330	$(CH_3)_3COH$	435
CCl_4	40	C_5H_5N	850
CH_2Cl_2	72	Thiophene	130
CH_3Cl	140	CH_2Cl-CH_2Cl	140

Hydrogen-bonding with the OH groups is involved for medium and strong interactions.

Silicas with the same silanol number should have the same isotherms for the adsorption of water because the silanol groups are the primary sites for adsorption. This idea was originally proposed by Kiselev (see for example, Kiselev, 1986), and was verified by experimental data (Zhuravlev, 1993). Using different large-pore and nonporous amorphous silicas (prepared from different syntheses), Kiselev and co-workers showed that their water isotherms are indeed the same at low relative pressures (<0.4), as shown in Figure 6.5.

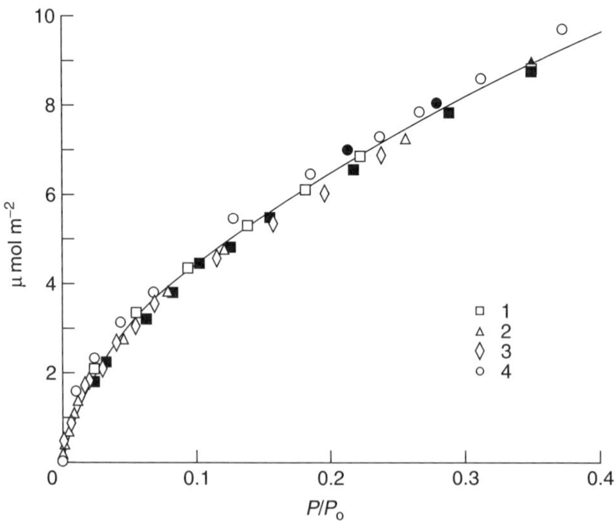

Figure 6.5. "Absolute" adsorption isotherms for water on different wide-pore amorphous silicas at room temperature (after the silicas were treated at 200 °C). The line shows the average of data from the literature (Kiselev, 1986; Zhuravlev, 1993).

6.4. MCM-41

Beck et al. (1992) succeeded in the synthesis of a new family of ordered, mesoporous silicate/aluminosilicate by hydrothermal formation of silica gels in the presence of surfactant templates. Quaternary ammonium surfactants were typically used. The surfactants self-assemble to form micellar templates with a three-dimensional, long-range order. The silicate precursors condense on the walls of the template. The organic templates are subsequently removed by air oxidation leaving behind a silicate structure. These materials are amorphous but exhibit simple X-ray diffraction patterns ($2\theta \sim 2^\circ$) that reflect the interplanar spacing of the regular mesoporous structure of the templates. These materials were named M41S, and the materials that have the honeycomb-shaped structures are named MCM-41. A schematic of the formation of MCM-41 is shown in Figure 6.6.

Different mechanisms for the interactions between the silicate precursors and the organic template, as well as the liquid crystal templating mechanisms, have been discussed (Ying et al., 1999; Tanev and Pinnavaia, 1995). Two other major types of M41S materials are MCM-48 (with 3-D pores) and MCM-50 (with a pillared layer or lamellar structure).

Among the M41S materials, MCM-41 has received the most attention because of its simple structure, as well as the ease of synthesis and tailoring of its structure and surface properties. The pore dimension is in the range of 20–100 Å and can be tailored by several different strategies (Rouquerol et al., 1999; Ying et al., 1999; Zhao et al., 1996). The first is to use surfactants with different chain lengths and pore diameters that can be controlled from near 15 to 45 Å (Huo, et al., 1996). The use of two surfactants can extend the pore sizes to 55 Å (Kaman et al., 1996). The addition of expander molecules, such as trimethylbenzene (Beck et al., 1992; Zhao et al., 1996), extends the pore size to near 100 Å. Good-quality, large-pore MCM-41 with pore sizes up to 65 Å can be made by post-synthesis hydrothermal restructuring (Huo et al., 1996; Sayari et al., 1997). Ordered, nanostructured materials with even larger pores (up to 300 Å) can be obtained by using block copolymers as the structure-directing agents (Zhao et al., 1998). Moreover, MCM-41 with heteroatoms (such as Al, Ti, B, and V) and many

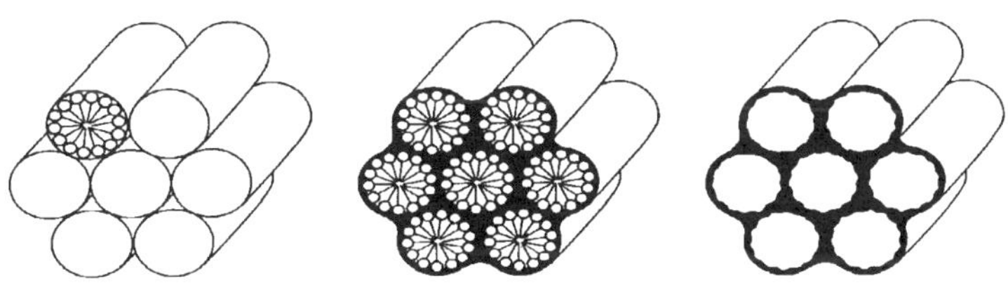

(a) (b) (c)

Figure 6.6. Schematic representation of the formation of MCM-41 by the liquid-crystal templating mechanism. (A) Hexagonal array of cylindrical micelles; (B) the same, with silicate species between the cylinders; and (C) hollow cylinders of MCM-41 after thermal elimination of organic material (Rouquerol et al., 1999, with permission).

non-silica materials with structures similar to MCM-41 can be synthesized (Ying, et al., 1999; Sun and Ying, 1997).

The unique feature of MCM-41 is the large uniform pore structure (and hence a remarkably large pore volume >0.7 cm^3/g). This unique feature has made these materials promising for both adsorption and catalysis (as discussed by Ying et al., 1999; Ravikovitch et al., 1998). However, this unique feature may not be attractive for gas adsorption because the interaction potentials are not enhanced within the pores, as discussed earlier. Consequently, relatively few studies have been made on MCM-41 as an adsorbent.

In order to obtain useful adsorption properties, MCM-41 needs to be modified in either surface chemistry or pore structure. Modification of silica surfaces will be discussed further in the next section. The same techniques of modification have been used for MCM-41. The reported modifications have been based mainly on the use of reactive silanes that contain organic groups (such as alkyls) and chloride. The formed MCM-41 has hydroxyl groups, and by reacting the hydroxyl group with silanes, that is, silanation, pore sizes can be reduced by the grafted silanes (Feng et al., 1997; Jaroniec et al., 1998). The grafting can be accomplished by many other compounds such as metal alkoxides and halides; hence the chemistry of the surface can be altered or functionalized (Moller and Bein, 1998; Jaroniec et al., 1998). Functionalization of the surface can be also accomplished before the final calcination step by directly displacing the surfactants with reactive silanes (Antochsuk and Jaroniec, 2000). A technique that reduces only the pore size at the opening regions has been suggested (Zhao et al., 1999) by grafting a controlled number of layers of silica in the opening regions. This procedure was performed by displacing the surfactant in the opening regions by H^+, followed by silanation with $Si(OEt)_4$ and hydrolysis, and finally calcination.

Only a few promising applications of the MCM-41 and its modified forms have been reported. Izumi and co-workers have reported the use of MCM-41 for VOC (Izumi, 1996) and SO_2 removal (Teraoka et al., 2000) by taking advantage of the weak bonds with the surface, hence ease in desorption. They also reported a low-temperature synthesis route for MCM-41 at very low pH (<1), which is a possible low-cost method. Feng et al. (1997) grafted the MCM-41 with a silane-containing thiol (−SH) group and produced a sorbent that is highly selective for binding heavy metal ions such as mercury, silver, and lead from wastewaters. The sorbent is also regenerable with HCl. Because of the large arrays of functionalities and pore sizes that can be achieved with the MCM-41 material, unique adsorption properties will no doubt be obtained. However, due to the cost, its potential use appears to be limited to specialized applications.

The syntheses of a large number of templated, mesoporous materials other than silica have also been reported. These materials are mainly of interest to catalysis and hence will be described only briefly here. A good summary of their synthesis routes and factors involved in the syntheses has been given by Ruckenstein and co-worker (Ruckenstein and Chao, 2001; Chao and Ruckenstein, 2002a and 2002b). Metals/elements such as Al, B, Ti, Zr, V, Pd, and Mn can be introduced by either grafting or isomorphous substitution onto the framework of the mesosilicates. Direct synthesis based on templating appears to be a more attractive route. Thus, pure oxides, such as those of Pb, Sb, Ti, Nb, Ta, Al, Mn, Ce, Ha, Zr, V, and W, and composites, such as V-Mg-O, Al-B-O, Al-P-O, V-P-O, Zr-P-O, and Al-P-V-O, have been prepared by the direct route. With V-Mg-O used as an example (Ruckenstein and Chao, 2001; Chao and Ruckenstein, 2002a and 2002b), the crystallinity as well as the morphology of the mesoporous product depend on the template surfactant, V/Mg ratio, and pH. Interestingly, fibrous and lamellar structures could be formed depending on the synthesis conditions.

6.5. CHEMICAL MODIFICATION OF SILICAS AND MOLECULAR IMPRINTING

Silicas are strong, mesoporous, and — more importantly — abundant in surface hydroxyl groups. These properties make them ideally suited for chemical grafting. The grafted sorbents are widely used for a number of applications. The most important application is for the stationary phases in a variety of chromatographies. The grafted silicas are also promising sorbents for selective adsorption. An example will be given for the potential use of a grafted silica as a selective sorbent for acid gas removal. The most commonly used grafting compound is 3-aminopropyltriethoxysilane:

$$(CH_3CH_2O)_3SiCH_2CH_2CH_2NH_2$$

Upon refluxing of silicas with a solution of the silane at mild temperatures (e.g., 70 °C), the surface is grafted by the following reaction:

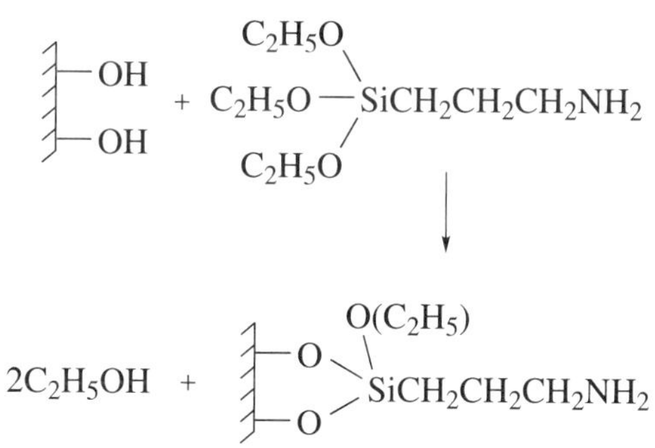

A large number of silanes carrying different functional groups have been used as the grafting agents. Due to the high separation power of chromatography, any small differences in interactions with the sorbent by the different components in the mixture will lead to resolution. New stationary phases and new applications are being developed rapidly, particularly for bioseparations (Belter et al., 1988; Garcia et al., 1999). Some of the silanes used for grafting are shown in Table 6.2.

Table 6.2. Silanes used for grafting silicas

Chlorosilanes:	Trichlorosilane, Cl_3SiH Methyltrichlorosilane, Cl_3SiCH_3 Dimethyldichlorosilane, $Cl_2Si(CH_3)_2$ Trimethylchlorosilane, $ClSi(CH_3)_3$ Propyltrichlorosilane, $Cl_3SiCH_2CH_2CH_2$ Octyltrichlorosilane, $Cl_3Si(CH_2)_7CH_3$ Octadecyltrichlorosilane, $Cl_3Si(CH_2)_{17}CH_3$
Alkenylsilanes:	Vinyltriethoxysilane, $(CH_3CH_2O)_3Si-CH{=}CH_2$ Methcryloxypropyltriethoxysilane, $(CH_3CH_2O)_3SiCH_2CH_2CH_2-O-CO-C(CH_3){=}CH_2$
Arylsilane:	Phenyltriethoxysilane, $(CH_3CH_2O)_3SiC_6H_5$
Epoxifunctional Silane:	3-Glyciodoxypropyltrimethoxysilane $(CH_3O)_3SiCH_2CH_2CH_2-O-CH_2CH-O-CH2$
N-Functional Silanes:	3-Aminopropyltriethoxysilane, $(CH_3CH_2O)_3SiCH_2CH_2CH_2NH_2$ 3-Aminopropyldiethoxymethylsilane, $(CH_3CH_2O)_3CH_2SiCH_2CH_2CH_2NH_2$ 3-Aminopropylethoxydimethylsilane, $(CH_3CH_2O)(CH_2)_2SiCH_2CH_2CH_2NH_2$ Cyanopropyltriethoxysilane, $(CH_3CH_2O)_3SiCH_2CH_2CH_2CN$ Hexamethyldisilazane, $(CH_3)_3SiNHSi(CH_3)_3$
S-functional Silane:	3-Mercaptopropyltriethoxysilane, $(CH_3CH_2O)_3SiCH_2CH_2CH_2SH$
Cl-functional Silane:	Chloropropyltriethoxysilane, $(CH_3CH_2O)_3SiCH_2CH_2CH_2Cl$

Courtesy of Vansant et al., 1995.

The organosilanes have the general formula R_N-Si-X_{4-N}, where R is an organic group and X is a hydrolyzable ligand. X may be an acyloxy, amine, halide, or alkoxy group. Chloride, ethoxy, and methoxy are the most-used X groups.

Silica gels are used in gel permeation chromatography, based on size exclusion of polymers from the pores. Silica gels have also found use as a polar column sorbent. Modification of the silica by silanes makes the stationary phase apolar. Because most interactions are reversed, compared with the pure silica, separations using these apolar columns are termed "reversed-phase chromatography".

Amine grafted silicas are promising sorbents for selective adsorption, especially for applications in environmental control and catalysis. For example, Lasperas et al. (1997) and Angeletti et al. (1988) used amine function-modified

MCM-41 and silica gel as efficient and reusable heterogeneous catalyst for the Knoevenagel Condensation reactions. Burwell and Leal (1974) first reported selective chemisorption of sulfur dioxide on amine-modified silica gel. Leal et al. (1995) studied carbon dioxide adsorption on amine surface-bonded silica gel, although their CO_2 adsorption amount (0.3 mmol/g-sorbent at 1 atm CO_2) was low. Further work by Huang et al. (2003) produced high-capacity and selective sorbents for both CO_2 and H_2S.

Their work is summarized briefly: Silica xerogels and MCM-48 were used. The xerogels were prepared from tetraethoxysilane (TEOS) by using standard technique, and the MCM-48 was prepared by the techniques developed by Schumacher et al. (2000). These silicas were used because of their high surface areas (816 m^2/g for xerogels, and 1389 m^2/g for MCM-48) and high silanol numbers ($\sim$5 and 8, respectively), as well as thermal stability (MCM-48 is more stable than MCM-41). 3-Aminopropyltriethoxysilane was grafted on these silicas.

The coverages of about 1.7 and 2.3 mmol—$CH_2CH_2CH_2$ NH_2 per gram of sorbent were obtained for silica xerogel and MCM-48, respectively. The Fourier transform infrared (FTIR) spectrum of the MCM-48 sample is compared with that modified with amine. As shown in Figure 6.7, the sharp peak at 3743 cm^{-1} on the unmodified MCM-48 is attributed to the free hydroxyl groups on the surface. Silanol groups with hydrogen-bonding interactions were also observed at 3540 cm^{-1} (broad band). The FTIR spectrum of the 3-aminopropyl-functionalized MCM-48 sample is shown in Figure 6.7. After the surface reactions with the 3-aminopropyltriethoxyl silane, the IR bands of the surface hydroxyl groups vanished, indicating that all the surface hydroxyl groups reacted with the silane. The new IR bands shown in the figure are those attributed to the

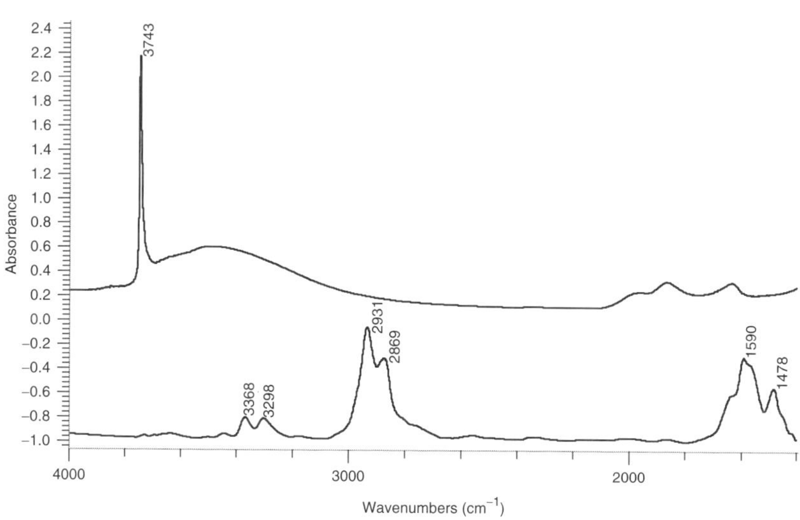

Figure 6.7. FTIR spectra of (A) MCM-48 (showing free and H-bonded hydroxyls on surface) and (B) amine-grafted MCM-48.

characteristic vibrations of NH_2 and CH_2 groups. The pair of IR bands at 3368 and 3298 cm^{-1} are due to the asymmetric and symmetric NH_2 stretchings. The band at 1590 cm^{-1} is attributed to NH_2 scissoring vibration. The bands at 2931 and 2869 cm^{-1} are due to CH stretching bands in the group $CH_2CH_2CH_2$ NH_2. The amine grafted silicas were stable, as no change was seen in the FTIR spectra upon heating to 350 °C.

The amine-grafted silicas adsorb CO_2 and H_2S at fairly high capacities and at high rates. The adsorption isotherms of CO_2 on the amine surface-modified MCM-48 and silica xerogels at 25 °C are plotted in Figure 6.8. The adsorbed CO_2 is completely desorbed at 75 °C, shown in Figure 6.9. H_2S is also adsorbed selectively by the amine-grafted silicas. Figure 6.10 shows the isotherms at room temperature. The adsorbed H_2S can be desorbed completely at 60 °C. The adsorption of methane (as for the sweetening of natural gas) and water vapor also carries practical interest. Only small amounts of methane are adsorbed by these grafted silicas, that is, ~3 cm^3 STP/g was adsorbed at 1 atm and room temperature. The presence of water significantly increases the adsorption of CO_2 (Leal et al., 1995; Huang et al., 2003) yet did not have any significant effect of the adsorption of H_2S. In the presence of water vapor, water is also involved in the sorption. The reactions with and without water are shown in Figure 6.11. In fact, the amount of CO_2 adsorbed in the presence of water was twice the amount without water, as shown by the reaction.

Molecular imprinting of microporous silica is an interesting idea, which could have new applications, particularly for bioseparation. The ease of polymerization of organosilanes (see Reactions 6.4 and 6.5) makes imprinting of microporous, amorphous silica possible. By this route, bulk silicas imprinted with desired

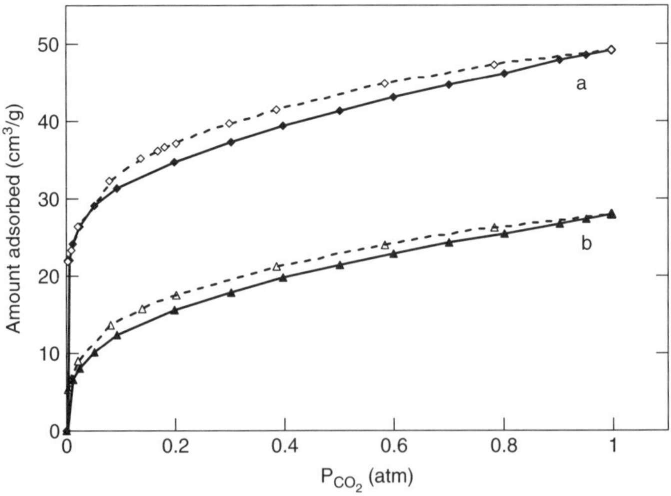

Figure 6.8. Adsorption/desorption isotherms of CO_2 on (A) amine-grafted MCM-48 and (B) amine-grafted silica xerogel at room temperature. Open symbols indicate desorption; solid symbols indicate adsorption.

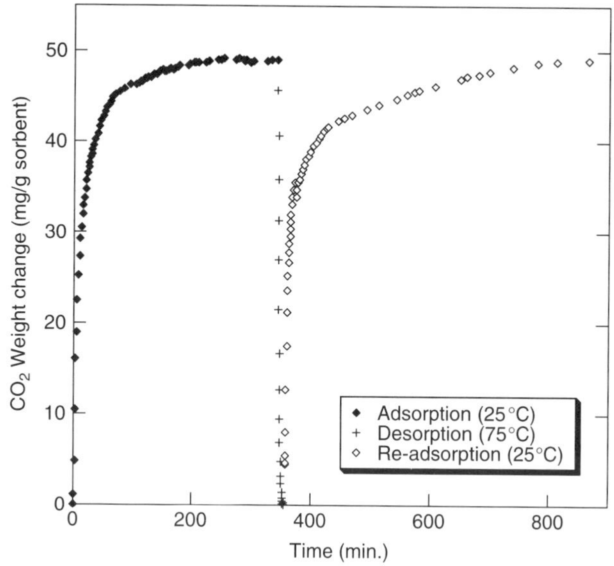

Figure 6.9. Cyclic CO_2 adsorption/desorption on amine-grafted MCM-48.

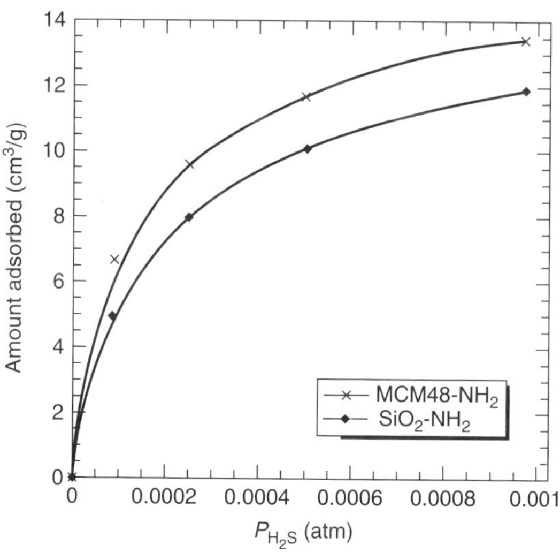

Figure 6.10. Adsorption isotherms of H_2S on amine-modified silicas at room temperature.

organic functional groups and microporosities could be prepared (Katz and Davis, 2000). The synthesis for a silica with three amine groups in each cavity (of the size of 0.8–1.0 nm) is shown in Figure 6.12. TEOS was used for polymerization in the sol-gel process. The imprint molecules contained one, two (in the para positions), and three (in the 1, 3, and 5 positions, as shown in the figure)

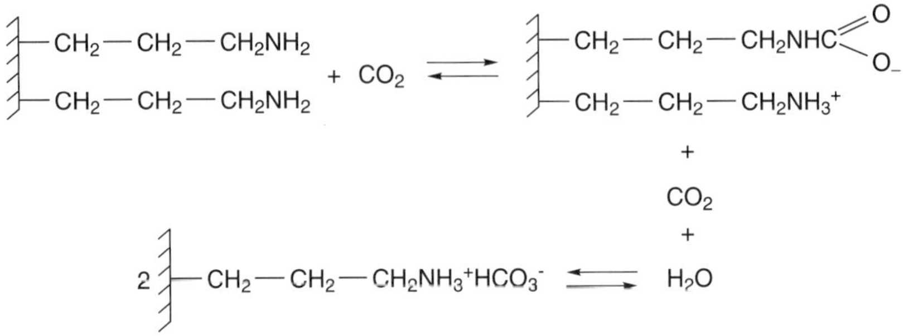

Figure 6.11. Adsorption of CO_2 on amine grafted silica with and without water vapor, showing that the amount adsorbed is doubled in the presence of water (Huang et al., 2003).

aminosilane groups. By this route, Katz and Davis could synthesize amorphous silicas with binding sites comprising one, two, or three propylamine groups that are covalently bonded to the silicon atom of the silica.

This covalent bonding is an advantage over imprinting with bulk polymers because it results in heterogeneity due to noncovalent bonding. The pore-size distribution of the imprinted silica was measured by adsorption of argon at 77 K. The pore volume indicated by the crosses in Figure 6.13 were those of the polymerized TEOS, with the imprint molecules remaining inside the structure. This was the pore volume of the micropores that was filled by the solvent used in the sol-gel process. Upon removal of the imprint, additional porosity was created, indicated by circles in the figure. This pore volume was confined to a narrow range of micropore sizes, between 0.8 and 1.0 nm. The differences in isotherms of Ar at 77 K are also shown in Figure 6.13 (inset).

This route can potentially lead to design and synthesis of silicas, not only with desired pore sizes, but, more importantly, imprinted with features modeled on those of enzymes and antibodies. Hence it is possible to mimic some of the essential features of binding and catalysis found in biological systems.

6.6. ACTIVATED ALUMINA

Commercial production of activated alumina is performed exclusively by thermal dehydration or activation of aluminum trihydrate, $Al(OH)_3$, or gibbsite (MacZura et al., 1977). The oldest form, which is still widely used, is made from Bayer α-trihydrate, which is a by-product of the Bayer process for aqueous caustic extraction of alumina from bauxite. The trihydrate, in the form of gibbsite, is heated or activated in air to $\sim$400 °C to form crystalline γ/η-alumina with a minor amount of boehmite and has a surface area of $\sim$250 m^2/g. Alternatively, the trihydrate is heated very rapidly at 400–800 °C to form an amorphous alumina with a higher surface area, 300–350 m^2/g. The major impurity in these products, besides water (typically 6%), is Na_2O at nearly 1%. The micropore volume is shown in Figure 6.1, but there is a considerable number of pores with sizes

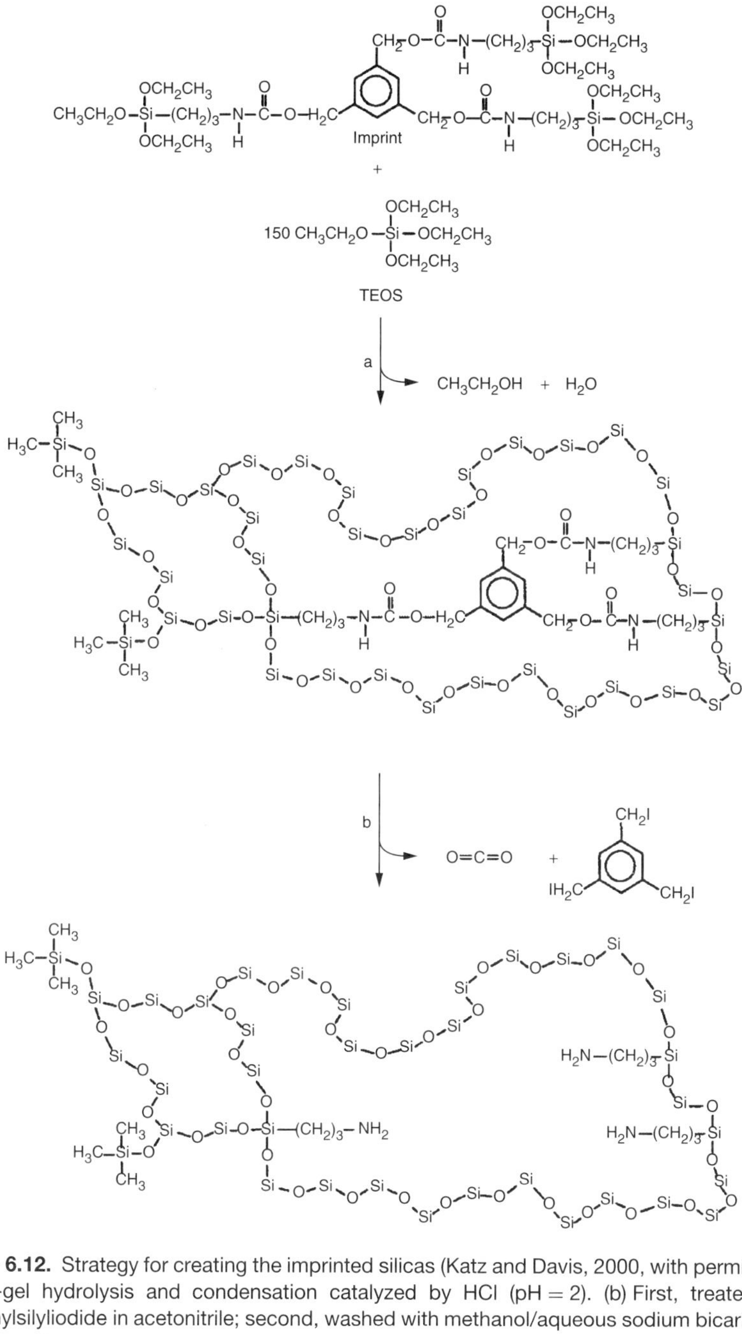

Figure 6.12. Strategy for creating the imprinted silicas (Katz and Davis, 2000, with permission). (a) Sol-gel hydrolysis and condensation catalyzed by HCl (pH = 2). (b) First, treated with trimethylsilyliodide in acetonitrile; second, washed with methanol/aqueous sodium bicarbonate to remove the core.

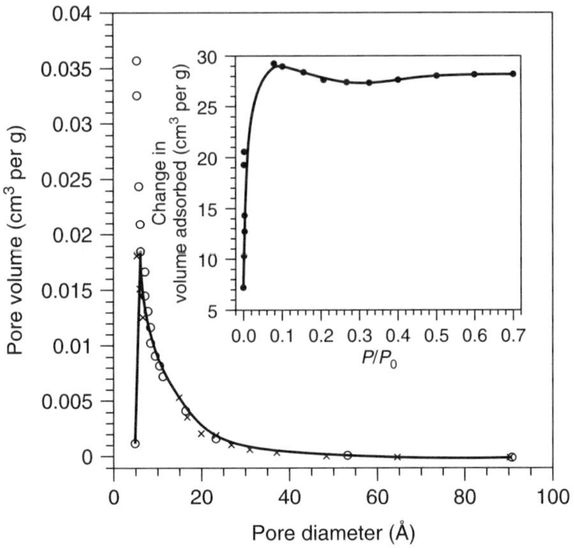

Figure 6.13. Crosses indicate pore-size distribution of the as-synthesized material after step a, and circles indicate that after imprint removal (step b). Inset: Differences of isotherm between these two materials with Ar at 77 K (Katz and Davis, 2000, with permission).

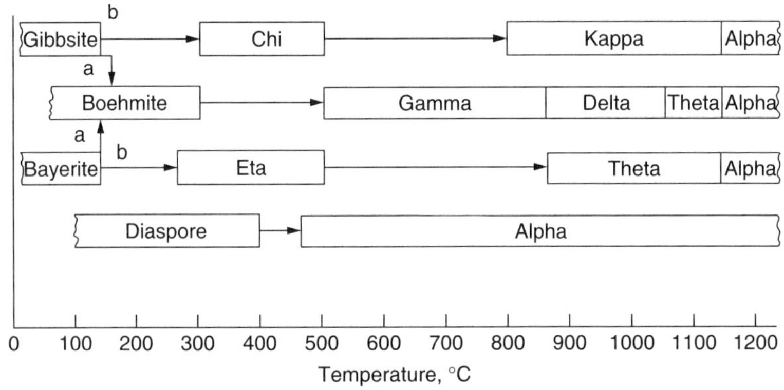

Figure 6.14. Decomposition of aluminum hydroxide and crystal transformation of alumina (from Wefers and Bell, 1972). Path a: >1 atm, moist air, heating rate >1 °C/min, >100 μm size. Path b: 1 atm dry air, <1 °C/min, <10 μm.

greater than 50 Å. A highly impure form of activated alumina is made by the thermal activation of bauxite, which contains alumina in the form of gibbsite.

The crystalline phase transformation of alumina is shown in Figure 6.14. Transformation is also dependent on the gaseous atmosphere, as well as the other conditions, as detailed in Figure 6.14. In this figure, gibbsite and bayerite are trihydroxide, $Al(OH)_3$, or $Al_2O_3{\cdot}3H_2O$. Boehmite and diaspore are AlOOH.

These natural forms of hydroxides have well-defined crystal structures (Wefers and Bell, 1972). Upon heat-treatment, they transform into different forms of alumina, shown by Figure 6.14. Among the aluminas, the γ-alumina is the most commonly used form for both adsorption and catalysis.

The pore structure of the activated alumina depends strongly on the conditions of the heat-treatment. This can be best illustrated by the results of Rouquerol et al. (1999), using controlled heating conditions and well-defined gibbsite as the starting material. Their results are summarized in Figure 6.15. The processes involved in the heat-treatment are obviously very complex. Dehydration, crystal transformation, gas evolution, and sintering are all involved. As shown in Figure 6.14, the moisture content in the atmosphere is also important.

The crystal structures and the surface chemistry of various aluminas have been discussed by Gates et al. (1979). Surface acidity is a most important property for both adsorption and catalysis. Unlike silicas, Lewis acid sites (i.e., sites that can

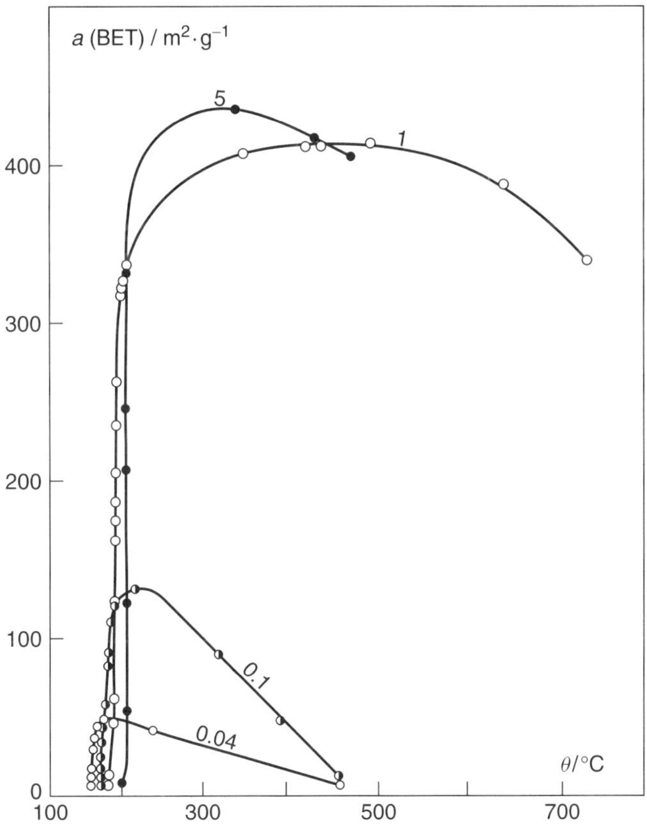

Figure 6.15. Development of surface area (BET-N_2) during dehydration and transformation of a fine gibbsite, 1 μm grain size, at a controlled rate of dehydration of 11.4 mg/h/g, and under pressures in mbar marked on the curves. (Rouquerol et al., 1999, with permission).

accept electrons) are usually abundant on aluminas. These are the Al^{3+} sites on the surfaces. There are both tetrahedral and octahedral Al^{3+} sites. Fully hydrated surfaces of aluminas also have Brønsted acid sites (i.e., $-OH$ groups that can donate protons). The high-temperature forms of alumina, for example, θ-alumina and δ-alumina, contain only Lewis acid sites. On γ-alumina and η-alumina, however, both types of acidity may exist, depending on the degree of hydration. Peri has shown models of surface of alumina as a function of its dehydration level (Peri, 1976).

The acidity of the surface can be increased by treatment with acid, such as HCl and HF. Partial conversion of a fully hydrated alumina surface by the acid treatment would form a surface that contains both Cl^- and OH^- groups (Gates et al., 1979):

$$\begin{array}{lccccccccccc}
 & OH^- & & OH^- & & OH^- & & OH^- & & OH^- & & OH^- \\
 & & Al^{3+} & & Al^{3+} & & Al^{3+} & & Al^{3+} & & Al^{3+} & \\
\Rightarrow & Cl^- & & OH^- & & Cl^- & & OH^- & & OH^- & & Cl^- \\
 & & Al^{3+} & & Al^{3+} & & Al^{3+} & & Al^{3+} & & Al^{3+} &
\end{array}$$

From the above discussion, it can be seen that both pore structure and surface chemistry of the activated alumina can be manipulated and controlled. As a result, activated alumina is a very versatile sorbent and can be tailored for specific applications.

6.7. ACTIVATED ALUMINA AS SPECIAL SORBENTS

Desiccation remains to be a major application for activated alumina. Like silica gel, activated alumina have higher ultimate capacities for water vapor than zeolites do. When high removals of water vapor (to very low dew points) are required, zeolites are used because their water capacities at very low concentrations are higher, as shown in Figure 6.16.

Activated alumina is a versatile sorbent that can be tailored for many special applications. New applications continue to be developed, mainly by the aluminum companies. Little is disclosed on the details of their modifications. However, the modifications follow simple general principles of surface chemistry, such as acid–base chemistry. Two methods are used for tailoring: (1) variation of the activation process, and (2) use of dopants. The following are proven applications of various tailored aluminas:

- Removal of HCl and HF from gases and liquids
- Removal of acidic gases (COS, CO_2, H_2S, CS_2) from hydrocarbons
- Removal of oxygenates and Lewis Bases
- Removal of polar organic compounds
- Removal of As^{5+}, $PO_4{}^{3-}$, Cl^-, and F^- from water

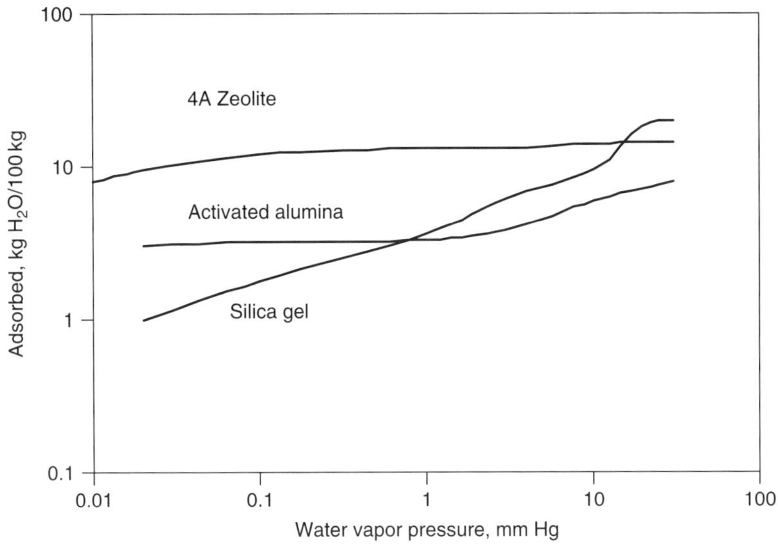

Figure 6.16. Comparison of water adsorption on various sorbents (Humphrey and Keller, 1999, with permission).

- Scavenger for organic process liquids
- Alkalized alumina for SO_2 removal

Acid Gas Removal. The aluminas that are used for acid gas removal obviously contain relatively high "soda" contents. Sodium is always contained in aluminas because Na(OH) is used in the Bayer process for extraction of alumina from bauxite. The typical content of Na_2O in activated aluminas is 0.3% (wt). The soda content can vary considerably. The soda content in the alkalized alumina is typically >1%. The BET (Brunauer-Emmett-Teller) surface areas of activated aluminas are generally in the range 200–500 m^2/g, with total pore volumes near 0.5 cc/g.

Oxygenates Removal. The alumina that is tailored for the removal of oxygenates (e.g., Alcoa Selexsorb CDO-200) has relatively low soda (0.3% Na_2O), high Lewis acidity (i.e., high-surface O vacancies), and very low Brønsted acidity. The surface areas are also low (200 m^2/g), apparently as a result of the heat-treatment. Such an alumina is selective toward Lewis bases, such as alcohols, aldehydes, peroxides, ketones, other carboxylic acids, ethers, ammonia, nitriles, and HCN. Its low Brønsted acidity minimizes unwanted side reactions caused by proton transfer upon adsorption of oxygenates. This alumina has proven to be particularly effective in the removal of oxygenated hydrocarbons from co-monomer streams (hexene-1 and octene-1) in polyethylene production processes (Alcoa, 2001).

Water Treatment. Water treatment is a useful application for alumina. Many studies have been performed for the removal of arsenic and fluoride from water. Alumina is particularly effective for adsorption of As^{5+}, which is in the form of anionic $H_2AsO_4^-$ in the aqueous solution. In aqueous media, the zero point charge (ZPC) is an important, measurable property of activated alumina (or of any sorbent). It is the pH value that is required to give zero net surface charge. The ZPC can be measured by acid–base titration (Berube and DeBruyn, 1968; Tewari and Campbell, 1976; James and Parks, 1982; Noh and Schwarz, 1990). At pH below the ZPC, the alumina surface is positively charged; whereas at pH above the ZPC, the surface is negatively charged. Therefore, at pH below ZPC, activated aluminas will adsorb anions; whereas at pH above ZPC, cations will be adsorbed. The typical values of ZPC for the commercial activated aluminas are in the range of pH = 8–10, depending on the grades. The selectivity sequence for a number of competing ions for activated alumina is (Clifford, 1999; Vagliasindi et al., 1996):

$$OH^- > H_2AsO_4{}^- > Si(OH)_3O^- > F^- > HSeO_3{}^- > TOC > SO_4{}^{2-} > H_3AsO_3$$

As^{5+} exists in water as H_2AsO_4. As^{3+} is hardly adsorbed because H_3AsO_3 is not charged. A number of plant treatment results are available (e.g., Wang et al., 2000), all indicating that with raw water containing arsenic in the 50–70 μg/L range, the use of fixed beds of activated alumina could produce treated water with less than 5 μg/L As. Similarly, numerous studies have been performed for the defluoridation of water (e.g., Singh and Clifford, 1981; Karthikeyan et al., 1994). In 12 test runs of defluoridation of drinking water, fluoride contents <1 mg/L were achieved from a raw water with 3 mg/L by using adsorption with activated alumina (Karthikeyan et al., 1994).

The adsorption isotherm of arsenic (as As(V)) on activated alumina is given in Figure 6.17 (Wang et al., 2000; Rosenblum and Clifford, 1984). Beside the oxidation state of As, pH and competing ions significantly affect the adsorption of As. As mentioned, the best pH is below 8.2. Below this pH, the surface of alumina has a net positive charge that can be balanced by adsorbing anions, such as $H_2AsO_4{}^-$ (for As^{5+}), and all other anions listed above. Several studies have shown the optimum range as 5.5–6.0 (Singh and Clifford, 1981; Rosenblum and Clifford, 1984; Clifford, 1999). The isotherm shown in Figure 4.16 can be fitted by the Freundlich model:

$$q = 0.1679C_e^{0.452}$$

where C_e is the concentration in micrograms per ligand and q is the amount adsorbed in mg/g, both for As(V). This isotherm is in agreement with other studies (Clifford, 1999).

Alkalized Alumina. Activated alumina can be alkalized by impregnation with alkali oxides, such as K_2O, Na_2O, or NH_4OH. Alkalized alumina is a commercial sorbent for flue gas desulfurization, developed since the early 1970s (e.g., Emerson et al., 1971). Alkalized alumina also has other interesting applications; two will be discussed here: CO_2 removal and NO_x removal.

The alkalized aluminas are prepared by impregnation with solutions of alkali salts, for example, alkali bicarbonates. The salts are typically decomposed by heat-treatment to 500 °C. The alkali oxide contents are typically 5% (wt), but a range of 1–10% has been reported. The surface areas are reduced

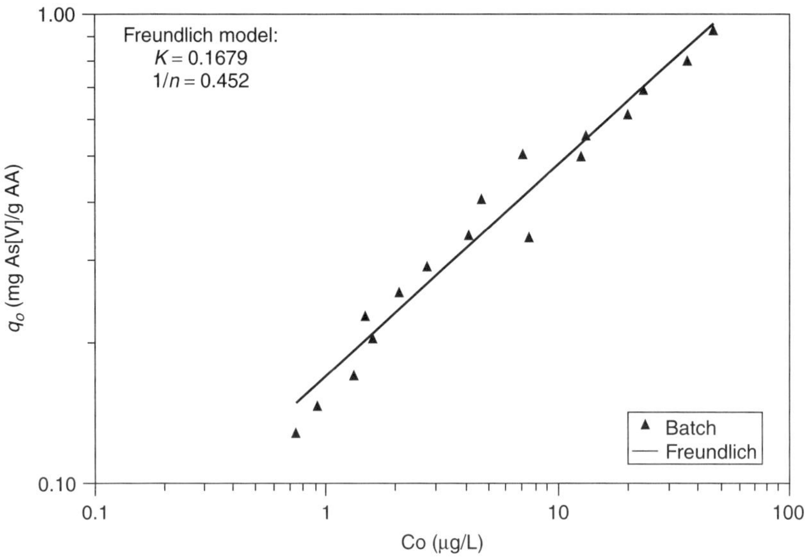

Figure 6.17. Adsorption isotherm of arsenic on activated alumina at pH 7.7. The adsorbed amount is in mg As^{5+} per *g* activated alumina, and the abscissa is concentration in μg/liter (Wang et al., 2000).

by impregnation, likely by a half, while the pore-size distributions are not significantly affected because only the micropores are blocked (Lee et al., 1999).

A very low tolerance for the CO_2 concentration is allowed in some applications. For example, in gaseous olefin containing streams, removal down to 1 ppm CO_2 is needed because CO_2 is a known catalyst inhibitor for olefin polymerization (Hogan et al., 1985). In order to avoid freezing, CO_2 also needs to be removed to the parts-per-million level from air prior to cryogenic air separation. Zeolites adsorb CO_2 strongly but are not useful for CO_2 removal, as they also strongly adsorb other components. Activated alumina, on the other hand, does not adsorb CO_2 strongly enough for these applications. It has been shown that the adsorption of CO_2 can be enhanced significantly when the alumina is alkalized (Slaugh et al., 1984; Hogan et al., 1985; Golden et al., 1997). Henry's constants of CO_2 on alkalized aluminas at room temperature are given below (Golden et al., 1997):

Adsorbent	Henry's Constants, mmol/g/atm
Act. Alumina	5.6
5% K_2CO_3	22.8
5% NaOH	15.1
5% $NaOH/NH_4HCO_3$	27.3
5% $(NH_4)_2CO_3$	5.8

The alkalized aluminas shown above were not fully calcined, as they were heated at 120 °C (Golden et al., 1997). After being fully calcined, Henry's constants obviously would have increased further.

Adsorption of NO_x on γ-alumina, as well as that alkalized with alkali carbonates and hydroxides, has been studied by Lee et al. (1998; 1999). It is known that NO does not adsorb effectively on activated alumina, whereas only NO_2 adsorbs strongly. The mechanism of adsorption of NO_2 on alumina, was studied by Nelli and Rochelle (1996). Their study suggested that the adsorption on alumina (and other oxide sorbents) follows the general mechanism:

$$H_2O + 3NO_2 \Rightarrow 2HNO_3 + NO$$

The reaction with the sorbed water of three NO_2 results in the formation of one NO. This mechanism was also verified by Lee et al. (1999). Lee et al. (1999) showed that the same mechanism was also followed for adsorption of NO_2 on alkalized aluminas and, more importantly, the alkalis significantly increased the adsorption of NO_2. As expected, the order of the increase followed: Cs > Rb > K > Na > Li.

REFERENCES

Alcoa Industrial Chemicals, MSDS Number 595 (2001).

Anderson, J. H. (1965) *Surf. Sci.* 3, 290.

Angeletti, E., Canepa, C., Martinetti, G., and Venturello, P. (1988) *Tetrahedron Letters* 29, 2261.

Antochsuk, V. and Jaroniec, M. (2000) *Chem. Mater*. 12, 6271.

Beck, J. S., Vartuli, J. C., Roth, W. J., Leonowicz, M. E., Kresge, C. T., Schmitt, K. D., Chu, C. T.-W., Olson, D. H., Sheppard, E. W., McCullen, S. B., Higgins, J. B., and Schlenker, J. L. (1992) *J. Am. Chem. Soc*. 114, 10834.

Belter, P. A., Cussler, E. L., and Hu, W.-S. (1988) *Bioseparations*. John Wiley and Sons, New York, NY.

Bergna, H. E. (1994) *The Colloid Chemistry of Silica*. American Chemical Society, Washington, DC, Chapter 8.

Berube, Y. G. and DeBruyn, P. L. (1968) *J. Coll. Interf. Sci*. 27, 305.

Brinker, C. J. and Sherer, G. W. (1990) *Sol-Gel Science*. Academic Press, New York, NY.

Burwell, R. L. Jr. and Leal, O., J.C.S. (1974) *Chem. Comm*. 342.

Chao, Z. S. and Ruckenstein, E. (2002a) *Langmuir* 18, 734.

Chao, Z. S. and Ruckenstein, E. (2002b) *Langmuir* 18, 8535.

Clifford, D. (1999) Ion exchange and inorganic adsorption. In *Water Quality and Treatment: A Handbook of Community Water Supplies*, 5th Edn., American Water Works Association (ed). McGraw-Hill, New York, NY.

De Boer, J. H. (1958) *Angew. Chem.* 70(13), 383.

De Boer, J. H. and Vleeskenns, J. M. (1958) *Proc. K. Ned. Akad. Wet. Ser. B* 61, 2.

Emerson, R. B. and Brian, W. A. Method or producing alkalized alumina and products produced thereby. U.S. Patent 3,557,025 (1971).

Feng, X., Fryxell, G. E., Wang, L.-Q., Kim, A. Y., Liu, J., and Kemmer, K. M. (1997) *Science* 276, 923.

Garcia, A. A., Bonen, M. R., Ramirez-Vick, J., Sadaka, M, and Vuppu, A. (1999) *Bioseparation Process Science*. Blackwell Science Inc., Malden, MA.

Gates, B. C., Katzer, J. R., and Schuit, G. C. A. (1979) *Chemistry of Catalytic Processes*. McGraw-Hill, New York, NY.

Golden, T. C., Taylor, F. W., Wang, A. W., and Kalbassi, M. A. Base-treated alumina in pressure swing adsorption. U.S. Patent 5,656,064 (1997).

Hair, M. L. (1967) *Infrared Spectroscopy in Surface Chemistry*. Dekker, New York, NY.

Hogan, J. P. and Rease, C. R. Removal of carbon dioxide from olefin containing streams. U.S. Patent 4,493,715 (1985).

Huang, H. Y., Yang, R. T., Chinn, D., and Munson, C. L. (2003) *Ind. Eng. Chem. Res.* in press.

Huo, Q., Mragolese, D. L, and Stucky, G. D. (1996) *Chem. Mater.* 8, 1147.

Iler, R. K. (1979) *The Chemistry of Silica*. John Wiley and Sons, New York, NY.

Izumi, J. Mitsubishi *VOC Recovery Process*. Mitsubishi Heavy Industries, Ltd. (1996) Cited from Ying et al. (1999).

James, R. O. and Parks, G. A. (1982) *Surface Colloid Sci.* 12, 119.

Jaroniec, C. P., Kruk, M., Jaroniec, M., and Sayari, A. (1998) *J. Phys. Chem.* R 102, 5503.

Jones, R. W. (1989) *Fundamental Principles of Sol-Gel Technology*. The Institute of Metals, London, UK.

Kaman, N. K., Anderson, M. T., and Brinker, C. J. (1996) *Chem. Mater.* 8, 1682.

Karthikeyan, G., Meenakshi, S., and Apparao, B. V. (1994) 20th Water, Engineering & Development Center (WEDC), Loughborough University, UK, p. 278.

Katz, A. and Davis, M. E. (2000) *Nature*, 403, 286.

Kiselev, A. V. (1986) *Intermolecular Interactions in Adsorption and Chromatography*. Vyssh, Shkola, Moscow.

Kiselev, A. V. and Lygin, V. I. (1975) *Infrared Spectra of Surface Compounds* (translated by N. and Kaner). John Wiley and Sons, New York, NY.

Klier, K. and Zettlemoyer, A. C. (1977) *J. Coll. Interf. Sci*. 58, 216.

Lasperas, M., Llorett, T., Chaves, L., Rodriguez, I., Cauvel, A., and Brunel, D. (1997) *Heterogeneous Catalysis and Fine Chemicals*, Vol. IV. Elsevier Science, Amsterdam.

Leal, O., Bolivar, C., Ovalles, C., Garcia, J. J., and Espidel, Y. (1995) *Inorganica Chimica Acta* 240, 183.

Lee, M. R., Allen, E. R., Wolan, T. J., and Hoflund, G. B. (1998) *Ind. Eng. Chem. Res.* 37, 3375.

Lee, M. R., Wolan, T. J., and Hoflund, G. B. (1999) *Ind. Eng. Chem. Res*. 38, 3911.

MacZura, G., Goodboy, K. P., and Koenig, J. J. (1977) In *Kirk-Othmer Encyclopedia of Chemical Technology*, Vol. 2, 3rd Edn. Wiley-Interscience, New York, NY.

Moller, K. and Bein, T. (1998) *Chem. Mater*. 10, 2950.

Nelli, C. H. and Rochelle, G. T. (1996) *Ind. Eng. Chem. Res*. 35, 999.

Noh, J. S. and Schwarz, J. A. (1990) *Carbon* 28, 675.

Peri, J. B. (1976) *J. Catal*. 41, 227.

Ravikovitch, P. I., Haller, G. L., and Neimark, A. V. (1998) *Adv. Colloid Interface Sci*. 203, 76.

Rosenblum, E. and Clifford, D. (1984) The Equilibrium Arsenic Capacity Of Activated Alumina. EPA-600/52-83-107. U.S. EPA, Cincinnati, OH.

Rouquerol, F., Rouquerol, J., and Sing, K. (1999) *Adsorption by Powders and Porous Solids*. Academic Press, San Diego, CA.

Ruckenstein, E. and Chao, Z. S. (2001) *Nano Lett*. 12, 739.

Sayari, A., Liu, P, Kruk, M., and Jaroniec, M. (1997) *Chem. Mater*. 9, 2499.

Schumacher, K., Ravikovitch, P. I., Chesne, A. D., Neimark, A. V., and Unger, K. K. (2000) *Langmuir* 16, 4648.

Singh, G. and Clifford, D. (1981) *The Equilibrium Fluoride Capacity of Activated Alumina* EPA-600/52-81-082, U.S. EPA, Cincinnati, OH.

Sun, T. and Ying, J. Y. (1997) *Nature* 389, 704.

Tanev, P. T. and Pinnavaia, T. J. (1995) *Science* 267, 865.

Teraoka, Y., Fukunaga, Y., Setaguchi, Y. M., Moriguchi, L, Kagawa, S., Tomonago, N., Yasutake, A., and Isumi, J. (2000) In *Adsorption Science and Technology*, (D. D. Do, ed.). World Scientific Publishers, Singapore, p. 603.

Tewari, P. H. and Campbell, A. B. (1976) *J. Coll. Interf. Sci*. 55, 531.

Unger, K. K. (1979) *Porous Silica, Its Properties and Use as a Support in Column Liquid Chromatography*. Elsevier, Amsterdam, The Netherlands.

Unger, K. K., Kumar, D., Schumacher, K., Du Fresne, C., and Grun, M. 2001 Paper 3M-03, Seventh International Conference on Fundamentals of Adsorption, May 20–25, Nagasaki, Japan, 2001.

Vagliasindi, F. G. A., Henley, M., Schulz., N., and Benjamin, M. M. (1996) *Proc. Water Quality Technol. Conf.*, p. 1829, American Water Works Association, Denver, CO.

Vansant, E. F., Van Der Voort, P., and Vrancken, K. C. (1995) *Characterization and Chemical Modification of The Silica Surface*. Elsevier, Amsterdam, The Netherlands.

Wang, L., Chen, A., and Fields, K. (2000) Arsenic Removal from Drinking Water by Ion Exchange and Activated Alumina. EPA/600/R-00/088, U.S. EPA, Cincinnati, OH.

Wefers, K. and Bell, G. M. (1972) Oxides and Hydroxides of Aluminum, Technical Paper No. 19, Alcoa Research Laboratories, Aluminum Company of America, Pittsburgh.

Yang, R. T. (1997) *Gas Separation by Adsorption Processes*. Butterworth, 1987; Boston, reprinted by Imperial College Press, London, UK.

Ying, J. Y., Mehnert, C. P., and Wong, M. S. (1999) *Angew. Chem. Int. Ed*. 38, 56.

Zhao, D., Feng, J., Huo, Q., Melosh, N., Frederickson, G. H., Chmelka, B. F., and Stucky, G. D. (1998) *Science* 279, 548.

Zhao, X. S., Lu, G. Q., and Hu, X. (1999) *Chem. Comm*. 1391.

Zhao, X. S., Lu, G. Q. and Millar, G. J. (1996) *Ind. Eng. Chem. Res.* 35, 2075.

Zhuravlev, L. T. (1993) *Colloids and Surfaces A*. 74, 71.

Zhuravlev, L. T. (1987) *Langmuir* 3, 316.

7

ZEOLITES AND MOLECULAR SIEVES

Zeolites are crystalline aluminosilicates of alkali or alkali earth elements, such as sodium, potassium, and calcium, and are represented by the chemical composition:

$$M_{x/n}[(\mathrm{AlO_2})_x(\mathrm{SiO_2})_y] \cdot z\mathrm{H_2O}$$

where x and y are integers with y/x equal to or greater than 1, n is the valence of cation M, and z is the number of water molecules in each *unit cell*. The primary structural units of zeolites are the tetrahedra of silicon and aluminum, SiO_4 and AlO_4. These units are assembled into secondary polyhedral building units such as cubes, hexagonal prisms, octahedra, and truncated octahedra. The silicon and aluminum atoms, located at the corners of the polyhedra, are joined by a shared oxygen. The final zeolite structure consists of assemblages of the secondary units in a regular three-dimensional crystalline framework. The tetrahedra can be arranged in numerous ways, resulting in the possibility of some 800 crystalline structures, less than 200 of which have been found in natural deposits or synthesized in laboratories around the world (Thompson, 1998).

Substitution (i.e., isomorphous substitution) of other elements for Al and/or Si in the zeolite framework can yield myriad *molecular sieves* (which are formally not zeolites). However, the main interest for synthesizing these new molecular sieve materials has been in catalysis for developing (1) large pores or channels and (2) catalytic sites other than acid sites. The largest windows in zeolites are the 12-membered oxygen rings (with an unobstructed diameter of 8.1 Å), which do not admit large hydrocarbon molecules that are of interest to the petrochemical industry. Since the development of VPI-5 (Davis et al., 1988), a crystalline aluminophosphate with unidimensional channels formed by 18-member oxygen rings (with a free diameter of 12.5 Å), a number of large-pore molecular sieves have been synthesized (Chen et al., 1994). These include $AlPO_4$-8, which contains

Adsorbents: Fundamentals and Applications, Edited By Ralph T. Yang
ISBN 0-471-29741-0

14-oxygen rings (Dessau et al., 1990) and Cloverite, a gallophosphate with a 20-oxygen ring and 8-oxygen ring dual pore system (Eastermann et al., 1991). [AlO_4] provides acid sites (as Lewis acid, or Brønsted acid when OH is bonded) for catalytic reactions. The addition of tetrahedra such as [TiO_4] provides oxidation sites for redox reactions.

Types A, X, and Y remain the dominant zeolites and molecular sieves that are in commercial use for adsorption and ion exchange. As the focus of this book is on sorbents, these zeolites will be the main subject for discussion. The basic principles on adsorption properties discussed below, however, are applicable to all other zeolites and molecular sieves. Potentially interesting adsorption properties of other zeolites and molecular sieves will also be included.

7.1. ZEOLITE TYPES A, X, AND Y

Unit cells of type A and type X zeolites are shown in Figure 7.1. The cations are necessary to balance the electric charge of the aluminum atoms in AlO_2, each having a net charge of −1. The water molecules can be removed with ease upon heating and evacuation, leaving an almost unaltered aluminosilicate skeleton with a void fraction between 0.2 and 0.5. The skeleton has a regular structure of cages, which are interconnected by windows in each cage. The cages can imbibe or occlude large amounts of guest molecules in place of water. The size of the window apertures, which can be controlled by fixing the type and number of cations, ranges from 3 to 8 Å. The sorption may occur with great selectivity because of the size of the aperture (and to a lesser extent due to the surface property in the cages)—hence the name molecular sieve. The windows of type A zeolite consist of 8-membered oxygen rings, or simply, 8-rings. Similarly, the windows of type X zeolite are referred to as 12-ring, which remain the largest windows in zeolites today.

The ratio of Si/Al in Type A zeolite is normally one, while those in types X and Y are typically one to five. The aluminum atom can be removed and replaced by silicon in some zeolites, thereby reducing the number of cations. The cations can also be exchanged. The inner atoms in the windows are oxygen. The size of the windows then depend on the number of oxygen atoms in the ring (4, 5, 6, 8, 10, or 12). The aperture size, as well as the adsorption properties, can be modified further by the number and type of exchanged cations. A description of the structures will be given for the zeolites, Type A and Types X and Y, important in gas separation. As mentioned, these types have dominated the commercial use of zeolites for gas separation and purification as well as ion exchange since their invention.

7.1.1. Structure and Cation Sites of Type A Zeolite

The structural unit in Type A zeolite (Linde Type A or LTA), as well as in Types X and Y (faujasite or FAU), is the truncated octahedron shown in Figure 7.1(a). This unit is also called sodalite cage or beta cage, as sodalite is formed by directly

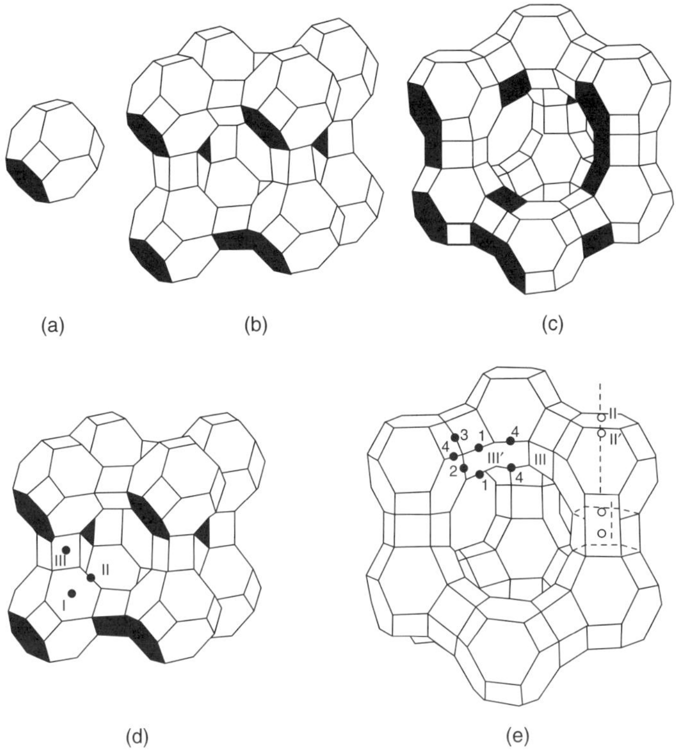

Figure 7.1. Line representations of zeolite structure: (a) sodalite cage, or beta cage or truncated octahedron; (b) type A zeolite "unit cell"; (c) "unit cell" of types X and Y, or faujasite; (d) cation sites in type A (there are eight I, three II, and twelve III sites per unit cell); (e) cation sites in types X and Y (16 I, 32 I', 32 II, 32 II', 48 III, and 32 III' sites per unit cell).

fusing the four-member rings of the units. The four-member rings of the sodalite units can also be linked through four-member prisms, as shown in Figure 7.1(b), which is Type A zeolite. The unit cell of Type A zeolite, as shown in this figure, contains 24 tetrahedra, 12 $A1O_4$ and 12 SiO_4. When fully hydrated, 27 water molecules are contained in the central cage or cavity (also called supercage) of the unit cell, and in the eight smaller sodalite cages. The free diameter in the central cavity is 11.4 Å, which is entered through six 8-member oxygen-ring apertures with an unobstructed diameter of 4.4 Å. There are 12 negative charges that are balanced by cations in each unit cell. The most probable locations for the cations are indicated in Figure 7.1(d). Type I is at the center of the six-member ring (with a free diameter of 2.8 Å, which is approximately the dimension of water) and thus sits at one of the eight corners of the cavity. Type II is at the eight-member aperture directly obstructing the entrance. Type III is near the four-member ring inside the cavity.

Type A zeolites are synthesized in the sodium form, with 12 sodium cations occupying all eight sites in I and three sites in II, plus one site in III. This is the commercial Type 4A zeolite, with an effective aperture size of 3.8 Å. The sodium form can be replaced by various other cations or by a hydrogen ion. The commercial Type 3A zeolite is formed by exchanging Na^+ with K^+, resulting in a smaller effective aperture size (of 3.0 Å) due to the larger K^+. The aperture size of the sodium form can also be increased by exchanging Na^+ with Ca^{+2} or Mg^{+2}, because 2 Na^+ are replaced by one divalent cation. The form of the exchanged Ca^{+2} or Mg^{+2} is Type 5A with rather unobstructed and larger apertures. The unobstructed apertures of 5A have a size of 4.3 Å (Breck, 1974).

A detailed discussion on the sites of important cations in zeolites A, X, Y, chabazite (cage-type with 8-oxygen ring window) and heulandite (channel-type) is given in 7.4.1.

7.1.2. Structure and Cation Sites of Types X And Y Zeolites

The skeletal structure of Types X and Y zeolites is the same as naturally occurring faujasite. The sodalite units are linked through 6-member prisms, as shown in the unit cell in Figure 7.1(c). Each unit cell contains 192 (Si, A1)O_4 tetrahedra. The number of aluminum ions per unit cell varies from 96 to 77 (i.e., Si/Al = 1 to 1.5) for Type X zeolite, and from 76 to 48 (Si/Al = 1.5 to 3) for Type Y zeolite. Loewenstein's rule forbids the formation of Al-O-Al bridges (Loewenstein, 1954). Thus, the maximum number of Al corresponds to a Si/Al ratio of 1. Kuhl (1987) reported a procedure for the synthesis of the low-silica X (LSX) zeolite (with Si/Al = 1). At Si/Al > 3, they are named USY (i.e., ultra-stable Y). The Si/Al ratio in the commercial USY zeolite can be very high, for example, 195. The framework of faujasite has the largest central cavity pore volume of any known zeolite, amounting to about 50% void fraction in the dehydrated form. The free diameter of the central cavity is 13.7 Å (Eulenberger et al., 1967). A unit cell, when fully hydrated, contains approximately 235 water molecules, primarily in the central cavity. The volume of the central cavity, however, accounts for only a small fraction (1/5–1/8) of the pore volume of the unit cell since there are portions of other central cavities from the neighboring unit cells, as well as window spaces that are also contained in the same unit cell (see Table 5.9 of Breck, 1974). The aperture is formed by the 12-member oxygen rings with a free diameter of approximately 7.4 Å. The size of the unobstructed 12-ring is approximately 8.1 Å (Breck, 1974).

Three major locations for the cations are indicated in Figure 7.1(e). The locations are center of the 6-member or hexagonal prism (I) and opposite to I and located in the sodalite cage (I′), similar to I and I′ but further from the central cavity (II and II′), and the 12-ring aperture (III and III′). The commercial 10X zeolite contains Ca^{+2} as the major cation, and Na^+ is the major cation for 13X zeolite. The distribution of Na^+, K^+, Ca^{+2}, other cations, and H_2O among the sites in X and Y zeolites has been discussed in detail by Barrer (1978). Cation sites for important cations in X and Y zeolites will be given in 7.4.1. The BET surface area measured with N_2 for zeolites falls in the range between 500 and 800 m^2/g.

7.1.3. Examples of Molecular Sieving

Separation and purification can be accomplished by molecular sieving, that is, by size exclusion. An example is drying or dehydration of gases or alcohols by 3A zeolite, which excludes all hydrocarbons, O_2, N_2, and essentially all permanent gases except ammonia. It is particularly useful for drying gases under reactive conditions. Isotherms of water on 3A are given in Figure 7.2, which also shows the approximate minimum condition for dehydration of zeolites. Another example is the large-scale commercial use of 5A zeolite in processes for the separation of normal paraffins from branched-chain (e.g., iso-) paraffins and cyclic hydrocarbons, and the Union Carbide isoSiv process is a prime example (Yang, 1987). The free aperture size of 5A zeolite is 4.3 Å, which admits only linear paraffins but not branched-chain paraffins and cyclic hydrocarbons. The branched-chain paraffins have higher octane numbers than their *n*-paraffin homologs as a gasoline product. It is known that temperature has a sizable effect on molecular sieving. The aperture size decreases with temperature. Thus, for some molecules, molecular sieving occurs only below certain temperatures. The temperature effect may be illustrated by the sieving of N_2/O_2 with 4A zeolite (Breck, 1974). The kinetic diameter of N_2 is approximately 0.2 Å larger than O_2. At temperatures below $-100\,^{\circ}C$, N_2 becomes essentially excluded, while Ar becomes excluded at below $-150\,^{\circ}C$. A schematic is given in Figure 7.3 that shows the relationship

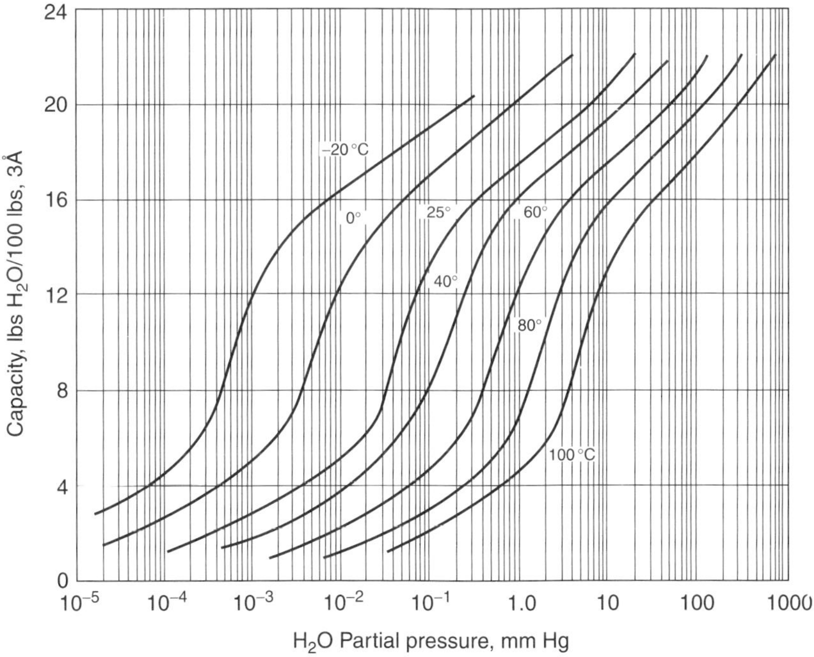

Figure 7.2. Equilibrium isotherms of water on 3A zeolite (taken from undated brochure of Grace Davison Division).

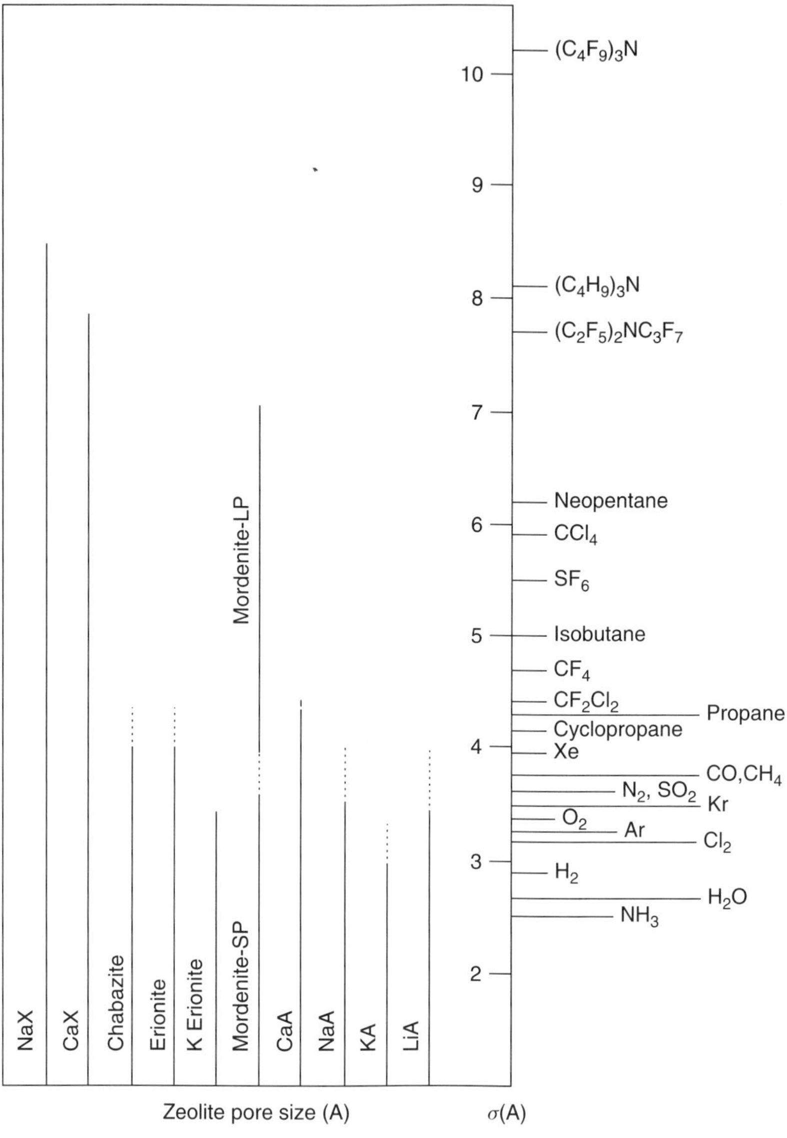

Figure 7.3. Correlation for molecular sieving of molecules (with kinetic diameter σ) in various zeolites with different effective pore sizes at temperatures of 77 K (solid lines) and 420 K (dotted lines) (from Breck, 1974, with permission).

between effective pore sizes of selected zeolites and the kinetic diameters of various molecules.

Separation and purification can also be accomplished by using differences in diffusivities, that is, kinetic separation (discussed in Chapter 3). 4A zeolite has been used in PSA separation for producing nitrogen for inert purge applications such as fuel tank purging in military airplanes, where pressurized air is readily available as the feed. A potentially useful approach for kinetic separation is to use partially ion-exchanged zeolites, that is, zeolites with mixed cations. For example, the diffusivities of CH_4, C_2H_6, and CO_2 in 4A zeolite can be reduced to desired values by partial exchange of Na with K to decrease the aperture size. The equilibrium isotherm and diffusivity of CO_2 in mixed NaKA zeolite are shown in Figure 7.4. From Figure 7.4, it is seen that the molecular sieving effects (for molecules smaller than CO_2) are fully achieved at less than 30% K^+ exchange. Thus, full K^+ exchange (which is costly) is not needed. This result was caused by the preferred occupation of site II by K^+ (Yeh and Yang, 1989). Similarly, the pore size of 4A can be increased by partial exchange of Na^+ with Ca^{2+}. At room temperature, propane, *n*-heptane and isobutane are all excluded from 4A. By exchanging 30–35% of the Na^+ with Ca^{2+}, the aperture size is opened up just enough to admit propane and *n*-heptane, while iso-butane remains excluded (Breck, 1974). Thus, mixed-cation zeolites offer a potential opportunity for tailoring sorbents for kinetic separation. Vansant (1990) discussed a number of approaches for the modification of aperture size as well as the framework structure, including ion exchange. A number of examples for tailoring the aperture size by partial ion exchange have been shown (Vansant, 1990). Silanation is an interesting approach. By silanation with SiH_4, Vansant (1990) showed that the pores in H-mordenite could be tailored from being fully open to Kr at 273 K to being fully closed at about 1.4 mmol/g of chemisorbed silane. Examples of molecular sieving for separations by $AlPO_4$ and $SAPO_4$ will be given in Section 7.2.

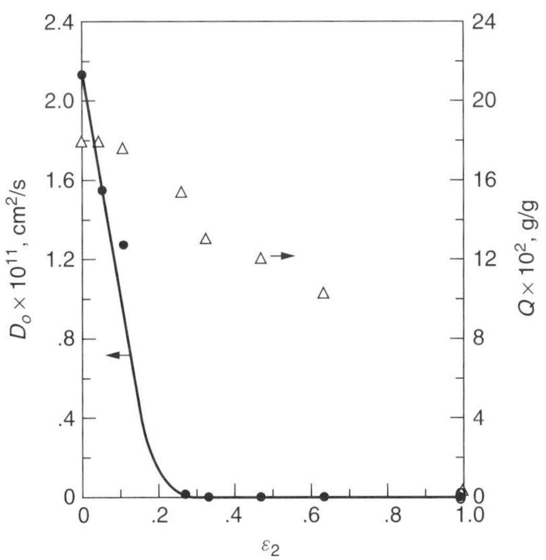

Figure 7.4. Equilibrium adsorption (Q) and diffusivity at Q = 0 (D_0) of CO_2 in NaKA zeolite at 25 °C with fraction of $K^+ = \varepsilon_2 = K^+/[K^+ + Na^+]$ (from Yeh and Yang, 1989, with permission).

7.2. ZEOLITES AND MOLECULAR SIEVES: SYNTHESIS AND MOLECULAR SIEVING PROPERTIES

At least 40 different types of naturally occurring zeolites have been found, beginning with the discovery of stilbite (STI) by the Swedish mineralogist Cronstedt in 1756, who also coined the term "zeolite." The principal natural zeolites are chabazite, $(Ca, Na_2)Al_2Si_4O_{12} \cdot 6H_2O$; gmelinite, $(Na_2, Ca)Al_2Si_4O_{12} \cdot 6H_2O$; mordenite, $(Ca, K_2, Na_2)Al_2Si_{10}O_{24} \cdot 6.66H_2O$; heulandite, $CaAl_2Si_6O_{16} \cdot 5H_2O$; clinoptilolite, $(Na_2, K_2, Ca)Al_2Si_{10}O_{24} \cdot 12H_2O$; levynite, $CaAl_2Si_3O_{10} \cdot 5H_2O$; and faujasite, $(Na_2, Ca, Mg, K_2)Al_2Si_5O_{14} \cdot 10H_2O$. More than 150 types of zeolites have been synthesized and are designated by a letter or group of letters (Type A, Type X, Type Y, Type ZSM, etc.). Standard three-letter structure codes have been adopted by the International Zeolite Association (IZA). (The website for IZA provides helpful information about the codes as well as crystallographic data.) Many more zeolite-like, crystalline molecular sieves have been successfully synthesized by using amine additives as "templating" agents. Early work in zeolite synthesis was performed largely by mineralogists using reaction conditions that were thought likely to have arisen naturally under volcanic circumstances. The commercial production of synthetic zeolites started with the successful development of low-temperature (25–100 °C) synthesis methods by using very reactive materials such as freshly co-precipitated gels or amorphous solids (Breck, 1974; Milton, 1959). Two comprehensive monographs by Barrer (1978) and Breck (1974) deal with all aspects of zeolites, including synthesis. The zeolites that have been synthesized more recently are discussed by Szostak (1998) and others (Chapters in Karge and Weitkamp, 1998; Jacobs and Martens, 1987; Dyer, 1988).

7.2.1. Synthesis of Zeolites A, X, and Y

Many alkali metal hydroxides and raw materials containing silica and alumina can be used in low-temperature synthesis. The steps involving the Na_2O-Al_2O_3-SiO_2-H_2O system, which is used in synthesizing zeolites of types A, X, and Y, are as follows (Breck, 1974):

$$NaOH(aq) + NaAl(OH)_4(aq) + Na_2SiO_3(aq)$$

$$\xrightarrow{T_1 \cong 25\,^{\circ}C}$$

$$[Na_a(AlO_2)_b(SiO_2)_c \cdot NaOH \cdot H_2O]gel$$

$$\xrightarrow{T_2 \cong 25 - 175\,^{\circ}C}$$

$$Na_x[(AlO_2)_x(SiO_2)_y] \cdot mH_2O + solution(zeolite\ crystal)$$

The first step involves gel formation between sodium hydroxide, sodium silicate, and sodium aluminate in aqueous solution at room temperature. The gel is probably formed by the copolymerization of the silicate and aluminate species by a condensation–polymerization mechanism. Expressed in moles per mole of Al_2O_3 for $Na_2O/SiO_2/H_2O$, typical compositions of the reactants are (at pH >10):

1. Type 4A zeolite, 2/2/35
2. Type X zeolite, 3.6/3/144
3. Type Y zeolite, 8/20/320

The gels are crystallized in a closed hydrothermal system at temperatures between 25 and 175 °C. Temperatures as high as 300 °C are used in some cases. The time for crystallization ranges from a few hours to several days. A large amount of work was directed at determining the reaction conditions (e.g., temperature, time, and degree of agitation) and compositional parameters that yield a single-phase, fully crystallized zeolite. Figure 7.5 shows one of the diagrams (Breck, 1974; Kouwenhoven and de Kroes, 1991). Many "reactant composition" diagrams for synthesizing a number of zeolites are given by Breck (1974).

As the synthesis proceeds at elevated temperatures, zeolite crystals are formed by a nucleation step, followed by a crystal growth step involving assimilation of alumino-silicate from the solution. The amorphous gel phase continues to dissolve, thereby replenishing the solution with alumino-silicate species. This process results in the transformation of amorphous gel to crystalline zeolite.

7.2.2. Organic Additives (Templates) in Synthesis of Zeolites and Molecular Sieves

The basic building block for zeolite types A and X is the sodalite cage or beta cage, as shown in Figure 7.1. The beta cages are connected by 4-prisms to form A zeolite and are connected by 6-prisms to form type X zeolite. Beside beta cage, 15 other cages are known (Gellens et al., 1982). All can be used as the basic building blocks for zeolite structures.

The zeolites that are built with cages have the topology of cages and cavities. These cavities are interconnected by windows, as shown in Figure 7.1. The other common topology is the tubular form. There are 10 basic tubular building units, with 4 shown in Figure 7.6. Gellens et al. (1982) showed the 10 tubular units as well as 16-cage building units.

The use of organic amines (mainly quaternary amines) as additives in the synthesis gel to influence the subsequent crystallization has been an exciting element and powerful tool in zeolite synthesis. These additives are often referred to as "templates" or "structure directing agents," although their roles in crystallization are far more complex than templating and are not well understood (Barrer, 1978; Breck, 1974; Szostak, 1998; chapters in Karge and Weitkamp, 1998). They are usually added after the gel solution is prepared, and after synthesis the residual hydrocarbons are removed from the crystals by air burning at 500–600 °C. In addition to the amine templates, HF or KF is frequently used to increase the crystallization rates, apparently by increasing the concentration of free silicate ions.

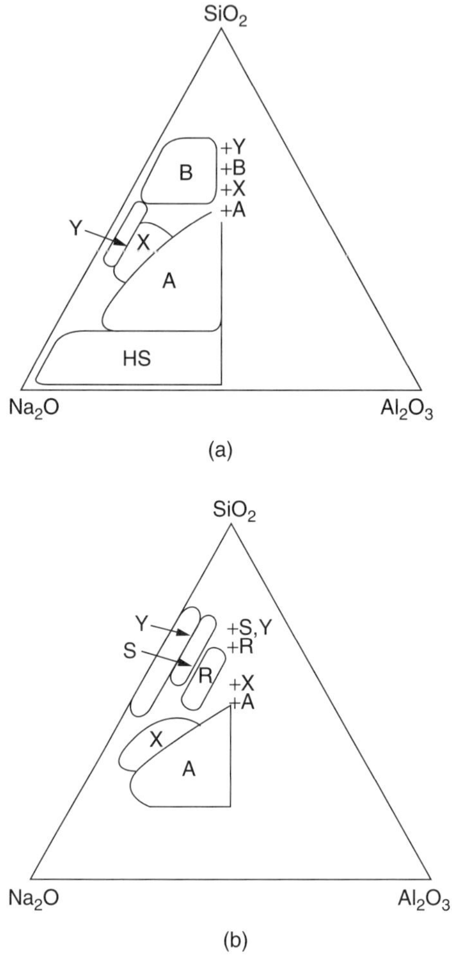

Figure 7.5. Compositional "synthesis windows" for the Na_2O-Al_2O_3-SiO_2-H_2O system at 100 °C and 90–98 mol% H_2O. Source of SiO_2 is (a) sodium silicate and (b) colloidal silica. The area enclosing a letter represents the composition that yields the corresponding phase, while the + marks the typical composition of the product. A, X, and Y = zeolites types A, X, and Y; B = zeolite P; R = chabazite; S = gmelinite; and HS = hydroxysodalite (from Breck and Flanigen, 1968, with permission).

Tetramethylammonium (TMA) cation was investigated as an additive in the synthesis of type A zeolite (LTA or Linde Type A) in 1961 (Barrer and Denny, 1961; Kerr and Kokotailo, 1961). Many new crystalline phases were found by the addition of TMA. It was soon found that the Si/Al ratio in LTA could be increased up to 2.5 by the use of TMA. An analog of these crystals is named ZK-4 (Kerr, 1966). The size of the TMA cation is approximately 6 Å, which is about the same size as that of hydrated Na^+. These cations fit well in the beta or sodalite cage. A typical composition of ZK-4 is $Na_8(TMA)_{1.2}[(AlO_2)_{9.2}(SiO_2)_{14.8}] \cdot 28H_2O$.

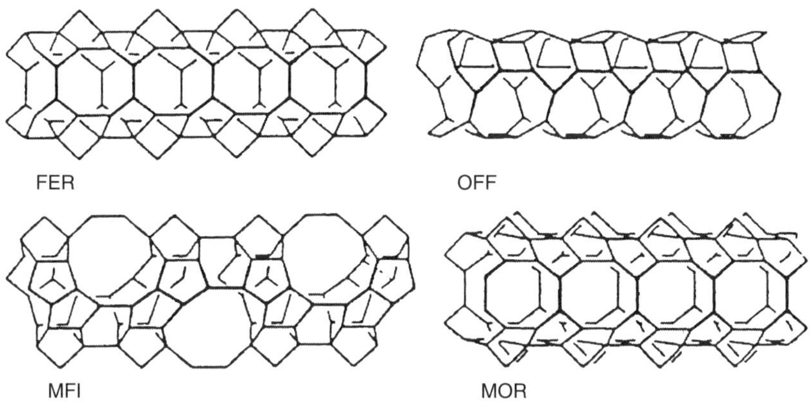

Figure 7.6. Four of the tubular building units: ferrierite (FER), offretite (OFF), ZSM-5 (MFI or pentasil), and mordenite (MOR). Six other tubular building units are shown in Szostak (1998), taken from Gellens et al. (1982).

Charnell used triethanolamine (2,2′,2″-nitrilotriethanol) as an additive in the synthesis of zeolites A and X and reported that crystals as large as 100 and 140 μm, for zeolites A and X, respectively, were formed (Charnell, 1971). Subsequent work showed that triethanolamine formed a complex with Al^{3+}, the presence of which would reduce the tendency for nucleation and crystal growth (Coker and Jansen, 1998).

While the work at Linde of Union Carbide was mainly on the development of new sorbents (resulting in the invention of zeolites A, X, and Y), the interest at Mobil was in developing new catalysts. With the use of amine additives, a series of zeolites named ZSM (zeolite secony mobil) were synthesized. (The ZSMs have different framework structures but share the common structure code simply because they were synthesized at Mobil.) The most useful one is ZSM-5 (Argauer and Landolt, 1972), which has been used as a shape-selective catalyst for xylene isomerization (to produce *p*-xylene, which can diffuse in ZSM-5), methanol-to-gasoline process (MTG), and several other commercialized catalytic reactions (Chen et al., 1994). ZSM-5 has two intersecting channels (one straight and one sinusoidal). The pore dimensions are shown in Figure 7.7, and are compared with that of X or Y zeolite. The channels of ZSM-5 are elliptical and their openings are 5.1 × 5.6 Å and 5.4 × 5.6 Å. The high-silica form of ZSM-5, developed independently at Linde, was named silicalite (Flanigen et al., 1978). The template for ZSM-5 was tetrapropyl ammonium (TPA) ion. Because the TPA ion fits so well within the voids of the tubular structure of ZSM-5, the successful synthesis of ZSM-5 provided major support for the templating theory. However, ZSM-5 was subsequently synthesized by many groups during the 1980s without using any templates (Szostak, 1998). A critical review and discussion of the vast literature on the subject was made by Szostak (1998).

Adsorption, diffusion, and reaction in ZSM-5 and silicalite have been studied extensively. For adsorption, it has been used as a selective sorbent for VOC

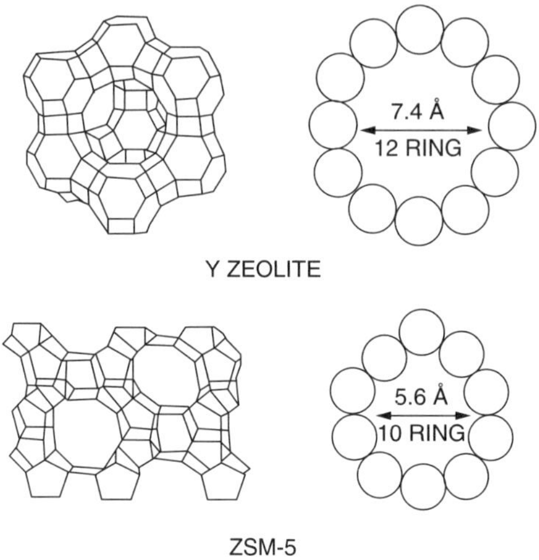

Figure 7.7. Channel-type framework structure and channel dimension (with 10-ring) of ZSM-5 (MFI) compared with the cage structure and window dimension (with 12-ring) of zeolite Y.

removal, including automotive cold start. It has a potential for interesting adsorption properties due to its small channel type pores. The small channel type pore offers maximum interactions with sorbate molecules of similar sizes. For example, *n*-paraffins and olefins can have a tight fit in the channels that gives rise to high heats of adsorption (Stach et al., 1984). The channel sizes of ZSM-5 and silicalite are comparable with the sizes of many important molecules such as *n*-paraffins, isoparaffins, aromatics, and their substitutes. Many unexpected and exciting nonlinear phenomena arise when the ratio of the channel size and the molecular size approaches one (Wei, 1994). These include high activation energies (June et al., 1990), “length selectivity” and occupancy effects for hydrocarbon diffusion (Wei, 1994). For future applications in environmental control where ultra-purification is required, a high Henry’s law constant for adsorption is important while the sorbent capacity is not a main concern. Because of its channel structure, the sorbent capacity of ZSM-5 is low compared with the cavity-type zeolites. For example, the pore volume of silicalite measured by n-hexane adsorption is 0.199 cm^3/g (Flanigen et al., 1978) or 0.185 cm^3/g (Ma, 1984). However, high Henry’s constants can be obtained by the tight fit. Using two silicalites with slightly different O-atom positions, the Henry’s constants of benzene adsorption in these two silicalites can differ by factors as high as 3.1 (Clark and Snurr, 1999; Li and Talu, 1993). Other potential applications such as its use as the sorbent for desulfurization of liquid fuels (e.g., removal of thiophene from benzene) will be discussed in Chapter 10.

By using templates, two major categories of crystalline (zeolite-like) molecular sieves were synthesized starting in the 1980’s. One category was molecular sieves containing transition metals in the framework, by replacing aluminum with Ti, V, Fe, etc. (Perego et al., 1998). Another category was aluminophosphates

($AlPO_4$s) and silicoaluminophosphates ($SAPO_4$s). Although the main interest in their synthesis was in catalysis, there are many potential applications for their use as sorbents, one of which has already been recently commercialized (i.e., ETS-4, a titanium silicate, for N_2/CH_4 separation, by Kuznicki et al., 2001).

The successful syntheses of transition metal-containing silicates have been accomplished by incorporating transition metal atoms in the pentasil (MFI) framework (shown in Figure 7.6). The large amount of literature on their syntheses has been discussed by Perego et al. (1998). The "mixed alkoxide" method appeared to be the most successful. For example, titanium-silicalite-1 (TS-1) was synthesized by controlled hydrolysis of an aqueous solution containing tetraethylorthosilicate (TEOS), tetraethylorthotitanate (TEOT), and tetrapropylammonium hydroxide (TPAOH) as the template (Taramasso et al., 1983). Different templates were used for the syntheses of TS-2, TS-3, etc., as reviewed by Perego et al. (1998). The Ti atom is coordinated tetrahedrally in these structures. Many forms of vanadosilicates and ferrisilicates have also been synthesized (Perego et al., 1998). Transition-metal containing silicates can also be prepared without the use of templates. Young claimed hydrothermal synthesis of titano- and zirconosilicate frameworks in 1967 (Young, 1967). Kuznicki (1990) was able to synthesize a small pore titanosilicate molecular sieve, named ETS-4, by reacting a solution of sodium silicate/$TiCl_3$/NaOH//KF at 150 °C. KF was used to increase the crystallization rate, and the pH of the solution was 10.5. ETS-4 has a Si/Ti ratio of 2.6 and pore dimensions of 3–5 Å, depending on the calcination temperature (Kuznicki et al., 2001). Its structure collapses near 350 °C. Unlike TS-1, the Ti atom in ETS-4 is octahedrally coordinated. A form of titanosilicate, named TAM-5, has very high selectivities for Cs^+ and Sr^{2+} over Na^+ (Anthony et al., 1994; Hritzko et al., 2000). Using this sorbent, Wang and co-workers have designed a highly efficient carousel process for removal of radioactive $^{137}Cs^+$ from a simulated nuclear waste (Hritzko et al., 2000).

The syntheses of aluminophosphate ($AlPO_4$) molecular sieves were first reported in 1982 (Wilson et al., 1982a; 1982b and 1984). These molecular sieves have a very narrow range of chemical composition (i.e., rather invariant ratio of P/Al compared with the wide range of Si/Al ratio in zeolites), but exhibit a rich diversity of framework structures. The chemical composition of $AlPO_4$ is

$$xR \cdot Al_2O_3 \cdot 1.0 \pm 0.2P_2O_5 \cdot yH_2O$$

where R is an amine or quaternary ammonium ion. The average of the ionic radii of Al^{3+} (0.39 Å) and P^{5+} (0.17 Å) is 0.28 Å, which is similar to the ionic radius of Si^{4+} (0.26 Å). This similarity apparently is responsible for the narrow range of the chemical composition (i.e., P/Al $\approx$ 1). A large number of amines and quaternary ammonium ions have been used as the templates for their syntheses. $AlPO_4$-5, $AlPO_4$-11, $AlPO_4$-17, and $AlPO_4$-20 were among the first synthesized. They are formed in both cage-type and channel-type framework structures, and the sizes of their pore apertures vary from the size of 6-ring (2.8 Å, in $AlPO_4$-20) to 7.3 Å of 12-ring (in $AlPO_4$-5). Tetrapropyl ammonium ion was a typical template used for $AlPO_4$-5. The structural diversity apparently reflects a dominant role of the templating agent in organizing and shaping structural voids during crystallization. The template in most cases remains entrapped in these voids and is removed by calcination. Some of the $AlPO_4$ types, such as $AlPO_4$-5, can be synthesized with many different templates. The $AlPO_4$s are formed by tetrahedrally coordinated AlO_4 and PO_4, and have no need for charge-balancing cations. A large number of silicoaluminophosphate ($SAPO_4$) analogs have also been synthesized. The syntheses and characterization of a high amount of $SAPO_4$ and $SAPO_4$ have been reviewed by Szostak (1998), Ernst (1998), and Hartmann and Kevan (1999). The $SAPO_4$ analogs are formed by silicon, aluminum, phosphorous, and oxygen atoms in tetrahedral coordination, with uniform pore channels in molecular dimension (Lok et al., 1984; Lok et al., 1988). $SAPO_4$'s have a framework with a net charge that varies depending on how the silicon is substituted into the aluminophosphate. That is, if silicon substitutes for aluminum, phosphorous, or both, the resulting net charge will be, respectively, +1, −1, or 0 (Djieugoue, et al., 1999). Usually the second and third substitutions occur during the crystallization process.

The rich variety of pore structures, both cavities and channels, as well as the cation sites that can be exchanged in the $SAPO_4$ analogs, offer promising opportunities for their use as new sorbents for separations. An example for such an application is the use of $AlPO_4$-14 for the separation of propane/propylene (Padin et al., 2000). The size of the channel aperture of $AlPO_4$-14 is 3.8 Å, which essentially excludes propane but admits propylene (which has a slightly smaller kinetic diameter than propane, i.e., 3.6 vs. 3.8 Å). Figure 7.8 shows the isotherms of propane and propylene on $AlPO_4$-14. More details of this application will be given in Chapter 10. An example of a possible application of $SAPO_4$ involves the selective adsorption of CO_2, shown in Figure 7.9. (Hernandez-Maldonado and Yang, 2002). $SAPO_4$-14, which has the structure of gismondine, shows a pore

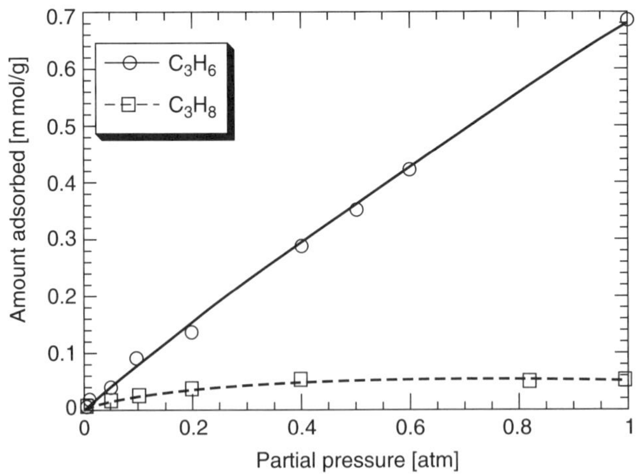

Figure 7.8. Equilibrium isotherms of propane (3.8 Å) and propylene (3.6 Å) on $AlPO_4$-14 (with a channel dimension of 3.8 Å) at 120 °C (Padin et al., 2000, with permission).

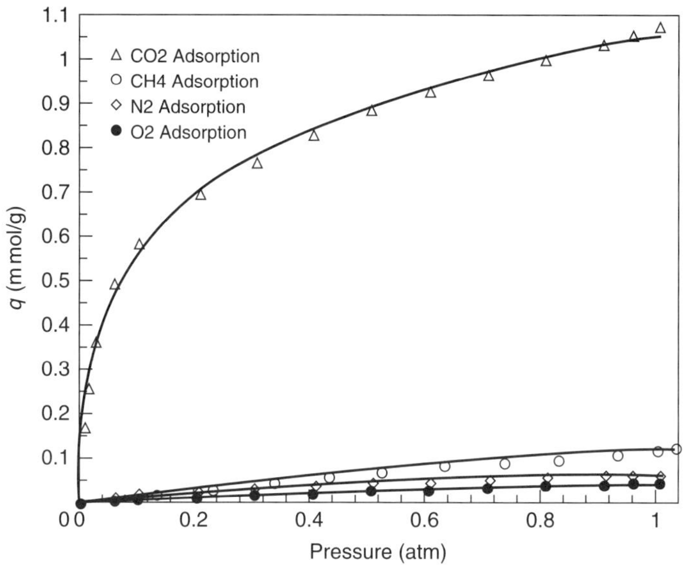

Figure 7.9. Equilibrium isotherms on $SAPO_4$-43 (Gismondine) at 25 °C (with 8-ring aperture 3.1 × 4.5 Å) (Hernandez-Maldonado and Yang, 2002, with permission).

opening by 8-rings with dimensions of 4.5 × 3.1 Å. This crystalline material was first synthesized in a pure phase by Akporiaye et al. (1996), using isopropylamine as the template. The template was carefully removed by calcination in air at 400 °C, which was the limiting temperature before the structure would collapse.

Because of the potential applications of $AlPO_4$ and $SAPO_4$ for separations by either size exclusion or by kinetic separation, the pore-opening dimensions of $AlPO_4$ are given in Table 7.1. The $SAPO_4$ analogs of $AlPO_4$ (e.g., $SAPO_4$-43 and $AlPO_4$-43) have the same framework structures and hence approximately the same channel or window sizes.

Finally, the current understanding about the roles of the templating agents in the syntheses of zeolites and molecular sieves warrants discussion. The early use (in the 1960s) of tetramethyl ammonium (TMA) ion increased the Si/Al ratio of the zeolites (e.g. zeolite A). Subsequent work showed evidence that the amine ions or neutral amines could stabilize the formation of structural subunits that were thought to be precursors of crystalline zeolite species. Myriad new crystalline phases of zeolites, as well as new molecular sieves, were synthesized with the aid of templates. Moreover, many different templates could lead to the same crystal phase. For example, about 23 different nitrogen-containing templates have been used to form the same $AlPO_4$-5 (Szostak, 1998). These additives have also played a major role in the search for large-pore zeolites and molecular sieves. The theory behind this was to find proper "void fillers" that would stabilize or template large voids. This led to the use of ever-larger quaternary ammonium cations in zeolite synthesis. The three-dimensional open pore topologies of zeolite beta (BEA), ZSM-5, and ZSM-11 were prepared, respectively, with tetraethyl

Table 7.1. Structures and aperture sizes of A1PO-*n* molecular Sieves

n	IZA* Structure Code	Pore Diameter/nm (Ring**)
	Large Pore	
5	AFI	0.73 (12)
36	ATS	0.75 × 0.65 (12)
37	FAU	0.74 (12)
40	AFR	0.43 × 0.70 (10)
46	AFS	0.64 × 0.62 (12) 0.4 (8)
	Intermediate Pore	
11	AEL	0.63 × 0.39 (10)
31	ATO	0.54 (12)
41	AFO	0.43 × 0.7 (10)
	Small Pore	
14		0.38 (8)
17	ERI	0.36 × 0.51 (8)
18	AEI	0.38 (8)
26		
33	ATT	0.42 × 0.46 (10)
34	CHA	0.38 (8)
35	LEV	0.36 × 0.48 (8)
39	ATN	0.4 (8)
42	LTA	0.41 (8)
43	GIS	0.31 × 0.45 (8)
44	CHA	0.38 (8)
47	CHA	0.38 (8)
	Very Small Pore	
16	AST	(6)
20	SOD	(6)
25	ATV	0.30 × 0.49 (8)
28		
	Very Large Pore	
8	AET	0.79 × 0.87 (14)
VPI-5	VFI	1.21 (18)

*International Zeolite Association.

**Number of oxygen atoms in the ring that forms the channel or window.

From Hartmann and Kevan, 1999, with permission.

ammonium, tetrapropyl ammonium, and tetrabutyl ammonium ions. A modified templating theory is the "lock-and-key" version, which envisions that the zeolite structure grows around the template, thus stabilizing certain pore/cavity structures or cages. In some cases, the additive acts not only as a template for a given structure to crystallize, but also prohibits another structure from forming during nucleation.

Organic molecules react with silica in aqueous solution-forming complexes and thus increase the solubility of silica in the solution. Hence another role of the organic additive is to act as a gel modifier that influences both the gelling and crystallization processes. An extensive critical review and discussion on the subject of templating has been given by Szostak (1998).

7.3. UNIQUE ADSORPTION PROPERTIES: ANIONIC OXYGENS AND ISOLATED CATIONS

Zeolites exhibit many unique adsorption properties, mainly because of their unique surface chemistry. The surface of the framework is essentially oxygen atoms, whereas Si and Al are buried or recessed in the tetrahedra of oxygen atoms. They therefore are not fully exposed and cannot be readily accessed by adsorbate molecules. Also, the anionic oxygen atoms are more abundant and are much more polarizable than the Al and Si cations. Therefore, the numerous anionic oxygen atoms dominate the van der Waals interactions with the sorbate molecules, that is, the $\phi_D + \phi_R$ (dispersion + repulsion) terms (see Chapter 2).

Besides the anionic oxygen, cations are located at certain sites. Some of these sites are hidden or inaccessible to the adsorbate molecules. However, some other cations are located above the oxide surfaces and are fully accessible. For adsorbate molecules with permanent dipoles and quadrupoles, the interactions with these exposed cations often dominate the total interaction potential (see, for example, Table 2.1 and discussion in 7.4).

The anionic surface oxygens carry negative charges. The charge depends on the location of the oxygen relative to the cation sites, and also on the cation. In Monte Carlo simulations, a constant charge is usually assigned to all surface oxygen atoms and the value is usually determined by fitting the experimental data (of isosteric heat of adsorption or the isotherm). For example, for each oxygen, a charge of $-1/3$ was used by Razmus and Hall (1991) and -1.2 was used by Mellot and Lignieres (1997).

It is instructive to compare the relative anion electronegativities of the zeolite framework with simple anions such as halides. Such a comparison can be made by calculating the net charges of the anions (or cations) by using molecular orbital theories. The Gaussian 94 Program in Cerius^2 molecular modeling software from Molecular Simulations, Inc. was used for the calculation (Takahashi et al., 2000; Yang and Yang, 2002). The calculations were performed at the Hartree–Fock (HF) and density functional theory (DFT) levels by using effective core potentials (ECPs). The LanL2DZ basis set was used for both geometry optimization and natural bond orbital (NBO) analysis. The net charges were calculated by using

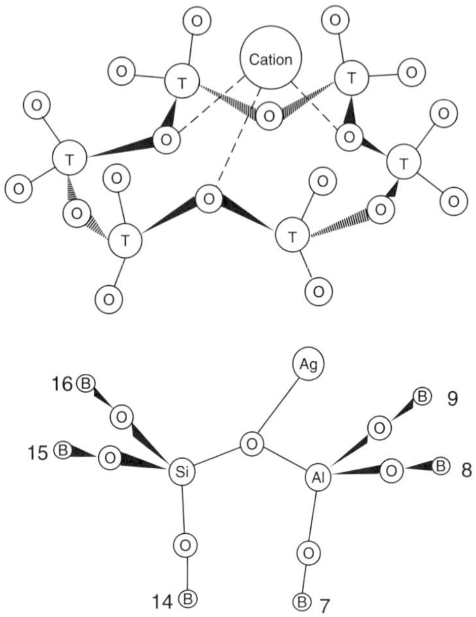

Figure 7.10. (Top) Site II cation on six-membered oxygen ring as the basic unit on A and X zeolites. T denotes Si or Al. (Bottom) Geometry-optimized cluster model to represent the chemistry of Ag-zeolite.

Table 7.2. Relative electronegativities of zeolite anion and halides. Comparison of anion net charges calculated by molecular Orbital Theory

	Anion Charge, Electronic Unit	Electron Occupancy in 5s Orbital of Ag^+
Ag^+Z^-	0.5765	0.142
Ag^+F^-	0.5111	0.295
Ag^+Cl^-	0.3404	0.357
Ag^+Br^-	0.3017	0.393
Ag^+I^-	0.2375	0.437

AgZ denotes Ag-Zeolite using the Model in Figure 7.10.

NBO. The zeolite model used in the calculation was the cluster model shown in Figure 7.10. The results are shown in Table 7.2. In the Ag-zeolite cluster model, H is used to terminate the structure and does not have a long-range influence on the bonding of Ag.

From Table 7.2, the zeolite anion is more electronegative than F^-. Also shown in Table 7.2 are the electron occupancies in the 5s orbital of the Ag^+, which is bonded to the anion. The 5s orbital is the orbital at the highest energy level of Ag and is the valence orbital. For a perfect anion, the Ag^+ to which it is bonded should have an empty 5s orbital. Again, it is seen that Ag^+ in AgZ has the lowest occupancy in its 5s orbital, indicating that the Z^- (i.e., zeolite framework anion) is the most electronegative anion.

The strong anionic nature of the zeolite framework and the correspondingly strong cations that are bonded to the framework zeolite are unique with zeolites. Furthermore, the cations and anions are not located closely to each other. Thus, these cations and anions exert strong electric fields and field gradients over the surface. The surfaces of zeolites are different from ionic crystals such as NaCl or cement (that has the same chemical composition as zeolites). On the surfaces of ionic crystals, the cations and anions are closely and periodically spaced. Thus, for an adsorbate molecule several angstroms in size, no net strong electric field and field gradient is exerted by the surface for interaction.

7.4. INTERACTIONS OF ADSORBATE WITH CATIONS: EFFECTS OF CATION SITE, CHARGE, AND IONIC RADIUS

In earlier discussions, the strong or dominating contributions of cation-dipole and cation-quadrupole interactions to the total bonding energy for adsorption on zeolites are already seen (see Table 2.1). Unfortunately, except ionic radii, information on the cation sites and the charges has not been well determined. Hence they are often treated as fitting parameters in molecular simulations of adsorption. However, the strong effects of these parameters on adsorption are well established.

7.4.1. Cation Sites

Zeolite frameworks usually have more sites for the number of charge-balancing cations that occupy them. The cations distribute themselves in a manner to minimize the free energy of the system. The distribution of the cations on the sites depends on (1) the temperature of heat-treatment, (2) the cationic species, and (3) the degree of hydration. X-ray diffraction has been the main tool for cation siting (Barrer, 1978). Neutron diffraction has advantages over X-ray diffraction because the X-ray scattering is nearly indistinguishable between Si and Al. Also, the X-ray scattering by small cations, such as Li, is too weak to detect (Hutson et al., 2000). Rietveld refinement, a trial-and-error procedure, is now the standard technique for determining the cation sites as well as the structure (Rietveld, 1967; Hutson, et al., 2000). It is difficult to determine all the cation sites in zeolites because of the relatively small number of cations compared with the large number of other atoms in the structure (Al, Si, and O). Other techniques, such as solid-state NMR (Engelhardt et al., 1994), infrared spectroscopy (Ozin et al., 1983), and diffusion studies (of probe molecules) (Ackley and Yang, 1991) have also assisted in cation siting.

Due to their importance, the most studied cation sites are for zeolites A, X, and Y. Results on cation siting in these zeolites will be summarized first. Cation sites in another main type of structure with cage-topology (i.e., chabazite) and that in a typical type with tubular topology (i.e., heulandite) will also be included. Adsorption of molecules with strong dipoles and/or quadrupoles (e.g., NH_3 and SO_2) is known to cause redistribution of some cations (Barrer, 1978). As mentioned, the remaining water molecules have a strong influence on cation sites. Because water is removed for adsorption applications, only highly dehydrated zeolites will be discussed.

Cation sites for zeolite A, with the common alkali and alkaline earth cations, are listed in Table 7.3. The cation sites in A and X zeolites from Figure 7.1 are shown again in Figure 7.11. The three sites in zeolite A, sites I, II and III, are only approximate indications of the actual sites. For example, some of the cations associated with 6-oxygen rings (i.e., site I) are extended into the large cavity; while some others, also referred to as being at site I, are recessed into the sodalite cage or are nearly in the plane of the 6-ring (e.g., Firor and Seff, 1979; McCusker and Seff, 1981). Although they are all referred to as being at site I, they are clearly different in terms of their abilities to interact with adsorbate molecules. The cation sites can be predicted by calculating the cation-lattice interaction energies (consisting of electrostatic, polarization, dispersion/repulsion, and charge-transfer energies), and the sites with the highest energies will be preferred (Ogawa et al., 1978). This technique can also be used to predict the sites for zeolites with mixed cations that are obtained by partial ion exchange. As a result, the site selectivities determined for cations are as follows: Li^+, Na^+, Ca^{2+}, and Sr^{2+} prefer 6-ring sites (site I); whereas K^+, Cs^+, and Ba^{2+} prefer the 8-ring sites (site II) (Ogawa et al., 1979). The predictions are in agreement with experimental data shown in Table 7.3.

Table 7.3. Cation site occupancies in dehydrated zeolite A

	Sites			
Zeolite	I	II	III	Others
Li-A[a]	8	3	1	—
Na-A[b,c]	8	3	1	—
NaCa-A (4Na + 4Ca)[d]	8	0	0	—
K-A[e]	6	3		3[e]
Ca-A[f]	5	1	0	—
Sr-A[f]	5	1	0	—
Ba-A[g]	4	2	—	—

[a] Vance and Seff, 1975.
[b] Reed and Breck, 1956; Smith and Dowell, 1968.
[c] Yanagida et al., 1973.
[d] Breck, 1974.
[e] Barrer, 1978.
[f] Firor and Seff, 1979.
[g] Dyer et al., 1971; Ogawa et al., 1978.
12 Monovalent cations per unit cell.

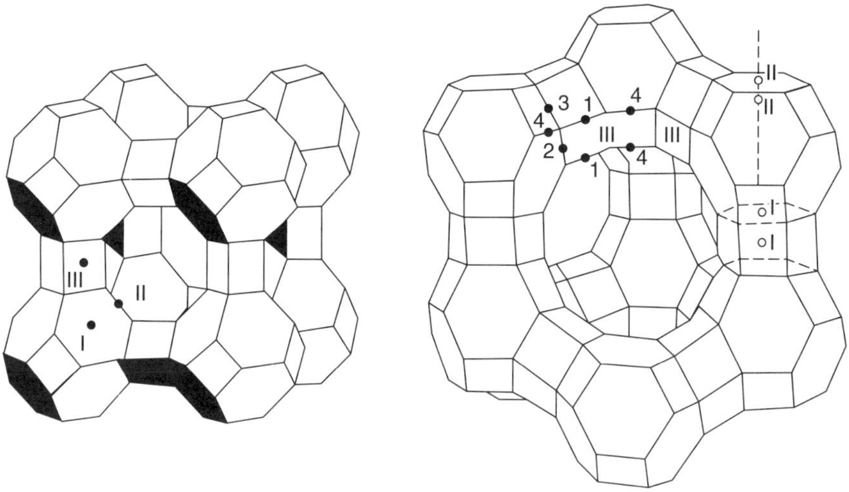

Figure 7.11. Cation sites in zeolites A (left) and X and Y (right).

The cation sites in zeolites X and Y are given in Table 7.4. Sites I, I', and II' are not exposed and are not available for interaction with adsorbate molecules, with the possible exception of water, which could fit in the 6-oxygen ring (with an opening of 2.8 Å). Thus, nearly one-half of the cations are not available for adsorption. Sites II and III are exposed to the cavity and are associated with, respectively, the 6- and 4-oxygen rings. Alkali metal cations occupy both of these sites but the alkaline earth metal cations tend to occupy site II only.

The basic building unit for zeolites A, X, and Y is the sodalite or beta cage. The largest window for this cage is the 6-oxygen ring, with a free opening of 2.8 Å. There are 7 known cages that contain 8-oxygen rings, and these cages are the building units of many zeolites, such as chabazite, levynite, and erionite (Gellens et al., 1982). The 8-ring window is elliptic (about 4 Å dimension) and is of interest for molecular sieving. The cation sites for chabazite are shown in Figure 7.12 (Calligaris and Nordin, 1982). The natural form of chabazite has been used for natural gas purification (Vaughan, 1988). Site I is associated with the 6-ring and is displaced into the ellipsoidal cavity. Site II is located near the center of the ellipsoidal cavity. Site III is found at the center of the hexagonal prism, and site IV is near the 8-ring window. Thus, with the exception of cations at site III, all cations are exposed to the ellipsoidal cavity and are available for interaction with adsorbate molecules.

Channel-type zeolites and molecular sieves also hold interest for gas separation. Cation sites in some of the $SAPO_4$ molecular sieves have been discussed by Hartmann and Kevan (1999). Cation sites in the channel-type zeolites have been determined for a number of topologies. The cations either fully block the channels or partially block them. The cation sites in clinoptilolite have been determined because of its practical importance (Alberti, 1975; Koyama and Takeuchi, 1977; Ackley and Yang, 1991).

Table 7.4. Cation site occupancies in dehydrated X and Y zeolites

Zeolite	Al/Unit Cell	Sites				
		I	I'	II	II'	III
Li-X (Li-LSX)[a]	95.8	—	27.2	33.9	—	32.4
Li-X[b]	85	—	25.6	30.4	—	11.5
Li-Y[c]	46	—	24.4	21.0	—	0
Na-X[d]	81	3.8	32.3	30.8	—	7.9
Na-Y[e]	53	3	15	35	—	—
Na-Y[f]	57	7.8	20.2	31.2	—	—
K-Y[g]	54.7	5.4	18.1	26.8	—	—
K-X[h]	87	9	13	26	—	38
Ca-X[g]	86	7.5	17.3	17.3	9.0	—
Sr-X[g]	86	11.2	7.0	19.5	4.2	—
$Ba_{36}Na_{16}$-X[i]	88	7.0 Ba	4.7 Ba	11.4 Ba	3.7Ba	Na
Ag-LSX[j]	96	8.5	23.4	32.0	0	19.2
Ag-Y[j]	56	10.9	12.4	27.5	4.5	0

[a]Hutson and Yang, 2000.
[b]Feuerstein and Lobo, 1998.
[c]Forano et al., 1989.
[d]Mortier, 1982.
[e]Engelhardt et al., 1994.
[f]Eulenberger et al., 1967a.
[g]Mortier et al., 1972.
[h]Hseu, 1972.
[i]Olson, 1968.
[j]Hutson et al., 2000.
Maximum = 96 monovalent cations per unit cell for Si/Al = 1.

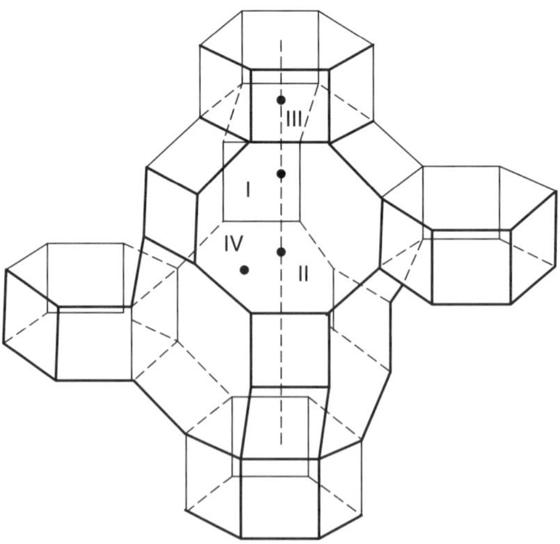

Figure 7.12. Structure and cation sites of chabazite (Calligaris and Nordin, 1982).

Clinoptilolite is a member of the heulandite group of natural zeolites. It has been used for radioactive waste disposal and ammonia recovery from sewage effluents (Vaughan, 1988). It has also been studied extensively in the separation of N_2/CH_4 for natural gas upgrading by molecular sieving or kinetic separation (Ackley and Yang. 1991; see chapter for details). The cation sites in clinoptilolite are shown in Figure 7.13.

The framework of clinoptilolite consists of three channels, A, B, and C, formed by 10- and 6-oxygen rings, as shown in Figure 7.13. Channels A and B are alternating and parallel, and channel C is on the same plane with A and B and

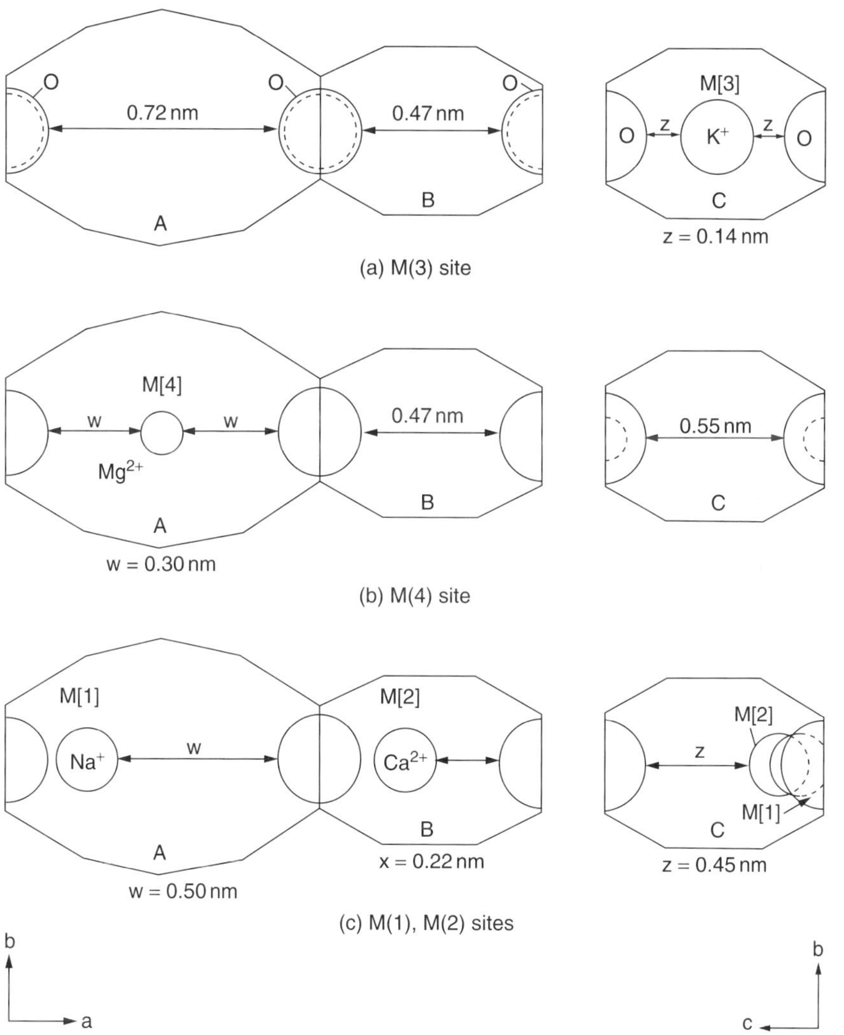

Figure 7.13. Cation sites in channel-type zeolite of clinoptilolite (heulandite), with dimensions of channel blockage (Ackley and Yang, 1991).

intersects them at an angle of 55°. Thus, the channel system is two-dimensional. Each unit cell of clinoptilolite contains 6 Al atoms, hence there are 6 monovalent cations or 3 bivalent cations. These cations are located at sites M(1), M(2), M(3), and M(4), as indicated in Figure 7.1. Each unit cell contains only 4 combined M(1)/M(2) sites, 4 M(3) sites, and 2 M(4) sites, or a total of 10 sites. Different cations have their own preferred sites as indicated in the figure. M(1)/M(2) sites are at the intersections of channels A/B with channel C. Na^+ and Ca^{2+} may occupy both M(1) and M(2) sites. The natural clinoptilolite has mixed cations. In fully K^+ exchanged form, 4 K^+ ions occupy the M(3) sites and the other two occupy M(2) sites. Thus, diffusion in the K- clinoptilolite is one-dimensional (since channel C is closed), whereas diffusion in all other pure cation forms is two-dimensional (since channel C is open) (Ackley and Yang, 1991).

7.4.2. Effects of Cation Sites on Adsorption

The effects of cation sites can be best illustrated by the important system of N_2/O_2 on type X zeolites. NaX (or 13X) has been used commercially for air separation since the 1970s. Li-LSX is the best sorbent that is commercially available today (Chao, 1989). Mixed-cation AgLi-LSX (with 1–3% Ag cations) has been shown to be even better than Li-LSX for air separation (Yang and Hutson, 1998; Hutson et al., 1999; Hutson et al., 2000).

As shown in Figure 7.1, there are 192 possible cation sites in a unit cell of faujasite (or X zeolite) and only a maximum of 96 cations to occupy them, i.e., LSX has 96 cations (when monovalent cations are used). Upon activation of the zeolite, i.e., heating at 350 °C, the cations migrate to the sites with the lowest energies. Migration is an activated process, which depends on the temperature, time, as well as the size of the cation. Unfortunately, the most stable sites (those at the lowest energies) are hidden and are not exposed to the supercage cavity. These are the sites with the maximum coordination. From Table 7.4, only about 1/3 to 1/2 of the cations are located at exposed sites.

By ion exchange of Na^+ with Li^+ in the LSX, Chao obtained significantly improved N_2/O_2 selectivity (Chao, 1989). This improvement is the result of the smaller ionic radius of Li^+ (0.68 Å) compared to that of Na^+ (0.97 Å). Since Li^+ and Na^+ have the same charge, N_2 interacts much more strongly with Li^+ due to a significantly higher $\phi_{\dot{F}Q}$ (electric field gradient - quadrupole) potential. However, no improvement is seen until over approximately 70% ion exchange is made. N_2 adsorption increases linearly with ion exchange beyond this threshold value (see Figure 7.14). (Figure 7.14 actually shows LiX with different Si/Al ratios, or different number of Li cations/unit cell. However, it illustrates the same phenomenon.) This point is discussed further in Chapter 10 (Figure 10.7). The reason for this significant phenomenon is that Sites I, I' and II' are lower-energy sites and are preferred by Li^+ (Chao et al., 1992; Coe, 1995). Sites II and III are exposed but have lower coordination and are less preferred. These exposed sites are most important for adsorption.

Although Ag^+ has a larger ionic radius (1.26 Å) than Li^+, a weak π-complexation bond can be formed between Ag^+ (in AgZ) and N_2 (Chen and Yang, 1996).

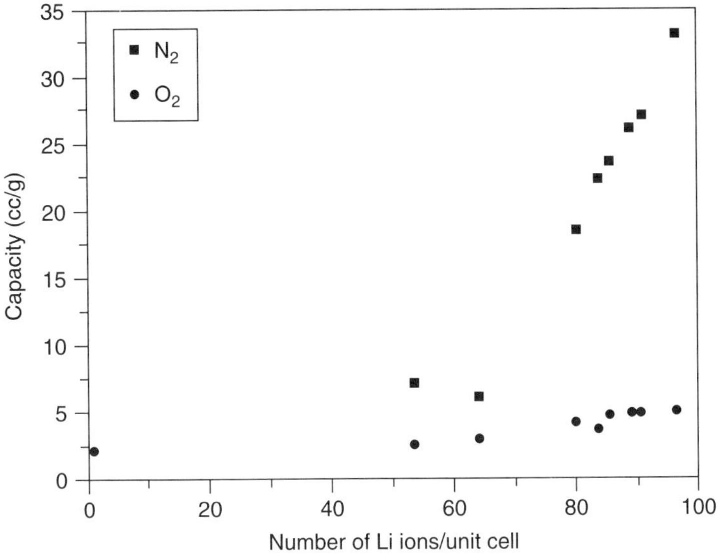

Figure 7.14. N_2 and O_2 adsorption capacities at 23 °C and 1 atm for Li faujasite with different Si/Al ratios (Coe, 1995, with permission; this result is similar to that given in Chao, 1989). This result illustrates that the first approximately 70 Li^+ are located at shielded sites that are not fully available for interaction with N_2 (or O_2).

This π-complexation bond, although weak, can significantly enhance the adsorption for N_2 (Yang and Hutson, 1998; Hutson et al., 1999; Hutson et al., 2000).

The pure Ag-LSX (Si/A1 = 1) adsorbs 22 nitrogen molecules per unit cell at 1 atm and 25 °C. The capacity depends on the temperature of heat-treatment as shown in Figure 7.15. X-ray photoelectron spectroscopy (XPS) results showed that some reduction occurs during heating from 350 to 450 °C in *vacuo* or in an inert atmosphere (Hutson et al., 2000). Moreover, a change in color from white to red occurs. This is the result of the formation of a trinuclear $Ag^+ - Ag^\circ - Ag^+$ cluster. A detailed neutron diffraction analysis has identified the site of the Ag cluster as shown in Figure 7.16 (Hutson et al., 2000). From Figure 7.16, it is seen that some of the Ag^+ originally located at site SII (after heating to 350 °C) are now located at site SII* (after heating to 450 °C). The cation at SII is significantly shielded by the six oxygen atoms of the 6-ring, and therefore are only sterically partially accessible to the adsorbate N_2. After heating to 450 °C, the Ag^+ located at SII* becomes less shielded by the 6 O atoms, and thus have more interactions (including weak π-complexation) with nitrogen.

The isosteric heats of adsorption of N_2 on Li-LSX and Ag-LSX are shown in Figure 7.17. As can be seen, the first N_2 molecule adsorbed in the unit cell of Ag-LSX has a bond energy of approximately 10 kcal/mol, decreased quickly to below 7 kcal/mol for other N_2 molecules. This difference of 3 kcal/mol is the result of bonding with Ag^+ at Site II*. The vertical distance between SII

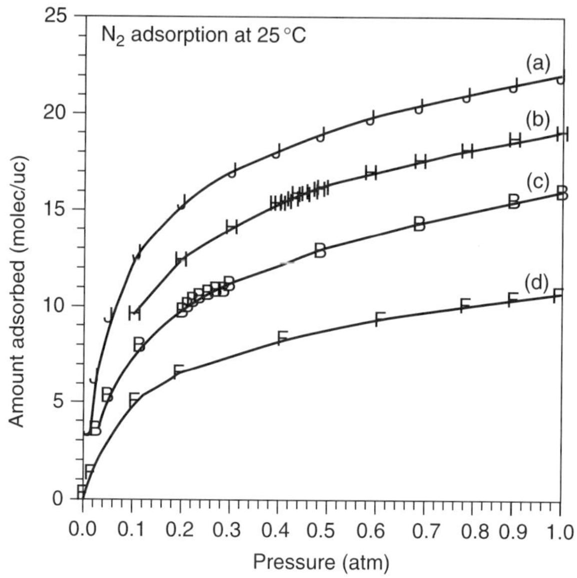

Figure 7.15. N_2 adsorption isotherms, measured at 25 °C, for Ag-LSX (a) after drying at room temperature followed by vacuum dehydration at 450 °C, (b) after drying at room temperature followed by vacuum dehydration at 350 °C, (c) after drying in air at 100 °C followed by vacuum dehydration at 350 °C, and (d) after drying in air at 100 °C in air followed by heat-treatment in air at 450 °C and finally vacuum dehydration at 450 °C (Hutson et al., 2000; Hutson, 2000, with permission).

and SII* is ∼0.75 Å (Figure 7.16). This small difference causes significantly less shielding (by O atoms) and consequently much stronger bonding with the sorbate molecule. The sorbate-cation bond energy also depends on the orientation of the sorbate molecule that is bonded. Calculations by Tielens et al. (2002) indicated that on the same site III of Li-LSX, the bond with N_2 is weaker when the N_2 molecule is orientated along the centerline of the 12-ring window.

Another example for illustrating the cation-shielding effect on adsorption is by comparing N_2 adsorption on NaY and LiY zeolites (Mellot and Lignieres, 1997). The type Y zeolite has the same framework structure as X, but with less than 76 cations per unit cell (due to higher Si/A1 ratios). In this case, Site II is the only exposed site for Na^+ and Li^+. LiY is expected to adsorb N_2 more strongly than NaY because of its smaller cations. The experimental isotherms are, however, the same (Mellot and Lignieres, 1997). The reason is that Li^+ in site II is more shielded by O atoms as evidenced by a shorter Li^+-O (framework) bond (of 2.07 Å) than the Na^+-O (framework) bond (of 2.48 Å) (Shepelev et al., 1990). Moreover, on the same X zeolite framework, it was necessary to assign different charges for the four Ca^{2+} cations (2, 1.2, 1.2, 1.2) on the four SII sites in order to account for the energetic heterogeneity for N_2 adsorption (Mellot and Lignieres, 1997).

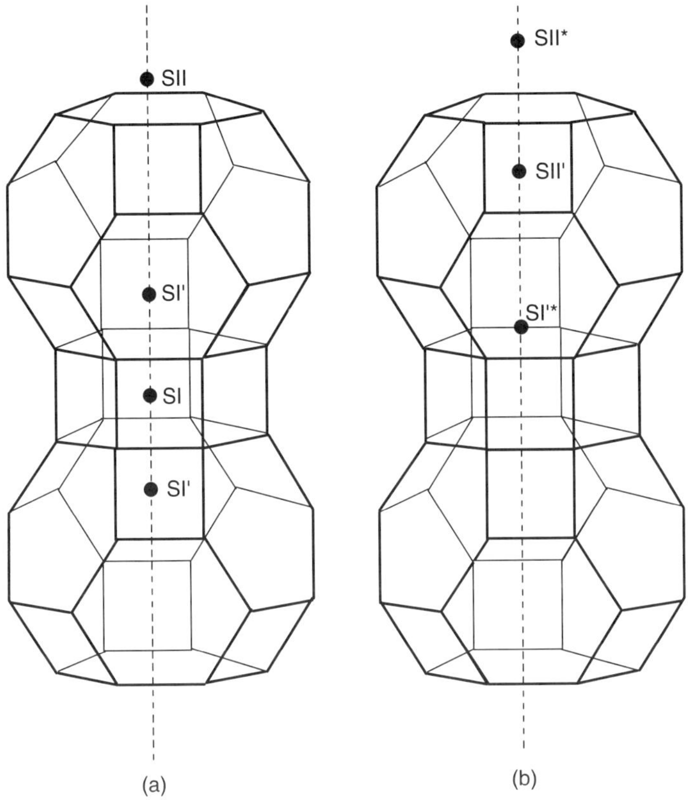

Figure 7.16. Sodalite and hexagonal prism site I and II cation locations in Ag- faujasites. Configuration (a) shows the normal cation locations with occupied sites at SI, SI', and SII. Configuration (b) shows cation sites that have resulted from cation and/or cluster migration upon vacuum thermal treatment at 450 °C. This configuration shows occupied sites at SI'*, SII', and SII* (Hutson et al., 2000; Hutson, 2000, with permission).

As discussed in 7.4.1, the sites associated with the 6-oxygen rings are not the same. Some are extended into the cavity, while others are recessed into the sodalite cage or are nearly in the same plane of the 6-ring (e.g., Firor and Seff, 1979; McCusker and Seff, 1981). For the exposed sites, the distance from the plane of the 6-ring also differs for different cations. Consequently, their interactions with the adsorbate molecule are different.

7.4.3. Effects of Cation Charge and Ionic Radius

The equilibrium distance between an interacting pair is the sum of the van der Waals or ionic radii of two atoms. Hence the ionic radius of the cation is important in all interactions, both nonspecific and electrostatic interactions. The ionic radii of important cations are listed in Table 2.3. The cationic charge, on the other hand, is important only to the electrostatic interactions.

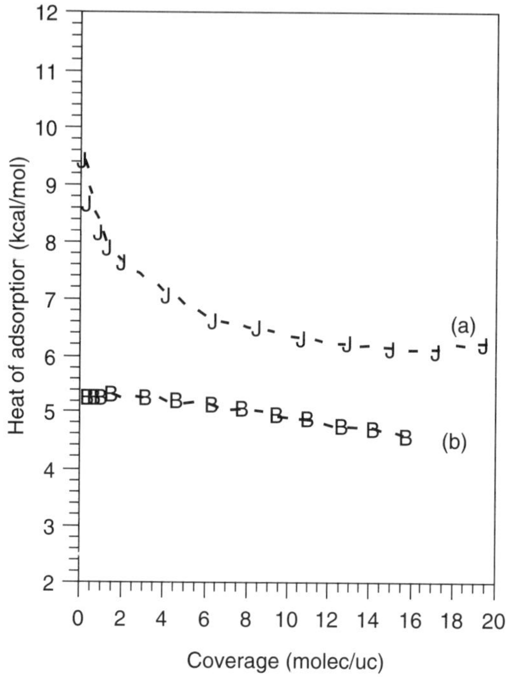

Figure 7.17. Heat of adsorption (kcal/mol) versus surface coverage (molecules/unit cell) for N_2 adsorption on (a) Ag-LSX-450 and (b) Li-LSX-450 (Hutson et al., 2000; with permission).

The effects of cation charge and ionic radius on the interaction energies are best seen in Eqs. 2.4–2.8. For electrostatic interactions, the following dependence holds:

Induction (Field-induced dipole):

$$\phi_{\text{Ind}} \propto \frac{q^2\alpha}{r^4} \tag{7.1}$$

Field-Dipole:

$$\phi_{F\mu} \propto \frac{q\mu}{r^2} \tag{7.2}$$

Field Gradient-Quadrupole:

$$\phi_{\dot{F}Q} \propto \frac{qQ}{r^3} \tag{7.3}$$

where:

$$r = r_i(\text{ionic radius}) + r_j(\text{adsorbate atom or molecule}) \tag{7.4}$$

q = electronic charge of ion, α = polarizability, F = electric field, μ = permanent dipole moment, and Q = quadrupole moment.

Table 7.5. Interaction energies (ϕ) between molecules and isolated cations

Molecule or Ion	r Å	$10^{24}\alpha$ cm^3	$10^{30}\chi$ cm^3/molc	Q esu	$-(\phi_D + \phi_R)$ kJ/mol	$-\phi_{\text{Ind}}$ kJ/mol	$-\phi_{\text{total}}$ kJ/mol
Ar	1.92	1.63	0				
O(O_2)	1.73	1.58	−1.3				
N(N_2)	1.89	1.74	−4.7				
Ar-Ion:							
Li^+	0.78	0.029	−0.99	0	0.21	21.3	21.5
Na^+	0.98	0.180	−6.95	0	0.67	16.0	16.6
K^+	1.33	0.840	−27.54	0	1.80	10.2	12.0
Ca^{2+}	0.99	0.471	−22.1	0	2.13	63.5	65.6
Sr^{2+}	1.13	0.863	−46.17	0	3.26	52.7	55.9
Ba^{2+}	1.35	1.560	−76.4	0	4.22	40.7	44.9
O_2-Li^+							32
O_2-Na^+							20
N_2-Li^+							51
N_2-Na^+							36

α = polarizability, χ = magnetic susceptibility, and Q = quadrupole moment.
Values for O_2 and N_2 are taken from Mellot and Ligniers (1997) and all others are from Barrer (1978). Van der Waals radius and ionic radius are denoted by r. N_2-ion and O_2-ion are in linear arrangements.

Table 7.5 shows the interaction energies of Ar, O_2, and N_2 interacting with isolated cations. For Ar, Eqs. 2.4–2.8 were used (Barrer and Stuart, 1959). For O_2 and N_2, the energies were calculated by Mellot and Lignieres (1997) from quantum mechanics that represent the sums of the L-J and electrostatic interactions. For the Ar-cation pairs, as the cation increases in size, the polarizability increases, hence $(\phi_D + \phi_R)$ also increases. The dispersion constant (A) in ϕ_D also increases with the magnetic susceptibility (χ), via the Kirkwood–Müller formula (Eq. 2.9). Hence the dispersion energies are higher for the divalent ions. The induction energy, in contrast, decreases sharply with the increasing size, as stipulated by Eq. 7.1 ($\phi_{\text{Ind}} \propto r^{-4}$). Here, α is fixed, which is for argon. The divalent cations are slightly bigger but have twice the amount of charge, hence the induction energies with the divalent cations are much larger than those with the monovalent cations ($\phi_{\text{Ind}} \propto q^2$). For N_2 and O_2 interacting with the same cation, the nonspecific $(\phi_D + \phi_R)$ and ϕ_{Ind} energies are about the same because their sizes, polarizabilities, and magnetic susceptibilities are quite similar. The main difference in the total interacting energies comes from $\phi_{\dot{F}Q}$ because N_2 has a much higher quadrupole moment than O_2. The substantial differences among the four pairs (O_2 – Li^+, O_2 – Na^+, N_2 – Li^+ and N_2 – Na^+) are clearly seen in Table 7.5. For O_2 interacting with Li^+ and Na^+, the difference of 12 kJ/mol is caused by the different sizes of the ions, hence the difference in the induction energy (see Eq. 7.1). For the same ion interacting with O_2 and N_2, the large difference is caused by the difference in the quadrupole moment. The dependence follows Eq. 7.3. In all cases, electrostatic interactions dominate (over the dispersion forces).

In the example above, none of the adsorbate molecules are polar. The following example illustrates interactions with polar molecules. Barrer and Gibbons (1965) performed calculations on the interaction potentials of CO_2 and NH_3 moving along the axis running through the center of the 12-ring window of the supercage or central cavity of zeolite X. CO_2 has a strong quadrupole but no dipole, and NH_3 has a strong permanent dipole but a weak quadrupole. The results are shown in Tables 7.6 and 7.7. The qualitative comparison with experimental data was remarkably good considering the calculations were made ca. 1965. In Table 7.6, it is seen that the field gradient- quadrupole interaction dominates the adsorption of CO_2 because it has no dipole but a strong quadrupole. The field gradient-quadrupole interaction energy is nearly proportional to r^{-3} (Eq. 7.3), showing the strong dependence on the ionic radius of the cation. For NH_3, the field-dipole interaction ($\phi_{F\mu}$) is clearly important (Table 7.7). The $\phi_{F\mu}$ term is proportional to r^{-2} (Eq. 7.2), hence the $\phi_{F\mu}$ term decreases sharply with increasing atomic weight. The strong dependence on the cation size is also clearly seen for the induction term, ϕ_{Ind} (proportional to r^{-4}) (Table 7.7).

Table 7.6. Components of interaction energies (ϕ, in kJ/mol) for CO_2 adsorbed on X zeolite with different cations

Component	Li^+	Na^+	K^+	Rb^+	Cs^+
$-\phi_D$ (Oxygens)	15.9	13.0	7.1	4.6	4.6
$-\phi_D$ (Cations)	0.4	0.8	3.3	4.6	9.2
$-\phi_{\text{Ind}}$	9.6	5.0	2.1	0.8	0
$-\phi_{\dot{F}Q}$	30.9	21.3	17.6	14.6	9.6

CO_2 is oriented along the ppp axis in the cavity (data taken from Barrer, 1978; Barrer and Gibbons, 1965). For CO_2: $\mu = 0$, $Q = -4.3$ esu and $\alpha = 2.91 \times 10^{-24}$ cm^3/molecule.

Table 7.7. Energy terms in kJ/mol for NH_3 in X zeolite with different cations

Cation	$-\phi_D$ (Oxygens)	$-\phi_D$ (Cations)	$-\phi_{F\mu}$	$-\phi_{\text{Ind}}$	$-\phi_{\text{total}}$	Expt'L ($-\Delta$H)
Li^+	47.2	1.7	50.6	23	77.7	76.5
Na^+	33.8	3.3	33.0	9.6	53.9	72.3
K^+	9.6	6.7	20.1	3.8	23.8	59.8
Rb^+	7.9	11.7	17.6	2.9	20.5	55.6
Cs^+	7.5	16.3	15.0	2.1	19.6	47.2

(Data taken from Barrer, 1978; Barrer and Gibbons, 1965). For NH_3: $\mu = 1.47$ debye, $Q = -1.0$ esu and $\alpha = 2.2 \times 10^{-24}$ cm^3/molecule.

A note needs to be made about the interactions with zeolites that have divalent (and higher valent) cations. The interaction energies of CO_2 with X zeolites that have different univalent cations follow the order that larger ions give lower heats of adsorption (Table 7.4 above, and Barrer, 1978). For divalent ions, the heats follow the reverse order of $Ba^{2+} > Sr^{2+} > Ca^{2+}$ (Barrer, 1978). This is also the case for N_2 adsorption on zeolites with different univalent and divalent cations (Mckee, 1964). Both N_2 and CO_2 are nonpolar but highly quadrupolar. For adsorption on zeolites exchanged with univalent cations, the $\phi_{\dot{F}Q}$ term dominates. With divalent cations, however, the large polarizabilities (Table 2.2) become important, and the dispersion and induction energies are significantly large, especially for Ba^{2+}. Hence all interaction terms need to be considered.

REFERENCES

Ackley, M. A. and Yang, R. T. (1991) *AIChE J.* 37, 1645.

Akporiaye, D. E., Dahl, I. M., Mostad, H. B., and Wendelbo, R. (1996) *Zeolites* 17, 517.

Alberti, A. (1975) *Tschermaks Min. Petr. Mitt.* 22, 25.

Anthony, R. G., Dosch, R. G., Gu, D., and Philip, C. V. (1994) *Ind. Eng. Chem. Res.* 33, 2702.

Argauer, R. J. and Landolt, G. R. U.S. Patent 3,702,886 (1972).

Barrer, R. M. (1978) *Zeolites and Clay Minerals*. Academic Press, New York, NY.

Barrer, R. M. and Denny, P. J. (1961) *J. Chem. Soc.* 971.

Barrer, R. M. and Gibbons, R. M. (1965) *Trans. Faraday Soc.* 61, 948.

Barrer, R. M. and Stuart, W. I. (1959) *Proc. Roy. Soc.* A249, 464.

Breck, D. W. (1974) *Zeolite Molecular Sieves*. Wiley, New York, NY.

Breck, D. W. and Flanigen, E. M. (1968) *Molecular Sieves*. Soc. Chem. Ind., London, U.K. p. 47.

Calligaris, M. and Nordin, G. (1982) *Zeolites*, 2, 200.

Chao, C. C. U.S. Patent 4,859,217 (1989).

Chao, C., Sherman, I. D., Mullhaupt, T. J., and Bollinger, C.-M. U.S. Patent 5,174,979 (1992).

Charnell, J. F. (1971) *J. Cryst. Growth* 8, 291.

Chen, N. and Yang, R. T. (1996) *Ind. Eng. Chem. Res.* 35, 4020.

Chen, N. Y., Degnan, T. F., Jr., and Smith, C. M. (1994) *Molecular Transport and Reaction in Zeolites*. VCH, New York, N.Y.

Clark, L. A. and Snurr, R. Q. (1999) *Chem. Phys. Lett.* 308, 155.

Coe, C. G. (1995) *Access in Nanoporous Material*. (T. I. Pinnavaia and M. E. Thorpe, eds.). Plenum Press, New York, N.Y., p. 213.

Coker, E. N. and Jansen, J. C. (1998) Approaches for the synthesis of ultra-large and ultra-small zeolite crystals. In *Molecular Sieves*. (H. G. Karge and J. Weitkamp, eds.). Springer, Berlin Germany and New York, N.Y. pp. 121–155.

Davis, M. E., Saldarriaga, C. H., Montes, C., Garces, J. M., and Crowder, C. (1988a) *Nature* 331, 698.

Davis, M. E., Saldarriaga, C. H., Montes, C., Garces, J. M., and Crowder, C. (1988b) *Zeolites* 8, 362.

Dessau, R. M., Schlenker, J. L., and Higgins, J. B. (1990) *Zeolites* 10, 522.

Djieugoue, M. A., Prakash, A. M., and Kevan, L. (1999) *J. Phys. Chem. B* 103, 804.

Dyer, A. (1988) *An Introduction to Zeolite Molecular Sieves*. Wiley, Chichester, U.K.

Dyer, A., Celler, W. Z., and Shute, M. (1971) *Adv. Chem. Ser.* 101, 436.

Eastermann, M., McCusker, L. B., Baerlocher, Ch., Merrouche, A., and Kessler, H. (1991) *Nature* 352, 320.

Engelhardt, G., Hunger, M., Koller, H., and Weitkamp, J. (1994) *Stud. Surf. Sci. Catal.* 84, 421.

Ernst, S. (1998) Synthesis of more recent aluminosilicates with a potential in catalysis and adsorption. In *Molecular Sieves*. (H. G. Karge and J. Weitkamp, eds.). Springer, Berlin Germany and New York, N.Y. pp. 63–96.

Eulenberger, G. R., Keil, J. G., and Shoemaker, D. P. (1967a) *J. Phys. Chem.* 71, 1812.

Eulenberger, G. R., Keil, J. G., and Shoemaker, D. P. (1967b) *J. Phys. Chem.* 71, 1817.

Feuerstein, M. and Lobo, R. F. (1998) *Chem. Matter.* 10, 2197.

Firor, R. L. and Seff, K. (1979) *J. Am. Chem. Soc.* 101, 3095.

Flanigen, E. M., Bennett, J. M., Grose, J. P., Cohen, J. P., Patton, R. L., Kirchner, R. M., and Smith, J. V. (1978) *Nature* 271, 512.

Forano, C., Slade, R. C. T., Andersen, E. K., Andersen, I. G. K., and Prince, E. (1989) *J. Solid State Chem.* 82, 95.

Gellens, L. R., Price, G. D., and Smith, J. V. (1982) *Mineral. Mag.* 45, 157.

Hartmann, M. and Kevan, L. (1999) *Chem. Rev.* 99, 935.

Hernandez-Maldonado, A. J. and Yang, R. T. (2002) Paper 9a, AIChE 2002 Spring National Meeting, March 11, New Orleans.

Hritzko, B. J., Walker, D. D., and Wang, N.-H. L. (2000) *AIChE J.* 46, 552.

Hseu, T. 1972 Ph.D. Thesis, University of Washington, Seattle, WA.

Hutson, N. D. (2000) Ph. D. Dissertation, University of Michigan.

Hutson, N. D., Rege, S. U., and Yang, R. T. (1999) *AIChE J.* 45, 724.

Hutson, N. D. and Yang, R. T. (2000) *AIChE J.* 46, 2305.

Hutson, N. D., Reisner, B. A., Yang, R. T., and Toby, B. H. (2000) *Chem. Mater.* 12, 3020.

Jacobs, P. A. and Martens, J. A. (1987) *Synthesis of High-Silica Aluminosilicate Zeolites*, *Studies in Surface Science and Catalysis*, 33, 15. Elsevier, Amsterdam, The Netherlands.

June, R. L., Bell, A. T., and Theodorou D. N. (1990) *J. Phys. Chem.* 94, 8232.

Karge, H. G. and Weitkamp, J., eds. (1998) *Molecular Sieves*. Springer, Berlin, Germany, and New York, NY.

Kerr, G. T. (1966) *J. Inorg. Chem.* 5, 1537; U.S. Patent 3,247,195 (1966).

Kerr, G. T. and Kokotailo, G. (1961) *J. Am. Chem. Soc.* 83, 4675.

Kouwenhoven, H. W. and de Kroes, B. (1991) *Stud. Surf. Sci. Catal.* 58, 497.

Koyama, K. and Takeuchi, Y. (1977) *Z. Kristallogr.* 145, 216.

Kuhl, G. H. (1987) *Zeolites* 7, 451.

Kuznicki, S. M. U.S. Patent 4,938,939 (1990).

Kuznicki, S. M., Bell, V. A., Nair, S., Hillhouse, H. W., Jacubinas, R. M., Braunbarth, C. M., Toby, B. H., and Tsapatsis, M. (2001) *Nature* 412, 720.

Li, J. and Talu, O. (1993) *J. Chem. Soc., Faraday Trans.* 89, 1683.

Lok, B. M., Patton, C. A., Gajek, R. T., Cannan, T. R., and Flanigen, E. M. U.S. Patent 4,440,871 (1984).

Lok, B. M., Vail, L. D., and Flanigen, E. M. U.S. Patent 4,758,419 (1988).

Loewenstein, W. (1954) *Amer. Mineralog.* 39, 92.

Ma, Y. H. (1984) *Fundamentals of Adsorption*. (A. L. Myers and G. Belfort, eds.). Engineering Foundation, New York, NY, p. 315.

McCusker, L. B. and Seff, K. (1981) *J. Am. Chem. Soc.* 103, 3441.

McKee, D. W. U.S. Patent 3,140,933 (1964).

Mellot, C. and Lignieres, J. (1997) *Physical Adsorption: Experiment, Theory and Applications*. (J. Fraissard and C. W. Conner, eds.). Kluwer Academic, Boston, MA.

Milton, R. M. U.S. Patent 2,882,243, to Union Carbide Corporation, (1959).

Milton, R. M. U.S. Patent 2,882,244, to Union Carbide Corporation, (1959).

Mortier, W. J. (1982) Proceedings 6th International Zeolite Conference (D. Olson and A. Bisio, eds.). Butterworth, Surrey, UK, p. 734.

Mortier, W. J., Bosmann, H. J., and Uytterhoeven, J. B. (1972) *J. Phys. Chem.* 76, 650.

Ogawa, K., Nitta, M., and Aomura, K. (1978) *J. Phys. Chem.* 82, 1655.

Olson, D. H. (1968) *J. Phys. Chem.* 72, 1400.

Ozin, G. A., Hugues, F., McIntosh, D. F., and Mattar, S. (1983) *ACS Symp. Ser.* 218, 409.

Padin, J., Rege, S. U., Yang, R. T., and Cheng, L. S. (2000) *Chem. Eng. Sci.* 55, 4525.

Perego, G., Millini, R., and Bellussi, G. (1998) Synthesis and characterization of molecular sieves containing transition metals in the framework. In *Molecular Sieves*. (H. G. Karge and J. Weitkamp, eds.). Springer, Berlin Germany, and New York, NY, pp. 187–228.

Razmus, D. M. and Hall, C. K. (1991) *AIChE J* 37, 769.

Reed, T. B. and Breck, D. W. (1956) *J. Am. Chem. Soc.* 78, 5972.

Rietveld, H. M. (1967) *Acta Crystallog.* 22, 151.

Shepelev, Y. F, Anderson, A. A., and Smolin, Y. I. (1990) *Zeolites* 10, 61.

Smith, J. V. and Dowell, L. G. (1968) *Z. Kristallogr.* 126, 135.

Stach, H., Thamm, H., Janchen, J., Fiedler, K., and Schirmer, W. (1984) Proceedings 6th International Zeolite Conference (D. Olson and D. Bisio, eds.). Butterworth, Guildford, Surrey, UK, p. 225.

Szostak, R. (1998) *Molecular Sieves*, 2nd Ed. Blackie Academic & Professional, New York, NY.

Takahashi, A., Yang, F. H., and Yang, R. T. (2000) *Ind. Eng. Chem. Res.* 39.

Taramasso, M., Perego, G., and Notari, B. U.S. Patent 4,410,501 (1983).

Thompson, R. W. (1998) Recent advances in the understanding of zeolite synthesis. In *Molecular Sieves*. (H. G. Karge and J. Weitkamp, eds.). Springer, Berlin Germany, and New York, NY, pp. 1–34.

Tielens, F., Baron, G. V., and Geerlings, P. (2002) *Fundamentals of Adsorption 7*. (K. Kaneko et al. eds.). IK International, Chiba City, Japan, p. 393.

Vance, T. B. and Seff, K. (1975) *J. Phys. Chem.* 79, 2163.

Vansant, E. F. (1990) *Pore Size Engineering in Zeolites* Wiley, New York, NY.

Vaughan, D. E. W. (1988) *Chem. Eng. Prog.* Feb., 25.

Wei, J. (1994) *Ind. Eng. Chem. Res.* 33, 2467.

Wilson, S. T., Lok, B. M., and Flanigen, E. M. U.S. Patent 4,310,440 (1982a).

Wilson, S. T., Lok, B. M., Messina, C. A., Cannan, E. R., and Flanigen, E. M. (1982b). *J. Amer. Chem. Soc.* 104, 1146.

Wilson, S. T., Lok, B. M., Messina, C. A., and Flanigen, E. M. (1984) Proc. 6th Intern. Conf on Zeolites. (D. H. Olson and A. Bisio, eds.). Butterworth, Guildford, UK, p. 97.

Yanagida, R. Y., Amaro, A. A., and Seff, K. (1973) *J. Phys. Chem.* 77, 805.

Yang, F. H. and Yang, R. T. (2002) Unpublished results.

Yang, R. T. and Hutson, N. D. Lithium-based zeolites containing silver and copper and use thereof for selective adsorption, U.S. Patent 60/114317 (December, 1998).

Yeh, Y. T. and Yang, R. T. (1989) *AIChE J.* 35, 1659.

Young, D. A. U.S. Patent 3,329,480 (1967) and U.S. Patent 3,329,481 (1967).

8

π-COMPLEXATION SORBENTS AND APPLICATIONS

The π-complexation bond is typically a weak bond that can be formed between the sorbent and sorbate. The sorbents that are used for separation and purification based on π-complexation are called π-complexation sorbents. Development of π-complexation sorbents began only recently. A number of such sorbents have already been used commercially, and tremendous potential exists for future applications in separation and purification, both for the chemical/petrochemical industry and environmental applications. For this reason, π-complexation sorbents are discussed in a separate chapter.

All major industrial adsorption processes are based on van der Waals and electrostatic interactions between the sorbate and sorbent. Chemical bonds have yet to be exploited in a significant way. Chemical complexation has been studied and used on a large scale in a number of other separation and purification processes by using mass separating agents (King, 1980). As suggested by King (1987), chemical complexation bonds are generally stronger than van der Waals interactions (thus giving rise to higher selectivities), yet many of them are weak enough to be reversible (i.e., to be broken by simple engineering means). This picture is well illustrated by the bond-energy-bond-type diagram of Keller (Humphrey and Keller, 1997). Indeed, a number of important separations have been proposed by King and co-workers, who used solvents with functional groups to form reversible chemical complexation bonds between the solute and solvent molecules (King, 1987). The π-complexation is a special class of chemical complexation. For π-complexation sorbents, it pertains to the main group (or *d*-block) transition metals (and there are 27 such elements). When interacting with a gas or solute molecule, these metals and their ions can form the usual σ bonds with their *s*-orbitals and, in addition, their *d*-orbitals can back-donate electron density to the antibonding π-orbitals of the molecule to be bonded. The π-complexation has been seriously considered for olefin/paraffin separation and purification by use of liquid

Adsorbents: Fundamentals and Applications, Edited By Ralph T. Yang
ISBN 0-471-29741-0

solutions containing silver (Ag^+) or cuprous (Cu^+) ions (Quinn, 1971; Ho et al., 1988; Keller et al., 1992; Blytas, 1992; Eldridge, 1993; Safarik and Eldridge, 1998). It has also been considered seriously for CO separation by using Cu^+ solutions, such as the COSORB process (Haase and Walker, 1974; Kohl and Reisenfeld, 1979). Although gas/solid operations can be simpler as well as more efficient than gas-liquid operations, particularly by pressure swing adsorption, the list of attempts for developing solid π-complexation sorbents was a short one until the recent work that will be described in this chapter. CuCl, which is insoluble in water, has been considered for olefin/paraffin separations by using CuCl powder as the sorbent (Gilliland et al., 1941; Long, 1972). The only successful solid sorbent of this nature, before our recent work, was $CuCl/\gamma\text{-}Al_2O_3$ for binding with the π bond of CO (Xie and Tang, 1990; Golden et al., 1992a; Golden et al., 1992b; Kumar et al., 1993). It should also be noted that the commercially available sorbents do not have significant selectivities for olefins (over corresponding paraffins) and that the use of these sorbents (e.g., 13X zeolite, with a olefin/paraffin separation factor of ~1.3, Da Silva and Rodrigues, 1999) would require additional, substantial operations (Kulvaranon et al., 1990; Jarvelin and Fair, 1993; Ghosh et al., 1993; Da Silva and Rodrigues, 2001).

Efficient solid π-complexation sorbents have been developed recently (Hirai et al., 1986b; Yokoe et al., 1987; Golden et al., 1992a; Yang and Kikkinides, 1995; Chen and Yang, 1996; Wu et al., 1999; Rege et al., 1998; Huang et al., 1999a, 1999b; Padin et al., 1999; Padin and Yang, 2000; Yang et al., 2001) for a number of applications in separation and purification. The bond between the sorbent and sorbate needs to be strong. However, excessively strong bonds would lead to either reaction or irreversible adsorption. Empirically, the adsorption is "reversible" when the bond is below 15–20 kcal/mol, that is, desorption can be achieved easily by simple engineering operations such as mild changes in pressure and temperature. The bonding strength between sorbate and sorbent depends on:

- Emptiness of the outer-shell *s*-orbital of the cation that is on the sorbent surface;
- The amount of π electrons in the target adsorbate molecule and the ease with which these π electrons can be donated to the *s*-orbital of the cation;
- The amount of *d*-orbital electrons of the cation and the ease with which they can be donated to the adsorbate molecule.

Molecular orbital theory has been used to study π-complexation (Chen and Yang, 1996; Huang et al., 1999a, 1999b; Yang et al., 2001; Takahashi et al., 2002). Molecular orbital theory can also serve as an ideal tool for designing sorbents for π-complexation for any given target adsorbate molecule. For this reason, a section (Section 8.2) will be devoted to the basics of molecular orbital theory calculations.

8.1. PREPARATION OF THREE TYPES OF SORBENTS

Although cations of many of the *d*-block metals can be used for π-complexation, Ag^+ and Cu^+ have been used most frequently. To prepare good sorbents, these cations need to be spread, at a high dispersion, on solid substrates that have high surface areas. Accordingly, there are three types of π-complexation sorbents:

a. Monolayer or near-monolayer salts supported on porous substrates
b. Ion-exchanged zeolites
c. Ion-exchanged resins

Among these three types, type (a) and type (c) are used for bulk separations, whereas type (b) is used for purification. Specific applications will be given in Sections 8.3 and 8.4.

8.1.1. Supported Monolayer Salts

It has been known for a long time that metal oxides and salts can be dispersed on solid substrates in a monolayer form, that is, as opposed to forming a stoichiometric compound with the support, or being dissolved in the support to form a solid solution. Russell and Stokes (1946) showed the first evidence for the monolayer dispersion of MoO_3 on γ-Al_2O_3. An extensive discussion on the subject has been given by Xie and Tang (1990). Monolayer dispersion of many ionic metal oxides and salts, particularly halides, has been accomplished by Xie and Tang (1990). Their substrates have included γ-Al_2O_3, silica gel, TiO_2, activated carbon, and a variety of zeolites.

There are two general approaches for dispersing monolayer or near-monolayer salts on porous supports: thermal monolayer dispersion and incipient wetness impregnation. Both will be discussed, although the latter has more practical use. The technique of spontaneous thermal dispersion has been described in detail in the literature (Xie and Tang, 1990; Xie et al., 1992). It was successfully applied to synthesize sorbents capable of π-complexation with olefins (Yang and Kikkinides, 1995; Cheng and Yang, 1995; Yang and Foldes, 1996; Deng and Lin, 1997; Rege et al., 1998; Padin and Yang, 2000). The work of Deng and Lin (1997) involved spreading salts by using microwave heating.

Thermal monolayer dispersion involves mixing a metal salt or oxide with a substrate at a predetermined ratio. This ratio is determined by the amount of salt that is required for monolayer coverage on the surface area of the substrate, assuming two-dimensional hexagonal close-packing. The BET surface area of the substrate is first measured. After the finely divided powders of the salt and substrate have been thoroughly mixed, it is heated at a temperature between the Tammann temperature and the melting point of the salt. The Tammann temperature is the point where the crystal lattice begins to become appreciably mobile, and it is approximately 1/2 T_m, where T_m is the melting point in absolute temperature. If the temperature is too low, the dispersion would take an unacceptably long time. A high dispersion temperature could cause the metal salt to oxidize or react with the substrate, and potentially deactivate the sorbent. A typical example of sample preparation is that for $AgNO_3/SiO_2$ (Padin and Yang, 2000). The SiO_2

had a surface area of 670 m^2/g, which would require ~0.55 g of $AgNO_3$ for monolayer close-packing. However, the surface area of the impregnated sample unavoidably decreases from that of the starting support. Hence the monolayer amount had to be determined empirically, which was 0.32 g/g. Thus, after thorough mixing of 0.32 grams of $AgNO_3$ per gram of SiO_2, the sample was heated in air at 200 °C for ~4 days to ensure complete dispersion. The BET surface area of this sample was 384 m^2/g, indicating some pore plugging as a result of $AgNO_3$ impregnation.

The thermal monolayer dispersion technique requires the use of fine powder and thorough mixing. Thus, pelletizing is needed when sorbents in the pellet form are to be used. An advantage of this technique is that salts that are insoluble in water can be dispersed directly, as in the case with the important salt CuCl. This technique is suitable for laboratory experiments.

The other technique involves incipient wetness impregnation, which is used at an industrial scale for catalyst preparation. It involves preparing a solution of the salt to be dispersed. The solution is then mixed with the substrate. It is then absorbed by the substrate due to incipient wetness. After the substrate has imbibed the solution containing the salt into its pore structure, the sample is heated to remove the solvent. Care must be taken when selecting solvents for use in this technique. First, the salt needs to be sufficiently soluble in the solvent to allow enough salt to be dissolved in the volume of solution that can be imbibed by the substrate pores. Second, the solvent selected needs to be able to wet the surface of substrate. The preparation of the $AgNO_3/SiO_2$ is used again to illustrate this technique (Padin and Yang, 2000). Because $AgNO_3$ is highly soluble in water, water was used as the solvent. Also, the high affinity of SiO_2 for water also ensures proper wetting of the surface. The pore volume and surface area of the SiO_2 were 0.46 cm^3/g and 670 m^2/g, respectively. A 1.2 M solution of $AgNO_3$ was first prepared. A volume of solution equal to the total pore volume of the SiO_2 support was mixed with the substrate such that an $AgNO_3/SiO_2$ weight ratio equal or close to 0.32 was achieved. The sample was then heated for 4 h at 105 °C in air to remove the water. The ratio of $AgNO_3/SiO_2$ in the resulting sample was calculated at 0.27. The BET surface area of this sorbent was 398 m^2/g, which indicated some pore plugging.

Water-Insoluble Salts. Many important salts for π-complexation are water-insoluble. The best examples are cuprous (Cu^+) salts, e.g., CuCl. The most practical technique for preparing monolayer CuCl is by a two-step process: incipient wetness impregnation of $CuCl_2$ followed by reduction to CuCl. This process also applies to other cuprous salts, as the cupric salts are generally water-soluble. Attempts have also been made for direct impregnation of CuCl. This could be accomplished by two ways: using acid or basic solutions or dissolving CuCl with the aid of ammonium chloride. These two techniques will be first briefly described, and the two-step process will be then discussed in more details.

Cuprous salts are generally soluble in acid or base solutions. Thus, one could impregnate a porous support with CuCl directly. For example, following the procedure of Hirai et al. (1986b), Tamon et al. (1996) have impregnated CuCl on activated carbon by using CuCl in 1 N HCl solution. However, thorough washing is required after impregnation to remove the HCl. The resulting sorbent is not fully covered by CuCl, and the bare surface of carbon will participate in adsorption of the gas mixture, thereby lowering the separation factor. Although washing was not mentioned by Hirai et al. (1986b), the CuCl loadings in their samples were also well below the amounts for monolayer coverage. A similar procedure was used by Yokoe et al. (1987). In their work, both activated carbon and activated alumina were used as the support. Before the activated alumina was impregnated with CuCl, it was first impregnated with a hydrocarbon solution followed by coking in nitrogen such that a layer of carbon (at 1–3% wt.) was deposited. Presumably, the carbon substrate provided a reducing environment that helped keep the CuCl in the reduced state. Cuprous-ammonium-salt aqueous solutions have been widely used commercially for CO absorption (Kohl and Riesenfeld, 1979; Safarik and Eldridge, 1998). The cuprous salts are solubilized by complexing with ammonia, forming $Cu(NH_3)_2{}^+$, which is soluble (Van Krevelen and Baans, 1950). Thus, the cuprous-ammonium-salt aqueous solution can be used directly for incipient wetness impregnation. Alternatively, non-aqueous double salts containing cuprous ion (Blytas, 1992; Safarik and Eldridge, 1998) can be used. Of particular interest is $CuAlCl_4$ in toluene (Long et al., 1972; 1973; 1979). Using the $CuAlCl_4$/toluene solution, Hirai et al. (1986a) successfully prepared $CuAlCl_4$ supported on activated carbon. At 1 atm and 20 °C, the adsorbed amounts on the original activated carbon were (in cc/g) 9.5 (CO), 21.5 (CH_4), whereas and 7.2 (N_2), while the corresponding amounts on the $CuAlCl_4$ impregnated carbon were 24.1 (CO), 4.2 (CH_4), and 0.8 (N_2). This result indicated that the surface of carbon was nearly fully covered by the salt. Hirai et al. (1985) also prepared cuprous ammonium chloride supported on resin by using the amino groups of the anion exchange resin. These sorbents showed good selectivities for ethylene over ethane. The disadvantage of using $CuAlCl_4$ and $Cu(NH_3)_2Cl$ is obvious. The loading of Cu^+ is substantially reduced due to the presence of the other salt, hence lowering the sorbent capacity.

CuCl/activated alumina and CuCl/activated carbon have been used commercially for CO separation and recovery by pressure swing adsorption (Golden et al., 1992a; 1992b; Kumar et al., 1993; Golden et al., 1998). These sorbents were prepared by incipient wetness impregnation of the support with an aqueous solution of a cupric compound with the aid of an "dispersant" (Golden et al., 1992a). Ammonium citrate was used as the dispersant, which stabilizes Cu^+ and should not be decomposed. The solvent was subsequently removed by heating in an inert atmosphere at an elevated temperature (e.g., 200 °C). For the activated carbon support, a pre-oxidation step was found helpful for CuCl dispersion. Oxidation can be performed by a number of means (see Chapter 5). Introduction of the surface oxide groups by oxidation increased the interactions with water, or wettability, which improved the adsorption of cupric salt and its dispersion.

Examples of the sample preparation were given by Golden et al. (1992a). For the activated alumina support, the sample was first heat-treated in air at

200 °C, followed by incipient wetness impregnation with an aqueous solution of $CuCl_2$ and ammonium citrate. For each kilogram of alumina, 0.49 liters of aqueous solution containing 0.31 kg $CuCl_2 \cdot 2H_2O$ and 0.0375 kg ammonium citrate was used. The resulting material was air-dried at 120 °C, followed by activating at 200 °C in nitrogen. The pore volume of activated alumina is generally near 0.3–0.4 cm^3/g (Figure 5.1). Hence the volume of the solution was in excess of the pore volume, since 0.49 cm^3/g of solution was used. For the activated carbon support, the sample was first pre-oxidized in wet (at 20% relative humidity) air at 120 °C. The pore volume of activated carbon is ~0.6 cm^3/g (Figure 5.1). Again, an excessive volume of aqueous solution was used. For each kilogram of activated carbon, 1 liter of aqueous solution containing 0.7 kg $CuCl_2 \cdot 2H_2O$ and 0.05 kg ammonium citrate was used. The resulting material was subjected to the same post-treatment as that used for the activated alumina sample.

The final activation step (i.e., in N_2 at 200 °C) described above (Golden et al., 1992a) was not adequate to fully reduce Cu^{2+} to Cu^+. A mixture of 25% CO and 75% H_2 was subsequently used as the reducing gas in the final reduction step (Golden et al., 1992b). Copper in the reduced states, with a valence of 1 or 0, can form π-complexation bonding with CO, while Cu^{2+} cannot form chemical bond with CO. Therefore, the extent of reduction can be measured by CO adsorption. Golden et al. (1992b) reported results of CO adsorption with sorbents treated under different conditions, shown in Table 8.1. The Cu^+ dispersion was also measured by CO chemisorption. From these results, it appears that reduction with CO/H_2 at 150 °C was the optimum treatment. However, an in situ (i.e., in the PSA system) activation step was described later (Golden et al., 1998) where 80–85% of the Cu^{2+} was reduced to Cu^+ by a synthesis gas (16% CO) at 90 °C.

The reduction behavior of $CuCl_2/\gamma\text{-}Al_2O_3$ has been studied by Takahashi et al. (2000) by using temperature programmed reduction (TPR) with hydrogen. As shown in Figure 8.1, the reduction took place in two steps: with peak temperatures at 270 °C and 410 °C. The first peak corresponded to reduction of Cu^{2+} to Cu^+, and the second was for Cu^+ reduction to Cu^0. The second step prolonged beyond 450 °C, indicated by the long tailing of the TPR curve at 450 to 700 °C. Thus, with 5.3% hydrogen, the optimum temperature window for reduction to CuCl

Table 8.1. CO working capacity

Reduction Condition	CO Working Capacity (mmol/g)	Cu^+ Dispersion (%)
Air-dried, 120 °C	0.02	1
25% CO & 75% H_2, 70 °C	0.20	29
25% CO & 75% H_2, 120 °C	0.52	54
25% CO & 75% H_2, 150 °C	0.63	72
N_2, 200 °C	0.45	56

Between 1 and 0.1 atm at 20 °C) for CuCl/alumina after different reduction treatments (Golden et al., 1992b.

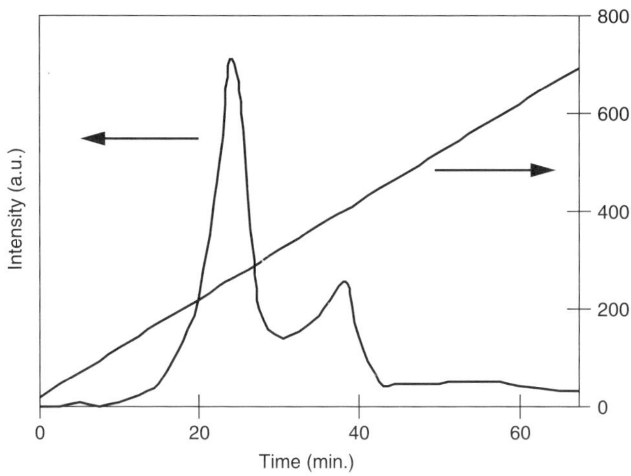

Figure 8.1. Temperature-programmed reduction of $CuCl_2/\gamma$-Al_2O_3 with 5.3% H_2 (in He) at a heating rate of 10 °C/min (Takahashi et al., 2000, with permission). Hydrogen consumption is expressed in arbitrary unit (a.u.).

Table 8.2. Empirical monolayer dispersion capacity on activated alumina

Salt	Dispersion Capacity ($g/100\ m^2$)
CuCl	0.095
$CuCl_2$	0.077
$AgNO_3$	0.083

was approximately 200–300 °C. A sorbent was subsequently prepared at 270 °C (Takahashi et al., 2000).

The empirical ratios for monolayer dispersion of CuCl, $CuCl_2$, and $AgNO_3$ on activated alumina are given in Table 8.2. These values are slightly lower than the theoretical values for close-packing (Xie and Tang, 1990). These values do not vary appreciably among different substrates (e.g., silica gel and activated carbon), and can be used as a general guide for sorbent preparation.

8.1.2. Ion-Exchanged Zeolites

Ion exchange is a chemical reaction. For uni-univalent ion exchange, it can be represented by:

$$A_S{}^+ + B_Z{}^+ \longleftrightarrow A_Z{}^+ + B_S{}^+ \tag{8.1}$$

where A and B are cations and S and Z denote the solution and zeolite phases. The selectivity for ion exchange between different cations depends on the free energy change of the reaction. Because ion exchange is usually performed in aqueous solutions, steric hindrance can be a factor when the size of the cation, or the hydrated cation, is larger than the aperture. This is particularly the case with type A and type X zeolites, where only small cations can penetrate the 6-oxygen rings into the sodalite cage. For zeolite A, large cations may also be excluded from the supercage by the 8-ring windows.

Cation exchange in zeolites has been discussed in detail by Breck (1974). The cation exchange behavior depends on (1) the nature of the cation, its size and charge; (2) the temperature; (3) the concentration of cations in solution; (4) the anions and the solvent; and (5) the structure of the zeolite. As a general rule, the equilibrium selectivity favors cations of a higher valence. The selectivity favors cations with a higher atomic weight for those with the same valence. The selectivity follows the relative order of free energies of reaction for different cations, favoring the reaction with the most negative free energy of reaction.

For type X and Y zeolites, below a level of 34 cations/unit cell (or, 40% ion exchange of a typical X zeolite with 86 cations per unit cell), the order of selectivity for univalent ions follow (Sherry, 1966; Breck, 1974):

$$\mathrm{Ag} \gg \mathrm{Tl} > \mathrm{Cs} > \mathrm{Rb} > \mathrm{K} > \mathrm{Na} > \mathrm{Li} \tag{8.2}$$

This series corresponds to occupancy of the most accessible cation sites (sites III and IV) within the supercage. At 50% exchange of X zeolite, which includes site II in the 6-ring adjacent to the supercage, the selectivity series is (Breck, 1974)

$$\mathrm{Ag} \gg \mathrm{Tl} > \mathrm{Na} > \mathrm{K} > \mathrm{Rb} > \mathrm{Cs} > \mathrm{Li} \tag{8.3}$$

These sites (II, III, and IV) are exposed to the supercage and hence important for adsorption.

The selectivity for Ca^{2+}, Sr^{2+}, and Ba^{2+} is similar to that of Rb^+ and K^+, whereas the selectivity of Ce^{2+} and La^{2+} is similar to that of Ag^+ (Sherry, 1967; Sherry, 1968; Breck, 1974). The selectivity for the important ion used in preparing π-complexation sorbents, Cu^{2+}, is not available. However, Cu^{2+} can be exchanged with ease for type X, Y, and ZSM zeolites (e.g., Huang, 1973; Rabo et al., 1977; Takahashi et al., 2001a).

In tailoring sorbents for π-complexation, both the cation–sorbate bond strength and the total number of cations are important. The density of cations depends on the cation exchange capacity of the zeolite. Table 8.3 provides useful information on the total cation capacities for a number of zeolites.

Ag^+ *Exchange.* Ag-Y has been shown to be an excellent sorbent for a number of purification processes, including the removal of dienes from olefins. This process has been used commercially. As discussed above, Ag^+ has a high selectivity for zeolites; hence it can be exchanged readily at the ambient temperature and low concentrations. A typical sample preparation procedure is described here for Ag^+ exchange in Y zeolites with different Si/Al ratios, and also for partial ion exchange for a given Y zeolite. (Takahashi et al., 2001b).

Table 8.3. Cation exchange capacity of various zeolites

Zeolite	Si/Al Ratio	Meq/g (Anhydrous)
Chabazite	2	5
Mordenite	5	2.6
Erionite	3	3.8
Clinoptilolite	4.5	2.6
Zeolite A	1	7.0
Zeolite X	1.25	6.4
Zeolite Y	2.0	5.0

Breck, 1974, with permission.

In the work of Takahashi et al. (2001b), four as-received zeolites were used: Na-type Y-zeolite (Si/Al = 2.43, or 56 Al atoms/unit cell), NH_4-type Y-Zeolite (Si/Al = 6, or 27.4 Al atoms/u.c.), H-type Y-zeolite (Si/Al = 15, 12 Al atoms/u.c.), and H-type ultra-stable Y-zeolite (H-USY, Si/Al = 95, 0.98 Al atoms/u.c.). All zeolites were in powder form (binderless). The zeolites were ion-exchanged by using excess amounts (five-fold cation-exchange-capacity) of 0.1 M $AgNO_3$ at room temperature for 24 h. Because Ag^+ has a higher selectivity compared with Na^+, this procedure ensured 100% exchange for Na-Y (Si/Al = 2.43) (Padin et al., 1999). The same procedure was applied to NH_4-Y (Si/Al = 6), H-Y (Si/Al = 15) and H-USY (Si/Al = 195). After the exchange, the zeolite suspension was filtered and washed with copious amounts of deionized water until no free ions were present in the filter water (i.e., no precipitation was observed upon treatment with Cl^-). The products were dried at room temperature and atmospheric conditions in a dark area.

To prepare mixed-cation zeolites by partial exchange, information on the ion-exchange equilibrium between Ag^+ and Na^+ is needed in order to control the Ag^+-exchange ratio. Thus, the ion-exchange isotherm for Ag-Na-Y-zeolite (Si/Al = 2.66) developed by Sherry (1966) was used by Takahashi et al. (2001b). The Na^+ and Ag^+ contents in the aqueous solution (0.1 M total) were pre-determined so the final ratio of $Ag^+/(Ag^+ + Na^+)$ in the zeolite was 21 or 49%, the desired ratios. These Ag^+-exchange ratios were chosen to match the Ag contents (or Ag^+/unit cell) in the fully Ag^+ ion-exchanged Ag-Y(Si/Al = 15) and Ag-Y(Si/Al = 6), respectively. $NaNO_3$ and $AgNO_3$ serve as the sources for Na^+ and Ag^+. Five-fold CEC of Ag^+ was added in the solution for full ion exchange. The Ag^+/Na^+ molar ratios in the solution were 0.04/0.96 for 21% Ag^+-exchange and 0.17/0.83 for 49% Ag^+-exchange. The procedure for mixed cations ion-exchange was exactly the same as the single ion-exchange procedure described above.

Preparation of Cu^+*-Zeolites.* Cu^+-zeolites are of particular interest as sorbents for π-complexation because Cu^+ can interact with CO and olefins (and other ligands) more strongly than Ag^+ (Huang et al., 1999a), and also for economic reasons. However, direct ion exchange with Cu^+ is not applicable because the cuprous salts are water-insoluble. Also, the cuprous compounds are fairly readily oxidized to cupric compounds in solution. Therefore, Cu^{2+} is exchanged first, followed by partial reduction of Cu^{2+} to Cu^+. Two approaches can be taken for the reduction step. The one studied most extensively involves the use of a reducing gas. The other approach used more recently is "auto-reduction," that is, without a reducing gas.

Cu(II)Y was first reduced with CO at 400 °C to Cu(I)Y by Naccache and Ben Taarit (1971), and the same procedure was subsequently used for preparing Cu(I)Y by others (Chao and Lunsford, 1972; Huang and Vansant, 1973; Pearce, 1988). Huang (1973) reported that by pre-adsorbing ammonia, Cu(II)Y could be reduced by CO completely to Cu(I)Y at substantially lower temperatures, as low as 100 °C. Rabo et al. (1977) reported the preparation of Cu(I)-ZSM-5 zeolite by a similar two-step process. In the reduction step, Rabo et al. (1977) reported that a mixture of 3% H_2O and 97% CO was the most effective reducing atmosphere and that the cuprous-form zeolite was obtained at 250–300 °C. Reduction of Cu(II)Y to Cu(I)Y by ethylene at 1 atm and 150 °C was reported by Cen (1989). In the work of Takahashi et al. (2001a), Cu(II)Y was reduced completely to Cu(I)Y with 75% CO in He at 450 °C and 12 h.

Auto-Reduction of Cu(II)Y *to* Cu(I)Y. Because Cu^{2+} is exchanged as $[Cu^{2+}OH^-]^+$, it is conceivable that it can be reduced in an inert atmosphere without the use of a reducing gas. This has indeed been proven and, as mentioned, this process is referred to as auto-reduction. Auto-reduction of Cu(II)-ZSM-5 to Cu(I)-ZSM-5 has been studied for the selective catalytic reduction of NO with hydrocarbon, where Cu-ZSM-5 is a catalyst (Iwamoto and Hamada, 1991; Sarkany et al., 1992; Larson et al., 1994; Shelef, M., 1994). More recently, Takahashi et al. (2001a) successfully prepared Cu(I)Y sorbent by auto-reduction and heating Cu(II)Y at 300–450 °C (for 1 h). In the work of Takahashi et al. (2001a), NaY was first exchanged to Cu(II)Y with 10-fold cation exchange capacity of 0.5 M $Cu(NO_3)_2$ at room temperature for 24 h. This procedure yielded 72% replacement of Na^+. The sample was subsequently heat-treated in He at 300–450 °C. Analysis of the resulting sample heat-treated at 450 °C showed that at least one-half of the copper existed as $[Cu^{2+}OH^-]^+$. In a study of Cu(II)-ZSM-5 that was subjected to auto-reduction, Larson et al. (1994) assumed that protons existed as the cations other than $[Cu^{2+}OH^-]^+$.

The mechanism of auto-reduction of Cu^{2+} in zeolites has been studied by a number of groups that used Cu-ZSM-5. Two main mechanisms are

(1) Mechanism by Larson et al. (1994)

$$[Cu^{2+}OH^-]^+ \longleftrightarrow Cu^+ + OH$$

$$[Cu^{2+}OH^-]^+ + OH \longleftrightarrow Cu^{2+}O^- + H_2O$$

$$2[Cu^{2+}OH^-]^+ \longleftrightarrow Cu^+ + Cu^{2+}O^- + H_2O$$

(2) Mechanism by others (Iwamoto and Hamada, 1991; Sarkany et al., 1992; Valyon and Hall, 1993)

$$2[Cu^{2+}OH^-]^+ \longleftrightarrow [CuOCu]^{2+} + H_2O$$

$$[CuOCu]^{2+} \longleftrightarrow 2Cu^+ + 1/2O_2$$

It appeared that the second mechanism was appropriate for the auto-reduction of Cu(II)Y to Cu(I)Y (Takahashi et al., 2001a).

8.1.3. Ion-Exchanged Resins

Many macroreticular polymeric resins are available commercially. A series of polystyrene cross-linked with divinyl benzene is also available commercially (Albright, 1986). These resins are functionalized to form cation-exchange or anion-exchange resins (Albright, 1986). The functional group for cation exchange is the sulfonate group in the form of $C_6H_5SO_3^-$. Usually the cation is H^+ (in the amberlyst resins) or Na^+ (in the DOWEX resins). These resins have high cation-exchange capacities (in the range of 4–5 meq/g), and the cations can be exchanged readily.

The resins are known to be hydrophobic as well as lyophobic, that is, with a low affinity for hydrocarbons. Although heavy hydrocarbons have high polarizabilities, the lyophobicity results from the weakness of electric fields on the surfaces of the polymeric resins. This property makes the resin an ideal candidate for modification as highly selective sorbents for separation and purification. After modification by ion exchange with a cation such as Ag^+, the sorbent will be highly selective for hydrocarbons that have π-electrons, while little of the hydrocarbons without π-electrons will be adsorbed.

Figure 8.2 shows the isotherms of ethane and ethylene on the cation exchange resin Amberlyst 15 (Yang and Kikkinides, 1995). Because of the lyophobicity of the resin, the amounts adsorbed were considerably lower than those on all other commercial sorbents, such as activated carbon, silica gel, and zeolites. However, upon ion exchange of H^+ by Ag^+, the amount adsorbed of ethylene exhibited a dramatic sevenfold increase, due to π-complexation between ethylene and Ag^+, while the adsorption of ethane was unaffected.

The procedure of sample preparation is given next (Yang and Kikkinides, 1995). Amberlyst 15 (from Rohm & Haas Company) was used as the cation exchange resin. It contained 20% divinylbenzene and was available as spherical beads in the size range of 16–50 U.S. mesh. The BET surface area was 55 m^2/g, and the cation exchange capacity was 4.7 meq/g. Its average pore diameter was given as 24 nm (Albright, 1986). Prior to ion exchange, the sample was washed successively with de-ionized water and methanol, followed by drying in air at 100 °C for 2 h. The sample was ready for ion exchange. The exchange was performed with a dilute (0.014 N) solution of $AgNO_3$ at room temperature. After repeated exchanges, the resin was subjected to successive washing with de-ionized water and methanol, followed by air drying at 100 °C. Methanol

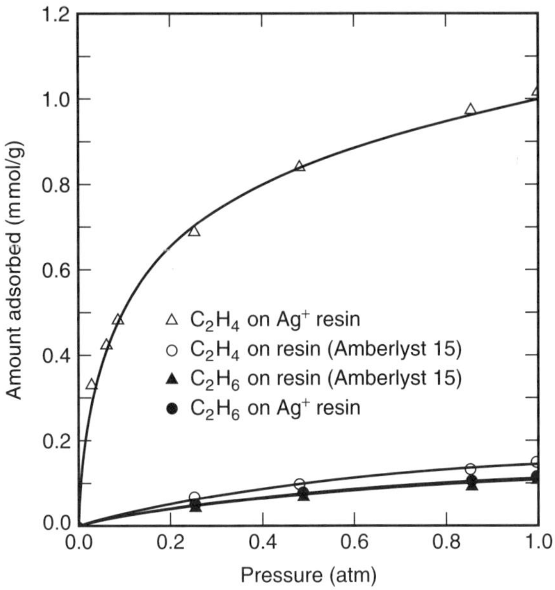

Figure 8.2. Equilibrium isotherms of C_2H_6 and C_2H_4 on Amberlyst 15 cation exchange resin (H^+ form) and the resin after 51.7% Ag^+ exchange, at 25 °C (Yang and Kikkinides, 1995).

was used to displace the water that remained in the polymer matrix and voids. The extent of ion exchange could be determined by the weight gain. After two exchanges, 51.7% H^+ was exchanged by Ag^+. This was the sample used in the results shown in Figure 8.2.

8.2. MOLECULAR ORBITAL THEORY CALCULATIONS

Molecular orbital (MO) theory is ideally suited for studies of sorbents for π-complexation. It provides a fundamental understanding of the bonding between the sorbate and sorbent. It can also be used to guide the design of π-complexation sorbents; that is, for a given target sorbate molecule, MO can be used to determine the best cation as well as the best companion anion that should be dispersed on the surface of the sorbent. Furthermore, in principle, the electron correlation term in MO calculation represents the dispersion energies (e.g., Sauer, 1989). Thus, MO has the potential to be used for predicting physical adsorption. However, we are concerned with only the π-complexation bonds for adsorption in this chapter.

8.2.1. Molecular Orbital Theory — Electronic Structure Methods

Electronic structure methods use the laws of quantum mechanics as their basis for computations. Quantum mechanics state that the energy and other related properties of a molecule may be obtained by solving the Schrödinger equation:

$$H\psi = E\psi \tag{8.4}$$

where H is the Hamiltonian operator, E is the energy of the particle (an electron or nucleus), and ψ is the wave function. The product of ψ with its complex conjugate ($\psi * \psi$, often written as $|\psi|^2$) is interpreted as the probability distribution of the particle (or electron).

Electronic structure methods are characterized by their various mathematical approximations to its solution, since exact solutions to the Schrödinger equation are not computationally practical. There are three classes of electronic structure methods: semi-empirical methods, density functional theory (DFT) methods, and *ab initio* methods.

8.2.2. Semi-Empirical Methods

Semi-empirical methods are parameters derived from experimental data to simplify the computation. They are used to solve an approximate form of the Schrödinger equation that depends on having appropriate parameters available for the type of chemical system under investigation. Different semi-empirical methods are largely characterized by their differing parameter sets. One of the most commonly used semi-empirical computer programs is MOPAC, developed by Stewart (1990). A commercial package based on MOPAC is available from Fujitsu Company of Japan. MOPAC includes the semi-empirical Hamiltonians MNDO (modified neglect of diatomic overlap), MINDO/3 (modified intermediate neglect of differential overlap), AM1 (Austin model 1), PM3 (parametric method 3), MNDO-d, and PM5. These methods have been calibrated by using experimental data for thermodynamic properties such as heats of formation (Pople et al., 1965; Baird and Dewar, 1969; Bodor et al., 1970; Murrell and Harget, 1972).

The advantage of semi-empirical methods is in the economy of computation. These programs can be performed with a personal computer. However, the bond energies calculated from these programs are substantially higher than the actual values. In many instances, the bond energies are divided by an empirical factor of 5 to give an indication of the real values. MOPAC provides the most accurate energy calculations among these methods, while the results in bond energies are still substantially over-estimated (e.g., Chen and Yang, 1997).

8.2.3. Density Functional Theory Methods

The approach of density functional theory (DFT) was developed in the 1960s by using mathematical functions, called functionals, to describe the electron density (Hohenberg and Kohn, 1964; Kohn and Sham, 1965). A summary of the theory is given by Foresman and Frisch (1996), and a detailed description is available from Parr and Yang (1989). DFT methods are attractive because they include the effects of electron correlation (i.e., electron–electron interaction) in the energy. Developments in DFT have led to nonlocal (gradient-corrected) functionals — BLYP (Becke-Lee-Yang-Parr), and to hybrid functionals — B3LYP. The nonlocal functionals account for the non-uniformity of the overall electron distribution. The hybrid functionals use a linear combination of the Hartree/Fock (HF) and the

DFT electron correlation, with parameters adjusted to provide the best fit with specific experimental data. The dynamic interaction between electrons is theoretically included by these density functional methods. This gives these methods the benefit of including electron–electron correlation for a computational expense similar to HF, giving DFT methods the major advantage of low computational cost compared to accuracy (Hohenberg and Kohn, 1964; Kohn and Sham, 1965; Parr and Yang, 1989; Foresman and Frisch, 1996).

8.2.4. *Ab Initio* Methods

Quantum mechanics provides a potential method for the complete description of the electronic properties of molecular systems, their structures, physical and chemical properties, and reactivities. Unlike semi-empirical methods, *ab initio* methods use no experimental parameters in their computations; they are based solely on the laws of quantum mechanics — the first principles referred to in the name *ab initio* (Foresman and Frisch, 1996). The computational difficulties encountered in the general case, as well as the magnitude of extraneous information generated by multi-electron wave functions, have been overcome by the development of entire conceptual frameworks, new computational methods, and more powerful computational machines. Progress in molecular orbital calculation has made it possible to make reliable predictions of molecular structures, relative energies, potential surfaces, vibrational properties, reactivities, reaction mechanisms, and so on. An increasing number of molecular orbital computer programs have become available, for example, Gaussian, GAMESS and AMPAC. Among all programs, Gaussian is most popular and has been applied successfully in many fields. It provides high-quality quantitative predictions for a broad range of systems. Gaussian 98 can handle jobs of more than 100 atoms on supercomputer systems.

Different *ab initio* methods can be characterized by their treatment of electron–electron interactions, that is, electron correlation. The first practical *ab initio* method was the HF method, which treats each electron as if it exists in a uniform field made from the total charge and space occupied by the other electrons. This treatment is only an approximation to the interactions between electrons as point charges in a dynamic system and excludes the contribution of excited electronic configurations. This neglect of electron correlation can lead to significant errors in determining thermochemical properties. It was theorized that the electron correlation was a perturbation of the wave function known as the Møller/Plesset perturbation (MP) theory, so the MP theory could be applied to the HF wave function to include the electron correlation. As more perturbations are made to the system, more electron correlation is included. These methods are denoted as MP2, MP3, and MP4. Another method is to calculate the energy of the system when electrons are moved into vacant orbitals, such as the QCISDT (quadratic configuration interaction with all single and double excitations and perturbative inclusion of triple excitations) method, which improves energy values but at greater computational costs (Clark, 1985; Hehre et al., 1986; Foresman and Frisch, 1996).

8.2.5. Basis Set

Both *ab initio* and DFT methods use sets of mathematical functions to represent the atomic orbitals. These are called the basis set. These mathematical functions are themselves made from a combination of simpler mathematical functions called primitives. Increasing the number of primitive functions and including contributions from valence orbitals imposes less restriction on the location of the electron and therefore more accurately models the exact molecular orbitals, but correspondingly increases the computational cost. The molecular orbitals are approximated as linear combinations of the basis functions. In the *ab initio* methods, a Gaussian-type atomic function is used as the basis function, which has the general form:

$$g(\alpha, \vec{r}) = cx^n y^m z^l e^{-\alpha r^2} \tag{8.5}$$

where vector $\vec{r}$ is the position of the electron, which is composed of coordinates x, y, and z, and α is a constant that determines the size (radial extent) of the function. The Gaussian function $e^{-\alpha r^2}$ is multiplied by powers (possibly zero) of x, y, and z and is normalized by constant c, so that:

$$\int_{\text{allspace}} g^2 = 1 \tag{8.6}$$

Linear combinations of primitive Gaussian functions shown in Eq. 8.5 are used to form the basis functions.

A set of standard basis sets has been devised to increase the comparability between researchers and to simplify the nomenclature when describing the model chemistry: STO-3G, 3-21G, 6-311G, 6-311G(*d*,*p*), 6-311 + G(*d*, *p*)... in the order of increasingly large basis sets. The smallest of these, STO-3G, is the abbreviated name for three Gaussian primitives (3G) to model a Slater-type orbital (STO). When each orbital is represented by two or more sizes of basis functions, we have split-valence basis set. The 3-21G uses two sizes of Gaussian primitives to represent each orbital, and a 6-311G uses three sizes of Gaussian primitives to represent each orbital. To allow orbitals to change shape, we can include contributions from orbitals of higher angular momentum, for instance, include one of more *d* orbitals on carbon. The notation for this involves adding a letter for the orbital type to the end of the abbreviation, for example, 6-31G(*2d,p*) adds two *d* orbitals to heavy atoms and one *p* orbital to hydrogen atoms. A further development in basis has been to add very large versions of orbitals to the basis set, called diffuse functions. This is denoted by adding a + to the abbreviation before the G, so 6-31+G adds diffuse functions to heavy atoms and 6-31++G adds diffuse functions to both heavy and hydrogen atoms. The best choice of basis set is largely dependent on the chemistry being studied.

8.2.6. Effective Core Potentials

Relativistic effects must be considered in the applications of *ab initio* molecular orbital calculations for the heavier elements; they have a significant influence on the physicochemical properties of molecules. The use of effective core potentials (ECP) has been a notable success in the molecular orbital calculations involving transition metals. Hence this method has been particularly useful for studies on π-complexation sorbents. ECP is simply a group of potential functions that replace the inner shell electrons and orbitals that are normally assumed to have minor effects on the formation of chemical bonds. Calculations of the valence electrons using ECP can be performed at a fraction of the computational cost that is required for an all-electron calculation, while the overall quality of computation does not differ much. In addition, the relativistic mass-velocity and Darwin terms, which are derived from all-electron relativistic HF calculations, are implicitly incorporated into the relativistic effective core potentials for heavier elements ($Z > 36$). Combined with the use of reliable basis sets, it appears to be a very powerful and economical method for dealing with molecules containing heavy transition metals. Recently, Hay and co-workers have shown that effective core potentials can be used reliably in density functional computations as well. The LanL2DZ basis set is a double-zeta basis set containing effective core potential representations of electrons near the nuclei for post-third row atoms. The reliability of this basis set has been confirmed by the accuracy of calculation results compared with experimental data as well as those from a more expensive all-electron basis set (Hay and Wadt, 1985; Gordon and Cundari, 1996).

8.2.7. Model Chemistry and Molecular Systems

Concepts of model chemistry and molecular system are required for *ab initio* molecular orbital calculation. Model chemistry refers to all theoretical aspects of calculation, whereas the molecular system refers to the molecules to be studied.

Model chemistry encompasses two elements: method + basis set, where method and basis set deal with Hamiltonian operator and wave function in the Schrödinger equation, respectively. Many methods and basis sets are available in the commercial *ab initio* molecular orbital calculation packages. The suitable combination of methods and basis set, as well as the selection of calculation level, is very important for a systematic calculation of a studied system. The higher the model chemistry, the more accurate the results. However, a highest model chemistry is to be avoided since the computational cost will increase with calculation level logarithmically. Using the minimal computational resources to achieve accurate enough results is a challenge for *ab initio* molecular orbital calculation (Foresman and Frisch, 1996).

Molecular system refers to the correct combination of atoms. The *ab initio* molecular orbital calculation is developed strictly for isolated molecules. Therefore, the correct extraction of a finite model from the infinite solid phase and the saturation of the boundaries of the model are crucial steps for calculations. Reviews on the application of *ab initio* molecular orbital calculation to the heterogeneous gas-solid systems are available (Sauer, 1989).

A suitable model chemistry may work well for a selected molecular system, but not for another. Therefore, there is a general procedure and criterion for the selection of model chemistry and molecular system. Usually, one selects a medium size of molecular model as a starting molecular system, and then performs a series of calculations with several different levels of model chemistries. After comparing the calculated results with experimental data, usually vibration frequencies, and if they match fairly well, one can expect that the similar/other molecular system (slightly enlarged) + model chemistry (a slightly higher level) can be considered valid and well representative. Subsequently, higher-level calculations may be performed to obtain more accurate results. Obviously, lowering the calculation level and decreasing the molecular system size lead to invalid results. Finally, there is a general notation for a given series of calculations (Foresman and Frisch, 1996):

Energy_Method/Energy_Basis Set//Geometry_Method/Geometry_Basis Set

where the model to the left of the double slash is the one at which the energy is computed, and the model to the right of the double slash is the one at which the molecular geometry was first optimized. For example, RHF/6-31+G//RHF/6-31G(*d*) denotes that the energy calculation was performed by using Hartree-Fock theory and the 6-31+G basis set on a structure previously optimized with Hartree-Fock theory and the 6-31G(*d*) basis set.

The saturation of boundaries for a model extracted from an infinite solid has been a confusing issue. There is no standard rule or criterion for boundary saturation. Point charge has been used to balance the boundary charges, but without any geometric and chemical meanings for the boundaries. Hydrogen replacement is widely used both to balance the charge and to terminate the boundaries, though it can introduce somewhat incorrect geometric and chemical environment to the boundaries. Regardless of the method that is used to saturate the boundaries, the final goal is to achieve the calculation results closest to the experimental data, usually the vibrational frequency data (Foresman and Frisch, 1996).

8.2.8. Natural Bond Orbital

The local atomic properties in a molecular system are often the most important properties from a chemical point of view, although they are not quantum mechanically observable. Usually there are several built-in schemes in a given molecular orbital calculation package, including the Gaussian 98 package, to partition the electron density among atoms in a molecular system and ultimately obtain certain atomic properties. The most popular scheme is Mulliken population analysis. Some other schemes, such as natural bond orbital (NBO) and Merz-Kollman-Singh analysis have also been used.

Atomic charge, orbital energy, and population are important pieces of information for determining electronic configuration, net charge association, and the nature of the bond. Mulliken population analysis is a widely used method in most *ab initio* molecular orbital calculations. However, reports about Mulliken population analysis that fail to yield reliable characterization of molecular systems have appeared. A more accurate method for population analysis, NBO, was introduced in 1983 (Glendening et al., 1995). The NBO method transforms a given wave function for the whole molecular structure into localized forms corresponding to one-center and two-center elements. The NBO method encompasses sequential calculations for natural atomic orbitals (NAO), natural hybrid orbitals (NHO), NBOs, and natural localized molecular orbitals (NLMO). It performs population analysis and energetic analysis that pertain to localized wave function properties. It is very sensitive for calculating localized weak interactions, such as charge transfer, hydrogen bonding, and weak chemisorption. Therefore, NBO is the preferred method for population analysis in studying adsorption systems involving weak adsorbate–adsorbent interactions (Mulliken and Ermler, 1977; Frisch et al., 1998).

8.2.9. Adsorption Bond Energy Calculation

Geometry optimization is the first step in all calculations. Calculations for all other parameters such as charges, orbital populations, and energies are all based on the geometrically optimized system. In geometry optimization, the geometry is adjusted until a stationary point on the potential surface is found, which means the structure reaches energy minimum. All adsorbate–adsorbent systems are subjected to geometry optimization first at the STO-3G level followed by the 3-21G or G-311G level (Chen and Yang, 1996; Huang et al., 1999a and 1999b). The bond lengths calculated by the 3-21G basis set deviates from experimental values by only 1.7% (e.g., by 0.016 for a 0.95 hydrogen bond). After geometry optimization, a number of higher-level basis sets, all including electron correlation, with NBO calculations are performed to obtain information such as energies, atomic charges, and orbital populations (occupancies) based on the same geometry-optimized system. Typically, energy and NBO calculations are performed on the B3LYP/3-21+G** level (Huang et al., 1999).

The energy of adsorption, E_{ads}, is calculated by using the optimized geometries by:

$$E_{ads} = E_{adsorbate} + E_{adsorbent} - E_{adsorbent-adsorbate} \tag{8.7}$$

where $E_{adsorbate}$ and $E_{adsorbent}$ are, respectively, the total energies of the adsorbate molecule and the bare adsorbent model, and $E_{adsorbent-adsorbate}$ is the total energy of the adsorbate/adsorbent system. A higher E_{ads} corresponds to a stronger adsorption.

The energies calculated by using different basis sets can vary widely. However, it needs to be stressed that the relative values are meaningful when *comparing* different sorbates/sorbents as long as *the same basis set is used.*

8.3. NATURE OF π-COMPLEXATION BONDING

The nature of π-complexation bonding between the adsorbate and adsorbent has been studied for a number of systems, including C_2H_4/Ag halides and C_2H_4/Ag-zeolite (Chen and Yang, 1996); C_2H_4/CuCl, C_2H_4/AgCl, CO/CuCl, and CO/AgCl (Huang, 1999a); C_2H_2/Ni halides (Huang and Yang, 1999), benzene/halides of Cu^+, Pd^{2+}, Ag^+, Au^+, and Pt^{4+} (Takahashi et al., 2000); thiophene/CuCl and AgCl (Yang et al., 2001); and thiophene/Ag-zeolite and Cu-zeolite (Takahashi et al., 2002). The results of adsorption of C_2H_4 on Ag halides and Ag-zeolite will be discussed first.

8.3.1. Understanding π-Complexation Bond through Molecular Orbital Theory

The geometries of C_2H_4, AgX (where X = halide), and Ag-zeolite are optimized first, using STO-3G and then at the 3-21G levels. The cluster model (Kassab et al., 1993; Hill and Sauer, 1995) is used to represent the chemistry of zeolite, shown in Figure 8.3. The optimized zeolite cluster shows a tilt of Ag toward the alumina tetrahedral. The adsorbate and adsorbent are then combined into a single molecule, thereby optimizing its geometry.

Using the NBO method, the results on electron occupancy (Oc) from population analysis of NAO are listed in Table 8.4 for the C atom in the adsorbate

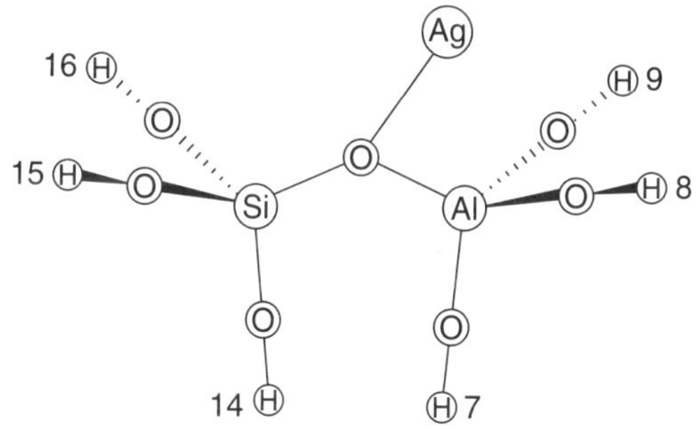

Figure 8.3. Geometry-optimized cluster model for Ag-zeolite.

Table 8.4. NAO (natural atomic orbital) electron occupancies in outer-shell orbitals of C and Ag

Atom	C				Ag					
Orbital	$2s$	$2P_x$	$2P_y$	$2P_z$	$5s$	$4d_{xy}$	$4d_{xz}$	$4d_{yz}$	$4d_{x2-y2}$	$4d_{z2}$
C_2H_4	1.0376	0.9977	1.2216	1.1578	—	—	—	—	—	—
AgF	—	—	—	—	0.1551	2.0000	1.9966	1.9966	2.0000	1.9615
AgCl	—	—	—	—	0.1223	2.0000	1.9986	1.9986	2.0000	1.9817
AgI	—	—	—	—	0.1947	2.0000	1.9991	1.9991	2.0000	1.9846
AgZ	—	—	—	—	0.0670	1.9936	1.9928	1.9999	1.9915	1.9966
C_2H_4-AgF	1.0597	0.9824	1.2573	1.2573	0.2820	1.9999	1.9963	1.9675	1.9996	1.9194
C_2H_4-AgCl	1.0602	0.9747	1.2576	1.1590	0.2427	1.9999	1.9982	1.9781	1.9997	1.9559
C_2H_4-AgI	1.0591	0.9735	1.2567	1.1591	0.3038	1.9999	1.9987	1.9808	1.9997	1.9627
C_2H_4-AgZ	1.0579	0.9829	1.2530	1.1605	0.1266	1.9920	1.9839	1.9967	1.9904	1.9937

Table 8.5. Changes in electron occupancies upon adsorption (ΔOc) in the outer-shell orbitals of Ag

Orbital	$5s$	$4d_{xy}$	d_{xz}	$4d_{yz}$	$4d_{x2-y2}$	$4d_{z2}$	$\Sigma 4d$	$\Sigma(5s+4d)$
C_2H_4-AgF	0.1269	−0.0001	−0.0003	−0.0291	−0.0004	−0.0421	0.0450	0.0819
C_2H_4-AgCl	0.1204	−0.0001	−0.0004	−0.0205	−0.0003	−0.0258	−0.0470	0.0734
C_2H_4-AgI	0.1091	−0.0001	−0.0004	−0.0183	−0.0003	−0.0219	−0.0410	0.0681
C_2H_4-AgZ	0.0596	−0.0016	−0.0089	−0.0032	−0.0011	−0.0029	−0.0177	0.0419

Chen and Yang, 1996.

molecule and for Ag in the adsorbents. Occupancies before and after adsorption are given. The changes in occupancy upon adsorption (ΔOc) are also given in Table 8.5. It is obvious from Tables 8.4 and 8.5 that, upon adsorption, the electron occupancy of the $5s$ orbital of Ag always increases, whereas the total occupancy of its $4d$ orbitals ($4d_{xy}$, $4d_{xz}$, $4d_{yz}$, $4d_{x2-y2}$, and $4d_{z2}$) always decreases. This result is caused by the σ-donation from the π-bond (i.e., mainly $2P_y$ and $2P_z$ orbitals) of the adsorbate molecule to the $5s$ orbital of Ag, and the d-π^* backdonation from the $4d$ orbitals of Ag to the π^*-bond ($2P_x^*$, $2P_y^*$, and $2P_z^*$ orbitals, also shown in the $2P$ orbitals) of the adsorbate. This is in line with the classic picture of Dewar (1951) and Chatt and Duncanson (1953) for π-complexation.

The net increase in occupancy indicates a net electron transfer from the adsorbate to the Ag-containing adsorbent. The net increase in the occupancy of the $5s$ orbital of Ag indicates the strength of the forward σ-donation bond. The net decrease in the occupancies of the $4d$ orbitals of Ag indicates the strength of the d-π^* backdonation bond. The ratios of ΔOc of $5s$ over the ΔOc of $4d$ are approximately 2 : 1 to 3 : 1, indicating the relative contributions of the σ-donation over the d-π^* backdonation to the overall bond.

It is interesting to compare the electron occupancies of the Ag atom in zeolite and different halides. The occupancy in the $5s$ orbital follows the order of electronegativities of the halides, that is, being the lowest with F^-, which has the highest electronegativity. The occupancy in the Ag-zeolite model is clearly lower than that in AgF, indicating that the zeolite framework anion is highly electronegative.

A careful examination of the occupancy changes in the $4d$ orbitals of Ag reveals an interesting pattern. The occupancy changes for $4d_{xy}$, $4d_{xz}$, and $4d_{x2-y2}$ are small, while the major changes occur in the $4d_{yz}$ and $4d_{y2}$ orbitals. The overlap between the $4d_{yz}$ orbital of Ag with the $2p^*$ orbitals of the adsorbate molecule follows the classic picture of π-complexation, illustrated in Figure 8.4. However, the large occupancy decreases of the $4d_{yz}$ orbitals are unexpected. The three other $4d$ orbitals ($4d_{xy}$, $4d_{xz}$, and $4d_{x2-y2}$) are positioned perpendicular in relation to $4d_{yz}$. Thus, there is no possibility for them to overlap with $4d_{yz}$. The $4d_{z2}$ orbitals, or the dumbbell-and-doughnut shaped orbitals, also shown in Figure 8.4, are in the vicinity of the spatial direction of the $4d_{yz}$ orbitals. As a result, the $4d_{z2}$ orbitals are the only ones that can overlap with the $4d_{yz}$

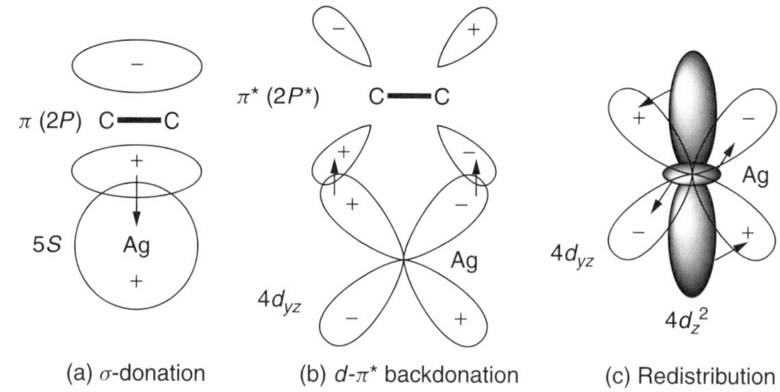

Figure 8.4. Schematic of the C_2H_4-Ag interactions by π-complexation, showing (A) donation of π-electrons of ethylene to the 5s orbital of Ag, (B) backdonation of electrons from the 4d_{yz} orbitals of Ag to the antibonding p^* orbitals of ethylene, and (C) electron redistribution. (C) depicts the possible electron redistribution from the 4d_{z2} orbitals to the 4d_{yz} orbitals (Chen and Yang, 1996, with permission).

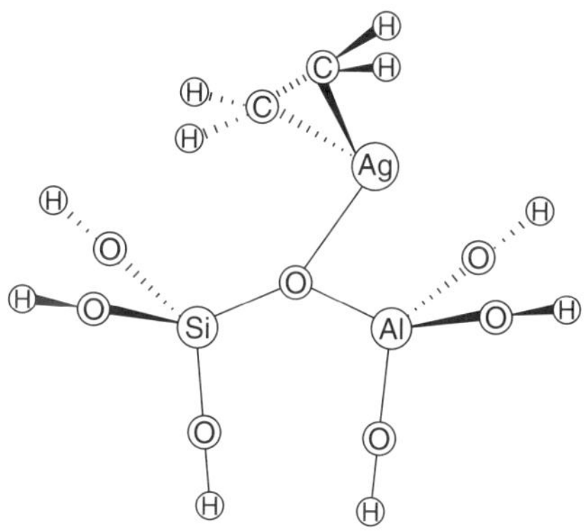

Figure 8.5. Geometry-optimized structure of C_2H_4 adsorbed on Ag-zeolite (Chen and Yang, 1996, with permission).

orbitals. This result indicates that there is considerable redistribution of electrons between the two 4d orbitals during the d-π^* backdonation. Taking the argument one step further, it seems that the electron redistribution (from the 4d_{z2} to the 4d_{yz} orbitals) enhances the d-π^* backdonation (from the 4d_{yz} orbital of Ag to the 2p^* orbitals of the adsorbate). The electron redistribution is also illustrated in Figure 8.4. In most of the adsorption pairs, as seen in Table 8.4, the net occupancy decreases are larger in the 4d_{z2} orbitals than those of the 4d_{yz} orbitals. Since only one Ag atom is used in the model, the electron redistribution is intra-atomic rather than interatomic in nature, and it contributes to the overall bonding.

The geometry-optimized model for ethylene adsorbed on Ag-zeolite is shown in Figure 8.5. The detailed structure is given in Chen and Yang (1996).

8.3.2. π-Complexation Bonds with Different Cations

The most important cations for π-complexation (for practical application) are Ag^+ and Cu^+; hence they are used for comparison (Huang and Yang, 1999). It is also interesting to compare the π-complexation between CO and ethylene, representing, respectively, C=O and C=C bonds. The results of electron occupancies in the outer-shell orbitals of Cu and Ag are shown in Table 8.6. The net change in electron occupancy of the outer-shell *s* orbital indicates the contribution by σ-donation, while that of the *d*-orbitals indicate the contribution by the d-π^* backdonation. It is seen that for Cu salts, the contribution by d-π^* backdonation (to π-complexation) is greater than that by σ-donation, while the opposite is true for Ag. The bonding with CO is stronger than that with C_2H_4. Moreover, the bonding with Cu salts is stronger than that with Ag.

The adsorption bond energies for these systems are shown in Table 8.7. Different basis sets lead to different energies. The relative values, using the same basis set, are in agreement with the experimental results. It is again seen that the bonding strength follows the order:

$$Cu^+ > Ag^+; \quad \text{and} \quad CO > C_2H_4$$

Adsorption of benzene on transition metal chlorides dispersed on silica gel was studied by Takahashi et al. (2000). A weak π-complexation bond was formed between benzene and these metal ions. Molecular orbital calculations for the bonding of benzene and chlorides of these metals were performed at the Hartree-Fock (HF) and density functional theory (DFT) levels using effective core potentials. The experimental values of heat of adsorption and the calculated bond

Table 8.6. Changes upon adsorption in electron occupancies in the outer-shell orbitals of Cu and Ag for adsorption of C_2H_4 and CO

Cu Electron Population Changes after C_2H_4 and CO Adsorption							
	$4s$(*)	$3d_{xy}$	$3d_{xz}$	$3d_{yz}$	$3d_{x2-y2}$	$3d_{z2}$	ΔOc(**)
CuCl-C_2H_4	0.052	0	0	−0.060	0	−0.019	−0.079
CuCl-CO	0.117	0	−0.051	−0.051	0	−0.027	−0.129
Ag Electron Population Changes after C_2H_4 and CO Adsorption							
	$5s$(*)	$4d_{xy}$	$4d_{xz}$	$4d_{yz}$	$4d_{x2-y2}$	$4d_{z2}$	ΔOc(**)
AgCl-C_2H_4	0.061	0	0	−0.029	0	−0.026	−0.055
AgCl-CO	0.101	0	−0.021	−0.021	0	−0.037	−0.079

*Indicating contribution of σ-donation.

**Total change in *d*-orbital electron occupancy upon adsorption (indicating contribution of d-π^* backdonation).

Huang and Yang, 1999, with permission.

Table 8.7. Energy of adsorption calculated by B3LYP/ 3-21+G basis set except for C_2H_4-AgZeolite**

Adsorbate	Adsorbent	Theoretical ΔH (kcal/mol)	Experimental ΔH (kcal/mol)
C_2H_4	AgCl	11.20	6.9
C_2H_4	CuCl	15.74	8.3
C_2H_4	AgZeolite	15.37* 19.63**	18.1
CO	AgCl	9.64	7.5
CO	CuCl	16.56	10.2

*Calculated by HF/3-21G basis set (Chen and Yang, 1996).
**Calculated by MP2/3-21G (Chen and Yang, 1966).
Huang and Yang, 1999.

Table 8.8. Energy of adsorption of benzene on MCl_x (in kcal/mol)

MCl_x	Theoretical ΔH (kcal/mol)	Experimental ΔH (kcal/mol)
CuCl	12.5	10.1 – 11.0
$PdCl_2$	10.8	9.3 – 10.9
AgCl	8.6	9.2 – 10.1
$AuCl_3$	6.5	8.8 – 10.1
$PtCl_4$	5.2	7.2 – 9.0

From Takahashi et al., 2000.

energies are compared in Table 8.8. The relative order in bond energies was predicted well by the natural bond orbital results.

8.3.3. Effects of Different Anions and Substrates

The effects of different anions on π-complexation have been studied for the adsorption of C_2H_4 and C_3H_6 on CuX and AgX (X=F, Cl, Br, I), by both experiment and molecular orbital theory (Huang et al., 1999b). The following trends of anion effects were obtained for the adsorption of C_2H_4 and C_3H_6 on the metal halides: $F^- > Cl^- > Br^- > I^-$. These trends were in excellent agreement with the experimental results. In addition, the theoretical metal-olefin bond energies are in fair agreement with the experimental data. The anion effects are illustrated in Figure 8.6 for the adsorption of C_3H_6. The effects of anions on the adsorption of C_2H_4 were similar.

The adsorption of acetylene on different nickel halides has been studied by *ab initio* molecular orbital calculations (Huang and Yang, 1999). The strengths of adsorption on different nickel halides were calculated by *ab initio* molecular

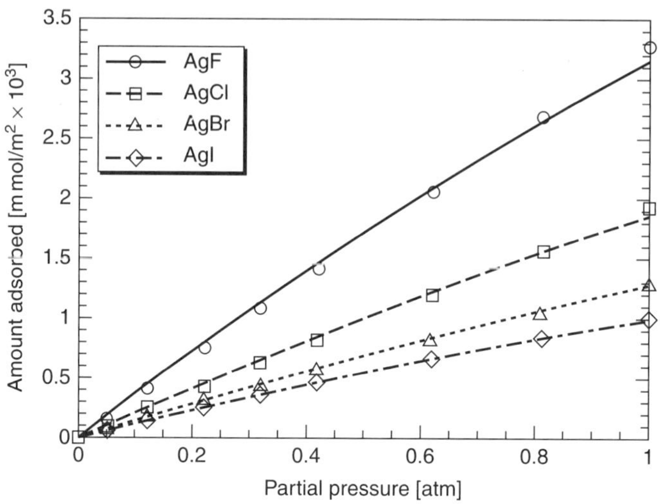

Figure 8.6. Normalized C_3H_6 adsorption isotherm on AgX (X=F, Cl, Br, I) salts at 0 °C (Huang et al., 1999b, with permission).

orbital calculation and follow the order NiF_2 > $NiCl_2$ > $NiBr_2$ > NiI_2. The calculated heats of adsorption were 25.97, 20.42, 18.24, and 16.42 kcal/mol, respectively. The calculations were performed on C_2H_2, NiX_2, and bonded C_2H_2-NiX_2, at the HF/3-21G level for geometry optimization and B3LYP/3-21+G** level for a detailed analysis of the electronic distribution by using NBO theory. The bonding between acetylene and NiX_2 involved three parts: (1) σ-donation (overlap of the $2p_x$ orbital of C with the $4s$ orbital of Ni), (2) electron redistribution (from the $4s$ orbital to the $3d_{xz}$ orbital of Ni), and (3) d-π^* back-donation (from the $3d_{yz}$ orbital of Ni to the $2p_y^*$, or π^*, orbital of C). The back-donation dominates the bonding. The three combined steps yield the minimum total energies.

The effects of substrates on π-complexation were studied by olefin adsorption on monolayer $AgNO_3$ supported on various substrates (Padin and Yang, 2000). The substrates selected were γ-Al_2O_3, SiO_2, and MCM-41. The following trend for olefin adsorption was observed for these substrates:

$$SiO_2 > \text{MCM-41} > \gamma\text{-}Al_2O_3$$

The silica surface (on both silica gel and MCM-41) provides a better substrate due to the lack of Lewis acid sites (unlike γ-Al_2O_3), and consequently the Ag atoms in these sorbents are more capable of forming π-complexation bonds with olefins. Although the effect of the physical characteristics of a substrate, such as surface area and pore size, would have on adsorption is clear, the effect of the electronic properties needs to be studied further.

The fundamental differences in interactions with different anions and cations can be further understood from the changes in orbital occupancies that give rise to the σ-donation bond and the d-π^* backdonation bond. The magnitudes of

Table 8.9. Summary of the NBO analysis of π-complexation between MX (metal halides) and C_2H_4

	C → M Interaction (σ Donation)	M → C Interaction ($d - \pi^*$ Back-Donation)	Net Change
	q1	q2	q1 + q2
$CuF\text{-}C_2H_4$	0.047	−0.089	−0.042
$CuCl\text{-}C_2H_4$	0.052	−0.080	−0.028
$CuBr\text{-}C_2H_4$	0.042	−0.077	−0.035
$CuI\text{-}C_2H_4$	0.030	−0.072	−0.042
$AgF\text{-}C_2H_4$	0.081	−0.073	+0.008
$AgCl\text{-}C_2H_4$	0.058	−0.053	+0.004
$AgBr\text{-}C_2H_4$	0.047	−0.049	−0.002
$AgI\text{-}C_2H_4$	0.032	−0.044	−0.011

q1 is the amount of electron population increase on valence *s* orbitals of the metal; q2 is the total amount of electron population decrease on valence *d* orbitals of the metal.
Huang et al., 1999b.

these two interactions for ethylene are shown in Table 8.9. An examination of Table 8.9 shows that in all cases the M–C interaction is a dative bond, that is, donation of electron charges from the π orbital of olefin to the vacant *s* orbital of metal and, simultaneously, back-donation of electron charges from the *d* orbitals of M to the π^* orbital of olefin. This can be interpreted in more detail. When the olefin molecule approaches M^+, some electronic charge is transferred from the C=C π orbital to the valence *s* orbital of M^+. At the same time, electrons in the filled *d* orbitals of the metal are transferred to the symmetry-matched π^* orbital of olefin. It can be seen from Table 8.9 that upon adsorption, the electron occupancies of the valence *s* orbitals of Cu and Ag always increase, whereas the total occupancy of their respective 3*d* or 4*d* orbitals always decrease. Obviously this is caused by the donation and back-donation of electrons between metal and olefin as stated above.

A comparison of the electron population changes in the *s* and *d* orbitals of M before and after adsorption shows that for the CuX-olefin complexes, the overall charge transfer is back-donation. The amount of back-donation is about double the amount of σ donation. This indicates that the Cu–C bonds contain more metal *d* than metal *s* character and that the strength of the covalent bonds depends mainly on the overlap of the metal *d* orbitals with the C hybrid orbitals. For the AgX-olefin complexes, quite differently, the back-donation is almost equal to the σ donation, which means the σ donation and back donation play equally important roles in the bonding of Ag–C. A comparison of the net changes of the electron occupation on the two different metals before and after adsorption shows greater net electron occupation changes on Cu than on Ag upon olefin adsorption. The amount of change indicates the extent of interaction. This is consistent with the conclusion that CuX has a stronger interaction with olefin than AgX.

8.4. BULK SEPARATIONS BY π-COMPLEXATION

Bulk separation/recovery of CO from synthesis gas by π-complexation has already been commercialized worldwide since 1989. π-Complexation is highly promising for other bulk separations such as olefin/paraffin and aromatic/aliphatic separations, either in vapor phase by PSA or in liquid phase by simulated moving bed processes. Before discussing these processes, problems of deactivation or stability of the π-complexation sorbents will be first addressed.

8.4.1. Deactivation of π-Complexation Sorbents

The deactivation behaviors of various π-complexation sorbents have been studied. Cu^+ salts such as CuCl and Cu(I) zeolites are oxidized quickly into Cu^{2+} state upon exposure to ambient air. Moisture is known to accelerate the oxidation process. Ag^+ salts and Ag^+-zeolites, on the other hand, are quite stable in ambient air with minimized light exposure (Hutson, 2000).

Because H_2 and H_2S are present in synthesis gas, cracked gases, and other gas streams encountered in industry, their effects on the π-complexation sorbents have been studied. The effects of exposure to 0.5 atm H_2 at various temperatures on $AgNO_3/SiO_2$ and AgY zeolite were discussed in detail by Jayaraman et al. (2001). Severe deactivation of both sorbents occurred at temperatures above 120 °C. X-ray photoemission spectroscopy (XPS) studies of the deactivated samples showed that the Ag^+ was reduced to Ag^0. However, these sorbents could be rejuvenated by oxidation with oxygen at 350 °C when the valence of Ag was restored to Ag^+. The π-complexation ability of the sorbent was tested by adsorption of ethylene, and the deactivation and reoxidation behaviors are shown in Figure 8.7.

The behavior of AgY zeolite in H_2S was studied by Takahashi et al. (2001b). At 25 to 120 °C, H_2S chemisorbed on AgY, while reaction with H_2S with continual weight gain was observed at 180 °C. XPS analysis showed the formation of Ag_2S. However, the adsorption capacities for 1-butene and 1,3-butadiene were only slightly lowered, indicating the π-complexation capability of Ag_2S.

The effects of H_2 and H_2S on Cu(I)Y were studied by Takahashi et al. (2001a). Unlike AgY, exposure to H_2 and H_2S at 120 °C showed no effect on CuY, demonstrating its excellent poison resistance toward H_2 and H_2S.

The deactivation behaviors of the Ag and Cu(I) sorbents are summarized in Table 8.10.

8.4.2. CO Separation by π-Complexation

Since 1989, CO separation/recovery by PSA using supported CuCl has been commercialized worldwide. PSA separation results are available in the literature (Kansai Coke & Chemicals Co., 1989; Chen et al., 1997; Golden et al., 1998).

CO is typically produced along with H_2 as synthesis gas, by steam reforming of methane or naphtha. Separation and recovery of CO has been accomplished by cryogenic processes. CO is used as a raw material for production of

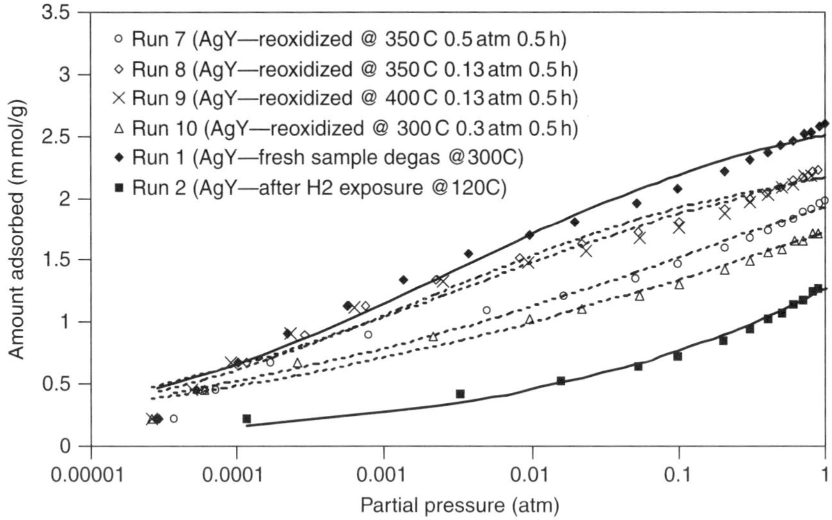

Figure 8.7. Deactivation by H_2 and rejuvenation by oxidation. Ethylene adsorption isotherms at 120 °C on different AgY samples, treated with H_2 (at 0.5 atm for 1 h) and O_2 for 0.5 h (Jayaraman et al., 2001, with permission).

Table 8.10. Deactivation/stability of π-complexation sorbents in various environments

	Air/Moisture	H_2	H_2S
CuCl and Cu(I)-zeolites	Deactivates quickly in ambient air	Stable at 120 °C	Stable at 120 °C
$AgNO_3$ and Ag-zeolites	Stable in ambient air	Deactivates at 70 °C	Slightly deactivates at 120 °C

polyurethane, polycarbonate, and other chemical products, and as "bottom blowing" gas for converters at steel mills. A low tolerance for methane impurity is required by chemical users to eliminate unwanted side reactions during synthesis of engineering plastics. The current level of tolerance for methane impurity is 25 ppm. This is not required for steel mill use. Depending on the end-use for CO, different sorbents can be used. Due to its high selectivity for CO over CH_4, $CuCl/Al_2O_3$ is used for CO separation when the low methane impurity is required. (Golden et al., 1998). CuCl/carbon or CuCl/coked-Al_2O_3 may be used when such limitation is not required (Yokoe, 1987; Kansai, 1989).

Because of the strong bond between CO and Cu^+, the isotherms of CO on these π-complexation sorbents are fairly steep (Hirai et al., 1986a and 1986b; Golden et al., 1992a; Tamon et al., 1996; Xie et al., 1996). The isotherm of Xie et al. (1996) is representative of the reported isotherms and is shown in

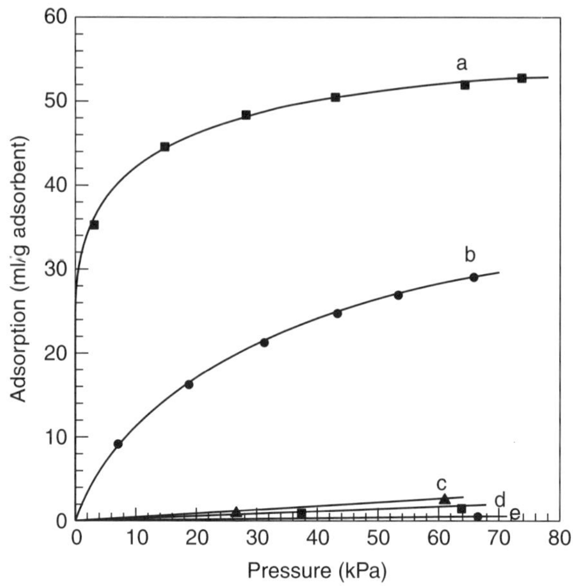

Figure 8.8. Isotherms of various gases at 30 °C on monolayer covered CuCl/NaY zeolite: (a) CO, (b) CO_2, (c) CH_4, (d) N_2, and (e) H_2 (Xie et al., 1996, with permission).

Figure 8.8. The sorbent was prepared by thermal dispersion of 0.554 g/g on NaY. Monolayer spreading of CuCl was confirmed by X-ray diffraction. Interactions between CuCl and the other gas molecules are weak due to the close proximity between Cu^+ and Cl^-, while each Cu^+ can bond one CO molecule. The steepness of the isotherm generally decreases at higher temperatures, at the expense of lower capacity. Hence a maximum working capacity (between two fixed working pressures) is achieved at an optimum temperature. Two strategies then become clear for the PSA operation: vacuum desorption and an increase in temperature. Indeed, 70 °C was the temperature of the PSA operation using the CuCl/NaY sorbent (Xie et al., 1996).

The PSA process using supported CuCl for CO separation/recovery has been described in the literature (Kansai, 1989; Chen et al., 1997; Golden et al., 1998). A detailed description was given by Golden et al. (1998). In this PSA process, CO is the strongly adsorbed component. In order to obtain a high-purity CO product, a purge step using the strong component is most effective (Yang, 1987, p. 247). This step follows the depressurization step. The feed mixture that is contained in the bed voids is displaced by the strong component, hence increasing the CO content in the bed. Thus, a CO rinse step is used in all PSA processes for CO recovery.

The other strategies used in the PSA processes are (1) H_2O/H_2S removal with a guard bed, (2) vacuum desorption, and (3) in situ activation of CuCl. Moisture and H_2S removal is necessary to avoid possible deactivation, and is accomplished by placing a layer of 3A zeolite and/or silica gel at the inlet of each bed to serve as the guard bed (Yang, 1987). Due to the steepness of the CO isotherm, the adsorption pressure is unimportant. The desorption pressure is, however, critically important. Generally, 0.1–0.3 atm pressure/vacuum is used for desorption. Golden et al. (1998) described an in situ sorbent activation technique for reducing Cu^{2+} to the Cu^{+} state with the feed mixture (which was a synthesis gas) at 90 °C. In all reports, the CO product purities were over 99 or 99.5% with CO recoveries over 80%. With CuCl/γ-Al_2O_3, the CH_4 impurity in the CO product was 30 ppm (Golden et al., 1998).

8.4.3. Olefin/Paraffin Separations

Olefin/paraffin separations by PSA will be discussed in Chapter 10. In this chapter, the isotherms of C_2H_4 and C_3H_6 on the best π-complexation sorbents will be given. The sorbents discussed here should also be suitable for separations of higher olefins from their corresponding paraffins.

Several new sorbents based on π-complexation have been prepared recently for selective olefin adsorption (Yang et al., 2002). These include Ag^+-exchanged resins (Yang and Kikkinides, 1995; Wu et al., 1999), monolayer CuCl/γ-Al_2O_3 (Yang and Kikkinides, 1995), monolayer CuCl on pillared clays (Cheng and Yang, 1995), monolayer $AgNO_3/SiO_2$ (Rege et al., 1998; Padin and Yang, 2000), and monolayer $AgNO_3$ supported on other substrates (Padin and Yang, 2000; Yang et al., 2002), particularly on acid-treated clays (Cho et al., 2001; Choudary et al., 2002). Among the different sorbents, monolayer $AgNO_3$ appears to give the best results, while CuCl/γ-Al_2O_3 appears to be the optimum when considering the costs of the sorbents.

The equilibrium isotherms for C_2H_4 and C_2H_6 at 70 °C on $AgNO_3/SiO_2$ are shown in Figure 8.9. The isotherms of C_3H_6 and C_3H_8 at 70 °C on the same sorbent are shown in Figure 8.10.

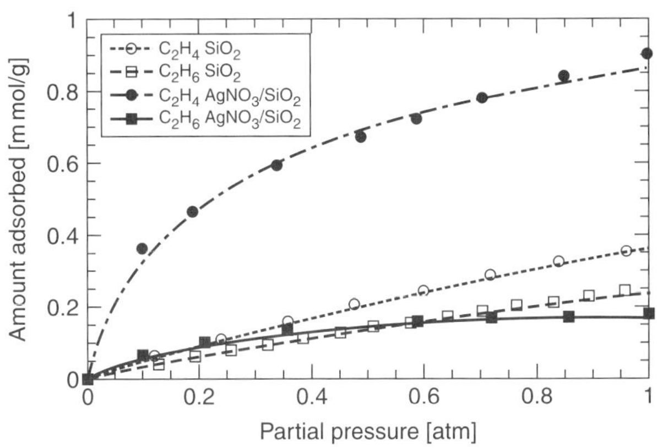

Figure 8.9. Equilibrium isotherms of C_2H_4 over C_2H_6 on $AgNO_3/SiO_2$ (by incipient wetness) at 70 °C (Padin and Yang, 2000; with permission).

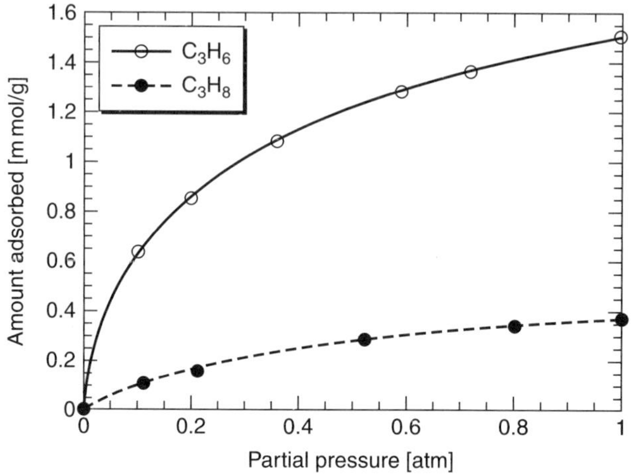

Figure 8.10. Equilibrium isotherms of C_3H_6 over C_3H_8 on $AgNO_3/SiO_2$ (by incipient wetness) at 70 °C (Padin and Yang, 2000).

From these results, it is seen that the sorbents have excellent selectivities and olefin capacities. The isotherms are also relatively linear. The linearity is desirable for cyclic processes such as PSA (Rege et al., 1998). Diffusion rates and isotherm reversibilities have also been measured on these systems, and they were all highly suitable for PSA (Rege et al., 1998).

8.4.4. Aromatics/Aliphatics Separation

Aromatics/aliphatics separation is accomplished by solvent extraction. A number of solvents have been used (Bailes, 1983). Although these separation processes are efficient, they are energy intensive, and more importantly, the solvents (such as sulfolane) increasingly pose as environmental hazards. Another possible separation technique is fractional distillation. It is, however, difficult because of the close relative volatilities. For benzene/cyclohexane, the mixture has a minimum azeotrope at about 53%. Therefore, acetone is added as an entrainer and a complex hybrid system (distillation combined with extraction in this case) can be used for separation (Stichlmair and Fair, 1998).

Because of the importance of aromatics/aliphatics separation and the problems associated with solvent extraction, possible alternatives have been studied. These include liquid membranes (Li, 1968; 1971; Goswami et al., 1985), pervaporation (Hao et al., 1997), and the use of liquid inclusion complexes (Atwood, 1984). No selective sorbents are known for aromatics/aliphatics separation. It is, however, certainly possible to develop such sorbents based on π-complexation. In the benzene molecule, the carbon atom is sp^2 hybridized. Hence, each carbon has three sp^2 orbitals and another P_Z orbital. The six P_Z orbitals in the benzene ring form the conjugative π bond. The P_Z orbitals also form the antibonding π^* orbitals, which are not occupied. When benzene interacts with transition metals, the π-orbitals of benzene can overlap with the empty outer-shell s orbital of the transition metal to form a σ-bond. Moreover, it is possible that the vacant antibonding π^*-orbital of benzene can overlap with the d-orbitals in the transition metal similar to that formed in the olefin-Cu^+ bond (Huang et al., 1999). Molecular orbital calculations indeed confirmed the π-complexation with benzene (Takahashi et al., 2000).

Takahashi et al. (2000) studied π-complexation sorbents with benzene and cyclohexane. Benzene and cyclohexane form an ideal pair of model compounds for developing selective sorbents for aromatics. These molecules have similar shapes and close boiling points (80 °C for benzene and 81 °C for cyclohexane). The kinetic diameter of benzene, which is calculated from the minimum equilibrium cross-sectional diameter, is estimated to be 5.85 Å compared with 6.0 Å for cyclohexane. The sorbents in that work were transition metal salts dispersed on high-surface-area substrates. Based on the results of selective olefin sorbents for olefin/paraffin separations, Cu^+, Ag^+, Pt^{4+}, and Pd^{2+} cations were the most promising sorbents due to their strong interactions with π-orbital to olefin molecules. The sorbent that yielded the highest benzene selectivity was $PdCl_2/SiO_2$. The pure-component isotherms are shown in Figure 8.11.

The pure component adsorption ratios and the separation factors for benzene/cyclohexane on these sorbents are shown in Table 8.11. The separation factors were calculated from mixed gas isotherms. Based on these figures, bulk separation with Ag and Cu salts is not promising. However, these sorbents are promising for purification, that is, removal of aromatics from aliphatics, since very high separation factors are obtained at low concentrations of benzene. Due to worldwide environmental mandates, refineries are required to decrease the contents of aromatics in gasoline and diesel fuels. The π-complexation sorbents

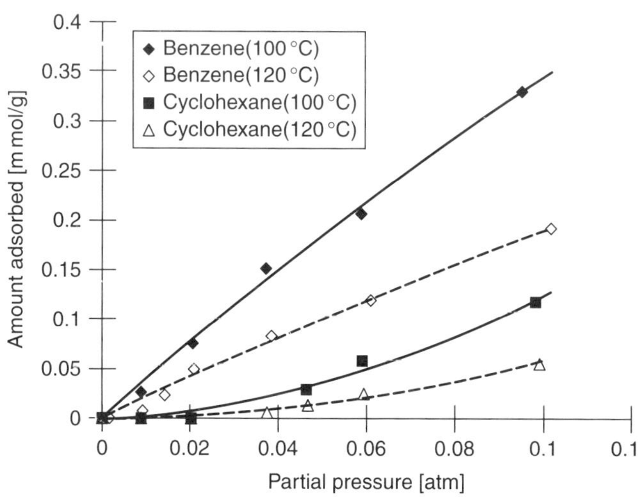

Figure 8.11. Pure-component equilibrium isotherms for benzene and cyclohexane $PdCl_2/SiO_2$ (0.88g/g) at 100 and 120 °C (Takahashi et al., 2000, with permission).

Table 8.11. Pure-component ratio and separation factor (based on mixture isotherms) for benzene/cyclohexane, all at 0.1 atm

Sorbent	Pure-Comp. Ratio at 100 °C	Pure-Comp. Ratio at 120 °C	Separation Factor at 120 °C
$PdCl_2/SiO_2$(0.88 g/g)	2.9	3.2	6.2
$AgNO_3/SiO_2$(0.33 g/g)	—	2.0	2.1
$CuCl/Al_2O_3$(0.50 g/g)	1.3	1.3	1.3

appear to be ideally suited for this application. This purification will be discussed in Chapter 10 (10.8).

8.4.5. Possible Sorbents for Simulated Moving-Bed Applications

Simulated moving-bed (SMB) processes have been widely used for difficult, liquid-phase separations (Ruthven, 1984; Humphrey and Keller, 1997; Juza et al., 2000). Sorbex is the generic name used by UOP for these processes. The most important application is the separation of the xylene isomers, named the Parex process. Other commercialized SMB separations include: n-paraffins/isoparaffins (Molex), olefins/paraffins (Olex), fructose/glucose (Sarex), and chiral SMB separations (Juza et al., 2000). A host of other separations have been demonstrated (Humphrey and Keller, 1997), although the commercial status of these applications is unknown. These demonstrated separations include: separation of hydroxyparaffinic dicarboxylic acids from olefinic dicarboxylic acids; removal of thiophene, pyridine, and phenol from naphtha; separation of unsaturated fatty acid methyl esters from saturated fatty acid methyl esters; and separation of saturated fatty acids from unsaturated fatty acid (Humphrey and Keller, 1997).

The sorbents used in the commercialized SMB processes are mostly cationic forms of type X or type Y zeolites, such as K-BaY, Sr-BaY, K-BaY, BaY, 5A, CaX, CaY, and SrX (Ruthven, 1984). The separation factors of the binary mixtures (i.e., extract and raffinate) on these zeolites are generally very low, typically around 2. For chiral separations, cyclodextrin-based chiral selective sorbents are used, with separation factors typically below 2 (Biressi et al., 2002). A separation factor much below 2 would make the SMB process economically unfeasible. Using a sorbent with a higher separation factor would obviously have many inherent advantages, such as higher throughput, higher product purity, and lower recycle rate.

From the available literature on SMB processes, the π-complexation sorbents have not been used. 5A zeolite is used for the separation of n-paraffins from branched and cyclo-paraffins, and the separation is accomplished by molecular size exclusion (of the branched and cyclo-paraffins). All other separations rely on alkaline-earth forms of zeolites. The interactions of the π-electrons of the aromatic or olefinic compounds with the alkaline earth cations are much weaker than those with the d-block metal cations such as Cu^+ and Ag^+. As a result, the separation factors on the π-complexation sorbents are significantly higher.

Therefore, the π-complexation sorbents are very promising candidates for some SMB separations. These separations include: olefins/paraffins (Olex); separation of hydroxyparaffinic dicarboxylic acids from olefinic dicarboxylic acids; removal of thiophene, pyridine, and phenol from naphtha; separation of unsaturated fatty acid methyl esters from saturated fatty acid methyl esters; and separation of saturated fatty acids from unsaturated fatty acid. Supported CuCl (such as $CuCl/\gamma\text{-}Al_2O_3$) and Ag^+ (such as $AgNO_3/SiO_2$) should be good sorbents for all of these separations, except for the purification of naphtha. For the latter application, Cu^+-zeolite and Ag^+-zeolites (such as CuY and AgY) are promising sorbents, as will be discussed in Chapter 10 (under *Desulfurization of Gasoline*).

8.5. PURIFICATION BY π-COMPLEXATION

A major difference exists between bulk separation and purification for sorbent design and selection. For bulk separation, isotherm linearity (hence high-working capacity) is needed. A steep isotherm (or high Henry's constant) is needed for purification. Thus, for π-complexation sorbents, supported salts with monolayer of full-surface coverage are desirable for bulk separation, whereas ion exchanged zeolites are suitable for purification.

The extent of π-complexation between the sorbate and sorbent depends, for a given sorbent, on the density of the π-electrons in the sorbate molecule. Thus, very strong bonds can be formed with molecules with more than two double bonds (e.g., dienes), triple bonds, and polynuclear aromatics. At the same time, for a given sorbate, the sorbent can be tailored to yield a desired bond strength, by choosing the appropriate cation.

For purification, particularly ultrapurification, a strong adsorption bond is needed. This means a high Henry's Law constant is needed. The product purity in the fluid effluent from a fixed bed adsorber depends on the Henry's constant, and the purity can be predicted with a mathematical model (Yang, 1987). For ultrapurification, e.g., when impurity levels of parts per billion (ppb) or parts per trillion (ppt) are required, no predictive models exist. This is the case for the removal of dioxins from effluent in incinerators (Yang et al., 1999).

For the reasons above, the π-complexation sorbents hold a tremendous potential for future applications in purification, some of which will be included for discussion. The removal of dienes from olefins by AgY and CuY has already been demonstrated and applied in the field (Padin et al., 2001). Other promising applications include:

- desulfurization of gasoline and diesel fuels
- removal of aromatics
- removal of CO from H_2 for fuel cell applications
- VOC removal
- dioxin removal
- removal of acetylene (by Ni^{2+} salts)

8.5.1. Removal of Dienes from Olefins

Normal α-olefins (NAO) are chemical intermediates used for making a variety of products. The largest uses for NAOs are in the production of alcohols (via oxo chemistry), as co-monomers for polyethylene production, and in the synthesis of poly (α-olefins) for synthetic lubricants. Also, oligomerization of *n*-butenes to more valuable octenes is an effective way of upgrading their value. A variety of catalysts are used for these reactions. The most common metal-based catalyst involves nickel, which is subject to poisoning by very low levels of 1,3-butadiene (C_4H_6). The selective butadiene hydrotreating process is one option to clean up unwanted butadiene in a mixed-C_4 stream (Meyers, 1986). Removal of dienes is also required for the production of higher α-olefins. Distillation is also being used for these purification processes.

Sorbents based on π-complexation for olefin purification have been developed recently in the author's laboratory (Padin et al., 1999; Jayaraman et al., 2001; Padin et al., 2001; Takahashi et al., 2001a and 2001b). AgY and Cu(I)Y are the best sorbents. Although only vapor phase isotherms are reported in the literature cited above, these sorbents have been demonstrated successfully for the liquid-phase feeds in the field. Diene impurities below 1 ppm can be readily achieved. The isotherms are shown in Figures 8.12 and 8.13, for 1,3-butadiene and 1-butene.

For purification, the undesired component or impurity is present at a low concentration or partial pressure. It is important that a significant amount of the

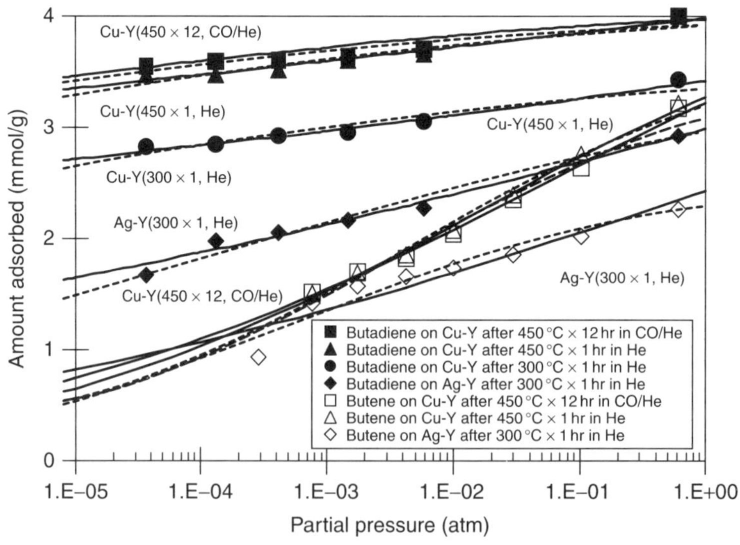

Figure 8.12. Pure-component equilibrium isotherms at 120 °C for 1,3-butadiene and 1-butene on Cu-Y and Ag-Y. Samples were prepared by reduction of Cu(II)Y by either CO or auto-reduction (in He) (Takahashi et al., 2001a, with permission). Lines are fitted isotherms with various models.

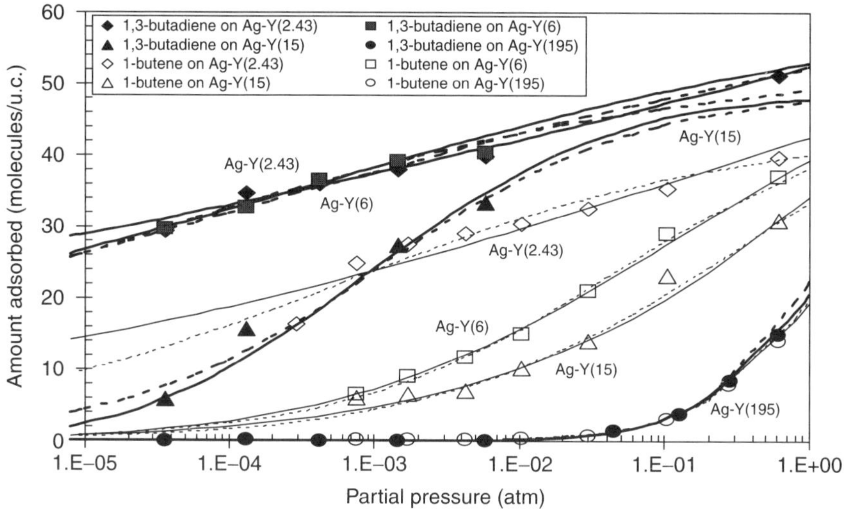

Figure 8.13. Isotherms of 1,3-butadiene and 1-butene on AgY with different Si/Al ratios (given in parentheses) at 120 °C (Takahashi et al., 2001b, with permission).

Table 8.12. Separation factors (α) of 1,3-butadiene (at pressures indicated) over 1-butene (at 1 atm) at 120 °C

Partial Pressure (atm)		Cu(I)Y	AgY
1,3-Butadiene	1-Butene		
0.001	1	1300	200
0.0001	1	10,000	1400
0.00001	1	77,000	10,000

Data taken from Takahashi et al., 2001a.

impurity can be adsorbed at a low partial pressure or concentration. Hence, in assessing the sorbent capability, the separation factor (α) is evaluated at low partial pressures of the impurity component, and at a high partial pressure for the main component. The capability of the sorbent for purification is reflected by the amount of adsorbed 1,3-butadiene at low partial pressures, that is, in the range of 10^{-4} to 10^{-5} atm in Figures 8.12–8.13. The separation factors thus evaluated are shown in Table 8.12. They were calculated by using mixed-gas isotherm models from single-gas isotherms. For any separation process using a given mass separating agent, the separation factor is a good indicator of the goodness of separation (King, 1980). For purifications of similar mixtures using liquid membranes, separation factors in the order of 10–100 yielded excellent results (Li, 1968; Li, 1971a and 1971b). The separation factors shown in Table 8.12 indicate that the π-complexation sorbents are excellent sorbents for purification of olefins.

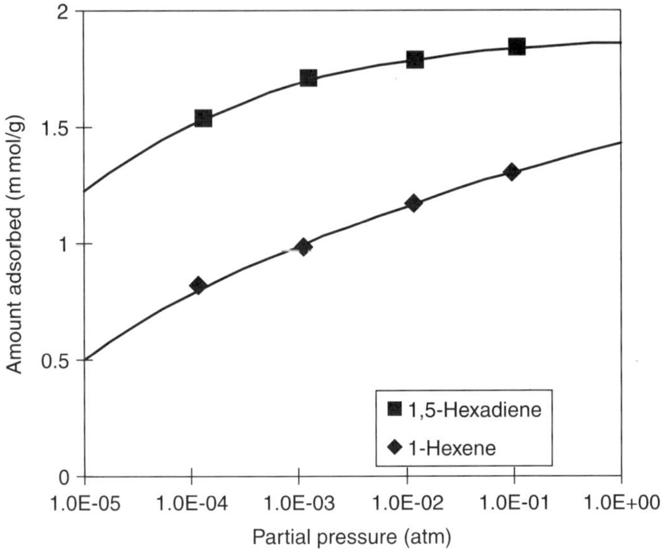

Figure 8.14. Isotherms of 1-Hexene and 1,5-Hexadiene at 180 °C on Ag-Y(Si/Al = 2.43).

For economic reason it is desirable to minimize the amount of Ag in the zeolite. Figure 8.14 shows the results of AgY with different Si/Al ratios. The amount of Ag in the AgY with Si/Al = 2.43 is 40 Ag/unit cell. The amounts for others are 28 Ag/unit cell for Si/Al = 6 and 12 Ag/unit cell for Si/Al = 15. The results showed that there was no noticeable reduction in the capacity for 1,3-butadiene at Si/Al = 6, and the zeolite at Si/Al = 15 was still quite effective. Results were also given for AgNaY (Si/Al = 2.43) with various degrees of Ag exchange (Takahashi et al., 2001a), and they were similar to that shown in Figure 8.13.

The purification of 1-hexene by removal of 1,5-hexadiene by AgY was also tested by vapor phase adsorption isotherms. The data in Figure 8.14. show the good capacity of the π-complexation sorbent for the removal of hexadiene at low concentrations.

8.5.2. Removal of Aromatics from Aliphatics

Purification of aliphatics by the removal of aromatics is important in the petrochemical industry as well as for pollution control. In a typical benzene removal process, a combination of extraction and distillation is used (Meyers, 1986). Improvements by other processes have been considered, such as pervaporation (Hao et al., 1997), liquid membranes (Li, 1968; Li, 1971a and 1971b), and adsorption by temperature swing adsorption (TSA) in the liquid phase (Matz and Knaebel, 1990). In the work of Matz and Knaebel, commercially available sorbents were used: silica gel, activated alumina, activated carbon, zeolite 13X, and polymeric resin (XAD-7). Among these sorbents, silica gel was considered the best due to its superior thermal-exchange capacity. However, the selectivities were low.

Takahashi and Yang (2002) studied adsorption of benzene and cyclohexane on various Y-zeolites. AgY showed superior benzene/cyclohexane selectivities to NaY, PdY, and H-USY. Separation factors as high as 10^4 were obtained with AgY at low concentrations of benzene. More details will be given in Chapter 10 (10.8).

REFERENCES

Albright, R. L. (1986) *Reactive Polymers*, 4, 155.

Atwood, J. L. (1984) *Separ. Sci. Tech.* 19, 751.

Bailes, P. J. (1983) Petroleum and petrochemicals processing — introduction and aromatics–aliphatics separation. In *Handbook of Solvent Extraction*. (T. C., Lo, M. H. I., Baird, and C., Hansen, eds.). Wiley, New York, NY p. 517.

Baird, N. C. and Dewar, M. J. S. (1969) *J. Chem. Phys.* 50, 1262.

Biressi, G., Mazzotti, M., and Morbidelli, M. (2002) *J. Chromatography* A 211, 957.

Blytas, G. C. (1992) Separation of unsaturates by complexing with nonaqueous solutions of cuprous salts. In *Separation and Purification 'Technology*. (N. N., Li and I. M., Clao, eds.), Chap. 2. Marcel Dekker, New York, NY.

Bodor, N., Dewar, M. J. S., Harget, A., and Haselbach, E. (1970) *J. Am. Chem. Soc.* 92, 3854.

Breck, D. W. (1974) *Zeolite Molecular Sieves.* Wiley, New York, NY.

Cen, P. L. (1989) *Third International Conference on Fundamentals of Adsorption*, Engineering Foundation, New York, NY, p. 191.

Chao, C. C. and Lunsford, J. H. (1972) *J. Phys. Chem.* 76, 1546.

Chatt, J. and Duncanson, L. A. (1953) *J. Chem. Soc.* 2939.

Chen, J., Gu, G. W., and Zhang, J. T. (1997) *Proceedings Fourth China-Japan-USA Symposium on Advanced Adsorption Separation Science and Technology*. (Z., Li and Z. H. Yeh., eds.). Guangzhou, China, p. 284.

Chen, N. and Yang, R. T. (1996) *Ind. Eng. Chem. Res*. 35, 4020.

Chen, S. G. and Yang, R. T. (1997) *Energy & Fuels* 11, 421.

Cheng, L. S. and Yang, R. T. (1995) *Adsorption* 1, 61.

Cho, S. H., Han, S. S., Kim, J. N., Choudary, N. V., Kumar, P., and Bhat, T. S. G., U.S. Patent 6,315,816 (2001).

Choudary, N. V., Kumar, P., Bhat, T. S. G., Cho, S. H., Han, S. S., and Kim, J. N. (2002) *Ind. Eng. Chem. Res.* 41, 2728.

Clark, T. (1985) *A Handbook of Computational Chemistry. A Practical Guide to Chemical Structure and Energy Calculation.* John Wiley & Sons, NY.

Da Silva, F. A. and Rodrigues, A. E. (1999) *Ind. Eng. Chem. Res.* 38, 2051.

Da Silva, F. A. and Rodrigues, A. E. (2001) *AIChE J.* 47, 341.

Deng, S. G. and Lin, Y. S. (1997) *Chem. Eng. Sci.* 52, 1563.

Dewar, M. J. S. (1951) *Bull. Soc. Chim. Fr.* 18, C71.

Eldridge, R. B. (1993) *Ind. Eng. Chem. Res.* 32, 2208.

Foresman, J. B. and Frisch, A. (1996) *Exploring Chemistry with Electronic Structure Methods*, 2nd Ed. Gaussian, Pittsburgh, PA.

Frisch, M. J., Trucks, G. W., Schlegel, H. B., Gill, P. M. W., Johnson, B. G., Robs, M. A., Cheeseman, J. R., Keith, T., Petersson, G. A., Montgomery, J. A., Raghavachari, K., Al-Laham, M. A., Zakrzewski, V. G., Ortiz, J. V., Foresman, J. B., Peng, C. Y., Ayala, P. Y., Chen, W., Wong, M. W., Andres, J. L., Replogle, E. S., Gomperts, R., Martin, R. L., Fox, D. J., Binkley, J. S., DeFrees, D. J., Baker, J., Stewart, J. P., Head-Gordon, M., Gonzalez, C., and Pople, J. A. (1998) *Gaussian 98, Revision A7*. Gaussian, Inc., Pittsburgh PA.

Ghosh, T. K., Lin, H. D., and Hines, A. L. (1993) *Ind. Eng. Chem. Res.* 32, 2390.

Gilliland, E. R., Bliss, H. L. and Kip, C. E. (1941) *J. Am. Chem. Soc.* 63, 2088.

Glendening, E. D., Reed, A. E., Carpenter, J. E., and Weinhold, F. (1995) NBO Version 3.1, Chemistry Department, University of Wisconsin, Madison, WI.

Golden, T. C., Guro, D. E., Kratz, W. C., Occhialini, J. M., and Sabram, T. E. (1998) *Fundamentals of Adsorption*. (F. Meunier, ed.). Elsevier, Amsterdam, The Netherlands.

Golden, T. C., Kratz, W. C., and Wilhelm, F. C. U.S. Patent 5,126,310 (1992a).

Golden, T. C., Kratz, W. C., Wilhelm, F. C., Pierantozzi, R., and Rokicki, A. U.S. Patent 5,175,137 (1992b).

Gordon, M. S. and Cundari, T. R. (1996) *Coord. Chem. Rev.* 147, 87.

Goswami, A. L. and Rawat, B. S. (1985) *J. Membrane Sci.* 24, 145.

Haase, D. J. and Walker, D. G. (1974) *Chem. Eng. Progress* 70(50), 74.

Hao, J., Tanaka, T., Kita, H., and Okamoto, K. (1997) *J. Membrane Sci.* 132, 97.

Hay, P. J. and Wadt, W. R. (1985) *J. Chem. Phys.* 82, 270.

Hehre, W. J., Radom, L., Schleyer, P. V. R., and Pople, J. A. (1986) *Ab initio Molecular Orbital Theory.* Wiley, New York, NY.

Hill, J. R. and Sauer, J. (1995) *J. Phys. Chem.* 99, 9536.

Hirai, H., Hara, S., and Komiyama, M. (1985) *Angew. Makromol. Chem.* 130, 207.

Hirai, H., Kurima, K., Wada, K., and Komiyama, M. (1985) *Chem. Lett. (Japan)* 1513.

Hirai, H., Wada, K., and Komiyama, M. (1986a) *Bull. Chem. Soc. Japan* 59, 1043.

Hirai, H., Wada, K., and Komiyama, M. (1986b) *Bull. Chem. Soc. Japan* 59, 2217.

Ho, W. S., Doyle., G., Savage, D. W., and Pruett, R. L. (1988) *Ind. Eng. Chem. Res.* 27, 334.

Hohenberg, P. and Kohn, W. (1964) *Phys. Rev.* 136, B864.

Huang, H. Y., Padin, J., and Yang, R. T. (1999a) *Ind. Eng. Chem. Res.* 38, 2720.

Huang, H. Y., Padin, J., and Yang, R. T. (1999b) *J. Phys. Chem. B.* 103, 3206.

Huang, H. Y. and Yang, R. T. (1999) *Langmuir* 15, 7647.

Huang, Y. Y. (1973) *J. Catal.* 30, 187.

Huang, Y. Y. and Vansant, E. F. (1973) *J. Phys. Chem.* 77, 663.

Huggahalli, M. and Fair, J. R. (1996) *Ind. Eng. Chem. Res.* 35, 2071.

Humphrey, J. L. and Keller, II, G. E. (1997) *Separation Process Technology.* McGraw-Hill, New York, NY.

Iwamoto, M. and Hamada, H. (1991) *Catal. Today* 10, 57.

Jarvelin, H. and Fair, J. R. (1993) *Ind. Eng. Chem. Res.* 32, 2201.

Jayaraman, A., Yang, R. T., Munson, C. L., and Chinn, D. (2001) *Ind. Eng. Chem. Res.* 40, 4370.

Juza, M., Mazzotti, M., and Morbidelli, M. (2000) *Trends Biotechnology* 18(3), 108.

Kansai Coke & Chemicals Co., (1989) Brochure on CO-PSA, Amagasaki City, Japan.

Kassab, E., Fouquet, J., Allavena, M., and Evleth, E. M. (1993) *J. Phys. Chem.* 97, 9034.

Keller, G. E., Marcinkowsky, A. E., Verma, S. K., and Wiliamson, K. D. (1992) Olefin recovery and purification via silver complexation. In *Separation and Purification Technology*. (N. N. Li and J. M. Calo, eds.). Marcel Dekker, New York, NY, p. 59.

King, C. J. (1980) *Separation Processes*, 2nd Ed. McGraw-Hill, New York, Chapter 14.

King, C. J. (1987) Separation processes based on reversible chemical complexation, In *Handbook of Separation Process Technology*. (R. W. Rousseau, ed.). Wiley, New York, NY, Chapter 15.

Kodde, A. J., Padin, J., van der Meer, P. J., Mittelmeijer-Hazeleger, M. C., Bliek, A., and Yang, R. T. (2000) *Ind. Eng. Chem. Res.* 39, 3108.

Kohl, A. L. and Reisenfeld, F. C. (1979) *Gas Purification*, 3rd Ed. Gulf Publishing, Houston, TX.

Kohn, W. and Sham, I. J. (1965) *Phys. Rev.* 140, A1133.

Kulvaranon, S., Findley, M. E., and Liapis, A. L. (1990) *Ind. Eng. Chem. Res*. 29, 106.

Kumar, R., Kratz, W. C., Guro, D. E., and Golden, T. C. A New Process for the Production of High Purity Carbon Monoxide and Hydrogen, Presented at the International Symposium on Separation Technology, University of Antwerp, Belgium (Aug. 22–27, 1993).

Larson, S. C., Aylor, A., Bell, A. T. and Reimer, J. A. (1994) *J. Phys. Chem.* 98, 11533.

Li, N. N. U.S. Patent 3,410,794 (1968).

Li, N. N. (1971a) *Ind. Eng. Chem. Proc. Dev. Des.* 10, 215.

Li, N. N. (1971b) *AIChE J.* 17, 459.

Long, R. B., (1972) Separation of unsaturates by complexing with solid copper salts, In *Recent Developments in Separation Science*. (N. N. Li, ed.), Vol. 1. CRC Press, Cleveland, OH, p. 35.

Long, R. B., Caruso, F. A., DeFeo, R. J., and Walker, D. G. U.S. Patent 4,141,960 (1979).

Long, R. B., Horowitz, H. H., and Savage, D. W. U.S. Patent 3,754,047 (1973).

Matz, M. J. and Knaebel, K. S. (1990) *Separ. Sci. Tech.* 25, 961.

Meyers, R. A., ed. (1986) *Handbook of Petroleum Refining Process.* McGraw-Hill, New York, NY.

Mulliken, R. S. and Ermler, W. C. (1977) *Diatomic Molecules Results of Ab initio Calculations.* Academic Press, New York, NY.

Murrell, J. N. and Harget, A. J. (1972) *Semi-empirical Self-Consistent-Field Molecular Orbital Theory.* Wiley-Interscience, London, UK.

Naccache, C. M. and Ben Taarit, Y. (1971) *J. Catal.* 22, 171.

Padin, J., Munson, C. L., and Yang, R. T. U.S. Patent 6,215,037 (2001).

Padin, J. and Yang, R. T. (2000) *Chem. Eng. Sci.* 55, 2607.

Padin, J., Yang, R. T., and Munson, C. L. (1999) *Ind. Eng. Chem. Res.* 38, 3614.

Parr, R. G. and Yang, W. (1989) *Density Functional Theory of Atoms and Molecules.* Oxford University Press, New York, NY.

Pearce, G. K. U.S. Patent 4,717,398 (1988).

Pople, J. A., Santry, D. P., and Segal, G. A. (1965) *J. Chem. Phys.* 43, S129.

Quinn, H. W. (1971) Hydrocarbon separations with silver (I) systems. In *Progress in Separation and Purification*. (Perry, ed.), Vol. 4. Interscience, New York, NY, p. 133.

Rabo, J. A., Francis, J. N., and Angell, C. L. U.S. Patent 4,019,879 (1977).

Rege, S. U., Padin, J., and Yang, R. T. (1998) *AIChE J.* 44, 799.

Russell, A. S. and Stokes, J. J. (1946) *Ind. Eng. Chem.* 38, 1071.

Ruthven, D. M. (1984) *Principles of Adsorption and Adsorption Processes*, Wiley, New York, NY, Ch. 12.

Safarik, D. J. and Eldridge, R. B. (1998) *Ind. Eng. Chem. Res.* 37, 2571.

Sarkany, J., d'Itri, J. and Sachtler, W. M. H. (1992) *Catal. Lett.* 16, 241.

Sauer, J. (1989) *Chem. Rev.* 89, 199.

Shelef, M. (1994) Chem. Rev. 95, 209.

Sherry, H. S. (1966) *J. Phys. Chem.* 70, 1158.

Sherry, H. S. (1967) *J. Phys. Chem.* 71, 780.

Sherry, H. S. (1968) *J. Coll. Interf. Sci.* 28, 288.

Stewart, J. J. P. (1990) QCPE Program No. 584; Quantum Chemistry Program Exchange, Indiana University, Department of Chemistry.

Stichlmair, J. G. and Fair, J. R. (1998) *Distillation: Principle and Practice.* Wiley-VCH, New York, NY.

Takahashi, A., Yang, F. H., and Yang, R. T. (2000) *Ind. Eng. Chem. Res.* 39, 3856.

Takahashi, A., Yang, F. H., and Yang, R. T. (2002) *Ind. Eng. Chem. Res.* 41, 2487.

Takahashi, A. and Yang, R. T. (2002) *AIChE J.* 48, 1457.

Takahashi, A., Yang, R. T., Munson, C. L., and Chinn, D. (2001a) *Langmuir* 17, 8405.

Takahashi, A., Yang, R. T., Munson, C. L., and Chinn, D. (2001b) *Ind. Eng. Chem. Res.* 40, 3979.

Tamon, H., Kitamura, K., and Okazaki, M. (1996) *AIChE J.* 42, 422.

Valyon, J. and Hall, W. K. (1993) *J. Phys. Chem.* 97, 7054.

Van Krevelen, D. W. and Baans, C. M. E. (1950) *J. Phys. Coll. Chem.* 54, 370.

Wu, S., Han, S., Cho, S. H., Kim, J. N., and Yang, R. T. (1999) *Ind. Eng. Chem. Res.* 36, 2749.

Xie, Y. C. and Tang, Y. Q. (1990) *Adv. Catal*. 37, 1.

Xie, Y. C., Xu, X., Zhao, B., and Tang, Y. (1992) *Catal. Lett.* 13, 239.

Xie, Y. C., Zhang, J., Qiu, J., Tong, X., Fu, J., Yang, G., Yan, H., and Tang, Y. Q. (1996) *Adsorption*. 3, 27.

Yang, R. T. (1987) *Gas Separation by Adsorption Processes*. Butterworth, Boston.

Yang, R. T. and Foldes, R. (1996) *Ind. Eng. Chem. Res.* 35, 1006.

Yang, R. T. and Kikkinides, E. S. (1995) *AIChE J.* 41, 509.

Yang, R. T., Long, R. Q., Padin, J., Takahashi, A., and Takahashi, T. (1999) *Ind. Eng. Chem. Res.* 38, 2726.

Yang, R. T., Padin., J., and Rege, S. U. U.S. Patent 6,423,881 (2002).

Yang, R. T., Takahashi, A., and Yang, F. H. (2001) *Ind. Eng. Chem. Res.* 40, 6236.

Yokoe, J., Takeuchi, M., and Tsuji, T. U.S. Patent 4,713,090 (1987).

9

CARBON NANOTUBES, PILLARED CLAYS, AND POLYMERIC RESINS

Three different types of sorbents are included in this chapter. Among them, polymeric resins and their derivatives have been used commercially for adsorption and ion exchange. The other two types have not been used commercially. However, they each have interesting and unique adsorption properties and are subjects of active research. For these reasons, they are included in this chapter.

9.1. CARBON NANOTUBES

The discovery of fullerenes and carbon nanotubes has opened a new chapter in carbon chemistry. Carbon nanotubes, in particular, hold tremendous potential for applications because of their unique properties, such as high thermal and electrical conductivities, high strengths, and high stiffness (chapters in Dresselhaus et al., 2001). Potential applications include: electron microscope tips, field and light emitters, microelectronic devices, nanoprobes and nanosensors, high Li capacitors for rechargeable Li batteries, composite materials, and replacing Si as the smallest computer chips.

A vast and rapidly growing volume of literature exists on carbon nanotubes. The coverage and discussion on this subject here will be limited to their syntheses, characterization, special adsorption properties, and potential applications as sorbents. Some terminologies are first defined. Single-wall nanotubes are denoted by SWNT, whereas MWNT stands for multiwall nanotubes. A SWNT is a seamless cylinder wrapped by a graphite sheet (or graphene sheet). The hexagonal honeycomb lattice of the graphene sheet can be oriented in many possible directions relative to the axis of the tube. Depending on the relative orientation (and size), a SWNT can be metallic or semiconducting (Louie, 2001; Yao et al., 2001). The relative orientation is referred to as the "helicity," expressed by a set of indices

Adsorbents: Fundamentals and Applications, Edited By Ralph T. Yang
ISBN 0-471-29741-0

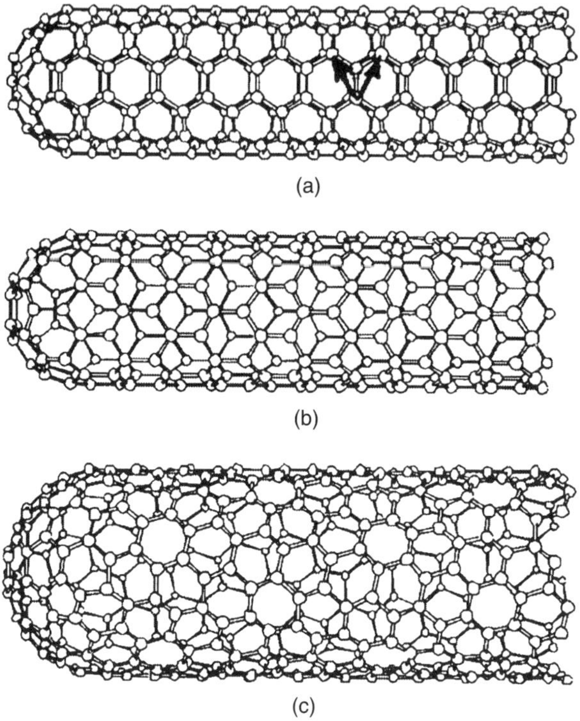

Figure 9.1. Schematic models for single-wall nanotubes with the nanotube axis normal to the chiral vector, $\vec{C} = n\vec{a}_1 + m\vec{a}_2$, where $\vec{a}_1$ and $\vec{a}_2$ are unit vectors shown by the two arrows in (a), which are both along the zigzag direction. (a) A (5, 5) nanotube, with tube axis having fivefold symmetry. (b) A (9, 0) nanotube (having threefold rotation symmetry). (c) A (n, m) nanotube (adapted from Dresselhaus and Avouris, 2001).

(n, m). The chiral vector of the nanotube, $\vec{C} = n\vec{a}_1 + m\vec{a}_2$, is normal to the tube axis, where $\vec{a}_1$ and $\vec{a}_2$ are unit vectors shown in Figure 9.1 (a). The nanotube diameter, d, can be calculated from n and m (Dresselhaus and Avouris, 2001):

$$d = \sqrt{3}a_{C-C}(m^2 + mn + n^2)^{1/2}/\pi = C/\pi \tag{9.1}$$

where a_{C-C} is the C–C bond length (1.42 Å), and C is the length of the chiral vector $\vec{C}$. Thus, a (n, n) nanotube is an "armchair" nanotube, that is, with its armchair lattice oriented along the circumference, whereas a $(n, 0)$ nanotube is a "zigzag" nanotube with its zigzag lattice along the circumference. A (n, m) nanotube is referred to as a "chiral" nanotube. These three types are illustrated in Figure 9.1. The helicity of nanotubes is important for applications in microelectronics and obviously less important for applications in adsorption.

Carbon nanotubes have been formed by many different methods, eight having been reviewed by Journet and Bernier (1998). Two of them have been studied extensively: catalytic decomposition of hydrocarbons and CO, and vaporization (and condensation) of graphite. The former has been the focus for commercial production (and is being used by various manufacturers). Both methods date back to the 1960's and 1970's. Some historical aspects of these two methods have been reviewed by Dresselhaus and Avouris (2001) and by Dresselhaus and Endo (2001). These two methods, along with their brief histories, are described separately.

9.1.1. Catalytic Decomposition

The formation of carbon filaments from decomposition of hydrocarbons was first reported by Schutzenberger (1890) in 1890, and the first substantive report on catalytic formation of carbon filaments was made by Radushkevich (as in the D–R equation) and Luk'Yanovich (1952). A tremendous amount of literature exists from studies performed by the carbon, catalysis, and metallurgy communities. A comprehensive review (including history) of the subject prior to 1978 is available in Baker and Harris (1978). Discussion on some of the later literature is available in Tibbetts (1990) and Yang and Chen (1989). Studies of carbon filaments paralleled the developments of the transmission electron microscope (TEM), which had become commercially available in the late 1940's and early 1950's, and had a resolution near 1 nm at the time. The formation of carbon filaments received considerable attention primarily because of its detrimental effects on catalyst deactivation and blast furnace operation (due to CO disproportionation). The research objective was to understand the formation process in order to minimize or alleviate this problem. It has long been known that Fe, Co, and Ni are the most active catalysts for forming filaments. The filaments typically had a well-defined hollow core. Figure 9.2 is a typical TEM image of the hollow graphite filament, by Hilbert and Lange (1958). With the TEM resolution high enough in the early 1970's to distinguish the graphitic layers, it was already known that some hollow filaments were formed by concentric layers of graphite with ~3.5 Å spacing. Indeed, such TEM images have been published (Baird et al., 1971; Baker and Harris, 1978; Fryer, 1979). One such example is shown in Figure 9.3. After 1993, the hollow carbon filaments acquired the name of MWNT. The term "carbon nanotube" was first used, in 1993, by Iijima and Ichihashi (1993) and by Bethune et al. (1993). In the 1980's, research on catalytic carbon filament growth was driven by a different impetus: to produce carbon fibers to replace those fibers made from PAN or pitch (Tibbetts et al., 1986; Dresselhaus et al., 1988; Tibbetts, 1990). This has not occurred because of the high costs of the catalytically grown carbon fibers.

In principle, carbon nanotubes can be grown from any gaseous hydrocarbons or CO, onto Fe, Co, or Ni particles dispersed on a substrate under appropriate reaction conditions. Higher temperatures and slower growth rates favor graphitic filament formation, while lower temperatures and fast rates lead to nongraphitic forms (Baker and Harris, 1978). Beside Fe, Co, and Ni, filaments can also be formed on other metals such as Pt and Cu. Acetylene is among the most reactive hydrocarbon precursors. Unsaturated hydrocarbons like propylene and butadiene are more reactive than the saturated hydrocarbons such as methane and

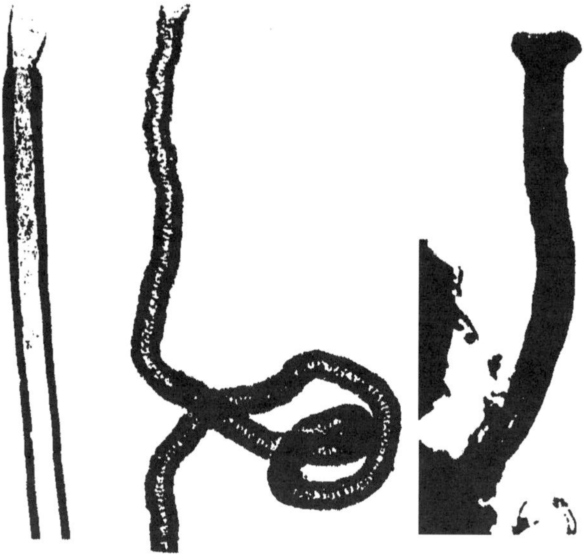

Figure 9.2. TEM images of hollow graphite filaments grown from n-heptane on Fe at 1100 °C, taken from Hilbert and Lange (1958). Magnification = 75,000 and 110,000 (far right).

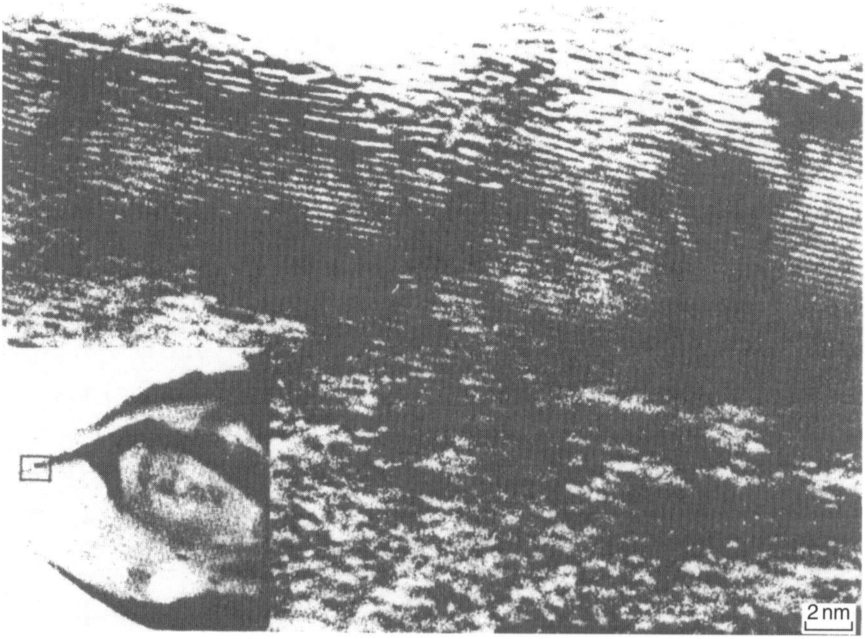

Figure 9.3. High-resolution TEM image of 0.35 nm graphitic layers in a hollow carbon filament grown from methane over Ni at 650 °C. Insert shows area at low magnification in square. Taken from Fryer (1979) and Baird et al. (1974). Baird et al. (1971) showed similar images of a hollow filament with graphitic sheets along the axis at 0.34 nm spacing.

propane. Thus, the appropriate reaction conditions depend on the gas as well as the activity of the catalyst. Nanotubes formed by a myriad of combinations of catalyst/support/gas/reaction conditions have appeared in the literature. Reviews for some of the reports are available (Ding et al., 2001; Dai, 2001). Ding et al. (2001) have given a list of detailed reaction conditions for a large number of them. A summary of these reports is given in Table 9.1.

The following mechanistic steps for filament growth have been generally accepted since the 1970's (Baker and Harris, 1978; Tibbetts, 1987). The hydrocarbon or CO first decomposes on the front-exposed surfaces of the metal particle to release hydrogen and carbon, which dissolves in the metal particle. The dissolved carbon diffuses through the particle and is precipitated on the rear faces to form the body of the filament. These steps are shown schematically in Figure 9.4. Burton (1974) has shown that the structure and melting point of a small metal particle differs significantly from that of the bulk metal, and that the melting point can be substantially lower than that of the bulk. Hence the supported metal particle acquires a liquid-like behavior. From in situ TEM studies, Baker has concluded that the metal particles (of sizes below a fraction of a micrometer) are generally shaped like a pear, with a truncated rear end, as illustrated in Figure 9.4. Thus, carbon solubility and diffusivity in the metal catalyst are prerequisites for filament growth. The solubilities of carbon in Ni are 0.29 at % (700 °C) and 0.37 at % (750 °C), and the diffusivity at 700 °C is 4.0×10^{-9} cm^2/s (Yang et al., 1990).

The carbon dissolution/diffusion/precipitation mechanism for filament growth has been studied in detail for hollow graphitic filaments formed on Ni, Co, and α-Fe from methane decomposition, by Yang and Chen (1989). The crystallographic orientations of the graphite/metal interface were examined with TEM/selected area electron diffraction. Epitaxial matching of the graphite lattice with different faces of the metals was identified. The structures of four different faces of Ni are shown in Figure 9.5. Their possible epitaxial matchings with the graphite lattice

Table 9.1. Catalytic formation of carbon nanotubes from reports after 1996

Gas	Reaction Conditions	Catalyst	Support
CO	*Typically 700 °C, 1–5 atm*	Fe, Co, Ni, Co-Mo, Fe-Mo, Co-Fe, Co-V, Fe-Ru, Ni-Cu, Ni-MgO, Fe-MgO, Ferrocene vapor ($Fe(C_5H_5)_2$), $Fe(CO)_5$ vapor	SiO_2 (most used), Al_2O_3, MgO, Al_2O_3-SiO_2, zeolite, clay
Unsaturated hydrocarbons (C_2H_2, C_2H_4, C_3H_6, C_4H_6, C_6H_6, acetone)	600–900 °C Carried in N_2 or another inert atmosphere at 1 atm		
Saturated hydrocarbons, mainly CH_4	700–1000 °C typically 900 °C, 1 atm		

Both SWNT and MWNT have been reported, depending on the particle sizes, see Figure 9.4.

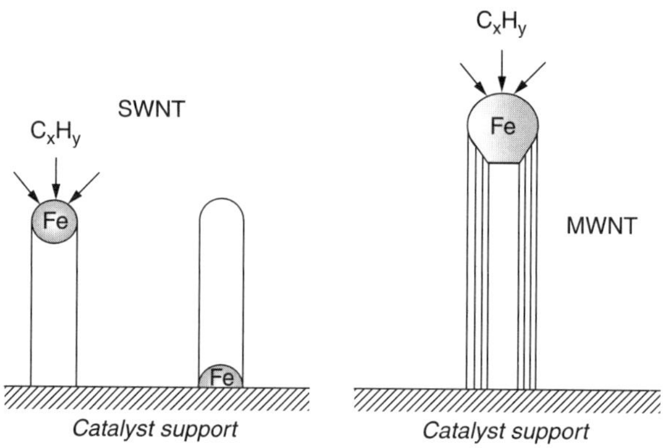

Figure 9.4. Growth modes of catalytically formed carbon nanotubes. Fe, Co, and Ni are the catalysts and hydrocarbons, and CO are the carbon precursors. SWNTs are grown on particles $<\sim$40–50 Å, whereas MWNTs are grown on larger particles. The outside tube diameter is the same as the particle diameter.

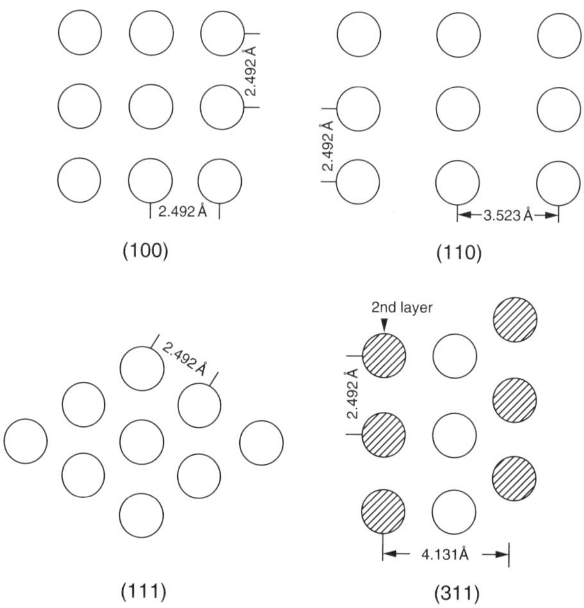

Figure 9.5. Structures and lattice distances of four Ni faces.

are illustrated by Figure 9.6. The distance between two parallel C–C bonds of the hexagonal ring of graphite is 2.46 Å. This distance is nearly perfectly matched with the lattice distance of 2.49 Å of Ni. Similarly, the lattice constant of Co (*hcp*) is $a = 2.51$ Å and that of α-Fe (*bcc*) is $a = 2.86$. These three metals are among those with lattice constants that are closest to 2.46 Å (Kittel, 1953).

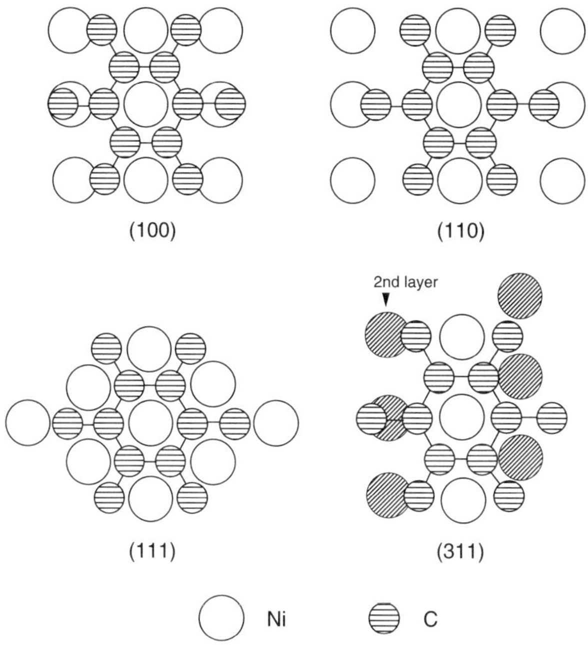

Figure 9.6. Epitaxial matching at the graphite/Ni interfaces on different faces of Ni where carbon precipitates (Yang and Chen, 1989).

Thus, epitaxy clearly plays an important role in the catalytic growth of carbon nanotubes, at least for MWNTs.

As illustrated by Figure 9.4, metal dispersion is the controlling factor for nanotube growth. Small particles lead to SWNTs, while large particles yield MWNTs. The uniformity of nanotube size depends on the dispersion of uniform particle size. Moreover, higher temperatures, and possibly post-annealing, will reduce the amount of defects in the resulting nanotubes.

Much has been learned about the dispersion of metals on various supports (Ruckenstein, 1987). Ruckenstein (1987) has laid the theoretical foundation for both the thermodynamics and kinetics of reaching equilibrium for metal/support systems. He also analyzed a large amount of experimental data. The thermodynamic equilibrium is based on the free energy of formation, which in turn can be determined from the following interfacial free energies: substrate/gas, crystallite/gas, and crystallite/substrate. The metal crystallite-support interactions play a crucial role. The dispersion is therefore influenced by a number of factors: metal and additive (i.e., alloys), support, temperature, and gas environment. For the same metal on different supports, the stability of dispersion follows the order: $Al_2O_3 > SiO_2 > C$. By varying these related factors, a large number of combinations have been employed in nanotube growths. Table 9.1 summarizes what have been used. Some empirical facts are known, but little is understood. It is instructive to discuss four cases in some detail below.

Pure metals (e.g., Fe, Co, or Ni) supported on alumina or silica gel lead to hollow graphite filaments with outside diameters around 0.1 μm. Higher dispersion is needed for growing smaller filaments or MWNTs. This can be achieved by using an additive to the metals, or by alloying the metals. Chen et al. (1997) were able to grow uniform and small MWNTs on Ni-MgO from catalytic decomposition of methane or CO. The optimal starting catalyst was a mixed oxide prepared by co-precipitation to form $Ni_{0.4}Mg_{0.6}O$. The catalyst was heated in H_2 at 650 °C before switching to CH_4 for nanotube growth (Yang, 2000). The catalyst particle was launched at the tip of the nanotube during its growth, and the catalyst could be removed by dissolving in nitric acid solution. Prior to the nanotube growth reaction, the catalyst was in the form of a solid solution of NiO and MgO, both having the same rock-salt crystal structure and nearly the same lattice constant. During H_2 pretreatment and nanotube growth, a small portion of the NiO was reduced to Ni^0. The effect of MgO was to inhibit, but not completely stop, the reduction of NiO. The small portion of reduced Ni metal formed small particles or clusters on the surface, which yielded MWNTs with small diameters. Thus, uniform-sized MWNTs with outside diameters of 15–20 nm were grown. CaO, on the contrary, did not have the dispersion effect on Ni (Chen et al., 1997), and yielded large MWNTs.

Using micrometer-size zeolite crystals as the support for MWNT growth did not present as an appealing idea due to the low surface area of the exterior surfaces on which the metal is supported. Surprisingly, they turned out to be excellent supports for Co and Fe, without added metals (Hernadi et al., 1996). Hernadi et al. (1996) compared a variety of supports for Co and Fe, including SiO_2, carbon, and different zeolites. Co supported on NaY resulted in the highest yield (27–40% yield) of uniform-sized MWNTs, which also had the highest BET surface area (653 m^2/g). The same group subsequently provided a recipe for the best results [Colomer et al. (1998)]. In their recipe, acetylene was decomposed on Co (2.5 wt %)/NaY at 600 °C. The zeolite support was removed by dissolution with HF solution, and the amorphous carbon was removed by either permanganate oxidation or air oxidation. This recipe has been widely used. Figure 9.7 shows a TEM image of MWNTs that were grown by using their recipe (Colomer et al., 1998). No study or discussion has been made on the mechanism of Co dispersion on zeolite. The Y zeolite (i.e., faujacite) is in the form of cubic crystals with a dimension in the order of a micrometer, with its pore size much smaller than the particle size of the Co metal. The zeolite is first impregnated with a Co salt (acetate) at room temperature, subjected to subsequent calcination. During calcination, water vapor, CO_2, and CH_4 will evolve from the internal pores. This may help the initial dispersion of Co. The external silicoaluminate surface is apparently better than both alumina and silica for the dispersion of Co. This phenomenon deserves study.

A direct comparison of the MWNTs prepared by the two different recipes (Chen et al., 1997, and Colomer et al., 1998) is as follows. MWNTs prepared from the former recipe gave a BET surface area of 155 m^2/g, with a pore-size distribution from 2.5 to 30 nm (with a peak size at 2.9 nm) (Long and Yang,

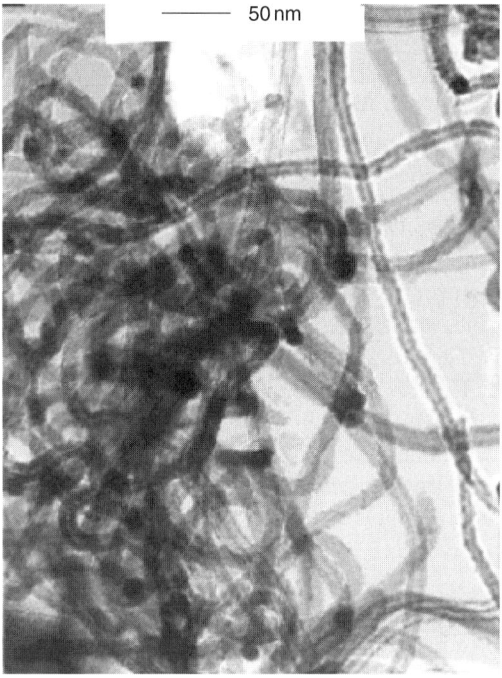

Figure 9.7. TEM image of MWNTs grown from methane decomposition on Ni-MgO catalyst at 650 °C, by Long and Yang (2001b) using the recipe of Chen et al. (1997). High-resolution TEM showed 0.35 nm graphitic layers along the tubes (Chen et al., 1997).

2001a), while that from the latter recipe yielded a surface area of 462 m^2/g and pore size distribution from 2.0 to 3.9 nm (Long and Yang, 2001b).

As illustrated in Figure 9.4, SWNT would grow when the metal particle is small enough, although no effort was made to image them prior to ca. 1996. Dai et al. (1996) reported the TEM images of SWNTs grown on Ni/Co and Mo supported on alumina, from CO disproportionation at 1200 °C. SWNTs with diameters from ~1 to 5 nm were seen, each capped with a metal particle of the same size as the tube diameter. However, the yield was low and the size distribution was wide. Resasco and co-workers (Kitiyanan et al., 2000; Alvarez et al., 2001) improved the technique by using Co/Mo supported on silica gel to increase the yield and, more importantly, to control the sizes to a uniform size of near 1 nm diameter. Their starting catalyst was in the form of mixed oxides of MoO_3 and Co_3O_4, with Co/Mo = 2–4 being the best (i.e., the highest yield and the highest selectivity for SWNT). The catalyst was pretreated in H_2 at 500 °C and SWNTs were grown from CO at 700 °C. The effects of the added MoO_3 to Co were more complex than those of MgO to Ni, as described above, but were somewhat similar. A small fraction of the Co was in the metallic state at the beginning of the nanotube growth reaction, while Mo was in the form of oxide. During the growth reaction, the fraction of the reduced Co increased with time, while Mo was gradually converted into Mo_2C (Alvarez et al., 2001). The growth ceased when all Mo converted to carbide and all Co was reduced. The small fraction of the reduced Co was apparently in the form of highly dispersed clusters (of the size of 1 nm), which were responsible for the most active SWNT growth period (i.e., at the beginning). In another work, Cassell et al. (1999) added Mo to Fe and showed a significantly increased yield of SWNTs by the addition of Mo. High yields of SWNTs were obtained from decomposition of CH_4 at 900 °C on these bimetallic catalysts. The optimal catalyst composition was: Fe/Mo $= 1/0.17$ supported on 1/1 ratio of mixed SiO_2 and Al_2O_3. Cassell et al. (1999) attributed the effect of Mo as a promoter for aromatization of methane, which in turn forms aromatic intermediates that facilitated the growth of nanotubes. However, from the discussion above, it is likely that Mo played a similar role in aiding dispersion of metallic Fe in the Fe/Mo system. TEM images of the SWNTs grown by Cassell et al. (1999) are shown in Figure 9.8. Using the catalytic route, SWNT strands as long as 20 cm have been grown (Zhu et al., 2002).

Aligned nanotubes (in the form of a carpet) are of interest for applications. This has been accomplished by a number of groups by using vapor phase catalysts such as ferrocene, $Fe(C_5H_5)_2$. In a recent example, Kamalakaran et al. (2000) prepared arrays of large (100–250 nm diameter) aligned MWNTs by pyrolyzing a spray solution of ferrocene and benzene in Ar at 850 °C.

The purification of nanotubes has been studied. It is relatively easier to purify the catalytically grown nanotubes than that from graphite vaporization. The metal catalysts and the inorganic support can be removed with HCl and HF solution. The transition metals can be dissolved in HNO_3 solution. The amorphous carbon can be removed with HNO_3 solution, and also by permanganate solution. Oxidation with air at mild temperatures is also effective in removing the amorphous carbon.

A closely related type of material to MWNT is graphite nanofibers (GNF), developed by Baker and Rodriguez (Rodriguez et al., 1995). GNF are prepared by catalytic decomposition of hydrocarbons on metals or metal alloys at temperatures

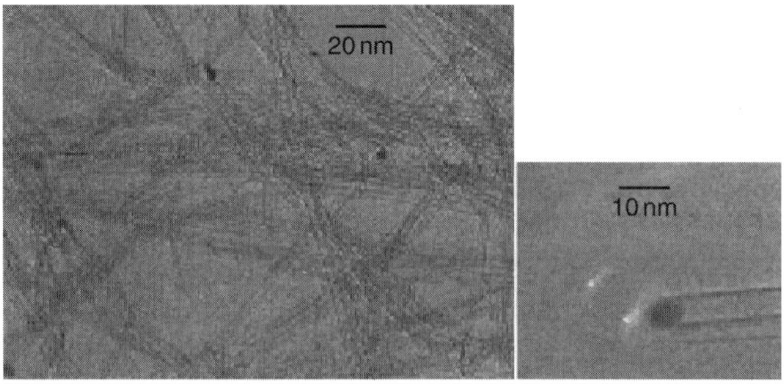

Figure 9.8. TEM images of SWNTs grown from methane decomposition on Fe-Ru/Al_2O_3 catalyst at 900 °C. Also shown is a typical image of the tip of a SWNT capped with a metal particle. SWNTs grown on Fe-Mo/Al_2O_3 are similar (from Cassell et al., 1999, with permission).

in the range between 450–750 °C. The fibers have dimensions of 30–500 Å^2 in cross-sectional area and 10–100 μm in length. They consist of platelets of graphite layers similar to the MWNTs. However, the layers are not parallel to the axial direction of the fiber, but instead form an angle with it. The structure has been described as "herring-bone." Thus, many graphitic edges are exposed on the surface of the fiber. These edges have unsaturated sp^2 bonds, or free sp^2 electrons. Hence much interesting chemistry can be exploited with the edges of GNFs. Of particular interest is the possibility of hydrogen storage (Rodriguez and Baker, 1997).

9.1.2. Arc Discharge and Laser Vaporization

Since the 1950's (Bacon, 1960), hollow graphite filaments have been found in deposits of high-temperature furnaces containing hydrocarbons or CO, with or without catalysts. The first substantive study of graphite fibers formed by arc discharge was performed by Bacon (1960). He used a dc arc generated by a moderate 75–80 volts and 70–75 amperes between two vertically aligned graphite electrodes. The arc provided the high temperature to vaporize the electrodes. The graphite vapor condensed on the cathode where graphite filaments were deposited. At the time, it was thought that the filaments would bc formcd at near the triple point of graphite, that is, near 100 atm and 3900 K. Thus, Ar at 92 atm was used as the inert atmosphere. A TEM with a resolution no better than 5 Å was employed. Selected-area electron diffraction/TEM analysis confirmed that the walls of the filaments were formed by cylindrical layers of perfect graphite sheets. Unfortunately, without high resolution, a "scroll" structure, rather than concentric tubes, was proposed. There is little doubt that what is now called MWNTs were formed in Bacon's samples.

Interest in the arc discharge technique was later revived by the discovery of the fullerenes in 1985 (Kroto et al., 1985). A detailed review on fullerenes (including their formation) is available in Dresselhaus et al. (1996). Kroto et al. (1985) identified the C_{60} molecule found in the soot, which was condensed from the carbon vapor generated by laser heating. Carbon vapor can be generated by a number of means, for example, arc discharge, laser ablation, resistive heating, or combustion (with deficient oxygen). It turned out that both MWNTs and fullerenes are formed in the condensed carbon. An interesting finding was that the formation of these materials depends on the pressure of the inert atmosphere. Also, a minimum pressure is required (Dresselhaus et al., 1996).

Table 9.2 is a summary of the main findings for the syntheses of carbon nanotubes and fullerenes by using the route of carbon vaporization/condensation. SWNT has never been found without the use of a catalyst. Fe, Co, and Ni are the main catalysts used in forming SWNTs.

The smallest SWNTs that have been grown are 4 Å in diameter. They have been grown by pyrolyzing tripropylamine filled in the 7.3 Å channels of $AlPO_4$-5 molecular sieve (Wang et al., 2000). An inner wall diameter of 4 Å of a MWNT formed by arc discharge has also been reported (Qin et al., 2000). Syntheses of 5 Å SWNTs have been reported as well (Peng et al., 2000; Sun et al., 2000).

Table 9.2. Formation of carbon nanotubes and fullerenes by graphite vaporization/condensation

Technique	Conditions/Observation	Reference
Arc discharge	Graphite electrodes at near triple point (in 92 atm Ar) Multiwall, cylindrical, graphitic whiskers formed on cathode, "scroll" structure was (mis)-identified by TEM	Bacon (1960)
Laser vaporization	1 atm He over graphite C_{60} (major) and C_{70} fullerenes formed	Kroto et al. (1985)
Ohmic vaporization	Graphite rod resistively heated in glass bell jar with He at >100 torr C_{60} fullerene formed	Kratschmer et al. (1990)
Arc discharge	Graphite electrodes in contact, 100 torr He C_{60} fullerene formed	Haufler et al. (1990).
Arc discharge	Similar to Bacon's, Ar at 100 torr was used. Multiwall nanotubes correctly identified	Iijima (1991)
Arc discharge with catalyst	(a) Fe placed on cathode to be co-vaporized, in 10 torr CH_4 + 40 torr Ar. (b) Co added to anode, in 100–500 torr He. Single-wall nanotubes (~1 nm) formed on cathode in both cases.	(a) Iijima and Ichihashi (1993); (b) Bethune et al. (1993)
Laser vaporization with catalyst	Mixed Co (1 at %) or Ni (0.6 at %) in graphite as target, in 500 torr Ar. Single-wall nanotubes	Guo et al. (1995)

The role of the catalyst and the mechanism for the formation of SWNTs during condensation of carbon vapor are not known. It is possible that the epitaxial matching of graphite lattice or rings on the faces of the transition metal plays a role, as illustrated in Figures 4–6. A rationalization for the growth process has also been given by Thess et al. (1996). In their rationalization, metal atom(s) is attached to the carbon atoms on the open end of the tube where growth occurs. These metal atoms prevent five-member rings from forming and keep the end open. The tube diameter is determined by competition between the strain energy of curvature of the graphite sheet and the dangling-bond energy of the open edge.

In all the studies since Bacon's work, "graphite rod" or "graphite disk" was used as the source of carbon for vaporization. No impurity data were given. The electrode graphite with the highest purity is used as anodes for aluminum smelting, with a total "ash" content of >0.1% (wt) (Yang, 1979). The commercially available electrode graphites have "ash" contents significantly higher than this level. The main impurity in electrode graphite is Fe. The question concerning the role of the Fe impurity in the formation of MWNTs (and possibly fullerenes) has not been addressed. With these impurities, SWNTs were likely to have been formed; however, they have not yet been found probably because of their scarcity.

9.1.3. Adsorption Properties of Carbon Nanotubes

The most unexpected and potentially most important adsorption property of carbon nanotubes is hydrogen storage. While controversy remains, intensive research efforts on this subject are on-going worldwide. This subject will be discussed separately in Chapter 10.

Carbon nanotubes have cylindrical pores. An adsorbate molecule interacts with the carbon atoms on the surrounding walls. As discussed in Chapter 2, the resulting potential in the cylindrical pore can be substantially higher than that in a slit-shaped pore with the same dimension. In addition, carbon nanotubes are highly graphitic (much more so than activated carbon). The surface of the nanotubes is highly aromatic and contains a high density of π electrons. With these two factors, it is *expected* that the carbon nanotubes can adsorb molecules much more strongly than activated carbon (which has slit-shaped or wedge-shaped pores). This expectation has indeed been shown by a number of simulation studies of adsorption for He, Xe, CH_4, and N_2. The general results showed that the interactions are of the order of two when compared with that on planar graphite, as to be discussed shortly.

Because SWNTs are grown in the form of bundles and ropes from both catalytic route (e.g., Colomer et al., 2000) and graphite vaporization (e.g., Thess et al., 1996), the inter-tube spaces (bounded by the outer surfaces of the tubes) are also important for adsorption. The SWNT bundles are arranged in a triangular lattice structure, held together by van der Waals forces. Hexagonal close-packed configuration without tube-tube contact has been used in most simulations. In such cases, the dimensions of the inter-tube pores are often smaller than that inside the tubes, thus, adsorption in the inter-tube spaces can be stronger. Adsorption in the inter-tube spaces has been assumed in many theoretical calculations, particularly in earlier work, when synthesis of small-diameter SWNTs were not reported. In such calculations, inter-tube spacing as small as 2.6 Å was used.

Since the first study on adsorption in carbon nanotubes by Pederson and Broughton (1992), most have been simulations. Few experimental studies have appeared, however. The observations from both simulations and experiments are discussed below.

Adsorption of Dioxin and Benzene. Ultrapurification is becoming an increasingly important topic due to environmental and health concerns. Dioxins are highly toxic and quite stable. Like dioxins, many of the highly toxic chemicals are low-volatile compounds. Removal of these compounds to the ppb or ppt level are often necessary. It has been a challenge to develop an experimental technique for measuring adsorption isotherms of low-volatile organics at low concentrations. A simple technique based on temperature programmed desorption (TPD) has been developed recently (Yang et al., 1999). This technique involves dosing the sorbate (as solution in a solvent such as DMF) at the inlet of the column that is packed with the sorbent, followed with TPD in a flow of an inert gas. From the peak desorption temperature as a function of heating rate, the activation energy of desorption is obtained, which is taken as the heat of adsorption. The heat of adsorption yields the Langmuir constant. The saturated amount is estimated from the monolayer amount, based on the molecular area of the sorbate. Thus, the Langmuir isotherm is obtained (Yang et al., 1999). Using this technique, Long and Yang (2001a) have measured the bond energy or heat of adsorption of dioxin on carbon nanotubes, and compared this value with that on other sorbents.

Dioxins and related compounds (e.g., polychlorinated dibenzofurans and biphenyls) are highly toxic and stable pollutants. Dibenzo-*p*-dioxins are a family of compounds consisting of two benzene rings joined by two oxygen atoms and have from zero to eight chlorine atoms attached around the rings. The dibenzofurans are a similar family, which differs only in the manner one of the bonds between the two benzene rings is bridged by oxygen. The toxicity of dioxins varies with the number of Cl atoms, with non- and monochloro dioxins being nontoxic, while becoming increasingly toxic with more Cl atoms. Dioxins are mainly generated from combustion of organic compounds in waste incinerators, such as municipal waste, medical waste, hazardous waste, and army stockpile (chemical agents). They are formed downstream the combustion zone with typical concentrations of 10–500 ng/Nm^3. Current regulations on dioxin emissions are complex, depending on the toxic equivalency of the actual compounds and O_2 concentration, and vary in different countries. Nonetheless, removal to well below 1 ng/Nm^3 is generally required (Harstenstein, 1993). Since 1991, activated carbon adsorption has been widely adopted for dioxin removal from waste incinerators in Europe and Japan (Harstenstein, 1993). Because of the higher bond energy between dioxin and activated carbon than other sorbents, the removal efficiency for dioxin by activated carbon is much higher than other sorbents, such as clays, pillared clays, g-Al_2O_3 and zeolites (Yang et al., 1999).

As mentioned, the desorption activation energy can be obtained from TPD by varying the heating rates. From the temperature dependence of the desorption peak temperature, one can calculate the desorption activation energy. For physical adsorption, the desorption activation energy is equal to the bond energy, or heat of adsorption (Yang, 1987). The bond energies for dioxin on three sorbents are compared in Table 9.3. The carbon nanotubes used in this work were MWNTs prepared from methane decomposition on Ni-MgO catalyst by using the recipe

Table 9.3. Peak desorption temperature of dioxin at different heating rates, activation energies for desorption, and Langmuir constants on different sorbents

	Peak Desorption Temperature (°C) at Different Heating Rates			Desorption Activation Energy	Langmuir Constant B at 25 °C	
Sorbent	2 °C/min	5 °C/min	10 °C/min	(kJ/mol)	(1/atm)	Ref.
Carbon nanotubes	588	609	620	315	2.7×10^{52}	a
ZX-4 activated carbon (Mitsubishi)	481	517	543	119	1.3×10^{18}	b
γ-Al_2O_3	306	353	394	47.9	4.5×10^{5}	b

(a) Long and Yang (2001a); (b) Yang et al. (1999).

of Chen et al. (1997). The surface area was 155 m^2/g and the peak diameter of the MWNTs was 2.9 nm. The ZX-4 activated carbon is commercially used for dioxin removal from incinerators. From the comparison, the bond energy of dioxin on the MWNTs is nearly three times that of dioxin on activated carbon. The Langmuir constant was obtained from the activation energy of desorption (Yang et al., 1999). The strong bonding between the dioxin molecule and the nanotube is obviously the result of the overlapping potentials with the surround walls. The position of the dioxin molecule is optimally aligned when it is collinear along the axis of the nanotube.

The results given in Table 9.3 indicate that for the same level of purification, much less sorbent is needed if the activated carbon is replaced by carbon nanotubes. Alternatively, if the same size adsorber is used, a much higher level of purification can be accomplished.

In the commercial operation of incinerators, activated carbon is used to adsorb both dioxins and Hg. The temperature of adsorption is near 150 °C. Adsorption at higher temperatures would be more economical; however, the temperature is limited by the sorbent capacity. By replacing activated carbon with carbon nanotubes, operation at higher temperatures would be possible.

Adsorption of Hg species on carbon nanotubes has not been studied. Such a study would be worthwhile. Likewise, studies on adsorption of hazardous polynuclear aromatic compounds and VOCs on carbon nanotubes would be of value.

The adsorption of benzene in/on single-wall nanotubes has been measured by Eswaramoorthy et al. (1999), as shown in Figure 9.9. The peak radius of the SWNTs was 1 nm. At 25 °C, a saturated amount of benzene of approximately 2.1 mmol/g was reached at their lowest relative pressure of 0.01. This amount corresponded to 30–40 Å^2 per molecule. Unfortunately, no heat of adsorption data were reported. However, from the available result, the bond energy between benzene and SWNTs is clearly very strong.

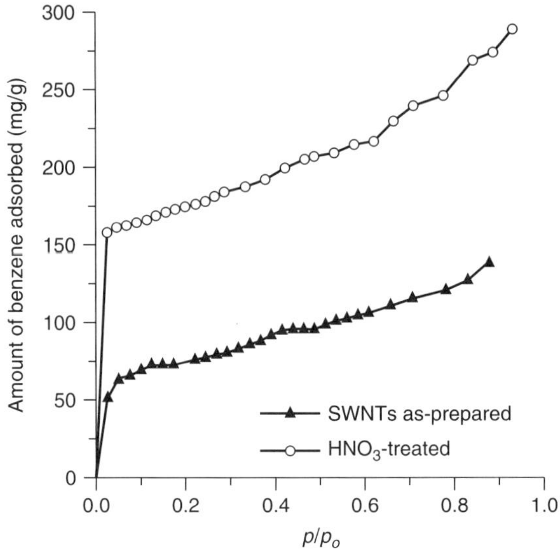

Figure 9.9. Adsorption isotherms of benzene on catalytically grown SWNTs at 25 °C. HNO_3 removed the Ni-YO_3 catalyst hence opened the tubes. (Eswaramoorthy et al., 1999).

***Selective Adsorption of* NO_x *Over* CO_2.** There has been a continuing search for selective sorbents for NO_x and SO_2 over CO_2 and H_2O. Such sorbents would be desirable for flue/combustion gas cleanup applications. It has been known that CO_2, like H_2O, is only very weakly adsorbed on styrenic resins (Kikkinides and Yang, 1991) and graphite (Chen and Yang, 1993). The styrenic resins are polystyrene cross-linked with divinyl benzene, and approximately 60% of their surfaces are benzene rings. CO_2 can have strong interactions with edge planes of graphite, but has only very weak interactions with the basal plane of graphite. This has been shown experimentally (by using graphite crystals with various basal/edge plane ratios) and by molecular orbital calculations (Chen and Yang, 1993). The potential energy curves of CO_2 approaching basal plane of graphite and that approaching the zigzag edges are compared in Figure 9.10. SO_2/graphite is also included in the comparison. It is clearly demonstrated that while SO_2 has a slight preference for the edge plane, the interaction of the edge plane with CO_2 is much stronger than that with the basal plane.

Unlike resins and graphite, activated carbons do not have high selectivities for NO_x and SO_2 over CO_2. The reason for lack of selectivity is that the activated carbon is not highly graphitic, and contains a high proportion of edge planes as well as some amorphous carbon.

Long and Yang (2001b) studied the adsorption of NO_x, SO_2, and CO_2 in the presence of O_2 on MWNTs. The MWNTs were prepared by decomposition of acetylene on 2.5 wt % Co/NaY zeolite, that is, the recipe of Colomer et al. (1998). The MWNTs had a BET surface area of 462 m^2/g, and an inner pore diameter of 2–4 nm. The single gas adsorption results are shown in Figure 9.11.

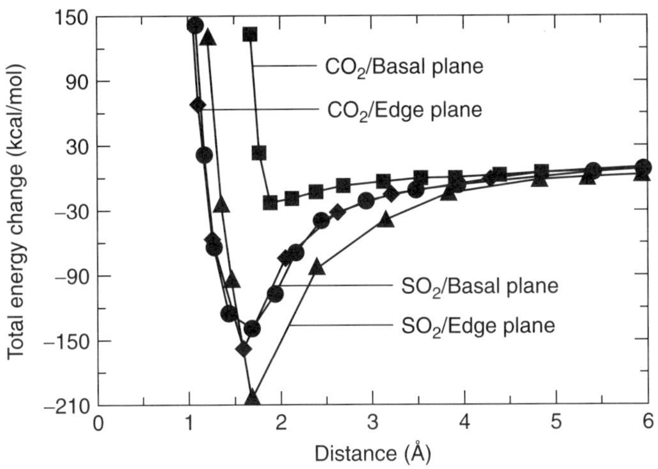

Figure 9.10. Potential energy as a function of internuclear distance from semi-empirical (INDO) molecular orbital theory calculation (see Chapter 8) (Chen and Yang, 1993, with permission). The absolute values from INDO are too high, but the relative values for different systems are correct. The edge plane is the zigzag edge.

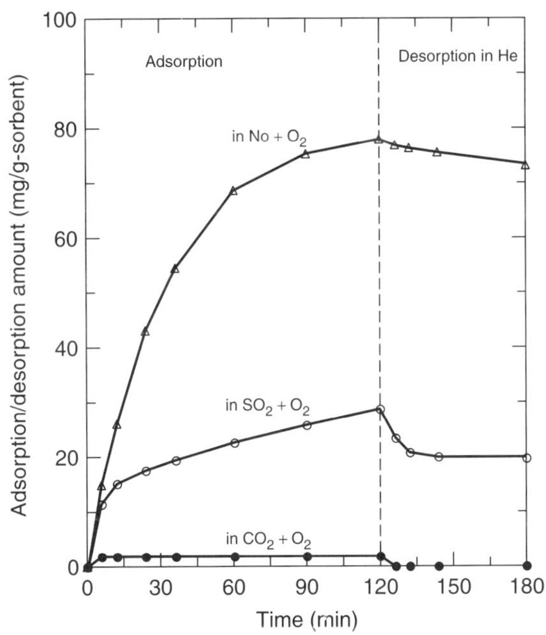

Figure 9.11. Adsorption/desorption of NO_x, SO_2, and CO_2 on MWNTs at 25 °C. Conditions for adsorption: 1000 ppm NO, 500 ppm SO_2, or 10% CO_2, each with 5% O_2 and balance He at 1 atm. (Long and Yang, 2001b).

The high selectivity for NO_x and SO_2 over CO_2 simply reflects the fact that the surfaces of the nanotubes are the basal plane of graphite. Subsequent measurements by Long and Yang (unpublished results) with activated carbon showed that the amounts adsorbed of NO_x and SO_2 were similar to those by MWNTs; however, the amount of CO_2 was also high, that is, activated carbon did not have the selectivity for NO_x and SO_2 over CO_2. Moreover, the amounts adsorbed of NO_x and SO_2 on MWNTs were significantly higher than those on activated carbon when normalized based on surface area, since the surface area of activated carbon was more than twice that of the MWNTs. The enhanced adsorption by MWNTs was most likely due to the cylindrical pore geometry.

Molecular Simulations and Adsorption of Other Gases. The first molecular simulation study on adsorption in carbon nanotubes was reported by Pederson and Broughton (1992). The potential curves were calculated for two HF molecules, approaching from the two opposite ends of a SWNT. Two tubes were subjected to calculations, with inner diameters of 8.2 and 10 Å. At the equilibrium distance of approximately 3 Å between the two HF, collinearly positioned along the center of the tube axis, a potential energy of approximately -0.3 eV was obtained. This energy is equivalent to approximately 17.6 kJ/mol. More interestingly, they arrived at an effective polarizability per unit area of the nanotube of 3.4 $Å^3/Å^2$. This value may be compared with the polarizability of the carbon atom at the ground state. Such a value is 1.76 $Å^3$ (see Table 2.2 in Chapter 2). Thus, the inner tube surface appears to have an enhanced polarizability.

GCMC (grand canonical Monte Carlo) simulations of adsorption of N_2 on both SWNTs (Maddox and Gubbins, 1995; Yin et al., 1999; Ohba et al., 2001; Mays et al., 2002) and MWNTs have been performed. Experimental data are also available (Inoue et al., 1998; Ohba et al., 2001; Mays et al., 2002). Interesting yet not unexpected results are seen. Maddox and Gubbins (1995) studied both adsorption of argon and nitrogen at 77 K on three nanotubes with 1.02 and 4.78 nm diameters. For the small tube, the isosteric heats of adsorption was approximately 16 and 17 kJ/mol for Ar and N_2, respectively. The heats of adsorption decreased to approximately 12 kJ/mol for the large tubes. For tubes shorter than 30 nm, strong end effects were predicted.

Simulations as well as experiments have been performed on the adsorption of ^{4}He on bundles of SWNTs (Teizer et al., 1999; Teizer et al., 2000; Cole et al., 2000; Talapatra et al., 2000; Gatica et al., 2000). From desorption measurements at $T > 14$ K, a large binding energy of 330 K was first reported by Teizer et al. (1999). This was 2.5X the value on planar graphite. They subsequently corrected it to 1.6X (Teizer et al., 2000). Adsorption isotherms and binding energies of Xe, CH_4, and Ne on bundles of SWNTs were measured by Talapatra et al. (2000). Again, higher binding energies than those on planar graphite were obtained. Remarkably, the percent increase in the binding energy relative to planar graphite, at about 75% for all three gases, was quantitatively the same. The temperatures of their measurements were up to 296 K for Xe, 195 K for CH_4, and 57 K for Ne. The binding energies were 222 meV for CH_4, 282 meV for Xe, and

52 meV for Ne. They actually calculated the isosteric heats of adsorption (q_{st}) from the temperature dependence first (at low loadings), and obtained the binding energies (ε) by $q_{st} = -\varepsilon + 2kT$, where k is the Boltzmann constant. From the ε values above, the heats of adsorption were approximately 23 kJ/mol for CH_4, 29 kJ/mol for Xe, and 7.4 kJ/mol for Ne. The SWNT bundles were reported as having closed ends. The results were interpreted as adsorption occurring on the outer surfaces and "ridges" of the bundles, not in the spaces between the tubes within the bundles. Simulation of Xe on the outer surfaces of SWNTs has been studied by Stan and Cole (1998), who concluded that the binding of Xe on the outer surface was weaker than that on planar graphite by about 20%. However, their estimated binding energy of 22.6 kJ/mol was close to the experimental data of Talapatra et al. (2000). Kuznetsova et al. (2000) measured the adsorption of Xe on both closed and open SWNTs at 95 K, using TPD. The desorption activation energy of Xe from saturated phase inside the tubes was 26.8 kJ/mol. This value was also in reasonable agreement with that obtained by Talapatra et al. and Stan and Cole. In a study of Muris et al., (2000), adsorption isotherms of CH_4 and Kr at 77–110 K were measured on closed bundles of SWNTs, with a tube diameter of 13.7 Å and intertubular distance of 17 Å. The isotherm of CH_4 on SWNTs had two steps, with isosteric heats of adsorption of 18.3 and 11.1 kJ/mol for the first and second steps, respectively. The heat of adsorption of the first step of adsorption for CH_4 on planar graphite was 14.9 kJ/mol. Their values seemed to be substantially lowered than the others. An earlier measurement of adsorption of CH_4 on MWNTs was performed on large tubes, ranging from 10 nm to 2 μm diameter (Mackie et al., 1997). In this study, typical wetting and capillary condensation behavior was seen.

An interesting result from the study of Kuznetsova et al. (2000) involved the opening of the tubes by heat-treatment under high vacuum. Evolution of CH_4, CO, H_2, and CO_2 gases began when heating the SWNTs to temperatures above 300 °C. Their final temperature was 800 °C. After such heat-treatment, the saturated amount of Xe increased by a factor of 23, indicating opening of the tubes. They postulated that the surface functionalities such as −COOH blocked the entry ports for adsorption at the nanotube ends and at defect sites on the walls. The thermal destruction of these functionalities would lead to opening of the pores. A similar postulation was actually made previously by Rodriguez and Baker (1997) from their studies of hydrogen storage on GNFs. This important point will be further discussed in detail in Chapter 10.

Isotope Separation. Separation of isotope mixtures (e.g., separations of tritium and deuterium from hydrogen) is a difficult task that requires energy-intensive processes such as diffusion, chemical exchange and laser isotope separation. Adsorption processes (mainly by chromatography) have attracted continuing interests; however, good sorbents have yet to be found.

Commercial sorbents have been considered for isotope separations. These include: D_2-H_2 on alumina at 77.4 K (Katorski and White, 1964; King and Benson, 1966), D_2-DT-T_2 on 13X zeolite at various low temperatures (Maienschein et al., 1992), and H_2-D_2-HD on NaA zeolite at 120-77 K (Stephanie-Victoire et al., 1998). The separation factors for these systems were all small (near or below 2).

The use of carbon nanotubes for isotope separations has been proposed recently by Scholl, Johnson and co-workers (Wang et al., 1999; Challa et al., 2001), based on "quantum sieving." Quantum sieving was first discussed by Beenakker et al., (1995) to describe the molecular transport in a pore that is only slightly larger than the molecule. In their description, a hard-sphere molecule with a hard-core diameter σ is adsorbed in a cylindrical pore with a pore diameter d. When d-σ is comparable with the de Broglie wavelength λ ($\lambda = h/mv$, where h is Planck's constant, m is mass, and v is the radial velocity), the transverse motion energy levels are quantized. For He at room temperature, $\lambda \approx 0.1$ nm. Hence, separation can be achieved by using differences in the quantum levels of the heavier and lighter molecules that are confined in the pore. From their hard-sphere theory, high selectivities were predicted for He isotopes at low temperatures and pressures in pores about 4 Å in diameter. Wang et al. (1999) employed path-integral Monte Carlo simulations with accurate potential models for studying isotope separations in SWNT bundles.

Wang et al. (1999) treated a low-pressure binary mixture in equilibrium with an adsorbed phase in the narrow pore. Because of the low density of adsorbate in the pore, the molecules undergo unhindered axial translational motion, and the transverse (radial) degrees of freedom are in their ground state and quantized. The chemical potential of the adsorbate molecule or atom can be expressed in terms of the ground-state energy (E) of its transverse wave function. By equating the chemical potentials of the adsorbate molecule and that in the gas phase, the selectivity can be calculated. The selectivity of component 1 over 2 is $S = (x_1/x_2)/(y_1/y_2)$, where $x_i(y_i)$ are the pore (gas) mole fractions. The selectivity approaching zero pressure (S_0) is given by:

$$S_0 = \frac{m_2}{m_1} \exp\left(-\frac{E_1 - E_2}{kT}\right) \tag{9.2}$$

where m is the molecular mass. This equation applies only when the ground state is populated. When the excited states (l) are also populated, Challa et al. (2001) obtained:

$$S_0 = \frac{m_2}{m_1} \left[\frac{\sum_l \exp(-E_1^l/kT)}{\sum_l \exp(-E_2^l/kT)} \right] \tag{9.3}$$

The selectivities calculated from these two equations for T_2/H_2 at 20 K were essentially the same (Wang et al., 1999; Challa et al., 2001). That calculated from Eq. 9.3 are shown in Figure 9.12. The results from the path integral MC calculations are also shown, and were nearly the same as that from the simple equations.

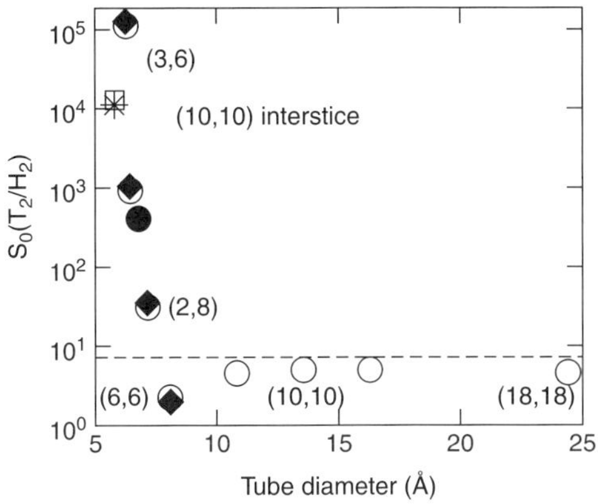

Figure 9.12. S_0 (equilibrium selectivity at pressure approaching zero) for T_2/H_2 as a function to tube diameter, from Eq. 9.3 (diamonds) and path integral Monte Carlo calculations (circles), (n, m) are tube indices, which are related to the tube diameter via Eq. 9.1, from Challa et al. (2001) with permission.

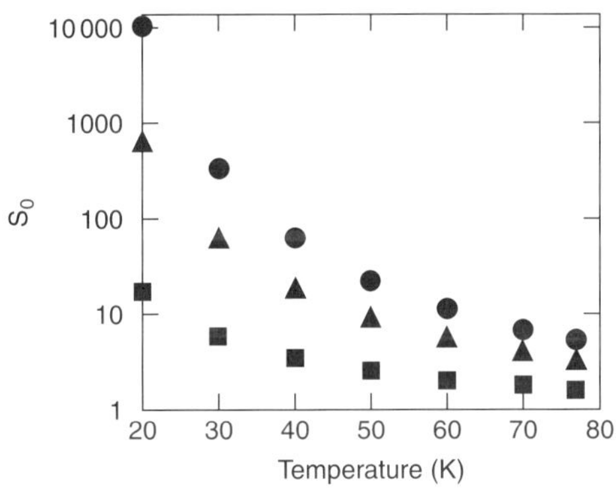

Figure 9.13. Selectivity S_0 in the SWNT interstices as a function of temperature for T_2/H_2 (circles), HT/H_2 (triangles), and T_2/HT (squares). From Challa et al. (2001) with permission.

The quantum sieving effects drop rapidly as the temperature increases, as shown in Figure 9.13. Using Eq. 9.3, Challa et al. (2001) calculated the selectivities for three different hydrogen isotope mixtures in the interstices of (10,10) SWNT bundles at different temperatures. The drop was due to the increased contribution from the excited states. Thirty energy states (in Eq. 9.3.) were used in their calculations. However, a selectivity of 5.2 for T_2/H_2 still remained at 77 K.

It should be noted that the quantum sieving effects are not limited to the carbon nanotubes. Small-pore molecular sieves such as $ALPO_4$-22 also could have such effects (Wang et al., 1999). However, carbon nanotubes have the advantage of being the most smooth and uniform pores. The interesting quantum sieving effects remain to be proven experimentally.

The interesting idea of separating enantiomers via chiral separation with carbon nanotubes was suggested and examined by Sholl and co-workers (Power et al., 2002). A Monte Carlo technique was used to calculate the isosteric heats of adsorption for enantiomers of *trans*-1,2-dimethylcyclopropane (DMCP) and *trans*-1,2-dimethylcyclohexane (DMCH) inside SWNTs. Sholl and colleagues used tubes of different diameters as well as different chiral angles. Tube diameters ranging from 9.5 Å to 28.7 Å and chiral angles from 34 to 54° were examined. Unfortunately, the isosteric heats of adsorption for the two pairs of enantiomers were negligible in all cases.

Kinetic Separations. As discussed in Chapter 5, carbon molecular sieves have already been used for gas separation that is based on differences in diffusivities of different gas molecules. The same separations should also be possible with carbon nanotubes. To this end, a number of simulation studies have been carried out. Mao and Sinnott (2000 and 2001) have reported molecular dynamics simulation results for diffusion of methane, ethane, *n*-butane, and isobutene, as well as their binary mixtures, in SWNTs and their bundles. As expected, diffusion of smaller molecules is faster, for example a factor of 25 was obtained for methane/isobutene in a (8,8) nanotube (Mao and Sinnott, 2001).

The difficult separation of N_2/CH_4 was studied with a SWNT of 13 Å diameter by Nicholson and Suh (2002) by using a Monte Carlo technique. The flux was expressed in the Fickian form to include both main-term and cross-term diffusivities, as well as a viscous contribution. Their results are summarized in Figure 9.14. In Figure 9.14, J_1/J_2 is the ratio of the total fluxes of CH_4 over N_2, which reflects the overall separation. D_{11} and D_{22} are the main-term Fickian diffusivities for CH_4 and N_2, respectively. The large deviation of D_{11}/D_{22} from J_1/J_2 reflects the significant contribution of the cross-term diffusivities.

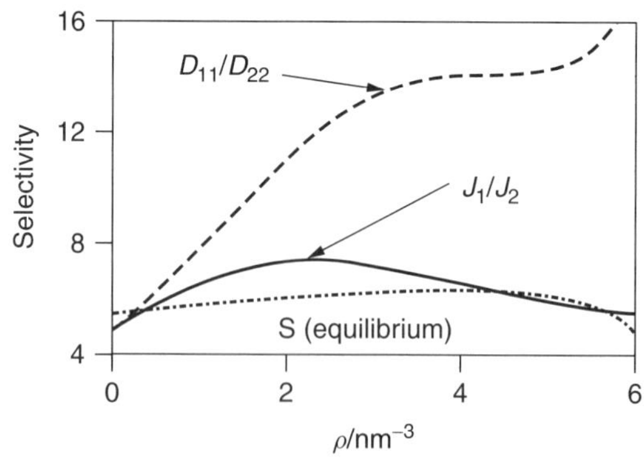

Figure 9.14. Comparison of equilibrium and kinetic selectivities of CH_4 over N_2 in a SWNT of 13 Å diameter at 25 °C. The fugacities of CH_4 and N_2 are equal and ρ is the total adsorbate density (Nicholson and Suh, 2002, with permission).

9.2. PILLARED CLAYS

Pillared interlayered clays (PILCs), or pillared clays, are a class of porous, high-surface area, two-dimensional materials that have been studied extensively for application as catalysts and, to a much lesser degree, as sorbents for gas separations. Comprehensive reviews on pillared clays and their catalytic properties are available (Pinnavaia, 1983; Burch, 1987; Figueras, 1988; Butruille and Pinnavaia, 1996; Gil et al., 2000).

There have been only a few studies on the adsorption properties of pillared clays, primarily since 1991 (Yang and Baksh, 1991). Some of the results have been reviewed by Yang and Cheng (1995).

The microporosity, that is micropore-size distribution of pillared clays has been correctly characterized only recently. For Al_2O_3-pillared clay, the main group of pores have sizes centered around 5.5 Å, with a minor group of pores centered around 4 Å (Gil and Montes, 1994; Hutson et al., 1998). These micropores can be controlled and tailored (Hutson et al., 1998) and could have potential applications as molecular sieves. PILCs are relatively hydrophobic when compared with zeolites, silica gel, and activated alumina (Malla and Komarneni, 1990). PILCs and acid-treated clays have been found to be excellent supports for the synthesis of π-complexation sorbents (Cheng and Yang, 1995b; Cho et al., 2001). In addition, synthetic clays are already produced commercially. Laponite is a synthetic hectorite, which along with montmorillonite (a member of smectite), can be easily pillared (Yang and Cheng, 1995). The availability of inexpensive clays of controlled quality is ensured by the availability of the synthetic clays. For the reasons above, further investigations of PILCs as possible sorbents for gas separation and purification are warranted.

9.2.1. Syntheses of PILCs

Clays are two-dimensional aluminosilicates with layer structures. Smectite is a family of clays that has been most studied as precursors for PILCs. Each layer of the smectite consists of three sub-layers: an alumina layer sandwiched between two silica layers. Two layers of smectite are depicted in Figure 9.15. The silica sub-layer is formed by a stable SiO_4 tetrahedral structural unit that has polymerized into a two-dimensional sheet. This occurs through sharing three of the oxygens at the corners of the tetrahedra. Like tetrahedra, the alumina octahedra can also polymerize in two dimensions to form a sheet; in this case, by sharing four oxygens. However, the aluminum atom is often substituted by Mg and Fe, referred to as isomorphous substitution. Substitution of the trivalent aluminum by divalent Mg or Fe imparts a net charge to the clay layer. Likewise, substitution of silicon by aluminum or magnesium by lithium could also impart a net charge. In order to achieve electroneutrality, the net charges must be compensated by interlayer cations such as Na^+, K^+, and Ca^{2+}. The interlayer cations are depicted in Figure 9.15. These cations are exchangeable. Pillared clays are prepared by exchanging the charge compensating cations between the clay layers with larger inorganic hydroxy cations, which are polymeric or oligomeric hydroxy metal

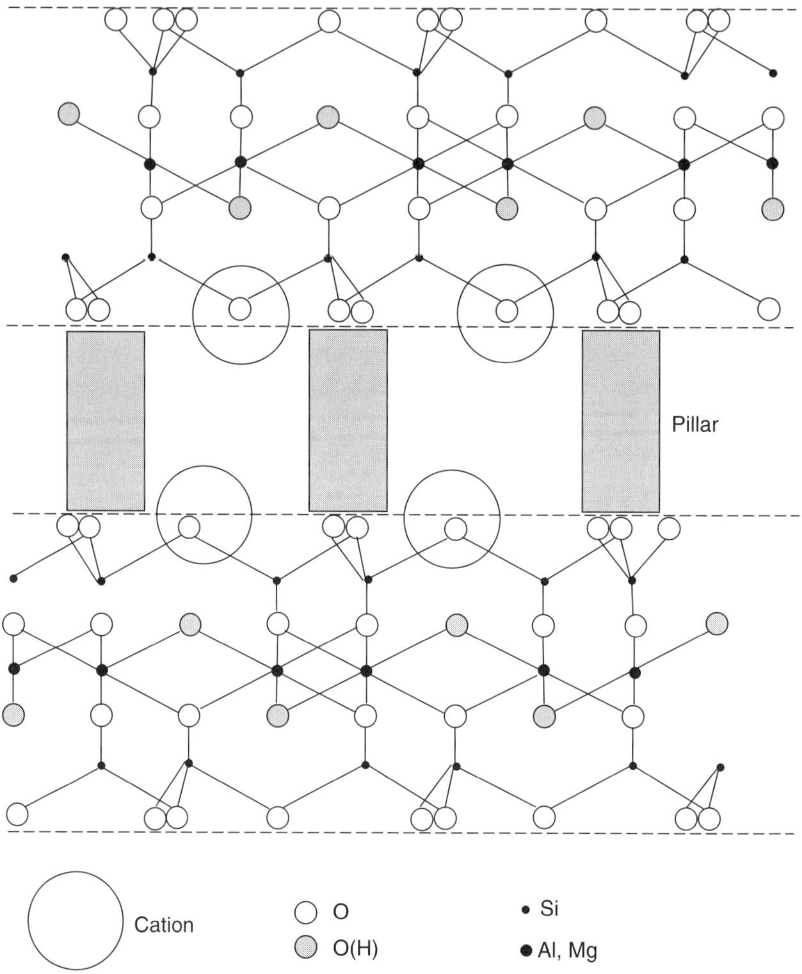

Figure 9.15. Schematic of structures of clay and pillared clay, with cation sites. The pillars are more closely spaced than depicted.

cations formed by hydrolysis of metal salts. Upon heating, the metal hydroxy cations undergo dehydration and dehydroxylation, forming stable metal oxide clusters which act as pillars keeping the clay layers separated and creating interlayer spacing (gallery) of molecular dimension. The steps for PILC synthesis are depicted in Figure 9.16.

Pillared clay development started in the mid 1950's by Barrer and co-workers (Barrer, 1978). They synthesized high-surface-area materials by pillaring montmorillonite clay with cations of $N(CH_3)_4^+$ and $N(C_2H_5)_4^+$. However, such materials have low thermal and hydrothermal stabilities and therefore have limited use as adsorbents and catalysts. Much interest and research have been directed toward the synthesis of pillared clays with high thermal and hydrothermal stabilities. The

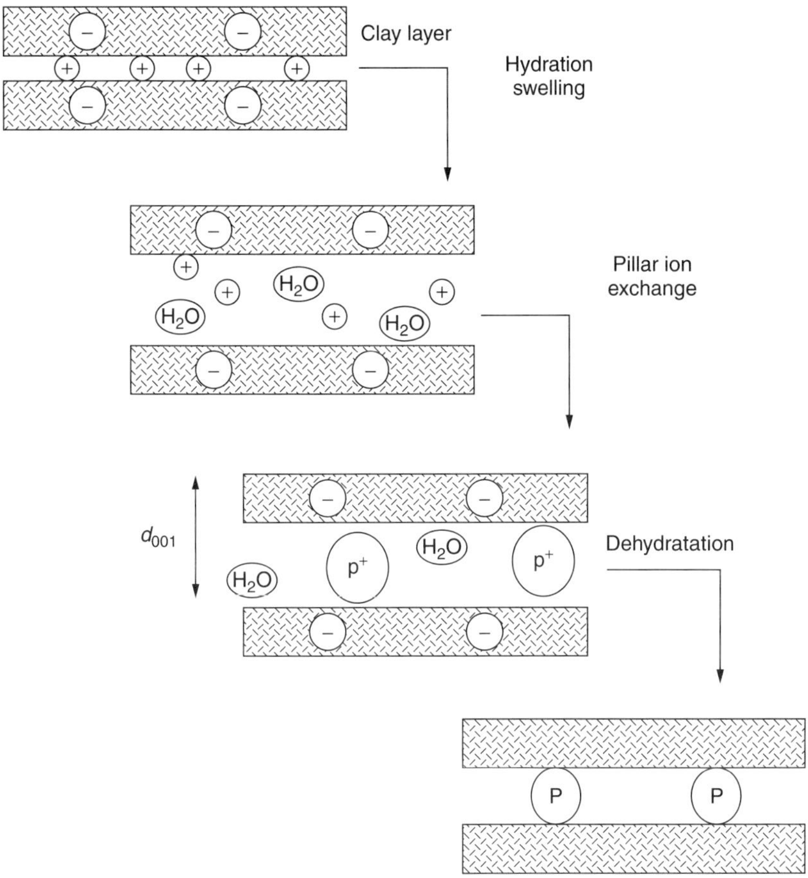

Figure 9.16. General scheme for synthesis of pillared clays (Butruille and Pinnavaia, 1996, with permission).

most promising ones for use as adsorbents and catalysts are as follows: ***Al-PILC*** (Vaughan et al., 1979, 1981a, b; Lahav et al., 1978; Shabtai et al., 1984a, b; Pinnavaia et al., 1984; Occelli, 1984; Sterte, 1991; Suzuki et al., 1988; Occelli and Tindwa, 1983; Pinnavaia, 1983; Burch, 1988; Figueras, 1988), ***Zr-PILC*** (Vaughan et al., 1979,1981; Yamanaka and Brindley, 1979; Figueras et al., 1989; Burch and Warburton, 1986; Bartley and Burch, 1986; Bartley, 1988; Yang and Baksh, 1991), ***Cr-PILC*** (Tzou and Pinnavaia, 1988; Carrado et al., 1986, Vaughan, 1987; Shabria and Lahari, 1980; Pinnavaia et al., 1985), ***Fe-PILC*** (Yamanaka and Hattori, 1988; Yamanaka et al., 1984; Burch and Warburton, 1987; Lee et al., 1989; Pinnavaia and Tzou, 1987; Rightor et al., 1991), and ***Ti-PILC*** (Sterte, 1986). The choices for hydroxy cations are not limited to those mentioned. In fact, any metal oxide or salt that forms polynuclear species upon hydrolysis (Baes and Mesmer, 1976) can be inserted as pillars, and all layered clays of the abundant phyllosilicate family as well as other layered clays can be used as the hosts (Clearfield, 1988; Drezdon, 1988, Sprung et al., 1990; Burch, 1988; Van Olphen, 1977; Fripiat, 1982,1988). In addition, several experimental parameters, such as the concentration of the metal ion, the basicity or degree of hydrolysis (given as r = OH/M), the temperature of preparation, the time and temperature of aging, the type of counter-ion, and the method of preparation, can strongly affect the degree of polymerization of the hydroxy-oligomeric cations in aqueous solution (Burch, 1987), and consequently the physicochemical properties of the pillared clays.

Typical chemical compositions of montmorillonite (bentonite) and major pillared clays are listed in Table 9.4. The clay was a purified form of bentonite from Fisher (Baksh et al., 1992).

9.2.2. Micropore Size Distribution

The main impetus for studying PILCs during the 1970's and 1980's was to develop large-pore catalysts for petroleum refining. X-ray diffraction was used as a guide for determining the pore sizes. The XRD patterns of the (001) reflection for the unpillared purified bentonite, and after pillaring with Al_2O_3, are shown in Figure 9.17. The 2θ angle of the (001) basal reflection was 7.1° for the unpillared purified bentonite and 5.1° for the Al_2O_3-PILC, which corresponds to d_{001} basal spacings of 1.21 nm for the unpillared clay and 1.69 nm for the Al_2O_3-PILC. Subtracting the thickness (0.93 nm) of the clay layer, the open spacing for the Al_2O_3-PILC is 0.76 nm. The open spacings for the PILCs reported in the literature are generally larger than this value. Depending on the preparation condition, open d-spacings as large as 20 Å have been reported. The BET surface areas of pillared clays are generally in the range of 200–400 m^2/g.

Table 9.4. Chemical compositions (in wt %) of clay and pillared clays

Oxides	Bentonite	Zr-PILC	Al-PILC	Cr-PILC	Fe-PILC	Ti-PILC
SiO_2	54.72	48.75	59.93	46.38	47.53	40.15
Al_2O_3	15.98	14.01	13.36	12.49	13.25	13.99
MgO	1.94	1.50	1.86	1.60	1.42	1.64
Fe_2O_3	2.93	2.49	3.05	2.57	1.58	2.90
TiO_2	0.12	0.12	0.19	0.12	0.13	0.12
Na_2O	2.04	0.22	0.17	0.23	0.07	0.19
CaO	0.82	0.09	0.07	0.064	0.087	0.089
K_2O	0.34	0.27	0.22	0.217	0.27	0.155
ZrO_2	—	17.65	—	—	—	—
Al_2O_3	—	—	10.46	—	—	—
Cr_2O_3	—	—	—	26.85	—	—
Fe_2O_3	—	—	—	—	27.20	—
TiO_2	—	—	—	—	—	30.17

(Baksh et al., 1992, with permission). Water (as balance) is not included.

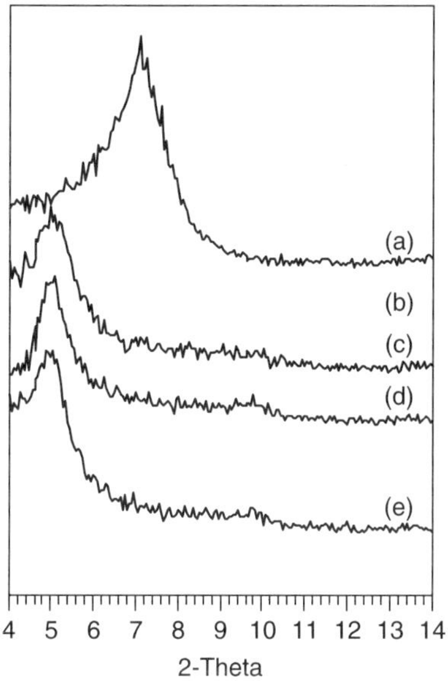

Figure 9.17. X-ray diffraction patterns for (a) unpillared purified bentonite, (b) unmodified Al_2O_3-PILC, (c) Li^+ -PILC, (d) Na^+-PILC, and (e) K^+-PILC (Hutson et al., 1998, with permission).

Using probe molecules as the sorbates, it was found that the pore sizes in PILCs were not limited by the interlayer spacing, but by the interpillar spacing (Yang and Baksh, 1991; Baksh et al., 1992). Furthermore, by using the low-pressure N_2 isotherm, Gil and Montes (1994) were able to correctly determine a bimodal micropore size distribution. For Al_2O_3-PILCs prepared with different Al/clay ratios, two groups of pores with sizes centered near 4 Å and 6 Å were observed for all samples. Hutson et al. (1998) studied the micropore size distributions of Al_2O_3-PILC and Al_2O_3-PILCs after ion exchanges with different alkali and alkaline earth metals. The N_2 isotherms of the clay and the Al_2O_3-PILC are shown in Figure 9.18. The low-pressure N_2 isotherm of the Al_2O_3-PILC is further shown in Figure 9.19. The hysteresis loop indicates the presence of mesoporosity. For the Al_2O_3-PILC shown in Figure 9.18, the mesopore volume was 0.11 ml/g. More importantly, the low-pressure isotherm provides information about the microporosity. The pore-size distribution of the micropores has been determined by using the Horváth–Kawazoe equation and its improved form by Cheng and Yang for slit-pore geometry (see Chapter 4 for details). The resulting micropore size distributions are shown in Figure 9.20. The distribution is in agreement with that determined by Gil and Montes. The total micropore volume for the Al_2O_3-PILC was 0.149 ml/g. Clearly these micropores are important since they account for 57% of the total pore volume.

Figure 9.20 also shows that the micropores could be modified by ion exchange of the PILC with different cations. The micropore volume for those pores in the lower distribution (<4.5 Å) increased with increasing ionic radius of the

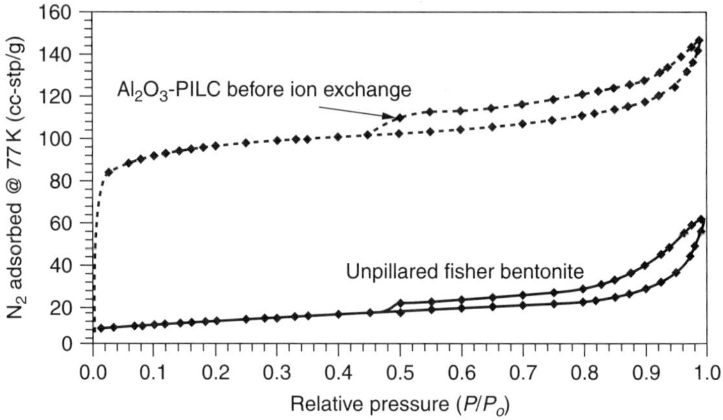

Figure 9.18. Nitrogen isotherms at 77 K for Al_2O_3-PILC, and unpillared purified bentonite (Hutson et al., 1998, with permission).

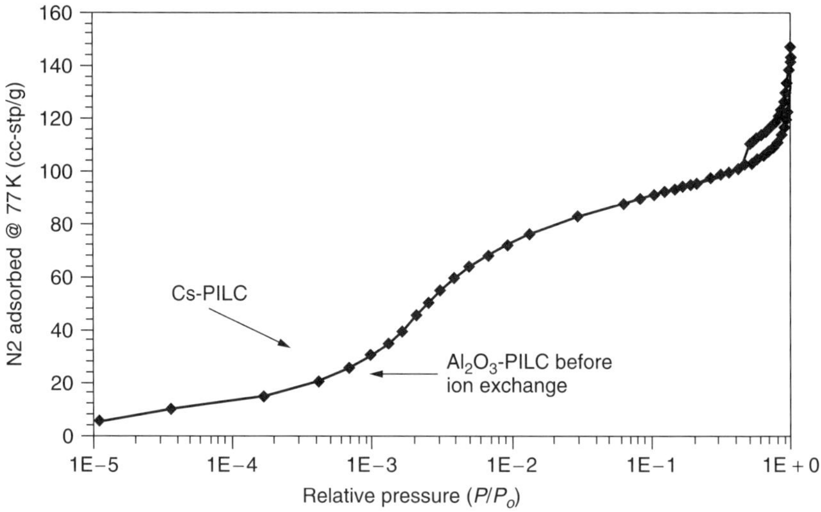

Figure 9.19. Low-pressure isotherms for Al_2O_3-PILC (Hutson et al., 1998, with permission).

charge-compensating cations used in the ion exchange. The micropore volume for those pores in the upper distribution (>4.5 Å) decreased with ion exchange.

9.2.3. Cation Exchange Capacity

The cation exchange capacity (CEC) is important for two reasons. First, like zeolites, by placing proper cations in the PILC, its adsorption properties can be tailored. Its pore structure can also be altered by ion exchange, albeit only

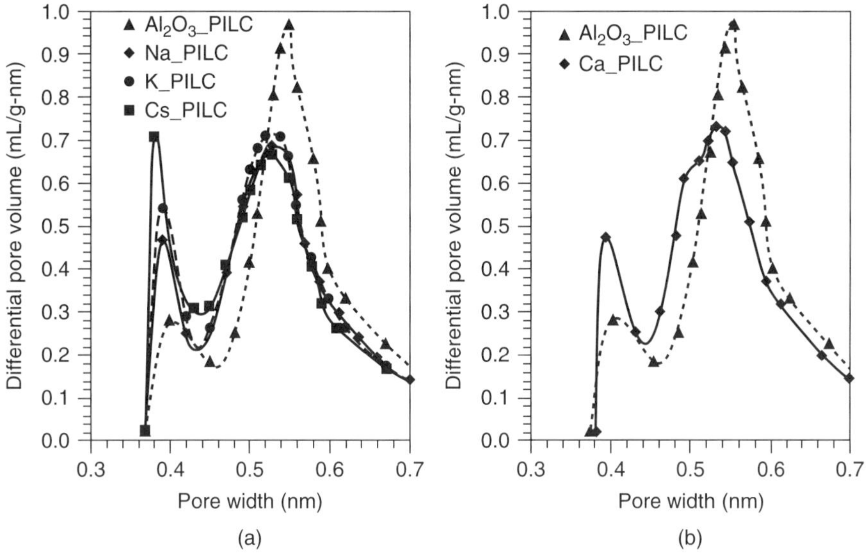

Figure 9.20. Overlays of the micropore-size distributions of the unmodified Al_2O_3-PILC and (a) monovalent cation-exchanged clays; and (b) divalent cation-exchanged clays (Hutson et al., 1998, with permission).

slightly, as discussed in 9.2.2. Second, PILC can be used as a stable, economical ion exchanger. It has been considered as a sorbent for the removal of heavy metal ions (such as Cu^{2+}, Cr^{3+} and Pb^{2+}) from aqueous media (Li et al., 1996).

The CEC of the precursor clay is equal to the total amount of the charge-compensating cations. Upon PILC synthesis, most of the CEC is lost; only about 10–20% remains. The CEC of the clay varies depending on the geological source. For the most used clay, montmorillonite (or bentonite), the CEC is 0.76 meq/g for Wyoming bentonite and is 1.40 meq/g for Arizona (Cheto) bentonite. The reason for the loss of CEC upon PILC synthesis will be discussed first, followed by methods for restoring it in the PILC.

In the scheme for PILC synthesis (Figure 9.16), the first step is to replace the charge-compensating cations in the clay by large polynuclear cations. The polynuclear cations are formed by hydrolysis of salts that contain these cations in proper ranges of pH, or OH/M ratio (Baes and Mesmer, 1976). Using Al^{3+} as an example, at r (OH/Al) ~2.4 (or, pH = 3.2–4.0), the main species in the pillaring solution is $[Al_{13}O_4(OH)_{24}(H_2O)_{12}]^{7+}$. This cation replaces the charge-compensating cations in the clay. The structure of this cation has been determined by Johansson (1960) to be the Keggin structure (after Keggin, 1934). The Keggin structure of Al is symmetric, built up by one tetrahedrally coordinated Al in the center and surrounded by 12 Al octahedra. After pillaring, the clay containing the polynuclear cation is calcined. Upon calcinations, the cation decomposes via:

$$[Al_{13}O_4(OH)_{24}(H_2O)_{12}]^{7+} \Rightarrow 6.5Al_2O_3 + 20.5H_2O + 7H^+ \qquad (9.4)$$

Thus, the original CEC is now taken up by the protons. Some or most of these protons migrate, at the calcination temperature, into the octahedral sheet of the clay, and toward the incompletely neutralized hydroxyl groups that are coordinated to magnesium, aluminum or other octahedral structural metal atoms. The migration of the cations into the octahedral layer is basically responsible for the CEC loss of the calcined product because these protons are no longer accessible for ion exchange.

It has been reported that approximately 80% of the initial CEC of the clay could be restored by treating the PILC with a base, such as ammonia, potassium carbonate or alkali solutions (Vaughan et al., 1981; Molinard et al., 1994a, b; Cheng and Yang, 1995a; Li et al., 1996).

The ammonia treatment can be accomplished by PILC exposure to a small partial pressure of ammonia at room temperature. For example, Molinard et al. (1994b) evacuated a desiccator that contained both the PILC and a beaker of ammonium solution. They were able to restore 80% of the CEC in 10 min. Their IR spectra showed the formation of the ammonium ions on the PILC. Thus, the restoration was apparently accomplished by retrieving, using ammonia, the proton from the octahedral layer to the surfaces in order to form ammonium ion.

Restoration of CEC by treatment with alkali solutions (such as NaOH) is also possible, but not well understood. Li et al. (1996) observed structural collapse of the PILC after such treatment. They also reported less restoration than that by ammonia treatment. The structural collapse was due to attack on the alumina pillars. Cheng and Yang (1995) have proposed the formation of −OH groups on the pillars where the proton is exchangeable.

9.2.4. Adsorption Properties

Although PILCs are aluminosilicates with cations, they are considerably less hydrophilic than zeolites and commercial desiccants. Earlier studies by Malla and Komarneni (1990) and by Yamanaka et al. (1990) indicated hydrophobicity of the PILCs. The capacity for water was increased by introducing Ca^{2+} into the interlayer spacing (Malla and Komarneni, 1990). The isotherms for water vapor on various PILCs are compared with that for activated carbon and 5A zeolite in Figure 9.21. The lack of a strong affinity for water is an advantage for applications.

Like zeolites, PILCs also show selectivity for N_2 over O_2. However, their capacities were substantially lower than that of zeolites. Cheng and Yang (1995a) corrected the earlier results of Baksh and Yang (1992) on the isotherms of cation exchanged PILCs for N_2 over O_2. In the work of Yang and Cheng (1995), the clay with the highest CEC, Arizona bentonite, was used as the starting clay. The PILC with the smallest pore sizes, Zr-PILC, was prepared from the Arizona bentonite. The smaller pores would provide the strongest force fields. After pillaring with ZrO_2, the sample was subjected to CEC restoration by treatment with ammonia. The resulting PILC was subsequently ion exchanged with alkali cations (Li^+, Na^+, K^+, Rb^+, Cs^+). The adsorption capacities for N_2 and O_2 on these ion

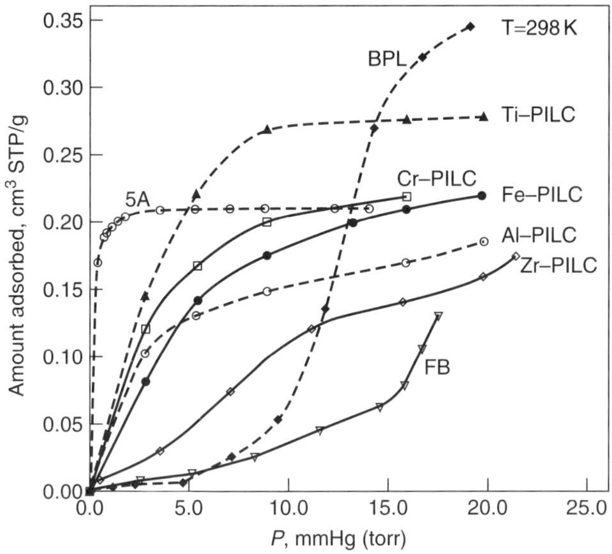

Figure 9.21. Adsorption isotherms of water at 25 °C on PILCs, BPL (activated carbon), FB (Fisher bentonite), and 5A zeolite (Baksh et al., 1992, with permission).

Table 9.5. N_2 and O_2 adsorption capacities of alkali ion-exchanged ZrO2-PILC at 25 °C and 1 atm

Adsorbent	Amount Adsorbed (mmol/g)	
	N_2	O_2
Li^+/Zr-PILC	0.076	0.024
Na^+/Zr-PILC	0.060	0.025
K^+/Zr-PILC	0.049	0.019
Rb^+/Zr-PILC	0.040	0.020
Cs^+/Zr-PILC	0.039	0.017
Sr^{2+}/Al-PILC	0.04	0.01
(0 °C, 0.45 atm)*	—	—

*Data point taken from Molinard and Vansant (1995).
From Yang and Cheng, 1995, with permission.

exchanged PILCs are shown in Table 9.5. The Li-exchanged form yielded the highest nitrogen capacity as well as the highest N_2/O_2 selectivity. The capacity for N_2 on Li-Zr-PILC is ~10% of that for 5A and 13X. Hence, if this PILC is used as the binder for zeolite pellets to be used for air separation, the N_2 capacity of the pellets would increase by about 2%.

The adsorption isotherms for a number of other gases on PILCs have been reported. The CO_2 isotherms on PILCs (Baksh and Yang 1992; Molinard and

Vansant, 1995) have similar shapes, however the equilibrium amounts are about 30–40% when compared to activated carbon. Occelli et al. (1985) studied the adsorption of normal paraffins on Al-PILC. Adsorption of nitrogen, water, n-butane, methanol, and neo-pentane on Al-PILC and Zr-PILC has been investigated by Stacey (1988).

Several strategies for PILC synthesis have been learned about achieving strong interactions with sorbate molecules. One needs small pore sizes and cations that have high valences as well as small ionic radii. The latter is for increasing the field-induced dipole, field-dipole, and field gradient-quadrupole interactions (see Chapter 2 and Chapter 7). To increase the density of cations, a clay with a high CEC should be used as the starting clay. Treatment with ammonia, after PILC synthesis and before cation exchange, will increase the CEC. The interlayer d-spacing is controlled mainly by the size of the polynuclear cation used for pillaring. Zr tetramers yield the smallest free interlayer spacing. Furthermore, the free interlayer spacing can be decreased by calcining at a higher temperature. For example, the d_{001} spacing of Al-PILC decreased by approximately 1 Å when the calcination temperature was raised from 400 to 600 °C (Cheng and Yang, 1997).

9.2.5. PILC and Acid-Treated Clay as Supports

In preparing catalysts and π-complexation sorbents, an inert support with a high surface area is often needed. The supports are, however, not always "inert." For example, strong metal-support interactions are known in catalysis (Ruckenstein, 1987). The effects of different supports (silica vs. alumina) for π-complexation by $AgNO_3$ have also been noted (Padin and Yang, 2000).

Pillared clays and acid-treated clays are two types of supports that have shown interesting properties. They are discussed separately below.

Cheng and Yang (1995b) compared the adsorption properties for π-complexation of CuCl spread on two different supports, γ-Al_2O_3 and TiO_2-PILC. TiO_2-PILC was chosen because it has the largest pore dimensions among the different PILCs. Both sorbents showed good olefin/paraffin selectivities. However, the isotherms of olefins were more linear or steeper when TiO_2-PILC was used as the support. This effect is shown in Figure 9.22. Isotherm steepness is desirable since the steep portion will yield a high working capacity for pressure swing adsorption.

Steepness of the isotherm beyond the "monolayer" region, that is, point "B" or where the "knee" is located, is a direct reflection of the surface energy heterogeneity. Mathematically, the factor "s" in the Unilam (i.e., uniform Langmuir) isotherm is an indicator of the steepness of this portion of the isotherm, and is hence referred to as the heterogeneity parameter. The Unilam isotherm has been discussed by Valenzuela and Myers (1989) and by Do (1998). It is given by:

$$q = \frac{q_m}{2s} \ln \left[\frac{c + Pe^{+s}}{c + Pe^{-s}} \right] \tag{9.5}$$

where q_m is the "monolayer" amount adsorbed, P is the pressure, and c and s are parameters. This isotherm is derived from the Langmuir isotherm by assuming

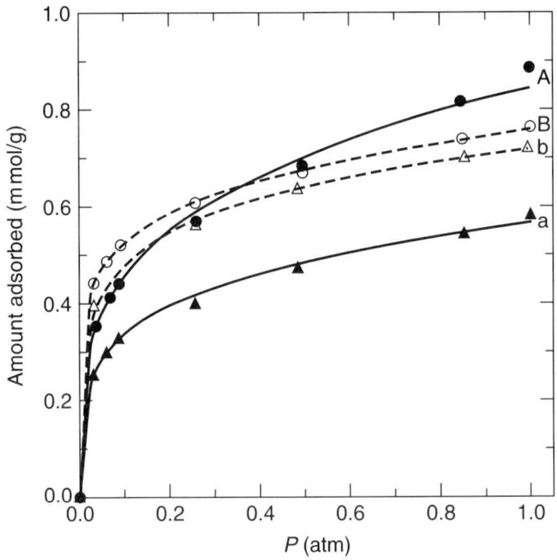

Figure 9.22. Isotherms at 25 °C for propylene on (A) $CuCl/TiO_2$-PILC, (B) $CuCl/\gamma$-Al_2O_3, and for ethylene on (a) $CuCl/TiO_2$-PILC, (B) $CuCl/\gamma$-Al_2O_3 (Cheng and Yang, 1995b with permission).

that the energy in the Langmuir constant follows a uniform distribution function, that is, a rectangular distribution with a spread of the energy between E_{max} and E_{min}. The factor s is proportional to the spread ($E_{max} - E_{min}$). When $s = 0$, the Unilam reverts to the Langmuir isotherm. Empirically, $s = 0$–14.

From the results of Cheng and Yang, it is clear that using the PILC support results in a higher value for s. This is likely caused by the heterogeneous nature of the PILC, because the PILC contains at least two different types of surfaces: that on the pillars and that on the clay layers.

Using Zr-PILC and Al-PILC as the supports for CuCl spreading, Engwall and Ma (2002) also showed very steep isotherms (with high s values) for olefins and paraffins. Since Zr-PILC and Al-PILC have the smallest pore sizes among the PILCs, the amounts of CuCl that were spread without blocking the micropores were very limited. In the work of Engwall and Ma (2002), less than 10% wt. CuCl could be spread, which were well below the monolayer spreading amounts.

"Acid activation" has long been known as a means for increasing (dramatically) the catalytic cracking activity of clays (e.g., montmorillonite) (Rupert et al., 1987). Upon acid treatment, the surface area of the clay is also increased. Typically, sulfuric acid is used in the treatment. The acid attacks and dissolves the octahedral layer that is sandwiched between two tetrahedral silica layers in the clay. The attack takes place uniformly on the edges of the octahedral layer, and eventually removes this layer. Thus, by optimal treatment (i.e., at a proper combination of acid concentration, temperature, and time) one can achieve a high surface area.

The chemical and structural evolutions of montmorillonite during acid treatment have been studied more recently by Rhodes and Brown (1993) and by Kumar et al. (1995). Kumar et al. (1995) followed the chemical changes and changes in pore structure after treatments of the clay with sulfuric acid at concentrations from 1 to 8 N at 80 °C for 2 h. After treatments with the acid at 1 and 2 N concentrations, alkali cations and Fe^{3+} in the tetrahedral layer were removed, while the surface area increased to 138 m^2/g. At higher concentrations, Mg^{2+} as well as Al^{3+} started being removed, with large gains in surface area, sometimes exceeding 300 m^2/g. The alumina octahedral layer was severely attacked only at concentrations >5N. The surface area remained at 370 m^2/g after treatments with higher concentrations. The pore-size distribution was also followed in their work (Kumar et al., 1995). Pores with dimensions of 2–4 nm were developed after treatment at low concentrations. Mesoporosity, with pore dimensions in the 4–10 nm range, was developed at high concentrations.

Using acid-treated clay as the support, Cho and co-workers (Cho et al., 2001; Choudary et al., 2002) developed an excellent sorbent for olefin/paraffin separation by spreading $AgNO_3$. Their C_2H_4/C_2H_6 isotherms on this sorbent are shown in Figure 9.23. The steep isotherms clearly reflect surface-energy heterogeneity. This sorbent and a comparison of it with other π-complexation sorbents will be discussed further in Chapter 10.

9.3. POLYMERIC RESINS

A broad range of synthetic, non-ionic polymers are available for use as sorbents, ion exchangers, and chromatographic column packings. The technology of designing and building porosity into polymers was accomplished in the late 1950's and early 1960's (Kunin et al., 1962; Millar et al., 1963; Albright, 1986). These macroporous polymers are termed macroreticular polymers. Building porosity can be accomplished by emulsion polymerization of the monomers in the presence of a solvent which dissolves the monomers but which is a poor swelling agent for the polymer. Although macroreticular polymers of acrylates and methacrylates are available, most commercial macroreticular polymers are based on styrene cross-linked by divinylbenzene (DVB). Over the years, these styrene/DVB copolymers have been produced with a wide range of porosities and macropore sizes.

The macroporous polymeric resins can be further reacted to attach functional groups to the benzene rings to generate functionalities for ion exchange. The resulting polymers are ion exchange resins. For example, polystyrene can be sulfonated by sulfuric acid resulting in an $-SO_3^-H^+$ group attached to the benzene ring, and the proton can be easily exchanged with other cations. Likewise, attaching ammonium or amine groups results in anion exchange resins. These polymeric resins and the functional groups are illustrated in Figure 9.24.

More recently, carbonaceous polymeric sorbents have been developed by partially pyrolyzing the styrene/DVB polymers and their sulfonated forms (Neely

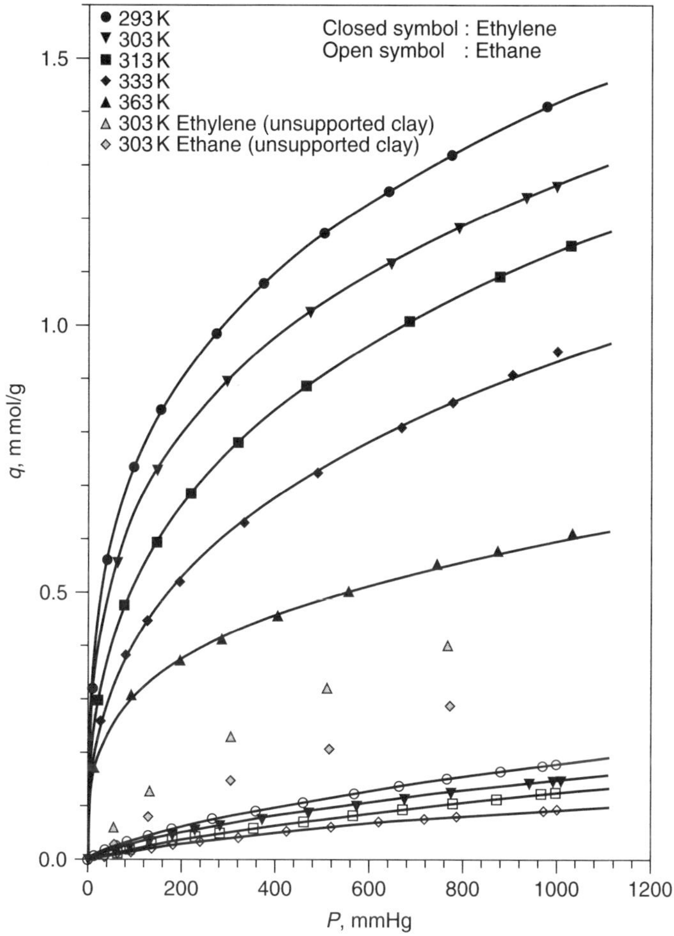

Figure 9.23. Isotherms of C_2H_4 and C_2H_6 on $AgNO_3$ supported on acid-treated clay (Choudary et al., 2002, with permission).

and Isacoff, 1982). These sorbents are particularly of interest for water purification because they have shown 5 to 10 times the capacity of granulated activated carbon (GAC) for low concentrations of some volatile organic compounds.

Polymeric resins have been widely used for water treatment as well as other applications. More recently, there has been a surge in applications designed to remove organic compounds from water. The main advantage of the polymeric resins lies in its ease of regeneration. An additional advantage is that the resins can be tailored for special applications such as that in the pharmaceutical and semiconductor industries. The major use for both polymeric sorbents and ion exchange resins involves water treatment; 75% of the resins were used for this purpose since 1987 (Albright, 1987). For such applications, the relatively high cost compared with GAC (about 10X) is justified.

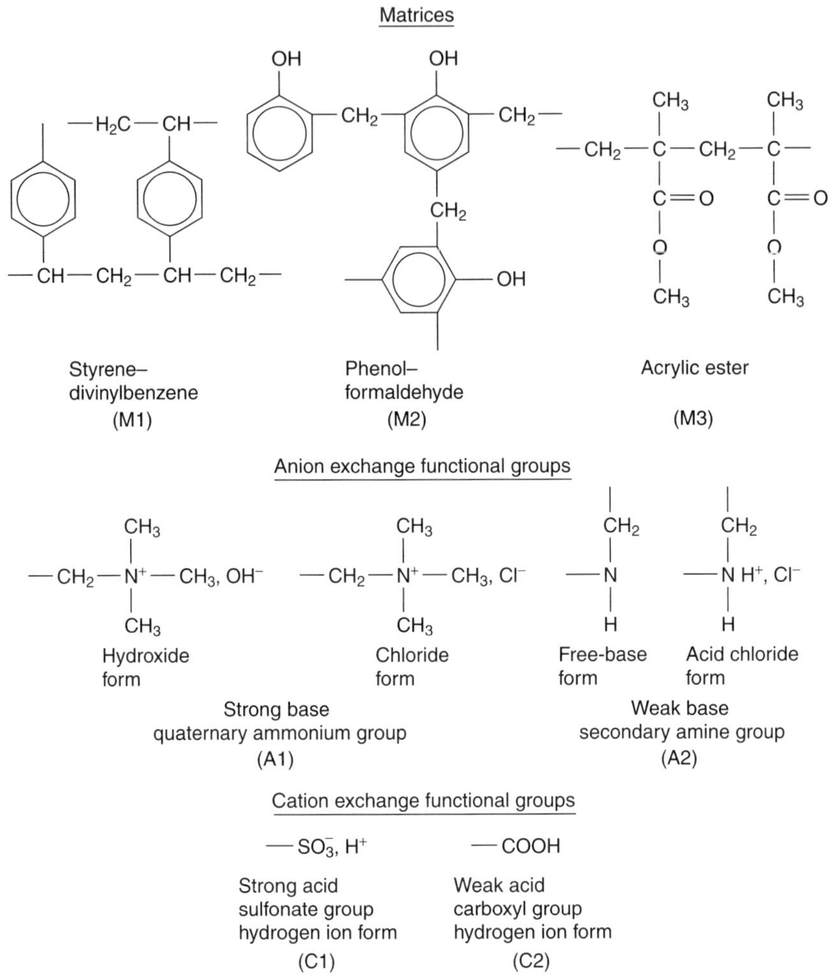

Figure 9. 24. Polymeric resins and functional groups that are attached to the benzene ring or matrix (Kim et al., 1976, courtesy of the Water Pollution Control Federation). Most commercial resins are based on Styrene/DVB.

9.3.1. Pore Structure, Surface Properties, and Applications

The polymer resins are in the form of spherical beads, usually in the size range of 20–60 mesh. Each spherical bead consists of an agglomeration of a large number of very small "microspheres." These microspheres are clusters of microgel particles ranging in size between 0.01 μm (or 100 Å) to 15 μm (Albright, 1986). Thus, the pore structure is comprised of inter-microsphere mesopores and the micropores within the microspheres. The latter depends directly on the degree of cross-linking, i.e., the amount of DVB. Most of the resins have 5–20% cross-linking; although resins with a very high degree of cross-linking are also

Table 9.6. Typical polymeric adsorbents

Resin Name	Chemical Nature	BET Surface Area (m^2/g)	Nominal Avg. Pore Diameter (Å)	Hydrophobicity
XAD-2	PS/DVB	300	90	Yes
XAD-4	PS/DVB	725	40	Yes
XAD-7	Acrylic ester	450	90	No
XUS-43493	PS/high DVB	1125	35	No
Dowex Optipore	PS/DVB/tert. amine	800	50	No
XE-563	Carbonaceous	550	38	No
XE-572	Carbonaceous	1100	30	No

Data from manufacturers. PS = polystyrene. DVB = divinyl benzene cross-linker. The XAD series is from Rohm and Haas under the generic name of Amberlite, and the XE series is from the same company under the name of Ambersorb. The XUS series is from Dow under the generic name of Dowex.

available. Cross-linking provides the high surface area as well as the rigidity and mechanical strength.

The general properties of some representative commercial resins are given in Table 9.6. The average pore sizes given by the manufacturers are not very meaningful because of the bi-modal pore distribution.

The surfaces of the resins are highly aromatic. Sizable fractions of the surfaces are benzene rings (Albright, 1986). For this reason, the unfunctionalized polymeric resins are hydrophobic. The ion exchange resins are, however, not hydrophobic. The hydrophobic resins need to be pretreated to become wetted prior to use in water treatment. The pretreatment involves immersion in a water-soluble solvent, such as acetone or methanol, to displace air from the pores, followed by displacement of the solvent by water or aqueous solution. The aromatic surfaces of the resins make them excellent sorbents for removal of organic compounds from aqueous solution, particularly those with low solubilities.

The polymeric resins and the carbonaceous polymers are significantly more hydrophobic than activated carbon. A comparison of water vapor isotherms is shown in Figure 9.25. With such highly hydrophobic surfaces, it is not clear whether the micropores are indeed wetted upon the pretreatment described above.

Ease in regeneration is a major practical advantage for the resins. Regeneration can be achieved with nonaqueous solvents or aqueous solutions of acids, bases, or salts. The nonaqueous solvents can also be regenerated for re-use, and the adsorbates may be recovered if desired. The ease in regeneration, or desorption, leads to another application: the purge and trap (P&T) technique for analytical purposes. This technique is popular for concentrating organic contaminants in water, in very low concentrations, for subsequent desorption and chemical analysis.

The major use for polymeric resins is water treatment (Faust and Aly, 1987). Commercial applications include removal of halogenated organic compounds,

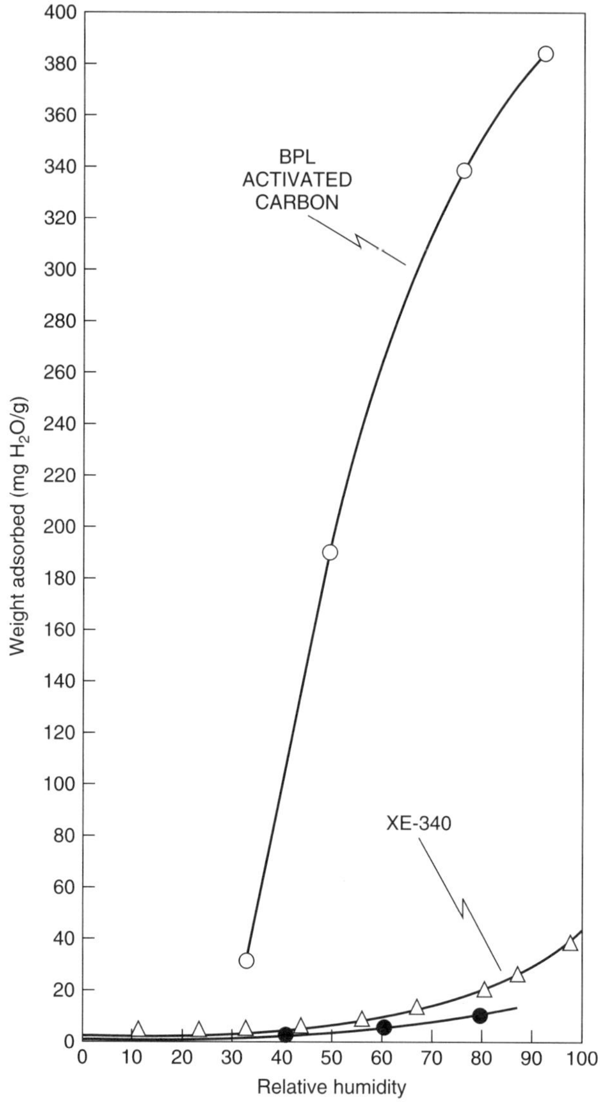

Figure 9.25. Water-vapor isotherms at room temperature on activated carbon and Ambersorb XE-340 (a carbonaceous polymeric resin) (Neely and Isacoff, 1982, with permission). Filled circles (•) are for XAD-16 styrene/DVB resin, with BET surface area = 760 m^2/g (Kikkinides and Yang, 1992).

phenols and pesticides from water, decolorization of effluents and dye wastes, removal of VOCs from air, and bioseparations. Adsorption of a large number of compounds by these sorbents has been discussed by Faust and Aly (1987).

Ion-exchange resins are used mainly for industrial and domestic water softening and deionization. As sorbents, they are also used for demineralization, dealkalization, desilicazation, and adsorption of ionic constituents from dilute solutions. For ion exchange, the selectivity increases with increasing valence of the ion, for example, with sulfonic acid group: $Th^{4+} > Al^{3+} > Ca^{2+} > Na^+$, and with increasing atomic number at constant ionic valence: $Cs^+ > Rb^+ > K^+ > Na^+ > Li^+$ and $Ba^{2+} > Sr^{2+} > Ca^{2+} > Mg^{2+}$.

Among the ion exchange resins, those based on weak bases and weak acids exhibit higher ion-exchange capacities than those based on strong bases and strong acids. However, regeneration of the strong-acid or strong-base resins is easier than that for the weak-acid or weak-base resins. For example, weak-acid resins require a two-step regeneration, first with a strong mineral acid, such as HCl, which exchanges H^+ for hardness, followed by NaOH neutralization to displace H^+ with Na^+.

Demineralization or deionization removes dissolved ionic impurities from water by a two-step process with cation- and anion-exchange resins. The former exchanges hydrogen for cations in solution. The acidic effluent passes through a column of anion-exchange resin that replaces the anions in solution with hydroxide. The H^+ from the cation-exchange resin are then neutralized by OH^- from the anion resin.

Since it was first proposed in the classic paper of Hollis (1966), the polymeric resins have been widely used as column packings for gas chromatography. For example, nonpolar styrene/DVB is used under the trade name Porapak Q, and its very polar derivative is used as Porapak T. These columns are used for analysis of various gas mixtures, including wet mixtures, based on differences of interactions with different gas molecules.

9.3.2. Comparisons of Resins and Activated Carbon

Many comparative studies have been performed on isotherms and column breakthrough performance between resins and activated carbon. The work of Weber and co-workers has been summarized elsewhere (Weber and van Vliet, 1981a and 1981b; Faust and Aly, 1987). Weber et al. have compared 10 different sorbents, including resins, activated carbons, and carbonaceous polymers. Data from selected sorbents, in terms of the two parameters in the D–R equation (see Chapter 3), are summarized in Table 9.7:

$$W = W_0 \exp[-(RT \ln(C_s/C))^2/B] \tag{9.6}$$

where W_0 is the maximum capacity, C_s is the saturated concentration, and B is the affinity coefficient.

The D–R isotherms of a number of compounds on activated carbon and resin are shown in Figure 9.26. The Calgon Filtrasorb 400 is an activated carbon for aqueous phase applications. XAD-2 and XAD-4, as mentioned earlier, are styrene/DVB resins. XAD-8 is cross-linked polyacrylic ester resin. XE-347 is a partially pyrolyzed resin.

The affinity coefficient, B, is an indication of the strength of the sorbent–sorbate interactions. From the isotherm data, two general facts become clear:

Table 9.7. Comparison of isotherm parameters

	Phenol		Carbon Tetrachloride		Dodecylbenzene Sulfonate	
Sorbent	*B*	W_0	*B*	W_0	*B*	W_0
F-400	462.1	33.78	184.3	41.94	1261.3	44.21
XAD-2	118.9	8.22	43.47	19.16	370.7	26.11
XAD-4	139.0	20.57	36.18	40.82	454.1	76.33
XAD-8	128.8	16.42	28.94	13.45	305.8	30.44
XE-347	549.9	16.85	214.6	1.74	739.0	3.707

Affinity coefficient, *B* in kJ/mol, and maximum capacity, W_0 in cc adsorbate/g, in the D–R Equation) for resins and activated carbon F-400. Data taken from Weber and van Vliet, 1981b.

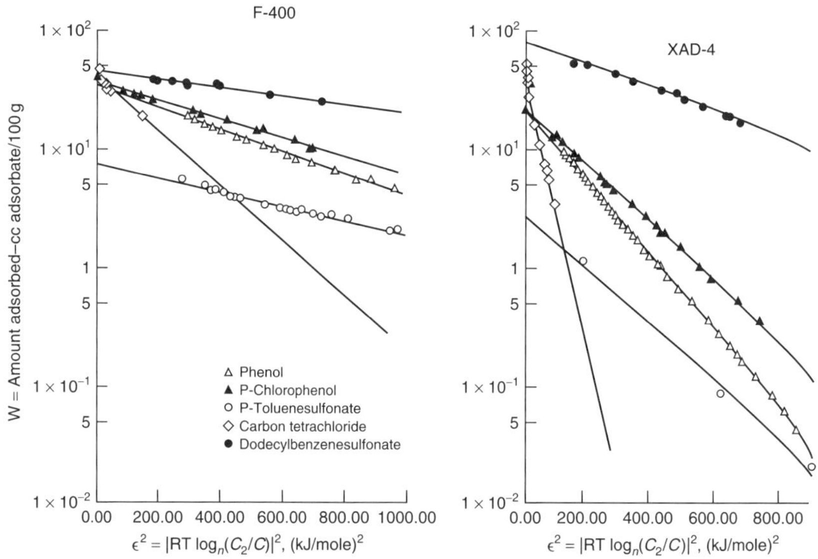

Figure 9.26. Comparison of isotherms in aqueous solution between activated carbon (Calgon F-400) and resin (XAD-4). Lines are D–R equation fits (Weber and van Vliet, 1981b, with permission).

(1) The sorbate–sorbent interactions are stronger on the carbonized resins. Hence, the carbonized resins are suited for achieving high purities or ultrapurification.

(2) The resins may have larger capacities at high concentration levels, particularly for large organic molecules, such as dodecylbenzene sulfonate (apparently by *absorption*). However, activated carbon generally compares well against resins in equilibrium isotherms. As mentioned, the ease of regeneration is a distinct advantage of the resins.

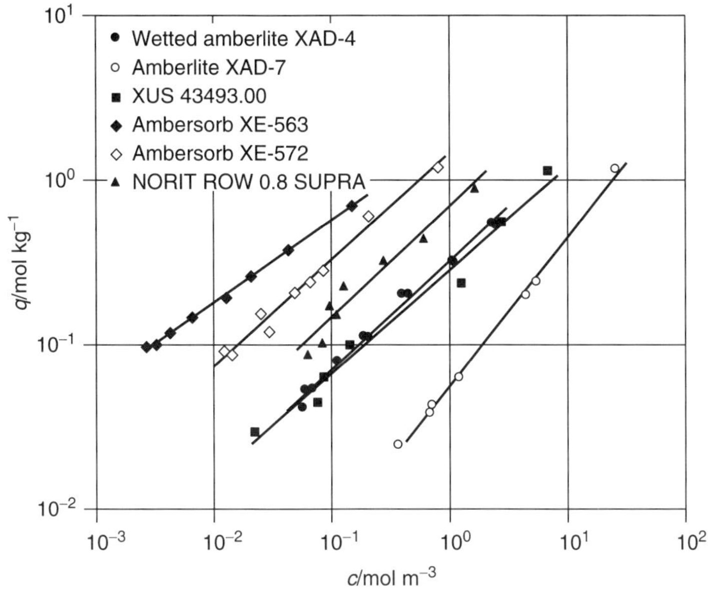

Figure 9.27. Isotherms for dichloromethane at 20 °C on polymeric resins (XAD and XUS), carbonized resins (XE), and activated carbon (Norit) (Rexwinkel et al., 1999, with permission).

Rexwinkel et al. (1999) studied the adsorption of a number of chlorinated hydrocarbons from aqueous solution on different resins, an activated carbon and carbonized resins. The comparisons are exemplified by adsorption of dichloromethane, shown in Figure 9.27. Again, the carbonized resins showed the highest capacities (by >5X compared with other sorbents) at dilute concentrations.

The general comparison shown above does not reflect the actual utility of the resins. For a target solute from the aqueous solution, a resin with high selectivity is obtained by a special functionality that is attached on the surfaces of the resin. A number of selective resins designed for special applications are available. High purity and selective resins are being used for potable water purification as well as by the food and pharmaceutical industries, for example, for purification of antibiotics and vitamins, recovery of products from fermentation, and purification of food products (sugars, glucose, citric acid, fruit juices, dairy products, and amino acids). Ultra-pure water for the semiconductor industry can be produced with impurities measured in the ppt level by using special resins.

9.3.3. Mechanism of Sorption and Gas-Phase Applications

The selectivities of specialty resins are provided by the functional groups that interact with the targeted solutes. Information and understanding on the interactions are scarce in the open literature. For example, resins containing basic groups can selectively adsorb carboxylic acids from complex solutions, and the capacities can exceed those of high-surface-area activated carbons (Kuo et al., 1987). The mechanism of the interactions between carboxylic acid and basic resins has been studied by Garcia and King (1989) by using acetic acid as prototype solute. The sorption was modeled by a 1 : 1 complex formation between the monofunctional acid and the basic functional groups on the resin. The complexation reaction leads directly to a Langmuir-type isotherm. Using data from a number of resins with different *pKa* (i.e., Gutmann donor number), Garcia and King (1989) showed that the Langmuir constant (K) can be correlated with *pKa*. They also showed that the same basic idea, that is, chemical complexation equilibria, may be used for selecting solvent used in regenerating the resin.

Uptake of organic molecules in polymer gels brings about swelling. Cross-linking builds macroporosity as well as rigidity. Solvent swelling occurs in resin beads that contain 1–5% of cross-links (Frechet and Farrall, 1977). Thus, the commercial resins are not swellable. However, large uptakes by resins of organic molecules with high molecular weights are known. The role of *absorption* in the uptake of organic molecules is not understood.

The polymeric resins are substantially more hydrophobic than activated carbon. The non-wetting types are subjected to the pre-wetting procedure described above (i.e., wetting with a solvent, such as methanol, followed by aqueous solution). The pores are presumably wetted (completely) upon this treatment. Undoubtedly, the mesopores (i.e., voids between the microspheres within each bead) are wetted. It is not clear, however, whether the micropores within each microsphere are completely wetted by this procedure.

An interesting study of adsorption on unwetted resins was reported by Rixey and King (1987; 1989a; 1989b). A number of polar organic compounds and four hydrophobic resins (XAD-2, XAD-4, Porapak Q and XE-340 carbonaceous resin) were used in their study. In using these compounds, it was observed that significant amounts of adsorption occurred when the resins were not wetted. The adsorption took place by the vapors of the solutes, and the rates could be modeled as being controlled by Knudsen diffusion of the vapors in the macropores. Selectivity within the pores for solute vs. water is enhanced by the lack of wetting. Moreover, separation of solutes based on difference in volatility by unwetted resins was proposed.

In a follow-up study, Rexwinkel et al. (1999) measured adsorption of seven common chlorinated hydrocarbons from aqueous solutions on wetted and un-wetted resins. These solutes are nonpolar, as opposed to the polar solutes studied by Rixey and King. Interestingly, the wetted and un-wetted resin (XAD-4) showed identical isotherms except for the heaviest compound, *2,4-dichlorophenol*. For *2,4-dichlorophenol*, adsorption on the wetted resin was slightly higher.

The gas-phase applications of polymeric resins are primarily used for gas chromatography and purification of air by removal of contaminants. The work of Hollis (1966) laid the foundation for the use of resins as packing materials for GC analysis. Hollis reported relative retention times of over 50 gas molecules on different EVB/styrene/DVB resins, using beads packed in a column. The same beaded forms are used today. The water vapor isotherm on the styrene/DVB resin

(XAD-16) is compared with that on activated carbon in Figure 9.25. Resins have the highest hydrophobicity among all sorbents. Based on the water capacity at low relative humidities, XAD-16 is of the order of 100 times more hydrophobic than activated carbon.

Such a high hydrophobicity gives resins a distinct advantage over activated carbon for removal of air contaminants from humid air. For activated carbon, the capacities for hydrocarbons are substantially reduced by humidity (see, for example, Doong and Yang, 1987). In addition, the resins should be easier to regenerate in the gas phase than activated carbon. A comprehensive study conducted by Mackenzie et al. (1994) included more than 100 sorbents in examining the removal efficiencies of humid air contaminated by chlorinated hydrocarbons. These sorbents included commercial polymeric resins, carbonaceous resins, commercial polymers and rubbers, and activated carbon. Based on the removal efficiency and ease of complete thermal regeneration, Ambersorb XE-563 (a carbonaceous resin) and Dowex Optipore were recommended as the best sorbents (for replacing activated carbon).

Another unique property of the resins is their low capacities for CO_2 (Chen and Pinto, 1990; Kikkinides and Yang, 1991). Based on this property, combined with hydrophobicity, removal and recovery of SO_2 and NO_x from combustion gases with resins has been suggested (Kikkinides and Yang, 1991). It should be noted, however, that basic ion-exchange resins are excellent sorbents for CO_2 (Yoshida et al., 2000).

REFERENCES

Ajayan, P. M. (1999) *Chem. Rev.* 99, 1787.

Albright, R. L. (1986) *React. Poly.* 4, 155.

Albright, R. L. and Yarnell, P. A. (1987) Ion exchange polymers. In *Encyclopedia of Polymer Science and Engineering*, Vol. 8. Wiley, New York, NY, p. 341.

Alvarez, W. E., Kitiyanan, B., Borgna, A., and Resasco, D. E. (2001) *Carbon* 39, 547.

Bacon, R. (1960) *J. Appl. Phys.* 31, 283.

Baes, C. F. and Mesmer, R. E. (1976) *The Hydrolysis of Cations*. Wiley, New York, NY.

Baird, T., Fryer, J. R., and Grant, B. (1974) *Carbon* 12, 591.

Baird, T., Fryer, J. R., and Grant, B. (1971) *Nature* 233, 329.

Baker, R. T. K. and Harris, P. S. (1978) The formation of filamentous carbon. In *Chemistry and Physics of Carbon*. (P. L. Walker, Jr. ed.). Dekker, New York, NY, pp. 83–165.

Baksh, M. S., Kikkinides, E. S., and Yang, R. T. (1992) *Ind. Eng. Chem. Res.* 31, 2181.

Baksh, M. S. A. and Yang, R. T. (1992) *AIChE J.* 38, 1357.

Barrer, R. M. (1978) *Zeolites and Clay Minerals as Sorbent and Molecular Sieves*. Academic Press, New York, NY.

Bartley, G. J. J. and Burch, R. (1985) *Appl. Catal.* 19, 175.

Bartley, J. J. (1987) Zirconium pillared clays. In Burch, R. loc. cit. p. 233.

Beenakker, J. J. M., Borman, V. D., and Krylov, S. Y. (1995) *Chem. Phys. Lett.* 232, 379.

Bethune, D. S., Kiang, C. H., deVries, M. S., Gorman, G., Savoy, R., Vazquez, J., and Beyers, R. (1993) *Nature* 363, 605.

Burch, R., Ed. (1987) Pillared clays. In *Catalysis Today*, Vol. 2. Elsevier, New York, NY.

Burch, R. and Warburton, C. I. (1986) *J. Catal.* 97, 503.

Burch, R. and Warburton, C. I. (1987) *Appl. Catal.* 33, 395.

Burton, J. J. (1974) *Catal. Rev.* 9, 209.

Butruille, J. R. and Pinnavaia, T. J. (1996) Alumina pillared clays. In *Supramolecular Chemistry*, Vol. 7 (G. Alberti and T. Bein, ed.). Pergamon Press, Oxford, UK, pp. 219–250.

Carrado, K. A., Suib, S. L., Skoularikis, N. D., and Coughlin, R. W. (1986) *Inorg. Chem.* 25, 4217.

Cassell, A. M., Raymakers, J. A., Kong, J., and Dai, H. (1999) *J. Phys. Chem. B* 103, 6484.

Challa, S. R., Sholl, D. S., and Johnson, J. K. (2001) *Phys. Rev. B* 63, 245419–1.

Chen, P., Zhang, H. B., Lin, G. D., Hong, Q., and Tsai, K. R. (1997) *Carbon* 35, 1495.

Chen, S. G. and Yang, R. T. (1993) *Langmuir* 9, 3258.

Chen, T. W. and Pinto, N. G. (1990) *Ind. Eng. Chem. Res.* 29, 440.

Cheng, L. S. and Yang, R. T. (1997) *Microporous Mater.* 8, 177.

Cheng, L. S. and Yang, R. T. (1995a) *Ind. Eng. Chem. Res.* 34, 2021.

Cheng, L. S. and Yang, R. T. (1995b) *Adsorption* 1, 61.

Cho, S. H., Han, S. S., Kim, J. N., Choudary, N. V., Kumar, P., and Bhat, T. S. G. U.S. Patent 6,315,816 (2001).

Choudary, N. V., Kumar, P., Bhat, T. S. G., Cho, S. H., Han, S. S., and Kim, J. N. (2002) *Ind. Eng. Chem. Res.* 41, 2728.

Clearfield, A. (1988) Recent Advances in Pillared Clap and Group IV Metal Phosphates, *NATO AS1 Ser., Ser. C* 231, 271.

Cole, M. W., Crespi, V. H., Stan, G., Ebner, C., Hartman, J. M., Moroni, S., and Boninsegni, M. (2000) *Phys. Rev. Lett.* 84, 3883.

Colomer, J. F., Piedigrosso, P., Willems, I., Journet, C., Bernier, P., Van Tendeloo, G., Fonseca, A., and Nagy, J. B. (1998) *J. Chem. Soc. Faraday Trans.* 94, 3753.

Colomer, J. F., Stephan, C., Lefrant, S., Van Tendeloo, G., Willems, I., Konya, Z., Fonseca, A., Laurent, Ch., and Nagy, J. B. (2000) *Chem. Phys. Lett.* 317, 83.

Dai, H. (2001) *Carbon Nanotubes*. (M. S. Dresselhaus, G. Dresselhaus, and Ph. Avouris, eds.). Springer, Berlin, Germany, and New York, NY, pp. 29–53.

Dai, H., Rinzler, A. G., Nikolaev, P. Thess, A., Colbert, D. T., and Smalley, R. E. (1996) *Chem. Phys. Lett.* 260, 471.

Ding, R. G., Lu, G. Q., Yan, Z. F., and Wilson, M. A. (2001) *J. Nanosci. Nanotech.* 1, 7.

Do, D. D. (1998) *Adsorption Analysis: Equilibrium and Kinetics*. Imperial College Press, London, UK, Chapter 3.

Doong, S. J. and Yang, R. T. (1987) *AIChE Symp. Series* 83 (No. 259), 87.

Dresselhaus, M. S. and Avouris, Ph. (2001) *Carbon Nanotubes*. (M. S. Dresselhaus, G. Dresselhaus, and Ph. Avouris, eds.). Springer, Berlin, Germany, and New York, NY, pp. 1–9.

Dresselhaus, M. S., Dresselhaus, G., and Avouris, Ph., Eds. (2001) *Carbon Nanotubes*. Springer Berlin, Germany, and New York, NY.

Dresselhaus, M. S., Dresselhaus, G., and Eklund, P. C. (1996) *Science of Fullerenes and Carbon Nanotubes*. Academic Press, San Diego, CA.

Dresselhaus, M. S., Dresselhaus, G., Suihara, K., Sapin, I. L., and Goldberg, H. A. (1988) *Graphite Fibers and Filaments*. Springer, Berlin, Germany, and New York, NY.

Dresselhaus, M. S. and Endo, M. (2001) *Carbon Nanotubes*. (M. S., Dresselhaus, G., Dresselhaus, and Ph. Avouris, eds.). Springer, Berlin, Germany, and New York, NY, pp. 10–28.

Drezdon, M. A. (1988) *Inorg. Chem.* 27, 4628.

Engwall, E. and Ma, Y. H. (2002) *Fundamentals of Adsorption 7*. (K. Kaneko, H. Kanoh, and Y. Hanzawa, eds.). IK International, Chiba City, Japan, p. 586.

Eswaramoorthy, M., Sen, R., and Rao, C. N. R. (1999) *Chem. Phys. Lett.* 304, 207.

Faust, S. D. and Aly, O. M. (1987) *Adsorption Process for Water Treatment*. Butterworths, Boston, MA.

Figueras, F. (1988) *Catal. Rev. Sci. Eng.* 30, 457.

Figueras, F., Mattrod-Bashi, A., Fetter, G., Thrierr, A., and Zanchetta, J. V. (1989) *J. Catal.* 119, 91.

Frechet, J. M. J. and Farrall, M. J. (1977) Functionalization of crosslinked polystyrene resins by chemical modification: a review. In *Chemistry and Properties of Crosslinked Polymers*. (S. S. Labana, ed.). Academic Press, San Diego, CA.

Fripiat, J. J. (1987) *Catal. Today* 2, 281.

Fripiat, J. J. (1982) *Developments in Sedimentology: Advanced Techniques for Clay Mineral Analysis*. (J. J. Fridat, ed.). Elsevier, New York, NY, Vol. 34.

Fryer, J. R. (1979) *Chemical Applications of Transmission Electron Microscopy*. Academic Press, New York, NY.

Garcia, A. A. and King, C. J. (1989) *Ind. Eng. Chem. Res.* 28, 204.

Gatica, S. M., Cole, M. W., Hartman, J. M., and Crespi, V. H. (2000) *Phys. Rev. B* 62, 9989.

Gil, A., Gandia, L. M., and Vicenti, M. A. (2000) *Catal. Rev., Sci. Eng.* 42, 145.

Gil, A. and Montes, M. (1994) *Langmuir* 10, 291.

Grim, R. E. (1968) *Clay Mineralogy*, 2nd Ed. McGraw-Hill, New York, NY.

Guo, T., Nikolaev, P., Thess, A., Colbert, D. T., and Smalley, R. E. (1995) *Chem. Phys. Lett.* 243, 49.

Harstenstein, H. U. (1993) Proceedings of International Specialty Conference, VIP-32; Municipal Waste Combustion Conference, Williamsburg, VA, March; Air and Waste Management Association: Pittsburgh, PA, 1993, p. 87.

Haufler, R. E., Conceicao, J., Chibanti, P. F., Chai, Y., Byrne, N. E., Flanagan, S., Haley, M. M., O'Brien, S. C., Pan, C., Xiao, Z., Billups, W. E., Ciufolini, M. A., Hauge, R. H., Margrave, J. L., Wilson, L. J., Curl, R. F., and Smalley, R. E. (1990) *J. Phys. Chem.* 94, 8634.

Hernadi, K., Fonseca, A., Nagy, J. B., Bernaerts, D., Fudala, A., and Lucas, A. A. (1996) *Zeolites* 17, 416.

Hilbert, M. and Lange, N. (1958) *Z. Kristallogr.* 111, 24.

Hollis, O. L. (1966) *Anal. Chem.* 38, 309.

Hutson, N. D., Gualdoni, D. J., and Yang, R. T. (1998) *Chem. Mater.* 10, 3707.

Iijima, S. (1991) *Nature* 354, 56.

Iijima, S. and Ichihashi, T. (1993) *Nature* 363, 603.

Inoue, S., Ichikuni, T., Uematsu, T., and Kaneko, K. (1998) *J. Phys. Chem.* B 102, 4689.

Johansson, G. (1960) *Acta Chem. Scand.* 16, 403.

Journet, C. and Bernier, P. (1998) *Appl. Phys. A.* 67, 1.

Kamalakaran, R., Terrones, M., Seeger, T., Kohler-Redlich, Ph., Ruhle, M., Kim, Y. A., Hayashi, T., and Endo, M. (2000) *Appl. Phys. Lett.* 77, 3385.

Katorski, A. and White, D. J. (1964) *Chem. Phys.* 40, 3183.

Keggin, J. F. (1934) *Proc. Roy. Soc. A (London)* 144, 75.

Kikkinides, E. S. and Yang, R. T. (1992) Unpublished data.

Kikkinides, E. S. and Yang, R. T. (1991) *Ind. Eng. Chem. Res.* 30, 1981.

Kim, B. R., Snoeyink, V. L., and Saunders, F. M. (1976) *J. Water Poll. Contr. Fed.* 48, 120.

King, J. and Benson, S. W. J. (1966) *Chem. Phys.* 44, 1007.

Kitiyanan, B., Alvarez, W. E., Harwell, J. H., and Resasco, D. E. (2000) *Chem. Phys. Lett.* 317, 497.

Kittel, C. (1953) *Introduction to Solid State Physics*. Wiley, New York, NY, p. 26.

Kratschmer, W., Fostiropoulos, K., and Huffman, D. R. (1990) *Chem. Phys. Lett.* 170, 167.

Kroto, H. W., Heath, J. R., O'Brien, S. C., Curl, R. F., and Smalley, R. E. (1985) *Nature* 318, 162.

Kumar, P., Jasra, R. Y. and Bhat, T. S. G. (1995) *Ind. Eng. Chem. Res.* 34, 1440.

Kunin, R., Meitzner, E. F., and Bortnick, N. M. (1962) *J. Amer. Chem. Soc.* 84, 305.

Kuo, Y., Munson, C. L., Rixey, W. G., Garcia, A. A., Frierman, M., and King, C. J. (1987) *Sep. Purif. Meth.* 16, 31.

Kuznetsova, A., Yates, J. T., Liu, J. and Smalley, R. E. (2000) *J. Chem. Phys.* 112, 9590.

Lahav, N., Shani, N., and Shabtai, J. (1978) *Clays Clay Miner.* 26, 107.

Lee, W. Y., Raythatha, R. H., and Tatarchuk, B. J. (1989) *J. Catal.* 115, 159.

Li, D., Scala, A. A., and Ma, Y. H. (1996) *Adsorption* 2, 227.

Long, R. Q. and Yang, R. T. (2001a) *J. Am. Chem. Soc.* 123, 2058.

Long, R. Q. and Yang, R. T. (2001b) *Ind. Eng. Chem. Res.* 40, 4288.

Louie, S. G. (2001) *Carbon Nanotubes*. (M. S., Dresselhaus, G. Dresselhaus, and Ph. Avouris, eds.). Springer, Berlin, Germany, and New York, NY.

Mackenzie, P. D., Sivavec, T. M., Grade, M. M., Horney, D. P., and Salva, J. J. (1994) Organic Sponges for Cost Effective CVOC Abatement, Report to DOE under contract DE-AC21-92MC29110, July, U.S. DOE, Morgantown Energy Technology Center, WV.

Mackie, E. B., Wolfson, R. A., Arnold, L. M., Lafdi, K., and Migone, A. D. (1997) *Langmuir* 13, 7197.

Maddox, M. W. and Gubbins, K. E. (1995) *Langmuir* 11, 3988.

Maienschein, J. L., Hudson, R. S., Tsugawa, R. T., Fearon, E. M., Souers, S., and Collins, G. W. (1992) *J. Vac. Sci. Tech. A* 10, 556.

Malla, P. B. and Komarneni, S. (1990) *Clays Clay Miner.* 38, 363.

Mao, Z. and Sinnott, S. B. (2000) *J. Phys. Chem. B* 104, 4618.

Mao, Z. and Sinnott, S. B. (2001) *J. Phys. Chem. B* 105, 6916.

Mays, T., Emmanuelle, A., and Yin, Y. F. (2002) *Fundamentals of Adsorption 7*. (K. Kaneko, H. Kahoh, and Y. Hanzawa, eds.). IK International, Chiba City, Japan.

Millar, J. R., Smith, D. G., Marr, W. E., and Kressman, T. R. E. (1963) *J. Chem. Soc.* 218.

Molinard, A., Cool, P., and Vansant, E. F. (1994a) *Microporous Mater.* 3. 149.

Molinard, A., Peters, K. K., and Vansant, E. F. (1994b) *Separation Technology*. (E. F. Vansant, ed.). Elsevier, Amsterdam, p. 445.

Molinard, A. and Vansant, E. F. (1995) *Adsorption* 1, 49.

Muris, M., Dufau, N., Bienfait, M., Dupont-Pavlovsky, N., Grillet, Y., and Palmari, J. P. (2000) *Langmuir* 16, 7019.

Neely, J. W. and Isacoff, E. G. (1982) *Carbonaceous Adsorbents for the Treatment of Ground and Surface Waters*. Dekker, New York, NY.

Nicholson, D. and Suh, S.-H. (2002) *Fundamentals of Adsorption 7*. (K. Kaneko, H. Kahoh, and Y. Hanzawa, eds.). IK International, Chiba City, Japan.

Occelli, M. L. (1984) Sorption of Normal Paraffins in a Pillared Clay Mineral. *Proceedings 8th International Congress in Catalysis* 4, 725.

Occelli, M. L., Innes, R. A., Hwu, F. S., and Hightower, J. W. (1985) *Appl. Catal.* 14, 69.

Occelli, M. L. and Tindwa, R. M. (1983) *Clays Clay Miner.* 31, 22.

Ohba, T., Murata, K., Kaneko, K., Steele, W. A., Kokai, F., Takahashi, K., Kasuyu, D., Yudasaka, M., and Iijima, S. (2001) *Nano Lett.* 1, 371.

Padin, J. and Yang, R. T. (2000) *Chem. Eng. Sci.* 55, 2607.

Pederson, M. R. and Broughton, J. Q. (1992) *Phys. Rev. Lett.* 69, 2689.

Peng, L. M., Zhang, Z. L., Xue, Z. Q., Wu, Q. D., Gu, Z. N., and Pettifor, D. G. (2000) *Phys. Rev. Lett.* 85, 3249.

Pinnavaia, T. J. (1983) *Science* 220, 365.

Pinnavaia, T. J. and Tzou, M. S., Pillared and Delaminated Clays Containing Iron. U.S. Patent 4,665,044 (1987).

Pinnavaia, T. J., Tzou, M. S., and Landau, S. D. (1985) *J. Am. Chem. Soc.* 107, 2783.

Pinnavaia, T. J., Tzou, M. S., Landau, S. D., and Raythatha, R. H. (1984) *J. Mol. Catal.* 27, 195.

Power, T. D., Skoulidas, A. I., and Sholl, D. S. (2002) *J. Am. Chem. Soc.* 142, 1858.

Qin, L., Zhao, X., Hirahara, K., Miyamoto, Y., Ando, Y., and Iijima, S. (2000) *Nature* 408, 50.

Radushkevich, L. V. and Luk'Yanovich, V. M. (1952) *Zh. Fiz. Khim.* 26, 88.

Rexwinkel, G., Heesink, B. B. M., and van Swaaij, W. P. M. (1999) *J. Chem. Eng. Data* 44, 1139.

Rhodes, C. N. and Brown, D. R. (1993) *J. Chem. Soc. Faraday Trans.* 89, 1387.

Rightor, E. G., Tzou, M. S., and Pinnavaia, T. J. (1991) *J. Catal.* 130, 29.

Rixey, W. G. and King, C. J. (1989a) *J. Coll. Interf. Sci.* 131, 320.

Rixey, W. G. and King, C. J. (1989b) *AIChE J.* 35, 69.

Rixey, W. G. and King, C. J. (1987) *Fundamental of Adsorption*. (A. I. Liapis, ed.). Engineering Foundation, New York, p. 503.

Rodriguez, N. M. and Baker, R. T. K. U.S. Patent 5,653,951 (1997).

Rodriguez, N. M., Chambers, A., and Baker, R. T. K. (1995) *Langmuir* 11, 3862.

Ruckenstein, E. (1987) The role of interactions and surface phenomena in sintering and redispersion of supported metal catalysts. In *Metal-Support Interactions in Catalysis, Sintering and Redispersion*. (S. A. Stevenson, J. A. Dumesic, R. T. K. Baker, and E. Ruckenstein, eds.). Van Nostrand Reinhold, New York, pp. 141–308.

Rupert, J. P., Granquist, W. T., and Pinnavaia, T. J. (1987) Catalytic properties of clay minerals. In *Chemistry of Clays and Clay Minerals*. (A. C. D. Newman, ed.). Mineralogical Society, Longman Scientific & Technical, Essex, UK.

Schutzenberger, P. and Schutzenberger, L. C. R. (1890) *Acad. Sci. [Paris]*. 111, 774.

Shabria, J. and Lahari, N. Cross-Linked Montmorillonite Molecular Sieves. U.S. Patent 4,216,188 (1980).

Shabtai, J., Maasoth, F. E., Tokarz, M., Tsai, G. M., and McCauley, J. (1984a) Characterization and Molecular Shape Selectivity of Cross-Linked Montmorillonite (CLM). *Proc. 8th Int. Congr. Catal.*, 4, 735.

Shabtai, J., Rosell, M., and Tokarz, M. (1984b) *Clays Clay Miner.* 32, 99.

Sprung, R., Davis, M. E., Kauffman, J. S., and Dybowski, C. (1990) *Ind. Eng. Chem. Res.* 29, 213.

Stacey, M. H. (1988) *Catal. Today* 2, 621.

Stan, G. and Cole, M. W. (1998) *Surf. Sci.* 395, 280.

Stephanie-Victoire, F., Goulay, A-M., and Cohen de Lara, E. (1998) *Langmuir* 14, 7255.

Sterte, J. (1991) *Clays Clay Miner.* 39, 167.

Sterte, J. (1986) *Clays Clay Miner.* 34, 658.

Sun, L. F., Xie, S. S., Liu, W., Zhou, W. Y., Liu, Z. Q., Tang, D. S., Wang, G., and Qian, L. X. (2000) *Nature* 403, 384.

Suzuki, K., Horio, M., and Mori, T. (1988) *Mater. Res. Bull.* 23, 1711.

Talapatra, S., Zambano, A. Z., Weber, S. E., and Migone, A. D. (2000) *Phys. Rev. Lett.* 85, 138.

Teizer, W., Hallock, R. B., Dujardin, E., and Ebbeson, T. W. (1999) *Phys. Rev. Lett.* 82, 5305; 84, 1844 (E) (2000).

Thess, A., Lee, R., Nikolaev, P., Dai, H., Petit, P., Robert, J., Xu, C., Lee, Y. H., Kim, S. G., Rinzler, A. G., Colbert, D. T., Scuseria, G. E., Tomanek, D., Fischer, J. E., and Smalley, R. E. (1996) *Science* 273, 483.

Tibbetts, G. G. (1990) Vapor-grown carbon fibers. In *Carbon Fibers, Filaments and Composites*. (J. L. Figueiredo, ed.). Kluwer, Netherlands, pp. 73–94.

Tibbetts, G. G. (1987) *Carbon* 25, 367.

Tibbetts, G. G., Endo, M., and Beets, C. P. (1986) Jr., *SAMPE J.* 22–5.

Tzou, M. S. and Pinnavaia, T. J. (1987) *Catal. Today* 2, 243.

Valenzuela, D. P. and Myers, A. L. (1989) *Adsorption Equilibrium Data Handbook*, Prentice Hall, Englewood Cliffs, NJ.

Van Olphen, H. (1977) *An Introduction to Clay Colloid Chemistry*, 2nd Ed. Wiley, New York, NY.

Vaughan, D. E. W. Multimetallic Pillared Interlayered Clay Products and Processes of Making Them. U.S. Patent 4,666,877 (1987).

Vaughan, D. E. W., Lussier, R. J., and Magee, J. S., Pillared Interlayered Clay Product, U.S. Patent 4,176,043 (1981a).

Vaughan, D. E. W., Lussier, R. J., and Magee, J. S. Stabilized Pillared Interlayered Clays. U.S. Patent 4,248,739 (1981b).

Vaughan, D. E. W., Lussier, R. J., and Magee, J. S. Pillared Interlayered Clay Materials Useful as Catalysts and Sorbents. U.S. Patent 4,176,090 (1979).

Wang, N., Tang, Z. K., Li, G. D., and Chen, J. S. (2000) *Nature* 408, 51.

Wang, Q., Challa, S. R., Sholl, D. S., and Johnson, J. K. (1999) *Phys. Rev. Lett.* 82, 956.

Warburton, C. I. (1987) Preparation and Catalytic Properties of Iron Oxide and Iron Sulphide Pillared Clays. *Catal. Today*, 2, 271.

Weber, W. J., Jr. and van Vliet, B. M. (1981a) *J. Amer. Water Works Assoc.* 73, 420.

Weber, W. J., Jr. and van Vliet, B. M. (1981b) *J. Amer. Water Works Assoc.* 73, 426.

Yamanaka, S., Doi, T., Sako, S., and Hattori, M. (1984) *Mater. Res. Bull.* 19, 161.

Yamanaka, S. and Brindley, G. W. (1979) *Clays Clay Miner.* 27, 119.

Yamanaka, S. and Hattori, M. (1987) Iron Oxide Pillared Clay., *Catal. Today*, 2, 271.

Yamanaka, S., Malla, P. B. and Komarneni, S. (1990) *J. Coll. Interf. Sci.* 134, 51.

Yang, R. T. (2000) *Carbon* 38, 623.

Yang, R. T. U.S. Patent 4,134,737 (1979).

Yang, R. T. and Baksh, M. S. A. (1991) *AIChE J.* 37, 679.

Yang, R. T. and Cheng, L. S. (1995) *Access in Nanoporous Materials*. (T. J. Pinnavaia and M. F. Thorpe, eds.). Plenum Press, New York, NY.

Yang, R. T. and Chen, J. P. (1989) *J. Catal.* 115, 52.

Yang, R. T., Goethel, P. J., Schwartz, J. M., and Lund, C. R. F. (1990) *J. Catal.* 122, 206.

Yang, R. T., Long, R. Q., Padin, J., Takahashi, A., and Takahashi, T. (1999) *Ind. Eng. Chem. Res.* 38, 2726.

Yao, Z., Dekker, C., Avouris, Ph. (2001) *Carbon Nanotubes*. (M. S. Dresselhaus, G. Dresselhaus, and Ph. Avouris, eds.). Springer, Berlin, Germany, and New York, NY.

Yin, Y. F., Mays, T., and McEnaney, B. (1999) *Langmuir* 15, 8714.

Yoshida, H., Oehlenschlaeger, S., Minami, Y., and Terashima, M. (2000) *Adsorption Science and Technology*. (D. D. Do, ed.). World Scientific, River Edge, NJ, p. 688.

Zhu, H. W., Xu, C. L., Wu, D. H., Wei, B. Q., Vajtal, R., and Ajayan, P. M. (2002) *Science* 296, 884.

10

SORBENTS FOR APPLICATIONS

The best sorbents for particular separations, both present and future, are discussed in the chapter.

10.1. AIR SEPARATION

Nitrogen and oxygen are, respectively, the second and third most produced chemicals. They are used in numerous chemical processing, refinery, metal production, and other industrial operations. For example, high-purity nitrogen is used for purging, blanketing, and providing atmosphere for metal treating and other purposes; while high-purity or enriched oxygen is used in chemical processing, steel and paper-making applications, wastewater treatment, and lead and glass production. Nitrogen and oxygen have been produced since 1907 when Carl von Linde built the first cryogenic distillation column for air separation in Buffalo, NY. Cryogenic processes are highly efficient, particularly for large-volume production. Cryogenics account for approximately 70% of the nitrogen and oxygen produced today (about 20% by adsorption and 10% by membrane and hybrid systems).

The polarizabilities of N_2, O_2, and Ar are nearly the same (1.74, 1.58, and 1.63 in units of 10^{-24} cm^3, respectively), and are all nonpolar. Consequently, they adsorb nearly the same on all sorbents except zeolites. The fact that zeolites can distinguish between N_2 and O_2 was observed as early as 1938 (Barrer, 1937; 1938). Barrer reported values for heats of adsorption of N_2 on chabazite as high as 8 kcal/mol. The high heats of adsorption were subsequently explained quantitatively in terms of the quadrupole–electric field gradient interactions (Drain, 1953; Kington and Macleod, 1959). The unique adsorption properties of zeolites derive from the fact that their surfaces are composed of negatively charged oxides with isolated cations that are located above the surface planes. Despite

Adsorbents: Fundamentals and Applications, Edited By Ralph T. Yang
ISBN 0-471-29741-0

the finding of N_2/O_2 selectivity by zeolites, no effort was made to separate air by adsorption until the 1960's, after the invention of synthetic zeolites types A and X, and the pressure-swing adsorption (PSA) cycles.

The invention of types A and X zeolites by Milton (1959) made it possible for these zeolites to be available at controlled quality and guaranteed supply. The inventions of PSA cycles by Skarstrom (1960) and by Guerin de Montgareuil and Domine (1964) have been discussed in detail by Yang (1987). Inspired by these inventions, dreamers in industry began contemplating the possibility of separating air *at ambient temperature* (as opposed to 77 K for cryogenic processes) by adsorption. 5A (CaA) and 13X (NaX) zeolites were used (and are still being used in some instances) as the sorbents. The PSA technology development encountered some challenges that were unexpected from small-diameter laboratory column experiments, for example, the "cold spot" problem (large temperature depression near the feed end, Collins, 1977; Yang, 1987) and the "creeping death" of beds (due to accumulation and freezing of water). The history of the PSA technology development may be reflected by the decline in the cost of O_2 from PSA as shown in Figure 10.1.

Prior to ca. 1980, PSA systems were used with both adsorption and desorption pressures well above atmospheric. These systems were low in capital (due to simplicity) but high in power consumption (since both N_2 and O_2 in the feed are compressed compared with vacuum swing adsorption (VSA) where only the waste gas is evacuated). The availability of improved sorbents and lower cost vacuum equipment led to the development of VSA, which is typically operated with adsorption pressure slightly above atmospheric and desorption pressure of typically 0.2 atm. A further breakthrough occurred in 1989, with the invention of LiLSX zeolite (low silica X, with Si/Al = 1) (Chao, 1989). The LiLSX is currently the best commercial sorbent for air separation and will be discussed

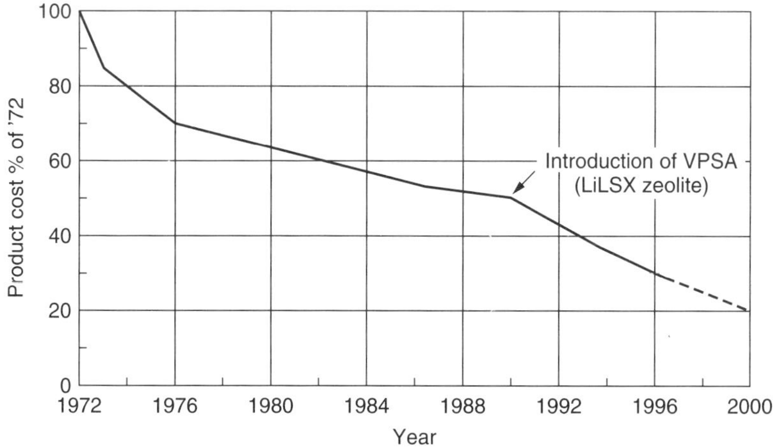

Figure 10.1. PSA oxygen product cost (in constant dollars). 5A and 13× zeolites were used before 1990, and LiLSX zeolite is used after 1990 (courtesy of J. P. Kingsley of Praxair, Inc.).

in some detail in this chapter. With the advances made in both sorbents and PSA/VSA cycle technology, PSA/VSA accounts for ~20% of the oxygen and nitrogen, and the size of a single PSA/VSA unit is approaching 250 tons/day oxygen, while these numbers continue to grow steadily. The sorbent productivity (or bed inventory), often expressed as "bed-size factor" (Leavitt, 1992), is well below 1000 lb of zeolite for per ton/day O_2 product (Notaro et al., 1999). The total power consumption is below 250 kWh per ton O_2.

10.1.1. 5A and 13X Zeolites

5A (CaA) and 13X (NaX) zeolites have been the most commonly used sorbents for air separation. The N_2/O_2 isotherms on 5A that are reported in the literature vary widely, depending on the cation composition in the zeolite. The typical commercial 5A used for air separation is made by exchanging ~70% of the Na^+ in NaA by Ca^{2+}.

Isotherms of N_2/O_2 on commercial 5A and 13X crystals are shown in Figure 10.2. High-pressure isotherms, including Ar, on 5A are shown in Figure 10.3. The isotherms of O_2 on zeolites are slightly higher than those of Ar at <1 atm, with the known exceptions of Ag-exchanged zeolites such as AgX (Hutson et al., 1999) and Ag-mordenite (Knaebel and Kandybin, 1993). The high-pressure isotherms on 13X are shown in Figure 10.4.

Mg-exchanged A zeolite has attracted interest due to its high N_2/O_2 selectivity or separation factor (α). Schollner et al. (1981) reported high N_2/O_2 selectivities on MgA based on chromatography data. Coe et al. (1994) measured isotherms of N_2 and O_2 on partially exchanged MgNaA with up to 81% exchange. Compared with 99% Ca-exchanged A, the N_2/O_2 separation factor (of air) at 1000 torr

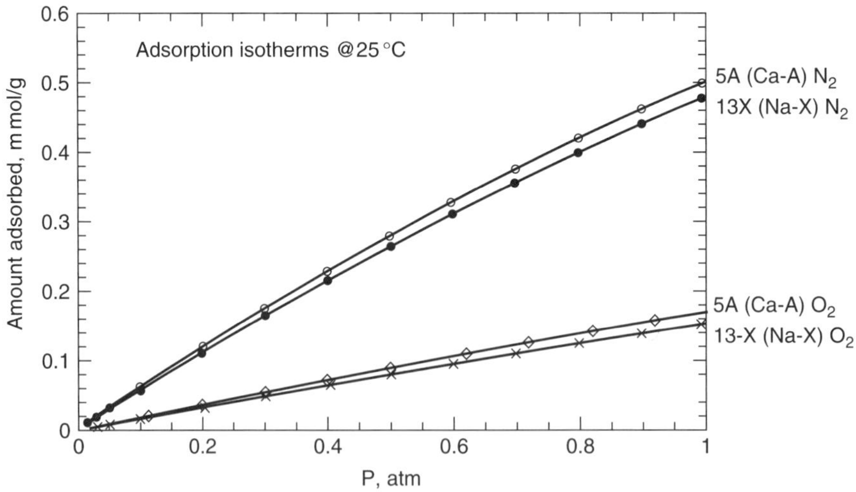

Figure 10.2. Equilibrium isotherms of N_2/O_2 on commercial 5A and 13X zeolite crystals at 25 °C.

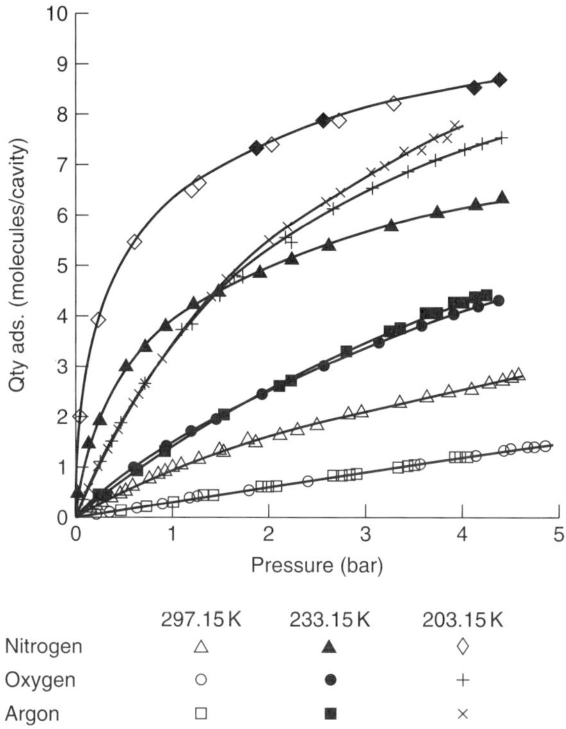

Figure 10.3. Equilibrium isotherms of N_2, O_2, and Ar on a 5A zeolite, $Ca_{4.5}Na_3[(AlO_2)_{12}(SiO_2)_{12}]$, that is, 75% Ca exchanged form of NaA. One molecule/cavity = 0.566 mmol/g anhydrous 5A crystals or 0.452 mmol/g sorbent with 20% inert binder (Miller et al., 1987, with permission).

(using the IAS model) increased with the level of Mg exchange. The α values were 4.8 (CaA), 6.6 (63% Mg in NaA), 7.6 (75% Mg in NaA), and 8.5 (81% Mg in NaA). Simulation results for VSA (adsorption at 1000 torr and desorption at 300 torr) using these sorbents were also reported. The highest oxygen product recovery as well as product throughput were obtained for the 75% MgNaA. The commercial status of using MgA in VSA is unknown. It is known, however, that Mg-exchanged zeolite is not stable because the framework structure tends to collapse upon calcination, due to the formation of MgO. For Mg-exchanged type A zeolite, it is not possible to obtain more than 81% Mg exchange.

10.1.2. Li-LSX Zeolite

As shown in Section 3.3 in Chapter 3, the PSA/VSA separation results may be correlated with the simple sorbent selection parameter S:

$$S = \frac{\Delta q_{N_2}}{\Delta q_{O_2}} \alpha_{N_2/O_2} \tag{10.1}$$

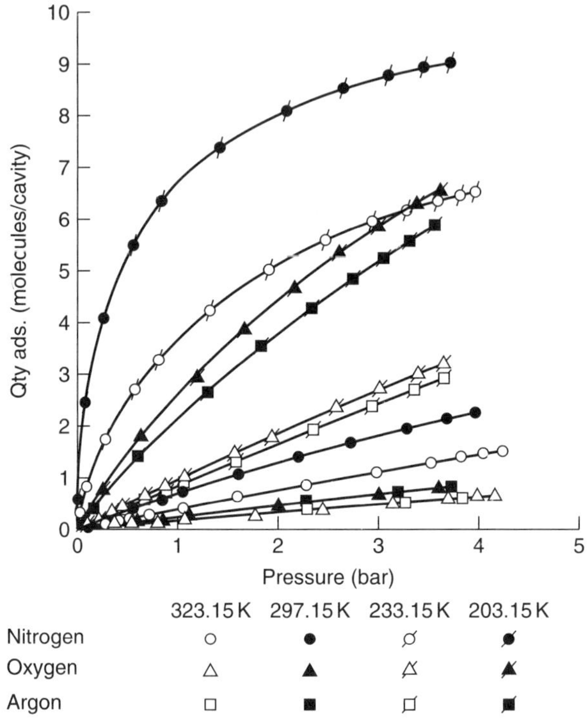

Figure 10.4. Equilibrium isotherms of N_2, O_2, and Ar on a 13X zeolite, $Na_{86}[(AlO_2)_{86}(SiO_2)_{106}]$. One molecule/cavity = 0.556 mmol/g anhydrous 13X crystals or 0.423 mmol/g sorbent with 24% inert binder (Miller, 1987, with permission).

where Δq_{N_2} represents the differential amounts of adsorbed N_2 during the pressure cycle, Δq_{O_2} represents that of O_2, and α is the separation factor at the adsorption condition. Thus, the performance of the sorbent depends not only on a high working capacity of N_2, but also a low delta loading in the O_2 capacity. In other words, a low capacity for O_2 is as important as increasing the N_2 capacity.

The adsorption of N_2 and O_2 on ion exchanged type X zeolite was studied, independently, by McKee (1964) and Habgood (1964). In both studies, the sorbents were commercial type X zeolite with Si/Al $\sim$ 1.25, and the zeolite was ion-exchanged with both alkali and alkaline earth metal ions. Li^+ was one of the ions included in both studies. These studies were performed at 1 atm pressure. The highest N_2/O_2 selectivities were obtained for: $Ba^{2+} > Sr^{2+} > Li^+ > Ni^{2+}$ (McKee, 1964).

The true potential of LiX for air separation was not understood until the invention of Chao (1989). The synthesis of stable LSX (Si/Al = 1) was accomplished in the early 1980s (Kuhl and Sherry, 1980; Butter and Kuznicki, 1986; Kuhl, 1987). Using the LSX, Chao found that (1) the N_2 adsorption capacity was significantly increased when the Si/Al ratio was near one; and (2) a threshold of $\sim$80% Li^+ exchange for X zeolite (Si/Al = 1.25) (or 70% for LSX) must be reached for increased N_2 adsorption, beyond which the amount of N_2 increased linearly with Li^+ content. Moreover, the O_2 capacity was decreased because the polarizability of Li^+ is lower than that of Na^+. Since Chao's invention, LiLSX has been the sorbent of choice for air separation.

Figure 10.5 shows the isotherms of N_2, O_2, and Ar on LiLSX. The high-pressure N_2 isotherms are shown in Figure 10.6. Figure 10.6 also shows the significant increase in N_2 capacity when the Si/Al ratio is decreased from 1.25 to 1. The high N_2 capacity, combined with the linearity of the N_2 isotherm, results in a large working capacity of N_2 for use in PSA/VSA. The low capacity for O_2 contributes equally to the improved performance for PSA/VSA separation.

The dependence of the N_2 capacity on the % Li exchange is shown in Figure 10.7. The reason for the threshold has to do with the site location of the Li^+ cations (Coe, 1995; Yoshida et al., 2001). The first 70% (out of a total of 96 cations/unit cell) of the Li cations are bonded to sites that are not fully exposed to the supercage where N_2 and O_2 are located. Beyond 70% exchange, the Li^+ ions begin to fill the sites that are more exposed, for example, SIII (see Figure 10.8 for site locations).

Herden et al. (1982) studied the cation sites in LiX and LiY by XRD. Feuerstein and Lobo (1998) studied the sites in LiLSX by neutron diffraction and solid-state NMR. Their results showed that only three sites were occupied: SI′, SII, and SIII, where SIII is at the center of the 4-oxygen ring of the sodalite cage, and the 96 cations were fairly evenly distributed among these three sites.

In faujasite zeolites, the cations in the beta-cages and the double six-ring (SD6R, the hexagonal prism) (i.e., at sites SI, SI″, and SII′) are sterically inaccessible to nitrogen, and so only the supercage cations (i.e., those at SII and

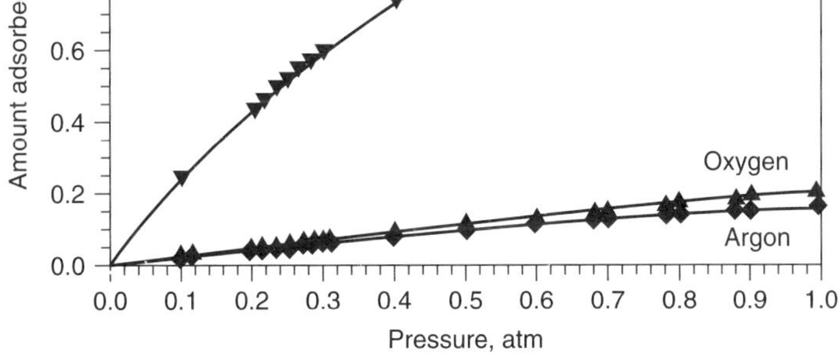

Figure 10.5. Adsorption isotherms for N_2, O_2, and Ar at 25 °C for $Li_{94.5}Na_{1.5}$-LSX dehydrated *in vacuo* at 350 °C (Hutson et al., 1999, with permission).

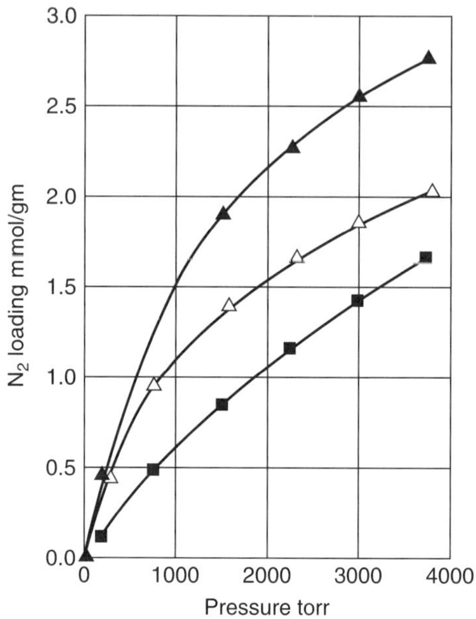

Figure 10.6. Nitrogen isotherms at 23 °C on: (▲) Li-LXS (99% Li, 1% Na, Si/Al = 1), (△) LiX (94% Li, 6% Na, Si/Al = 1.25), and (■) NaX (Si/Al = 1.25), from Chao, 1989.

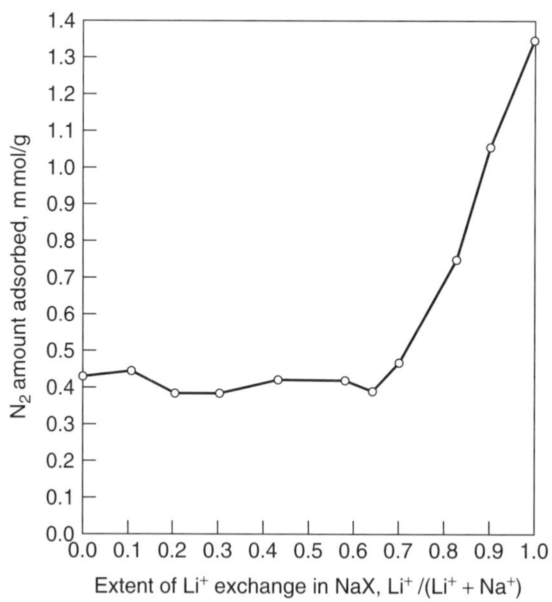

Figure 10.7. Dependence of nitrogen adsorption at 23 °C at 1 atm on fractional Na^+ exchange by Li^+ in LSX. The threshold value is at 80% Li for X zeolite with Si/Al = 1.5 (Chao, 1989). The data on LSX are from Coe et al., 1992.

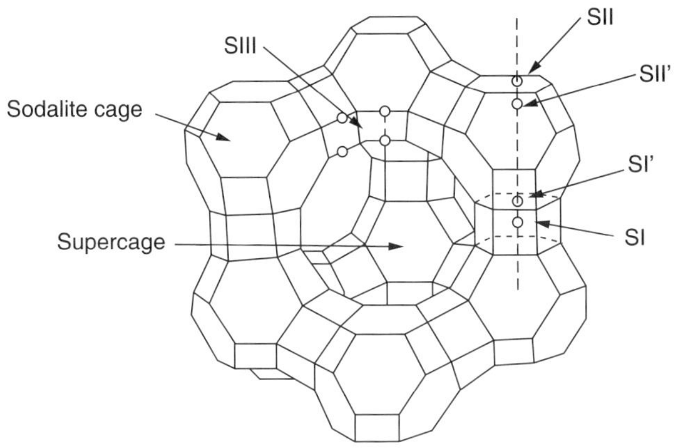

Figure 10.8. Unit cell of faujasite-type (X and Y) zeolites, including cation sites.

SIII) are available to interact with nitrogen. However, the electric fields around these supercage cations are partially shielded by the surrounding oxygen atoms. Because of this shielding, the electrostatic and induction interactions are expected to be lower than those of an isolated ion. Further, the dispersion forces acting on the molecule will be higher because adsorbate molecules also interact with oxygen atoms of the zeolite. Due to the small size of the lithium cation it can sit crystallographically very low in the face of the single six-ring (SR6, the SII position), allowing the electric field to be nearly completely shielded by the surrounding framework oxygen. This explains why one must exchange in excess of 64 lithium cations into the X zeolite before there is any increase in the N_2 adsorption capacity. Only the SIII Li^+ cations interact with the N_2 molecules. Na^+ and larger cations do interact from the SII locations, although the SIII locations are much more accessible and have higher energies because they are in a lower coordination.

It is obviously advantageous to lower the (70%) threshold of Li^+ exchange for increased N_2 adsorption. Li^+ is mobile in the faujasite structure (Herden et al., 1981), and the mobility is the highest during calcination. Lowering the threshold could possibly be accomplished by (1) low-temperature dehydration of the partially exchanged zeolite, or (2) filling the low-energy sites (SI, SI′, SII′, and SII) with large and inexpensive cations. Further research is needed to develop a viable technique for lowering the required Li threshold.

The performance of the LiLSX zeolite used in VSA has been compared with that of NaX in Figure 10.9. The standard five-step cycle (see Chapter 3) was used in the simulation (Rege and Yang, 1997). The O_2 product recovery was optimized for the two sorbents at different pressure ratios, keeping the product purity and throughput nearly constant. It is seen that the LiLSX outperforms NaX by a significant margin. More interestingly, it was found that in the case of LiLSX, it was possible to operate at a pressure ratio as low as 2, compared with the limit of 4 for NaX. Due to the large heat of adsorption for N_2 on LiLSX

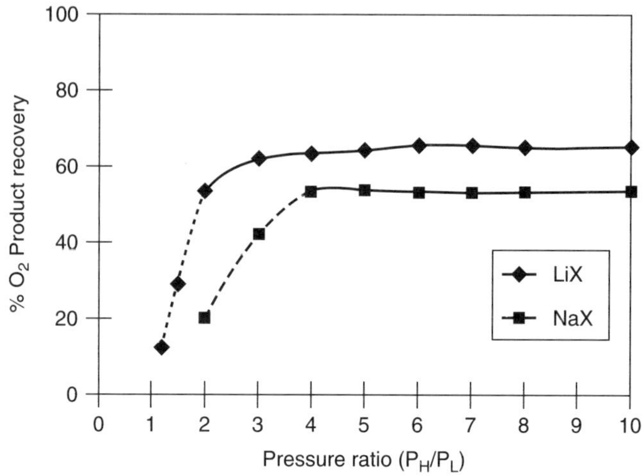

Figure 10.9. VSA O_2 product recovery (%) at different (Ads/Des) pressure ratios for LiLSX and NaX sorbents. The adsorption pressure was fixed at 1 atm, and the product throughput was fixed at 0.0267 kg O_2/kg sorbent/h. The O_2 product purity was near 95.2% for all runs (Rege and Yang, 1997, with permission).

(~5.5 kcal/mol), the temperature excursion during the VSA cycle was four times that in the NaX beds. Rege and Yang (1997) showed that by introducing 5–10% (v/v) of inert high-heat-capacity particles (such as iron), at the same total bed volume, the O_2 product recovery could be increased by 2% (at the same product purity and throughput).

10.1.3. Type X Zeolite with Alkaline Earth Ions

The work or McKee (1964) and Habgood (1964) showed that type X zeolite with divalent cations yielded the highest N_2/O_2 selectivity as well as the highest N_2 capacity at atmospheric pressure. Most of the sorbent development studies in the 1980's followed this line of thinking. For example, Coe and Kuznicki (1984) showed that CaX (followed by SrX) gave the highest N_2/O_2 selectivities and also N_2 capacities at 1 atm and 30 °C. Sircar et al. (1985) showed that mixed SrCaX (approximately 90% Sr and 10% Ca) exhibited higher N_2 capacities at 3 atm than pure CaX and SrX, without adverse effects on N_2/O_2 selectivity or a large increase in heat of adsorption.

The oxygen capacities on these divalent cation-containing X zeolites are also high, due to the higher polarizabilities of these cations when compared with the univalent cations. Consequently, the delta loading of O_2 in a VSA cycle is higher for the X zeolites with divalent cations. Table 10.1 illustrates this point. Comparing LiX and CaX (both with Si/Al = 1.25), the delta loading of N_2 (Δq_{N_2}) (or working capacity) is higher for CaA. However, the delta loading of O_2 is substantially higher for CaX. Thus, much work needs to be done by the CaX in O_2 adsorption and desorption. The result is poor separation performance by CaX.

Table 10.1. Sorbent selection parameter ($S \propto \Delta q_{N_2}/\Delta q_{O_2}$) for VSA/PSA performance for air separation, comparing LiX and CaX (Si/Al = 1.25 for both)

	LiX	CaX
N_2		
q @ 1.2 bar, mmol/g	0.94	1.28
q @ 0.24 bar, mmol/g	0.37	0.68
Δq_{N_2}	0.57	0.60
O_2		
q @ 0.3 bar, mmol/g	0.048	0.090
q @ 0.06 bar, mmol/g	0.010	0.019
Δq_{O_2}	0.038	0.071
$\Delta q_{N_2}/\Delta q_{O_2}$	15.0	8.5

P(adsorption) = 1.5 bar, P(desorption) = 0.3 bar
T = 300 K, $Y_{N_2} = 0.8$ and $Y_{O_2} = 0.2$.
Data courtesy of C. C. Chao, Praxair, Inc.

Following Chao's work, both Chao et al. (1992) and Coe et al. (1992) studied LSX zeolite containing mixed Li^+ and divalent cations. Both groups showed that LSX containing cations mixed at about 90% Li and 10% Sr or Ca are good sorbents. Fitch et al. (1995) reported good N_2/O_2 selectivity and N_2 capacity with mixed LiAlX zeolite (i.e., using Al^{3+} as the nonframework charge-compensating cation). However, LiLSX with near 100% Li exchange is the best sorbent used today for O_2 production.

10.1.4. LSX Zeolite Containing Ag (AgLiLSX)

Silver cation (Ag^+) exhibits very strong (but reversible) interactions with N_2. It has been shown that LiLSX mixed with only 1–3% Ag^+ can out-perform pure LiLSX in O_2 production from air by PSA/VSA. Upon heat-treatment, Ag^+ undergoes "auto-reduction," resulting in unique cation sites that are favorable for interactions with gas molecules. The interesting properties of Ag^+-containing zeolites are potentially useful for air separation as well as for other applications. Because of this, some details will be given for these zeolites.

Chemical State and Sites of Ag^+ In Faujasite. Silver is known to have very strong effects on the adsorption characteristics of zeolites (Habgood, 1964; Huang, 1974). Yang et al. (1996) reported the synthesis of a mixed lithium-silver (80/20) ion-exchanged X-type zeolite (Si/Al = 1.25 with ~17 Ag^+ per unit cell), and discussed its possible superior properties for air separation. This sorbent utilized the very strong adsorptive properties of the Ag^+ ion, which provided for increased capacity over that of Li-X while maintaining some degree of the advantageous isotherm linearity seen with Li-X.

Ab initio molecular orbital calculations showed the adsorption of nitrogen was enhanced by weak chemical interaction (through a classical π-complexation bond) with the Ag^+ cation on the zeolite framework (Chen and Yang, 1996).

Numerous attempts have been made to reduce transition metal ions in zeolites in order to form highly dispersed metallic clusters for use as catalysts. These attempts have typically been completed via treatment at elevated temperatures and/or in reducing atmospheres (e.g., sodium vapor, hydrogen gas, carbon monoxide gas). However, color changes upon vacuum dehydration of silver-exchanged A-type zeolites were found to be related to the formation of metallic clusters within the sodalite cage or the 6-prism of the zeolite (Kim and Seff, 1978a, 1978b; Jacobs et al., 1979). Using volumetric sorption techniques and temperature-programmed desorption, Jacobs et al. (1979) could relate these color changes to an autoreductive process involving framework oxygen. Autoreduction is the reduction of the transition metal ion and the oxidation of water or lattice oxygen. This has been observed for both Ag^+ and Cu^{2+} ions in zeolites A, X, and Y. Autoreduction of Cu^{2+} is discussed in Chapter 8. Autoreduction of Ag^+ has been shown to occur by two mechanisms in two clearly defined temperature regions (Jacobs et al., 1979; Baker et al., 1985):

(i) *autoreduction in the presence of zeolite water (25–250°C)*

$$2(Ag^+ - Z - O^-) + H_2O \rightarrow 2Ag^0 + (1/2)O_2 + 2Z - OH \qquad (10.2)$$

(ii) *autoreduction by oxygen from the zeolite lattice (127–380°C)*

$$2(Ag^+ - Z - O^-) \rightarrow 2Ag^0 + (1/2)O_2 + Z - O^- + Z^+ \qquad (10.3)$$

Kim and Seff (1978a, 1978b) proposed the formation of octahedral hexasilver metal clusters stabilized by coordination to six silver ions ($(Ag^+)_6(Ag)^6$) from X-ray structural determinations of a dehydrated silver-exchanged zeolite A. However, Jacobs et al. (1979) suggested that the formation of such large metal clusters is improbable because color changes are seen even at low temperatures. Also, low silver loadings where extensive migration of neutral silver atoms and subsequent sintering into Ag_6 metal clusters are highly unlikely. Alternatively, Jacobs et al. (1979) suggested, based on structural studies of Ag-A zeolites, the formation of linear $(Ag_3)^{2+}$ charged clusters ($Ag^+ - Ag^0 - Ag^+$) upon thermal dehydration of the zeolite.

The location of the extraframework silver in relation to the aluminosilicate framework has primary importance for elucidating the effect of silver clustering on the adsorptive characteristics of the zeolite. This is not a trivial endeavor. Numerous studies have been undertaken to identify the location of Ag^+ ions and Ag-clusters in argentiferous zeolites. These have mostly been for Ag-A and have included X-ray diffraction methods (Kim and Seff, 1978a; 1978b; Gellens et al., 1981a; 1981b) and far-infrared spectroscopy (Ozin et al., 1984; Baker et al.,

1985). It was found that, for dehydrated and fully Ag^+-exchanged faujasite-type zeolites, the silver molecules were distributed among the six-ring sites (SI, SI′, and SII for faujasites) and, for samples with high Al content, in the SIII locations. Gellens et al. (1981b) and Baker et al. (1985) showed the simultaneous occupancy of sites SI and SI′ by linear ($Ag^+ - Ag^0 - Ag^+$) clusters. Further information (prior to 1994) can be found in a comprehensive review of silver clusters and chemistry in zeolites by Sun and Seff (1994).

The detailed cation site locations in AgX, AgY, and AgLSX heat-treated under various conditions have been determined more recently by using powder neutron diffraction and Rietveld refinement (Hutson et al., 2000a; 2000b). The cation site distribution is summarized in Table 10.2 and the site locations are illustrated in Figure 10.10. X-ray photoemission spectroscopy confirmed partial reduction of $Ag^+ \rightarrow Ag^0$. A new site, named SII*, was identified, which is more elevated above the plane of the 6-ring of the sodalite cage, and hence can form a strong bond with N_2. Structural characterization, along with valence bond calculations, revealed the presence of cations in site II*, which are more active in Ag-LSX samples that were vacuum-dehydrated at 450 °C, as compared with those that were vacuum-dehydrated at 350 °C.

Air Separation by AgLiLSX. Mixed Li,Ag-LSX with different contents of Ag were prepared and characterized by Hutson and Yang (2000a). These samples were heat-treated *in vacuo* at various temperatures and their structures (including cation sites) were determined with powder neutron diffraction. The cation sites are summarized in Table 10.3. Structural characterization revealed the presence of cations at site II* in mixed Li,Ag-LSX zeolites that were vacuum-dehydrated at 450 °C. Cations in this site II* are more interactive with the atmospheric sorbates of interest than silver at the conventional site II location. Vacuum dehydration at 450 °C induced thermal migration of Ag^+ from site II to site II*. This is clearly seen in Table 10.3. Furthermore, the number of Ag cations at site II* per unit

Table 10.2. Site occupancies for silver exchanged faujasites in units of silver/unit cell determined from powder neutron diffraction

Site	Ag-Y-450	Ag-X-450	Ag-LSX-350	Ag-LSX-450
I	10.9	1.9	8.5	8.5
I′	12.4	14.0	23.4	23.0
I′b	—	13.9	—	—
II′	4.5	—	—	—
II	27.5	25.8	25.3	25.0
II*	—	6.2	6.7	7.0
III	—	13.4	19.2	20.2
Ag Found	55.3	75.2	83.1	83.7
Ag Predicted	56	86	96	96

Ag-X-450 denotes Ag-X heat-treated at 450 °C in vacuo. The sites are shown in Figure 10.9. Hutson et al., 2000a, with permission.

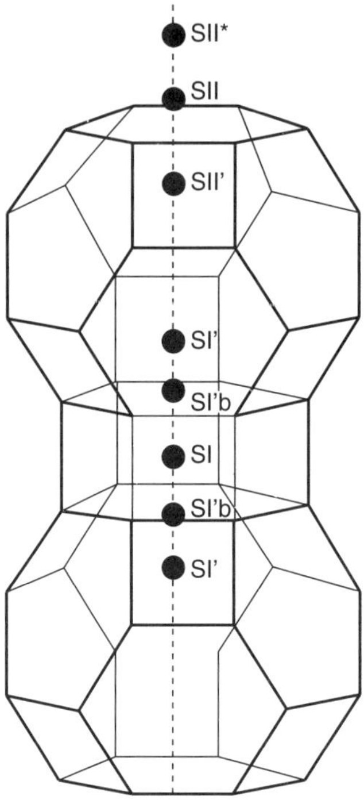

Figure 10.10. Extraframework sites for Ag^+ in the faujasite structure. SI, SI′, and SII are the same as that shown in Figure 10.7. From Hutson et al. (2000a) with permission.

Table 10.3. Site occupancies for extra framework species in mixed Li,Ag-LSX in units of silver/unit cell

Cation	$Li_{95.8}Na_{0.2}$ -LSX-450	$Li_{54.0}Ag_{41.8}Na_{0.2}$ -LSX-450	$Li_{93.3}Ag_{2.0}Na_{0.7}$ -LSX-350	$Li_{93.3}Ag_{2.0}Na_{0.7}$ -LSX-450
Li(I′)	27.2	23.4	28.5	29.3
Li(II)	33.9	40.3	34.9	34.9
Li(III)	32.4	—	23.0	25.3
Li found	93.5	63.7	86.4	89.5
Li predicted	95.8	54.0	93.3	93.3
Ag(I′)	—	1.9	—	—
Ag(II′)	—	—	1.7	0.9
Ag(II)	—	2.0	—	—
Ag(II*)	—	1.7	—	0.9
Ag(III)	—	5.1	—	—
Ag(III′)	—	17.9	—	—
Ag Found	—	28.6	1.7	1.8
Ag Predicted	—	41.8	2.0	2.0

450 and 350 indicate the temperature in °C of treatment in vacuo; Hutson and Yang, 2000a, with permission.

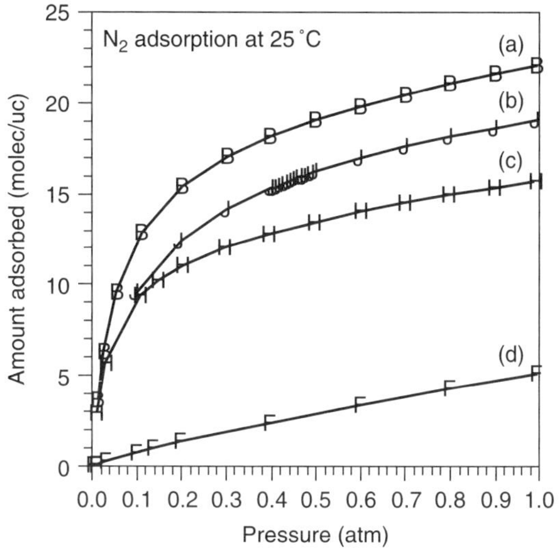

Figure 10.11. N_2 adsorption isotherms (in molecules/unit cell) at 25 °C, for (a) Ag-LSX-450 (treated at 450 °C in vacuo), (b) Ag-LSX-350 (treated at 350 °C in vacuo), (c) Ag-X, and (d) Ag-LSX treated at 100 °C (Hutson and Yang, 2000a; with permission).

cell is limited. As seen from Table 10.3, with 2 Ag per unit cell, 0.9 Ag was at site II*, whereas only 1.7 Ag were at this site in the sample with 41.8 Ag per unit cell.

Figure 10.11 shows the N_2 isotherms on Ag-LSX heat-treated at 350 and 450 °C. A significant increase in N_2 adsorption was seen as the heat-treatment temperature was raised from 350 to 450 °C. This increase in N_2 adsorption was caused by the Ag^+ at site II*, which interacted more strongly with N_2.

The N_2 isotherms on mixed Li_xAg_y-LSX zeolites dehydrated in vacuo at 450 °C are given in Figure 10.12. The results showed that the addition of increasing amounts of Ag resulted in a change in the general aspect of isotherm toward that of the nearly fully Ag^+-exchanged zeolite. For practical applications in VSA/PSA cycles, the relative linearity of the isotherms given by zeolites with 1-3 Ag^+ per unit cell is desirable.

The isosteric heats of adsorption of N_2 on almost fully Li and Ag exchanged LSX are shown in Figure 10.13, along with that of LSX with 1.1 Ag and 94.2 Li. The first N_2 molecule apparently adsorbed on Ag^+ at SII* with higher interactions than with Li^+ and Ag^+ at other sites. The results with mixed Li,Ag-LSX zeolites clearly pointed to the potential advantage of using the mixed form with only 1-3 Ag/unit cell for air separation. In Figure 10.14, the N_2/O_2 isotherms are compared for Li-LSX and that with about 1% Ag cation. The amount of N_2 adsorbed at 25 °C and 1 atm was significantly increased with the addition of 1.1 Ag/u.c. The increased amount was approximately 12%. The O_2 isotherms were both very low and hence did not significantly impact the VSA performance.

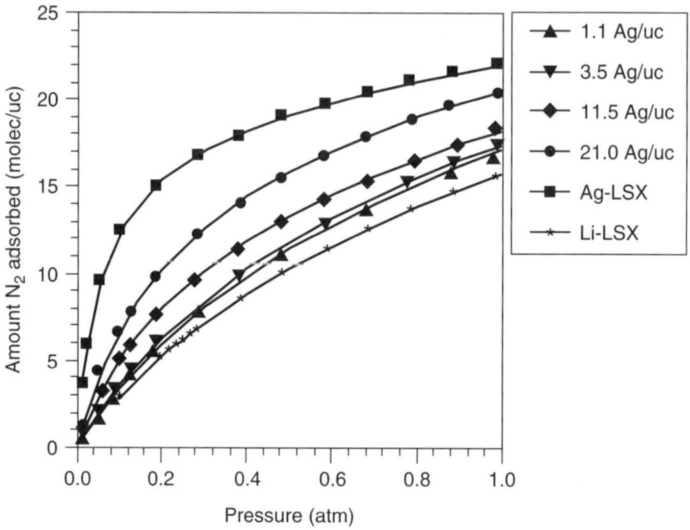

Figure 10.12. N_2 adsorption isotherm (in molecules/unit cell) at 25 °C, for Li_xAg_y-LSX zeolites dehydrated in vacuo at 450 °C. This shows that the addition of increasing amounts of Ag results in a change in the general aspect of isotherm toward that of the nearly fully Ag^+-exchanged material (Hutson and Yang, 2000a; with permission).

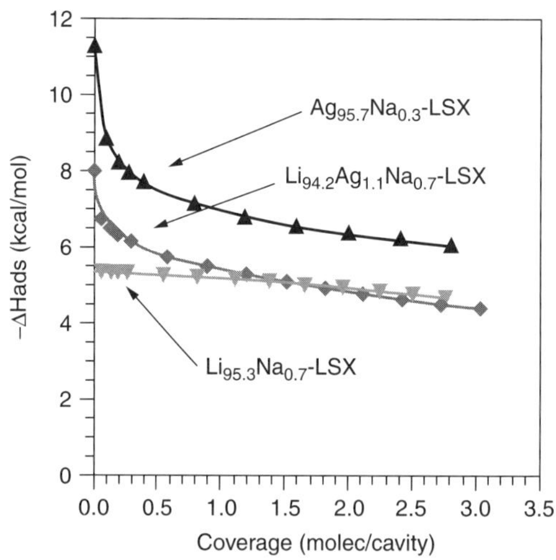

Figure 10.13. Isosteric heats of adsorption of N_2 on nearly fully Li/Ag exchanged LSX and LiLSX containing about 1% Ag (Hutson and Yang, 2000a; with permission).

Using the isotherm and heat of adsorption data, the VSA performance using these sorbents was evaluated by simulation (Hutson et al., 1999). The standard five-step PSA cycle used for air separation (see Chapter 3) was used in the simulation. The isotherms of the two sorbents to be compared are given in Figure 10.14.

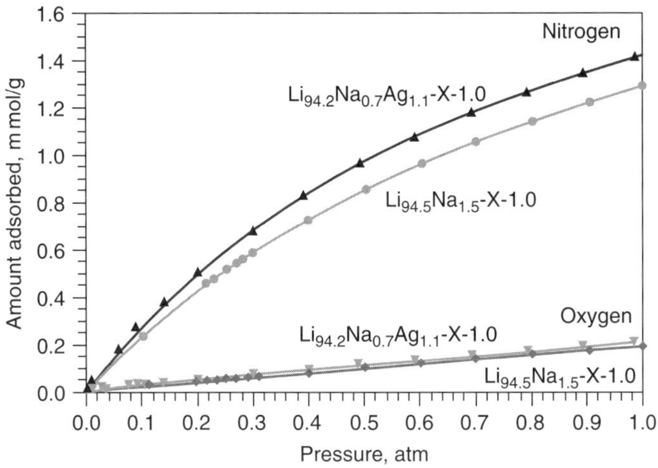

Figure 10.14. N_2 and O_2 isotherms for $Li_{94.2}Na_{0.7}Ag_{1.1}$-LSX dehydrated in vacuo at 450 °C and for $Li_{94.5}Na_{1.5}$-X-1.0 dehydrated in vacuo at 350 °C. All isotherms were measured at 25 °C (Hutson and Yang, 2000a, with permission).

Table 10.4. PSA/VSA simulation operating conditions and results for two sorbents

Sorbent	P_H (atm)	P_L (atm)	P_{CD} (atm)	U_H (m/s)	U_L (m/s)	O_2 Product Purity (%)	O_2 Product Recovery (%)	Product Throughput (kg O_2/h/kg Adsorbent)
Run 1								
Li_{96}-LSX	1.0	0.33	0.70	0.48	0.38	96.11	62.03	4.84×10^{-2}
$Li_{95}Ag_1$-LSX	1.0	0.33	0.69	0.60	0.42	96.42	62.74	5.40×10^{-2}
Run 2								
Li_{96}-LSX	1.2	0.4	0.70	0.40	0.38	90.68	78.02	6.31×10^{-2}
$Li_{95}Ag_1$-LSX	1.2	0.4	0.71	0.50	0.38	90.83	78.48	7.01×10^{-2}

P_H = high-pressure, P_L = low-pressure, P_{CD} = pressure of co-current depressurization, U_H = interstitial velocity in high-P step, U_L = interstitial velocity during desorption.
From Hutson et al., 1999.

In the simulation, the O_2 product purity and recovery were kept nearly the same. The adsorption and desorption pressures were fixed at typical values used in industry. The other cycle conditions were optimized. The results are summarized in Table 10.4. The performance of the given sorbent can be judged by the O_2 product throughput. In both runs shown in Table 10.4, the product throughput given by the sorbent containing 1.1 Ag/u.c. was higher by approximately 12%. The product throughput can be further increased by including pressure equalization steps (Yang, 1987), which were not used in the simulation. The simulation results for LiLSX are similar to VSA results obtained in industry.

Introducing only 1 Ag per unit cell can significantly improve the PSA separation. Cost estimates have been made that showed that a significant gain can be obtained by the use of Ag-containing Li-LSX zeolite.

10.1.5. Oxygen-Selective Sorbents

Since the N_2/O_2 ratio in air is approximately 4, much less work is needed to separate air by using an O_2-selective sorbent. This is indeed practiced in industry by kinetic separation (in a PSA cycle) using carbon molecular sieve (CMS) (Yang, 1987; Coe, 1995). CMS has been widely used for nitrogen generation by PSA (not VSA). 4A zeolite has also been used for nitrogen generation, mainly for small-volume fuel tank blanketing. Oxygen-binding cobalt complexes have been of long-standing interest and will be discussed here.

Carbon Molecular Sieves. Excellent carbon molecular sieve materials have been developed in industry. On these sorbents, the O_2, N_2, and Ar isotherms are approximately equal because they all adsorb by van der Waals interactions, and their polarizabilities are approximately the same. But the diffusivity ratio for O_2(Ar)/N_2 is approximately 30 due to differences in molecular size (Chen et al., 1994).

Details of the CMS preparation and its adsorption equilibrium and kinetics for air separation are given in Chapter 5 (5.7.2). PSA with CMS is widely used for N_2 production from air. The simple Skarstrom cycle (see Chapter 3) has been used, and the typical feed pressure is 8 atm, whereas the desorption pressure is near ambient. The maximum N_2 purity is 99.5% (Notaro et al., 1999). The adsorbent inventory is well below 500 lb per ton/day of N_2 capacity, and the largest existing single PSA unit is 100 tons/day for N_2 purity of 95% (Notaro et al., 1999). The power consumption is around 300 kWh per ton N_2. The low bed inventory as well as the low energy requirement while using the rather inefficient PSA cycle reflects the fact that air contains only 21% O_2, which is the adsorbed component.

4A Zeolite. NaA (4A) zeolite has also been used in small volume, enriched N_2 generation by PSA, for fuel tank blanketing of military aircrafts (particularly helicopters). A high-pressure air is readily available from the pressurized engine bleed or shaft-driven compressor air. The feed pressure is regulated to about 25 psig before feeding to the PSA unit. Fast PSA cycles, typically 2.5 s/cycle, are used. These on-board units generate a gas containing less than 12% O_2, than that which is the required limit for fuel tank blanketing. The N_2 enrichment is also accomplished by kinetic separation due to faster diffusion of O_2 in 4A zeolite.

Oxygen-Binding Cobalt Complexes. Since the 1940's, considerable work has been devoted to the study of oxygen-binding transition metal complexes, mostly in attempts to mimic important biological oxygen carriers, such as hemoglobin and myoglobin. Because of their similarity to the natural heme proteins, O_2-binding iron complexes have received the most attention. However, it is the oxygen-carrying complexes of cobalt that have proven the most promising as potential oxygen sorbents for air separation. Several thorough reviews are available on this subject (Jones et al., 1979; Niederhoffer et al., 1994; Li and Govind, 1994).

Pfeiffer described a compound of composition cobaltous bis-salicylaldehyde ethylenediamine that turned from a reddish color to black when exposed to air (Li and Govind, 1994). Tsumakei then showed that the blackening was due to adsorption of oxygen from the air (Li and Govind, 1994). He also showed that the sorption was reversible and that the oxygen could be driven off by heating in carbon dioxide. These results then stimulated a tremendous amount of work by Calvin and co-workers who synthesized and characterized a large number of cobalt chelates capable of binding oxygen (Calvin et al., 1946; Bailes et al., 1947; Calvin and Martell, 1952). Among these compounds were cobaltous bis-salicylaldehyde ethylenediamine (Co(salen) or salcomine) and cobaltous bis(3-fluoro-salicylaldehyde) ethylenediamine (Co(fluomine)). These compounds, shown in Figure 10.15, represent the most extensively studied of the oxygen-binding cobalt complexes (Calvin and Martell, 1952; Li and Govind, 1994). Co(salen) has reversible oxygen-binding capability, and there have been several attempts to use it to develop a system for oxygen production from air. The U.S. Air Force has attempted to develop the material for onboard oxygen support systems. Co(salen), however, is quickly deactivated by the presence of moisture. Many of the drawbacks of Co(salen) were reduced by using the compound Co(fluomine), which is stable in moisture and was studied extensively by the U.S. Air Force in the mid-1970's for potential use in breathing air systems for crews of military aircraft (Boscola, 1974; Adduci, 1975). As with the Co(salen) material, the commercialization of Co(fluomine) has been hampered by the long-term chemical instability of the complex.

Although the O_2-binding transition metal complexes have been studied extensively, the nature of their bond formation has long been the subject of some controversy. Vaska (1976) showed that almost all currently known transition metal dioxygen complexes can be divided into two types according to the characteristics of the dioxygen ligand. They are the (type I) superoxo (O_2^-) and the (type II) peroxo (O_2^{2-}) complexes. These complexes are further classified

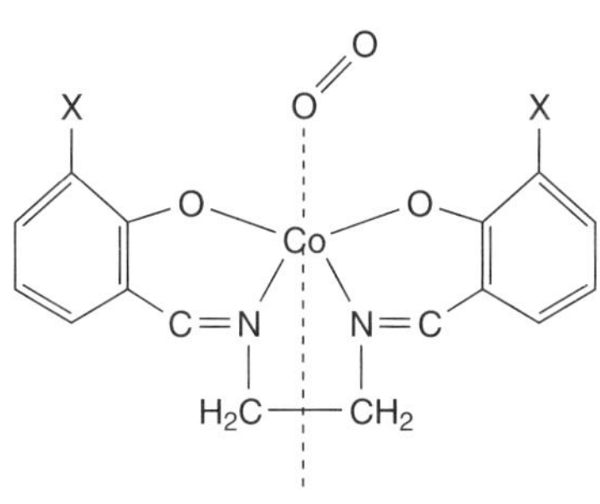

Figure 10.15. The Co(salen) [for X = H] and Co(fluomine) [for X = F] molecule.

according to whether the dioxygen is bound to one metal atom (type a), or bridges two metal atoms (type b).

Chen and Martell synthesized and characterized a large number of O_2-binding cobalt Schiff base complexes (Chen and Martell, 1987; 1989). Dzugan and Busch characterized new oxygen-binding macrocyclic cobalt complexes (Dzugan and Busch, 1990). Ramprasad et al. (1995) reported a group of solid-state crystalline cyanocobaltate complexes with reversible and very high oxygen-binding capacity. Specifically, lithium cyanocobaltate, $Li_3Co(CN)_5 \cdot 2DMF$, was obtained by calcination at 160 °C in N_2, as opposed to the same compound with 4DMF that was synthesized by others previously. The compound with 2DMF provided more voids and higher surface area, hence higher O_2 diffusion rates. The O_2 capacity was 2.2 mmol/g at 1 atm O_2 and 25 °C, and the isotherm was quite steep at pressures below 0.1 atm. O_2/N_2 cycling tests at 25 °C showed that the O_2 capacity declined steadily (linearly) to 85% of its initial value after 550 cycles (in 410 hr). Also, it deactivated rapidly in the presence of moisture. None of these materials, however, has shown the necessary combination of reversibility, capacity, and stability needed for use in industrial gas separations.

Recent research efforts have focused on steric hindrance as a means of protecting the oxygen-binding complex from oxidation and dimerization. The most promising approach has been to prepare the oxygen sorbent by entrapment or encapsulation as a solid-state metal complex within the cages of a synthetic zeolite (Lunsford, 1975; Howe and Lunsford, 1975; Imamura and Lunsford, 1985; Herron, 1986; Drago et al., 1988; Taylor et al., 1989; Taylor et al., 1992). None of these efforts, however, proved effective for separating oxygen from nitrogen due to instabilities and/or inadequate O_2-binding capacity.

Although numerous transition metal complexes with oxygen-binding ability have been reported, none of these materials has achieved commercial success as a sorbent for air separation. All these materials have suffered from one or more of the following drawbacks that have prevented commercialization: (1) chemical instability, (2) unacceptable adsorption characteristics, and/or (3) unacceptable cost.

An interesting idea that has not been pursued is to immobilize the oxygen-binding complexes on a solid support. In Wang's classic experiment (Wang, 1970), a heme diethyl ester was embedded in a matrix of polystyrene and 1-(2-phenylethyl)imidazole. The matrix not only prevented close approach of two heme but also provided a hydrophobic environment. Reversible oxygen uptake was observed (Wang, 1970).

More recently, Hutson and Yang (2000b) synthesized complexes of known oxygen-binding ability by using a modified synthesis technique that resulted in complexes that were attached (immobilized) to the surface of several porous substrates. Co(salen) and Co(fluomine) were the complexes used. The O_2-binding capacity and the stability of these resulting sorbents were then characterized. The modified procedure involved two steps. The first was to bond Co^{2+} on anion sites of a substrate, by ion exchange, to form a stable ionically bonded Co^{2+}. This was followed by attaching ligands coordinatively to Co^{2+} in order to give it the oxygen-binding ability. The rationale for this approach is given first and some of the results follow.

A basic understanding of the atomic orbitals will help in the understanding of the stability problem. Comparing the ionic salts of Co, Co^{2+} is more stable than Co^{3+} since the outer-shell orbitals for Co^{2+} are $3d^7$, $4s^0$ whereas that of Co^{3+} are $3d^6$, $4s^0$. However, the orbital occupations differ when ligands are attached to the carbon to form a coordination complex. In order for Co^{2+} to coordinate six ligands, six electrons are in three of the $3d$ orbitals, to evacuate two 3d orbitals to hybridize with one 4s and three 4p orbitals to form six d^2sp^3 hybridized orbitals. These six hybridized orbitals form coordination bonds with six ligands. The seventh electron in the $3d$ orbitals in Co^{2+} is excited to the $5s$ orbital, the next higher available orbital. Consequently, this electron can be easily lost, which is the reason that Co^{3+} is more stable than Co^{2+} in the coordination complex.

With this understanding, Hutson and Yang (2000b) first ion-exchanged Co^{2+} on anion sites of a substrate (i.e., a cation exchanger) to form stable Co^{2+}, and subsequently attached ligands (usually four) to Co^{2+} in order to give Co^{2+} the O_2 binding ability. Three different substrates were used: LSX zeolites, mesoporous MCM-41, and ion-exchange resin.

The adsorption/desorption isotherms of O_2 on Co(salen) are shown in Figure 10.16. These isotherms display a very noticeable and interesting hysteresis. The adsorption isotherm shows that very little O_2 is adsorbed until the O_2 pressure reaches a "threshold" at approximately 0.2 atm. The adsorption isotherm then sharply rises to nearly the full O_2-binding capacity of the complex. A very low pressure was then required to release the bound oxygen. The

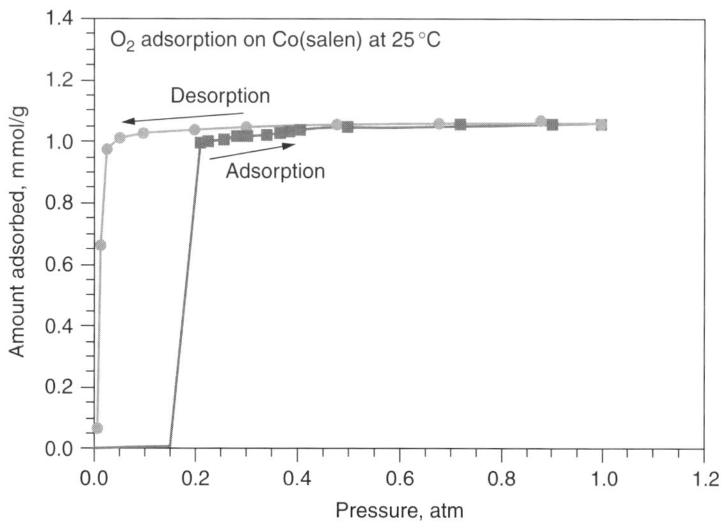

Figure 10.16. Oxygen adsorption and desorption isotherms, measured at 25 °C for Co(salen) (Hutson and Yang, 2000b, with permission).

O_2-binding capacity for the Co(salen) material was 1.06 mmol/g, approximately 70% of the theoretical value of 1.55 mmol/g based on a 1 : 2 (O_2 : Co^{2+}) adduct.

The O_2 adsorption and desorption isotherms, measured at 25 °C, for Co(fluomine) are shown in Figure 10.17. These curves are somewhat similar to those of the Co(salen), but do not have a similar hysteresis. The material strongly binds O_2 at very low pressures, immediately adsorbing the full capacity at pressures <0.1 atm. However, at 25 °C, the O_2 could not be desorbed even at very low O_2 pressures. The O_2-binding capacity for the Co(fluomine) material was 1.13 mmol/g, approximately 80% of the theoretical value of 1.39 mmol/g based on a 1 : 2 (O_2 : Co^{2+}) adduct.

The O_2 and N_2 adsorption and desorption isotherms for Co(fluomine)-MCM-41 are shown in Figure 10.18. The MCM-41 used in this study contained a Si/Al ≈ 5. These curves differ significantly from those of the free Co(fluomine) with nearly complete reversibility and much lower capacity (0.107 mmol/g at 1 atm). As with the free Co(fluomine), O_2 was strongly bound at very low pressures. However, unlike Co(fluomine), it did not immediately attain the full capacity, but rather maintained a positive slope through 1 atm. The positive slope yields a sorbent working capacity and is useful for application. The theoretical capacity for this material is 0.49 mmol/g for a 1 : 1 (O_2:Co^{2+}) and 0.25 mmol/g for a 1 : 2 (O_2:Co^{2+}) adduct. These values assume 3.6 molecules of Co(fluomine) per unit cell of the MCM-41. The actual capacity of this material at 1 atm oxygen pressure and 25 °C is less than the theoretical amount for both the 1 : 2 adduct and the 1 : 1 adduct. Although the pore structure of the MCM-41 is quite open, it may not be possible for a Co(fluomine) molecule to attach at every ion-exchange site. Also, significantly more oxygen may be bound at higher pressures (than 1 atm). The

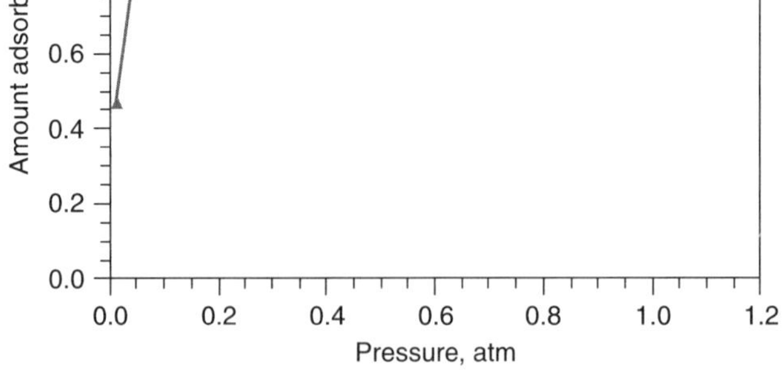

Figure 10.17. Oxygen adsorption and desorption isotherms, measured at 25 °C for Co(fluomine) (Hutson and Yang, 2000b, with permission).

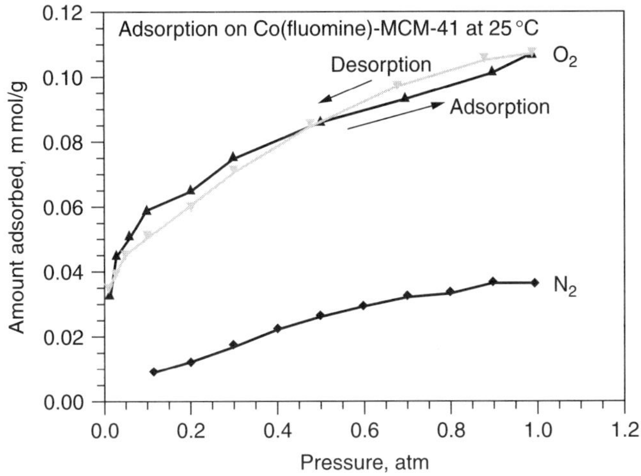

Figure 10.18. Oxygen and nitrogen adsorption isotherms, measured at 25 °C, for CoFluomine-MCM-41 (Hutson and Yang, 2000b, with permission).

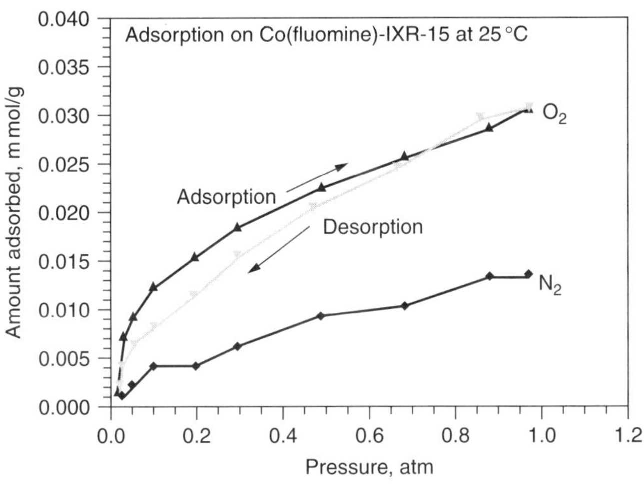

Figure 10.19. Oxygen and nitrogen adsorption isotherms, measured at 25 °C, for Co(fluomine) on ion exchange resin amberlyst-15 (Hutson and Yang, 2000b, with permission).

O_2-binding of this material was much more reversible than the neat Co(fluomine) complex. This may be an indication of the formation of 1 : 1 adducts since the free Co(fluomine) forms 1 : 2 adducts, which is very difficult to desorb.

The O_2 and N_2 adsorption and desorption isotherms, measured at 25 °C, for Co(fluomine)-IXR (ion exchange resin) are shown in Figure 10.19. These curves closely resemble those of the Co(fluomine)-MCM-41 with nearly complete reversibility but with much lower capacity (0.03 mmol/g at 1 atm). As with all other materials, O_2 was strongly bound at very low pressures. However, as with all of the other immobilized samples, it does not immediately attain the full capacity, but instead has a positive slope through 1 atm. The N_2 adsorption isotherm was also measured for this Co(fluomine)-IXR material and is shown in Figure 10.19. The theoretical capacity for this material is about 1.2 mmol/g for a 1 : 1 (O_2:Co^{2+}) and 0.6 mmol/g for a 1 : 2 (O_2:Co^{2+}) adduct. These values assume 100% ion-exchange of Co^{2+} with the ion-exchange resin (with a cation exchange capacity of 4.7 meq/g) and subsequent conversion to Co(fluomine). The actual capacity of this material, at 0.03 mmol/g at 25 °C and 1 atm, is considerably lower than the theoretical amount.

The immobilized samples were found to be very stable when exposed to atmospheric moisture. There was no noticeable decrease in the binding capacity after exposure to atmospheric oxygen and moisture. In order to test the stability to repeated oxygenation and deoxygenation, the samples were subjected to an aggressive stability test. Each of the samples was subjected to oxygenation and deoxygenation cycles in a thermogravimetric analyzer (TGA). The samples were first activated by heating to 120 °C in a pure Ar atmosphere. The samples were then alternatively exposed to 1 atm O_2 and 1 atm Ar at 60 °C for 15 min each. After 90 cycles, the samples were removed from the TGA and reactivated at 120 °C in a vacuum. The oxygen adsorption and desorption isotherms were then measured, and the sample was then returned to the TGA for continued cycling. The results are summarized in Figure 10.20. In each free and immobilized Co(fluomine) synthesis, piperidine was added as the last step. This was done in order to follow previously reported procedures (Bailes and Calvin, 1947), although samples prepared with and without this step did not show any difference (Hutson and Yang, 2000b). The results of the accelerated test for stability showed that the immobilized samples still lacked stability. However, they seemed to stabilize at about 40% of their initial capacities.

Attention has been given to the possibility of air separation based on magnetic properties. Oxygen is paramagnetic, while nitrogen and argon are not. Hence oxygen might be repelled magnetically from a diamagnetic superconductor, that is, the Meissner effect. Makarshin et al. (1997) reported that the BET surface area of a superconducting material measured from O_2 adsorption was about 6 times lower than that from N_2. Gordon and Cussler (1999a) measured the isotherms of N_2 and O_2 on two superconducting sorbents and compared them with those on graphite and SiO_2. No noticeable difference in the isotherms was observed. They also examined the possibility of air separation by using a membrane fabricated from a superconducting material (Gordon and Cussler, 1999b). The results of these experiments were complicated by the condensation of both O_2 and moisture, which explained earlier results as well. The conclusion was that the possibility of air separation with superconducting materials was not promising. A related study was performed by Ozeki et al. (2000), who measured the adsorption of O_2 on sorbents with external magnetic fields (from 1 T to 6 T). Like the experiment of Gordon and Cussler, the isotherms were measured at 77 K, below the onset of superconductivity of $YBa_2Cu_3O_{7-x}$. Interestingly, with 1 : 1 mixture of

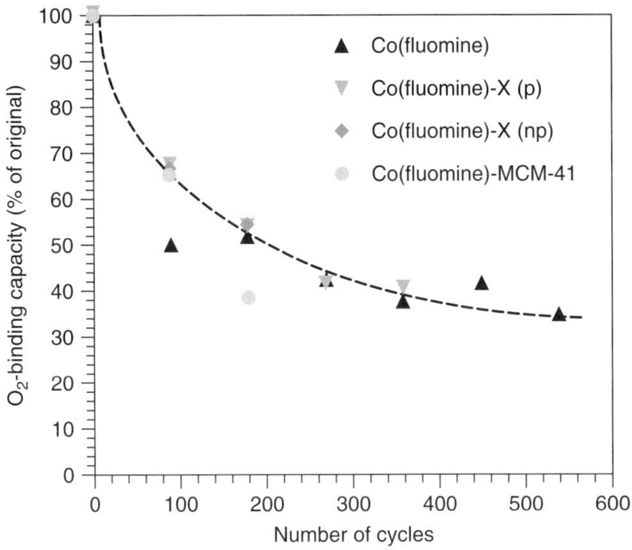

Figure 10.20. Stability curves for various O_2-binding materials after oxygenation and deoxygenation cycling. The materials were each cycled for 15 min in 1 atmosphere O_2 and in Ar at 60 °C (to accelerate the decay). The materials are Co(fluomine), Co(fluomine)-LSX synthesized with the use of piperidine (p), Co(fluomine)-LSX synthesized without the use of piperidine (np), and Co(fluomine)-MCM-41 (Hutson and Yang, 2000b, with permission).

5A zeolite and $YBa_2Cu_3O_{7-x}$, O_2 desorbed when the magnetic field was turned on, and adsorbed when it was turned off.

10.2. HYDROGEN PURIFICATION

Hydrogen purification is a necessary step in the production of hydrogen, which at present is done by catalytic steam reforming of natural gas, naphtha, or refinery gases. When hydrogen is the only desired product, the effluent from the steam reformer is processed further through a water–gas shift reactor to maximize the hydrogen content. Hydrogen purification is also performed to recover hydrogen from diverse petroleum-refinery and petrochemical streams, from coke ovens, and chemical plants such as ethylene plant effluent gas and ammonia plant purge gas. The typical hydrogen concentrations in these mixtures range from 50 to 99%. The remainder contains: N_2, CO, CH_4, CO_2, water, Ar, C_2 and higher hydrocarbons, sulfur containing compounds, and ammonia. The available pressures are 10 to 60 atm, which are well-suited for PSA processing. The required product purity is generally in the range of 99.99%–99.9999+%.

Hydrogen purification was the first large-scale application of PSA technology. The first commercial PSA hydrogen purification unit was installed in conjunction with a steam reformer, in Toronto around 1966. The standard five-step PSA cycle (with a co-current depressurization step) is used. Three or more pressure equalization steps are employed to maximize the product recovery. A review of the technology development prior to 1986 was given in Yang (1987).

Because of the wide range of adsorptive properties of the gas molecules in the feed, it was recognized from the early development that more than one sorbent was needed for the separation. Hence, layered beds were used from the beginning. Typically, the first layer (at the feed end) is activated carbon, which is followed by a zeolite (e.g., 5A). The reasoning for such layering becomes obvious from the equilibrium isotherms, shown in Figures 10.21 and 10.22. The most strongly adsorbed components are adsorbed in the activated carbon bed, while the other components are separated in the zeolite bed.

With activated carbon along, it is difficult to achieve a very high product purity with a reasonably high product recovery (Yang and Doong, 1985; Doong and Yang, 1986). With zeolite (5A), however, 99.9999% purity can be easily achieved with a high recovery (Doong and Yang, 1987). It was also shown by Doong and Yang (1987) that it was possible to predict product purity to the ppm level by standard numerical simulation. As in other PSA systems, a guard bed of activated alumina or silica gel (Yang, 1987) is normally used for water, sulfur-containing compounds and C_2+ hydrocarbons.

Numerous studies have been undertaken on the use of layered beds consisting of different sorbents for cyclic adsorption/ion exchange (Klein and Vermeulen, 1975; Frey, 1983; Wankat and Tondeur, 1985; Chlendi and Tondeur, 1995; Watson et al., 1996; Pigorini and LeVan, 1997). For hydrogen purification using layered activated carbon and zeolite, Chlendi and Tondeur (1995) used the

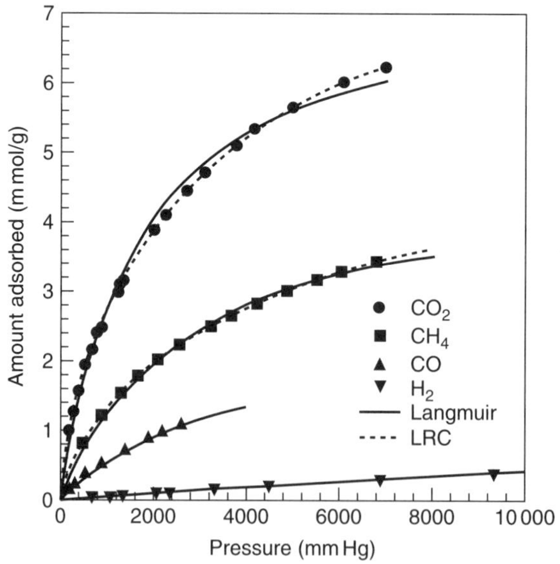

Figure 10.21. Equilibrium isotherms on activated carbon (Calgon BPL) at 25 °C (Park et al., 1998, with permission).

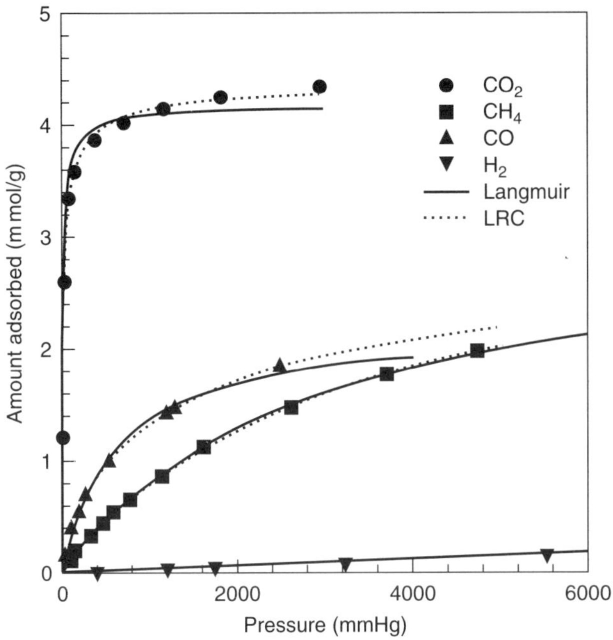

Figure 10.22. Equilibrium isotherms on 5A zeolite (pellets) at 25 °C (Park et al., 1998, with permission).

pseudo-characteristics to illustrate the interesting "refraction" phenomena of the different concentration wavefronts as they cross the interface of two layers. The basic principle of layered beds is to use a weak sorbent at the inlet, followed by stronger sorbents, in order to maximize the use of each sorbent and hence minimize the adsorber size. Optimization is needed to determine the ratio of the two beds (Chlendi and Tondeur, 1995; Park et al., 1998; Lee et al., 1999; Jee et al., 2001). The optimal design of layered beds can be made with the aid of a PSA model, based on the criterion that simultaneous breakthrough takes place in the two layers, that is, for the strong component (CO_2) in the weaker sorbent (carbon) and the weaker component (CO or CH_4) in the strong sorbent (zeolite). The optimal layering for a given gas-solid system depends on the feed composition and feed velocity (Park et al., 1998).

For a typical effluent from a water–gas shift reactor, the CO_2 content is in the 20–25% range. Thus, about 75% of the bed is activated carbon. The remaining bed of 5A zeolite adsorbs mainly CO and CH_4.

10.3. HYDROGEN STORAGE

Carbon nanotubes are the sorbents that are currently receiving the most attention. Hydrogen storage in carbon nanotubes is a rapidly evolving area, and also a subject of considerable controversy. For these reasons, it is only possible to present the current understanding on this subject. An attempt will be made to provide a fair assessment of its prospect. In addition to carbon nanotubes, metal hydrides and other sorbents are also included.

The four general approaches for on-board vehicle applications (compressed hydrogen gas, liquefied hydrogen, metal hydrides, and carbon sorbents) have been discussed in detail elsewhere and will not be repeated here (Department of Energy, 1999; Hynek et al., 1997; Schlapbach and Züttel, 2001).

10.3.1. Metal Hydrides

Extensive research on metal hydrides and their role in hydrogen storage have been conducted over 30 years (Alefeld and Volkl, 1978; Schlapbach, 1988; Sastri et al., 1998; Zaluska et al., 2001). Myriad metal hydrides have been studied, yet none are practical or commercially successful. However, some promising approaches are on the horizon and remain the subject of active research.

The most important basic properties expected in metal hydrides for hydrogen storage are (1) high hydrogen content, (2) fast hydrogen charging and discharging rates, and (3) moderate temperature for hydrogen desorption (discharge), preferably below 100 °C. The desorption rates are generally slower than the absorption rates, and are associated with hysteresis.

In addition to these basic requirements, metal hydrides are prone to the following common problems: (1) deactivation or poisoning by contaminants in hydrogen, such as CO, H_2O, O_2, CO_2, and H_2S; (2) deterioration upon absorption/desorption cycling — both capacity and rates decay upon cycling.

In order to discharge hydrogen at a moderately low temperature and to have acceptable rates (i.e., diffusion rates), the H-bond energy with the metal atoms must be low. Thus, the enthalpy of formation of the hydride should be below ~12 kcal/mol. The metals that are known to form hydrides all have high enthalpies of formation. This problem was circumvented by using *intermetallic* compounds, from the early work of Libowitz et al. (1958) and Reilly and Wiswall (1968). The intermetallic compounds are formed by alloying the hydride-forming metals with other metals (usually transition metals, Fe, Ni, Co, Cr, etc.) that do not form hydrides. This second metal greatly lowers the enthalpy of formation while retaining the hydride-forming capacity of the first metal.

Over the past three decades, a great number of binary and ternary alloys were developed for hydrogen storage. They are designated AB_x or A_xB type, where A denotes the hydride-forming metal and B denotes the non-hydride-forming metal. There are four classes of alloys that have been developed: (1) AB-type based on Ti as the hydride forming metal, (2) AB_5-type based on rare earth metals as the hydride forming metals, (3), A_2B-type based on Mg as the hydride forming metal, and (4) AB_2-alloys using Zr as the hydride forming metal (Sastri et al., 1998).

The thermodynamic equilibria of H_2/metal systems are described by pressure-composition isotherms, shown as an example in Figure 10.23 for $LaNi_5$ (Schlapbach and Züttel, 2001). At a constant temperature, at low pressures, the metal

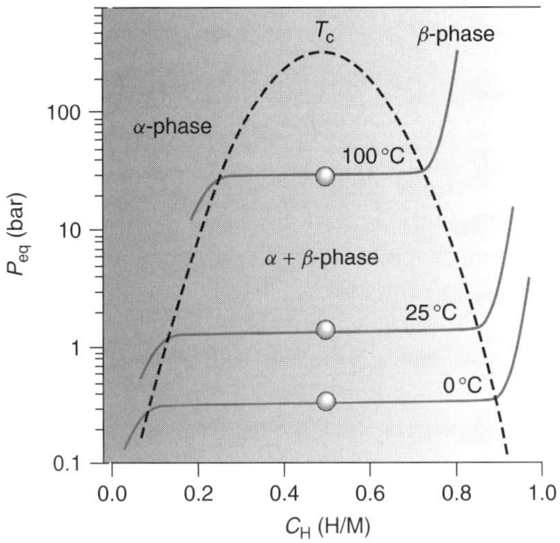

Figure 10.23. Pressure-concentration-temperature equilibrium curves for H_2/$LaNi_5$, forming $LaNi_5H_6$ at 25 °C (Schlapbach and Zuttel, 2001, with permission).

Table 10.5. Intermetallic compounds and their hydrogen storage properties

Type	Metal	Hydride	wt %	P_{eq}, T
Elemental	Pd	$PdH_{0.6}$	0.56	0.02 bar, 298 K
AB	FeTi	$FeTiH_2$	1.89	5 bar, 303 K
AB_5	$LaNi_5$	$LaNi_5H_6$	1.37	2 bar, 298 K
A_2B	Mg_2Ni	Mg_2NiH_4	3.59	1 bar, 555 K
AB_2	ZrV_2	$ZrV_2H_{5.5}$	3.01	10^{-8} bar, 323 K
Body-centered cubic	TiV_2	TiV_2H_4	2.6	10 bar, 313 K
AB	NaAl	$NaAlH_4$	8.0	90 bar, 403 K

Data taken from Schlapbach and Zuttel, 2001, except NaAl, from Zaluska et al., 2001.

dissolves some hydrogen-forming solid solution (or α phase). As the hydrogen pressure is increased, nucleation and growth of the hydride phase (or β phase) takes place, and coexists with the α phase. At still higher pressures, little increase in hydrogen concentration can occur. The width of the plateau on the phase diagram indicates the amount of hydrogen storage. From the isotherms, a van't Hoff plot would yield the enthalpy of formation, or −7.2 kcal/mol H_2 for $LaNi_5$. The plateau pressure at room temperature is near 2 atm and hence is convenient for application. The best known binary alloys are listed in Table 10.5. The mechanism of H_2 absorption starts from dissociation into H atoms, followed by diffusion into the metal lattice. In desorption, the process is revered, and 2 H atoms combine to form H_2.

It became clear by 1990 that metal hydrides would not be practical for hydrogen storage without new breakthroughs (Zaluska et al., 2001). The developments in the past decade are mainly on sodium alanates (complex hydrides of sodium and aluminum), the use of catalyst (such as Pd), and ball-milling the samples.

The thermodynamic equilibrium of the alanate hydride at 403 K is given in Table 10.5. However, without adding a catalyst, the system cannot be rehydrogenated. Bogdanovic and Schwickardi (1997) showed that by wet impregnation of a Pd catalyst, the system could be made reversible, and the kinetics also improved. Zidan et al. (1999) showed that "dry" addition of the catalyst was also effective. With these improvements, the hydrogen storage capacity was still well below the thermodynamic limits (Table 10.5).

Nano-crystalline metal particles have been developed more recently (Gleiter, 2000). The grain size is generally below 100 nm, and the range of 5–50 nm is more typical. These materials have better properties for hydrogen storage than their conventional counterparts.

Pd catalyst has also been used for other metal hydrides, such as Mg_2Ni and FeTi (Zaluska et al., 2001). Without the catalyst, an "activation" process is generally needed before hydrogen absorption, for example, FeTi needs to be activated by a series of long-term annealings at temperatures >400 °C. Moreover, if the material is subsequently exposed to air or water vapor, the activation process has to be repeated. With the addition of Pd nanoparticles, the FeTi nanoparticles could absorb hydrogen without activation.

The rates of absorption of hydrogen in all compounds listed in Table 10.5 are slow, that is, in the order of tens of minutes for completion in nanoparticles. For example, for absorption in nanoparticles of $LaNi_5$ at 298 K, 60 min for near completion is required. Approximately 100 min is necessary for Mg_2Ni at 200 °C (Zaluska et al., 2001). Desorption is generally slower than absorption, and the slow rates are detrimental to applications. The kinetics can be improved the use of a catalyst. Figure 10.24 shows the effects of Pd on the hydrogen uptake rates on $LaNi_5$. It has been found that ball-milling the samples could reduce the grain size of the crystals and consequently accelerates the rates for both absorption and desorption by an order of magnitude (Zaluska et al., 2001). The effects of ball-milling on the rates for Mg_2Ni are shown in Figure 10.25. The smaller grain size reduced the diffusion distance for H atoms. The effects of ball-milling have been studied on a number of metal hydrides; however, they are not limited on the grain size reduction alone. The effects of ball-milling are not understood, although mechanochemistry is clearly involved. Titanium has been found to be an effective catalyst for $NaAlH_4$ and Na_3AlH_6 in both dehydriding and hydriding kinetics, although at the expense of H-capacity (Sandrock et al., 2002). Ritter and coworkers have shown evidence of TiAl alloy formation which is possibly responsible for the catalytic effects (Riggleman et al., 2002).

10.3.2. Carbon Nanotubes

The U.S. department of Energy has set 6.5% (wt.) and 62 kg H_2/m^3 as the targets for on-board hydrogen storage in fuel cell applications in vehicles (U.S. DOE, 1998; Hynek et al., 1997), at ambient temperature. The pressure is not specified,

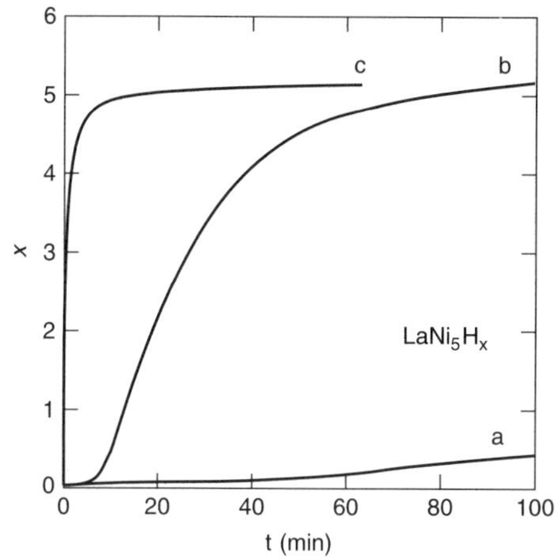

Figure 10.24. Rates of hydrogen absorption by $LaNi_5$ at 298 K for (a) polycrystalline $LaNi_5$, (b) nano-crystalline $LaNi_5$, and (c) nanocrystalline $LaNi_5$ with Pd nanocrystalline catalyst attached (Zaluska et al., 2001, with permission).

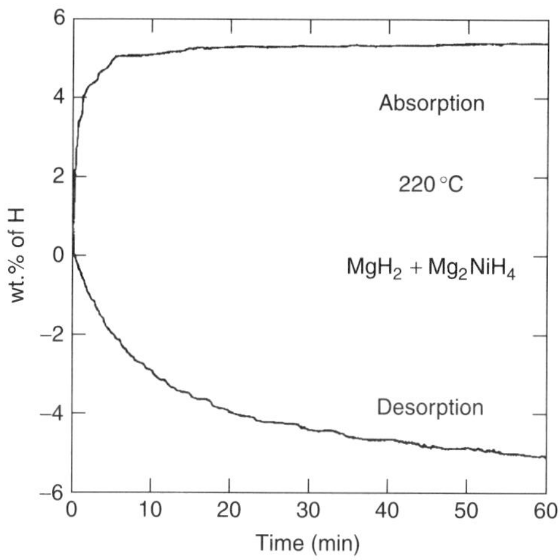

Figure 10.25. Rates of absorption and desorption at 220 °C for a mixture of hydrides (65% MgH_2 and 35% Mg_2NiH_4) after ball-milling in the hydrogenated state (Zaluska et al., 2001, with permission).

but 100 atm has been a nominal pressure for research. As a reference, for a compact passenger vehicle powered by fuel cell, 4 kg H_2 is needed for a driving range of 400 km.

The history, syntheses, and general adsorption properties of carbon nanotubes that have been reported are given in Chapter 9. A number of reviews on hydrogen storage in carbon nanotubes have appeared (Dresselhaus et al., 1999; Cheng et al., 2001; Ding et al., 2001; Dillon and Heben, 2001; Darkrim et al., 2002; Simonyan and Johnson, 2002).

Experimental Techniques and Pitfalls. The volumetric technique is the most commonly used method for measuring high-pressure isotherms (Yang, 1987). The apparatus involves a sample cell and a reservoir section. A pressure drop is measured when the pressurized reservoir is connected to the evacuated sample cell. The dead volumes of both compartments are measured by helium displacement using ideal gas law. Any additional pressure drop over that of He is attributed to adsorption. The amount adsorbed can be obtained from the dead volumes and a P-V-T relationship. Desorption can be measured by reversing the process.

This technique is prone to artifacts, particularly for H_2. Leakage through fittings and valves is a particular problem for H_2 because of its small molecular size. A number of pitfalls were discussed by Tibbetts et al. (2001). In a typical high-pressure apparatus (at 100 bar), the leakage of a few psi is equivalent to 1 wt % adsorption. A second source of error is temperature variation, by either ambient temperature fluctuation or heat of adsorption. In a typical setup, an artifact of 2.6 wt % adsorption can be caused by a 1 °C temperature rise (Tibbetts et al., 2001). Another source of artifact stems from the thermodynamic principle that a chamber being pressurized experiences a temperature rise. The temperature drop upon returning to the initial temperature can cause an artifact in adsorption. These errors could be associated with several reported results (Tibbetts et al., 2001). A similar temperature effect could be caused by the exothermic heat of adsorption (Darkrim et al., 2002). Here the temperature rises quickly, followed by a slow return to the ambient temperature (and the associated pressure drop), which could be erroneously taken as adsorption. Using a low value for the heat of adsorption for H_2, the temperature rise due to adsorption would amount to a couple degrees.

The gravimetric technique is relatively trouble-free because it measures the weight changes. However, because of the high sensitivities, careful calibration is necessary for changes in buoyancy and friction loss from gas flow upon changes of gas composition and temperature. High-pressure gravimetric (TGA) systems are available, but costly. With flow systems, contaminants in the supply gas could lead to artifacts because they could accumulate on the sorbent. This would not be a problem with static or volumetric systems due to the finite amount of gas that is in contact with the sorbent.

Temperature-programmed desorption (TPD) was used by Dillon et al. (1997) and Hirscher et al. (2001). This technique is also relatively trouble free, although its accuracy depends on the detector. Mass spectrometry was used in the work cited above.

Activated Carbon. Among the commercially available sorbents, activated carbon adsorbs the largest amounts of hydrogen. Activated carbon has been considered for on-board applications (Schwartz, 1993; Chahine and Bose, 1994; Hynek et al., 1997; Dillon and Heben, 2001). The isotherms of hydrogen on a typical activated carbon are shown in Figure 10.26. The low-temperature isotherms on a super-activated carbon are shown in Figure 10.27 (Zhou and Zhou, 1998). These amounts ($\sim$0.5 wt % at 100 atm and ambient temperature) are clearly not useful for meeting the DOE targets for on-board storage.

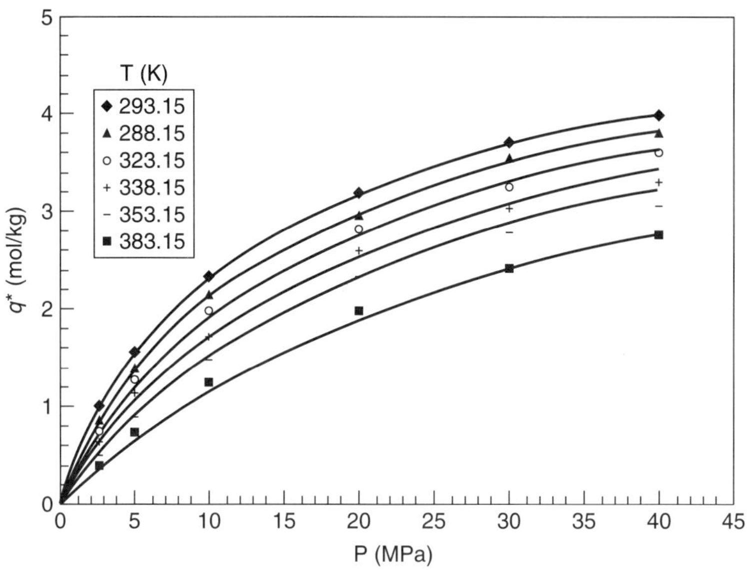

Figure 10.26. Hydrogen adsorption isotherms on activated carbon, CECA AC35 (BET surface area ~1000 m^2/g, from Lamari et al., 2000, with permission).

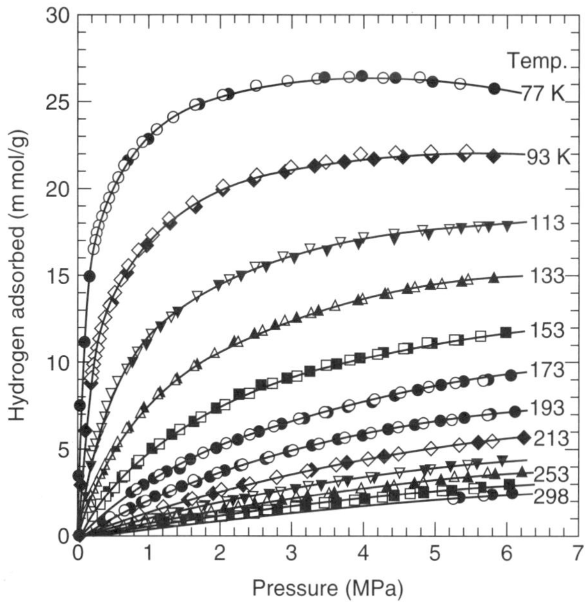

Figure 10.27. Hydrogen adsorption/desorption isotherms on super-activated carbon Anderson AX-21 with a BET surface area = 2800 m^2/g (see 10.4 on this carbon), solid symbols = adsorption, open symbols = desorption (from Zhou and Zhou, 1998, with permission). (BET surface area from Benard and Chahine, 2001).

The isotherm at 77 K exhibits a peak. This occurs when the absolute adsorbed density approaches saturation and the bulk density of the supercritical fluid begins to rise faster with increasing pressure (Menon, 1968; Benard and Chahine, 2001).

Single-Wall Carbon Nanotubes. The first report on hydrogen storage in SWNTs was by Dillon et al. (1997). A summary of the subsequent reports on experimental data on hydrogen storage in SWNTs, MWNTs and GNF is given in Table 10.6.

The storage capacity of Dillon et al. (1997) was estimated from TPD experiment. A small sample (of ~1 mg) of soot from arc discharge of Co/graphite was the sorbent. The sample was placed in a Pt foil and exposed to 300 torr H_2 for 10 min at 273 K, followed by 3 min at 133 K (in H_2). The sample was then cooled to 90 K while being evacuated prior to TPD. The evolved H_2 was detected by a mass spectrometer. For SWNT soot samples pretreated at 970 K in vacuum, broad H_2 desorption peaks were detected in the temperature range from 170 to 450 K, with peaks near 300 K. Using TPD curves from different heating rates, the activation energy for desorption (assuming first-order desorption kinetics) was calculated at 19.6 kJ/mol. From TEM images of the soot, the content of SWNTs was estimated as 0.1–0.2%. From these data, an estimate for H_2 storage capacity of 5–10% was obtained. The possible chemisorption of hydrogen by Pt was ruled out by calibration runs. The soot samples contained approximately 20% Co nanoparticles. The possible chemisorption on Co was also ruled out (Dillon et al., 1997; Dillon et al., 1999).

Ye et al. (1999) measured the isotherms of SWNTs (purified samples produced by laser-vaporization) by using volumetric technique. The samples were pretreated by heating in vacuo at 220 °C for 1 h. The highest adsorption was obtained on purified crystalline ropes, that is, 8 wt % at ~40 atm and 80 K. However, in contrast to the results of Dillon et al. (1997), high adsorption capacities were not observed at 300 K and pressures below 1 atm.

High adsorption capacities on SWNTs at 298 K were reported by Liu et al. (1999). Volumetric technique was used. The samples contained ~50–60% SWNTs with diameters near 1.8 nm. Mixed Ni/Co/Fe was used as the catalyst to produce the SWNTs by arc discharge. The samples were pretreated at 773 K in vacuo for 2 h. The uptake was very slow (~70% completion in 1 h). The highest capacity was 4.2 wt % at 10 MPa and 298 K. Approximately 80% of the adsorbed hydrogen could be desorbed at room temperature.

The SWNTs produced by laser-vaporization and arc discharge are usually capped by fullerene-like structures. Those from catalytic decomposition are usually capped by metal particles. Ye et al. (1999) and Dillon et al. (2000) both used ultrasonication to open the tubes. In the work of Dillon et al. (2000), it was reported that by using the ultrasonic probe of Ti-alloy (with 9 wt % Ti, 6 wt % Al, and 4 wt % V), the alloy ($TiAl_{0.1}V_{0.04}$) was incorporated into the SWNTs as contamination. The maximum adsorption capacity was ~7 wt %, and upon TPD, two desorption peaks occurred: approximately 2.5% evolved at 300 K, while the remainder evolved between 475–850 K. It was suspected that the alloy contaminant acted as a catalyst that stimulated the adsorption and desorption. This

Table 10.6. Reported hydrogen storage in single-wall nanotubes (SWNT), multiwall nanotubes (MWNT), and graphite nanofibers (GNF)

Material	H_2 wt%	T(K), P(MPa)	Notes	Ref.
SWNT	~5–10	133, 0.04	by TPD, see text	Dillon et al., 1997
MWNT	11.26	300, 9	See text	Chambers et al., 1998
GNF	67.55	300, 12	See text	Chambers et al., 1998
GNF	35	300, 11	Treated @ 1000 °C, in Ar	Park et al., 1999
Li/MWNT	20	473–673, 0.1	Flow system (TGA)	Chen et al., 1999
K/MWNT	14	300, 0.1	Flow system (TGA)	Chen et al., 1999
Li/MWNT	2.5	473–673, 0.1	Flow system (TGA), gas pre-dried	Yang, 2000
K/MWNT	1.8	300, 0.1	Flow system (TGA), gas pre-dried	Yang, 2000
K/MWNT	1.3	300, 0.1	Flow system (TGA), gas pre-dried	Pinkerton et al., 2000
SWNT	~8	80, 11.2	Treated at 220 °C in vacuo	Ye et al., 1999
SWNT	4.2	300, 10.1	Treated at 500 °C in vacuo	Liu et al., 1999
MWNT	3.98	300, 10	Treated in Ar at 2200 °C (2 h)	Li et al., 2001
GNF	<0.7	296, 11	Untreated or treated in H_2 at 500 °C	Tibbetts et al., 2001
GNF	0.7	295, 10.5	Treated at 1300 K in He flow for 1 h.	Poirier et al., 2001
SWNT	1.5	300, 0.08	Ti-contaminated during sonication	Hirscher et al., 2001
SWNT	0.05	296, 3.59	Treated @ 500 °C, H_2	Tibbetts et al., 2001
GNF	~15	300, 12	Pretreated at 150 °C in vacuo	Gupta and Srivastava, 2001
GNF	6.5	300, 12	Same as Gupta et al.	Browning et al., 2002
MWNT	0.65	300–373, 0.1	Residual $MgNiO_x$ catalyst, 800 °C in H_2	Lueking and Yang, 2002
MWNT	3.6	298, 7	Residual $MgNiO_x$ catalyst, 800 °C in H_2	Lueking and Yang, 2003

result was refuted by Hirscher et al. (2001), who obtained a maximum hydrogen capacity of 1.5 wt % H_2 equivalent by using D_2 at 0.08 MPa and room temperature. This amount was attributed to hydride formation of Ti metal.

In the tests of Tibbetts et al. (2001), two samples of SWNTs were included: one from MER Corporation and one from Tubes@Rice. The latter has been used by a number of groups. The adsorption capacities at 296 K and 3.59 MPa were 0.028 wt % for the MER sample and 0.050% for the Rice sample. The MER sample contained only about 10–15% SWNTs, while the Rice sample was quite pure with only a small amount of soot. The samples were also subjected to pretreatment by exposure to H_2 at 3.6 MPa while heating to 500 °C. Apparently the pretreatment was done in a closed, static system and no time of exposure was given. Tests similar to those of Tibbetts et al. were performed by Zuttel et al. (2002), and similar results were obtained. Zhu et al. (2001) reported 3.5 wt % hydrogen capacity for well-aligned SWNT bundles and 0.5 wt % capacity for random-ordered SWNTs, both at 290 K and 10 MPa. The extra amount for the well-aligned SWNTs was attributed to the inter-tube spaces, as discussed in detail by Dresselhaus et al. (1999). The question concerning a possible leakage of the system used by Zhu et al. (2000) and others was raised by Tibbetts et al. (2001).

Graphite Nanofibers. As defined by Baker and Rodriguez (Rodriguez et al., 1995; Chambers et al., 1998), graphite nanofibers are fibrous carbon consisting of graphite platelets that are arranged parallel, perpendicular, or at an angle with respect to the fiber axis. The GNF is grown by catalytic decomposition, discussed in detail in Chapter 9. The platelets could be 30–500 $Å^2$ in cross-sectional area. There is little or no distinction between MWNTs and GNF when the platelets are parallel to the fiber axis. When they are at an angle, the GNF is referred as having a "herring bone" structure. It was on this type of GNF that the highest hydrogen capacities were reported by Rodriguez and co-workers (Rodriguez and Baker, 1997; Chambers et al., 1998; Park et al., 1999). Their work will be described briefly below.

The catalysts for their GNF growth were based on Ni-Cu, although some Fe was also added in some cases. $C_2H_4/CO/H_2$ mixture was used as the hydrocarbon source. The GNF were grown at 500–700 °C. The resulting fibers were partially purified in "mineral acids." There is no doubt that some catalyst remained. The pretreatment before hydrogen adsorption was an important step. The samples were pretreated in an inert atmosphere (typically Ar) at 800–1100 °C. A strong dependence of the hydrogen storage capacity on the pretreatment temperature was reported (Park et al., 1999). Pretreatment at 1000 °C yielded the best result, about 35 wt % hydrogen uptake at 1600 psi and ambient temperature.

The sorption was envisioned as hydrogen entering the platelets. Thus, the pretreatment was intended to remove the surface functionalities on the edges of the graphite. The functionalities on carbon have been discussed in Chapter 5. Upon the removal of the functionalities, hydrogen could enter the interlayer spacing or be adsorbed on the edges.

The TPD of surface oxides from graphite has been studied extensively (see for example, Tremblay et al., 1978; Chen et al., 1993). Typically, a strong desorption peak occurs between 900 and 1000 °C. The desorption requires breakage of the C–C bonds to free CO. TPD of chemisorbed hydrogen (from the edge sites) has also been studied (Redmond and Walker, 1960), and even higher temperatures are required. These facts add credence to the explanation of Baker/Rodriguez et al.

Yet, a number of groups have failed to reproduce the high hydrogen capacities (Ahn et al., 1998; Strobel et al., 1999; Poirier et al., 2001). However, it is interesting that, when compared with other carbon materials (activated carbon, SWNTs, and MWNTs), GNF seems to yield the highest hydrogen capacities (e.g., Tibbetts et al., 2001). When expressed on a per surface area basis, GNF clearly yielded the highest values.

In contrast, two groups have reported high capacities by following the work of Baker/Rodriguez. Browning et al. (2002) reported 6.5 wt % hydrogen uptake at 12 MPa and 300 K. In their work, Fe/Ni/Cu catalyst was used for GNF growth. The samples were pretreated at 1000 °C in flowing Ar for 36 h. They interpreted the results by hydrogen dissociation (at room temperature) on the edge sites of graphite to form H atoms which were subsequently adsorbed in the platelets. Gupta and Srivastava (2001) reported ~15 wt % uptake in GNF under similar conditions, with a pretreatment temperature of only 150 °C in vacuo.

Alkali-Doped MWNTs. Alkali metals can intercalate graphite and form stable intercalated compounds. It was reported that up to 0.137 L(STP)/g hydrogen could be adsorbed between the layers of alkali-intercalated graphites (Wanatabe et al., 1971; Akuzawa et al., 1990; Ichimura and Sano, 1992). Rao et al. (1997) showed that doping by alkali metals resulted in change transfer to the SWNTs and significantly increased the ionic character of the doped SWNTs. The charge in the SWNTs can increase the hydrogen adsorption (Simonyan et al., 1999). With this background, the report of Chen et al. (1999) that large amounts of hydrogen were adsorbed in Li-doped and K-doped MWNTs at 1 atm pressure generated considerable excitement.

The adsorption was measured gravimetrically by using a flow system at the ambient pressure. The sample weight was monitored while the temperature was varied. A weight gain of 20% was observed for the Li/MWNTs while the temperature was lowered from 670 to 470 K. Likewise, the weight gain was 14% for K/MWNTs when the temperature was near 300 K.

Because laboratory gas cylinders of hydrogen, even at the ultra-high purity grades, are notoriously contaminated by moisture, Yang (2000) performed identical experiments with wet and dry hydrogen. With drying, the weight gains were substantially lower, at near 2%. With wet hydrogen, the weight gains were identical to those reported by Chen et al. The weight gains were caused by the formation of $Li(OH){\cdot}H_2O$ and K(OH). The formation of alkali hydroxides in these systems was further confirmed by Pinkerton et al. (2000). With stringent dehydration, 1.3 wt % adsorption was measured for K/MWNTs, of which 0.2 wt % was cyclable. These amounts were measured at the ambient pressure. Since these amounts are significant, the effects of alkali doping certainly merits further investigation.

Monte Carlo Simulations. Following the initial reports of hydrogen storage in carbon nanotubes, a large number of rigorous molecular simulations were published (Stan and Cole, 1998; Darkrim et al., 1998; Rzepka et al., 1998; Darkrim and Levesque, 1998; Wang and Johnson, 1999a and 1999b; Simonyan et al., 1999; Gordon and Saeger, 1999; Williams and Eklund, 2000; Darkrim and Levesque, 2000; Meregalli and Parrinello, 2001; Simonyan and Johnson, 2002). Reviews and summaries of these publications are available elsewhere (Dillon and Heben, 2001; Meregalli and Parrinello, 2001; Simonyan and Johnson, 2002) and are not repeated here.

Grand canonical Monte Carlo simulations using empirical and reliable potentials for hydrogen–carbon and hydrogen–hydrogen interactions were used. Adsorption of H_2 molecules is allowed to take place both inside and outside the tubes, that is, in a bundle. All possible tube diameters and inter-tube spacings have been assumed. As expected, adsorption potentials are enhanced within a SWNT over that in a slit pore of the same size. However, all simulations predicted less than 2.4 wt % storage at 298 K and 100 atm. The highest amount of 2.4 wt % was obtained with bundles having a tube diameter of 2 nm (Wang and Johnson, 1999a). Wang and Johnson (1999b) showed that the DOE target of 6.5% cannot be reached even by tripling the fluid-wall potential at ambient temperature.

Charging the SWNT can in increase the potential between SWNT and H_2. Thus, H_2 adsorption on Li-doped SWNTs (bundles) have been investigated by Simonyan and Johnson (2002) and Meregalli and Parrinello (2001). In the work of Simonyan and Johnson, both H_2-Li_2 interactions and H_2-charge interactions were included. However the predicted enhancement of adsorption was only about 40%, for example, at 298 K and 120 atm, the hydrogen capacity was increased from 1.25 wt % for a pure SWNT bundle to 1.9 wt % for Li-doped bundle ($Li/C = 0.1$).

Molecular Orbital Calculations. A common feature that exists in all reported experimental results, whether at ambient temperature or at cryogenic temperatures, is the *slow* uptake of H_2. In all cases, the uptakes were slow and equilibria were reached after a time span of the order of hours. Meanwhile, the adsorbed hydrogen could not be desorbed completely in vacuo. For example, 21–25% of the adsorbed hydrogen remained and had to be desorbed at 473 K in vacuo (Liu et al., 1999). Both results are clear indications of chemisorption. In all processes for synthesis of nanotubes and GNF, one or more of the metals from Co, Ni, and Fe must be used. Regardless of the processes that are used to purify the products, significant amounts of metals remain. Since metals were contained in the nanotubes in all cases, dissociation of hydrogen to atomic H was likely to have occurred. In addition, hydrogen dissociation is known to occur on the edge sites of graphite with a free sp^2 electron per carbon site (e.g., Ishikawa et al., 1975), although such a process would be much slower than that on metals at ambient temperature. Consequently, hydrogen dissociation followed by spillover and chemisorption on the surface of nanotubes is an entire possibility.

The first step in examining such possibility is to see whether H atoms can adsorb on the basal plane of graphite. This would involve electron transfer or formation of a chemical bond. Monte Carlo simulation cannot be used for this study; rather, molecular orbital theory calculations can provide a full description of the process (Chapter 8). Molecular orbital calculations on the chemisorption of atomic hydrogen on graphite have been reported since the early development of quantum chemistry, that is, the calculations of H/graphite by Sherman and Eyring (1932). The subsequent calculations on H/graphite have been mostly performed with semi-empirical methods. Chen and Yang (1989) performed such calculations and showed that the chemisorption was most stable on the zigzag edges, followed closely by the armchair edges, while the adsorption on the basal plane was meta-stable. More recently, Yang and Yang (2002) have undertaken ab initio calculations using Gaussian. The ab initio method is capable of energy predictions. Their results are summarized below.

Figure 10.28 shows the models used. H atoms are adsorbed on different edge sites of graphite (armchair, zigzag and basal planes). The energies of adsorption are listed in Table 10.7. It is seen that the energies are negative for all models, meaning that they are stable and the adsorption is exothermic for all cases. The strength of the bond is higher on the edge planes than the basal planes, and follows the order:

Zigzag Edge > Armchair Edge > Basal-Plane

For adsorption on the basal plane sites, the energy of adsorption is substantially lower when the H atoms are occupying adjacent sites. Comparing Model F with

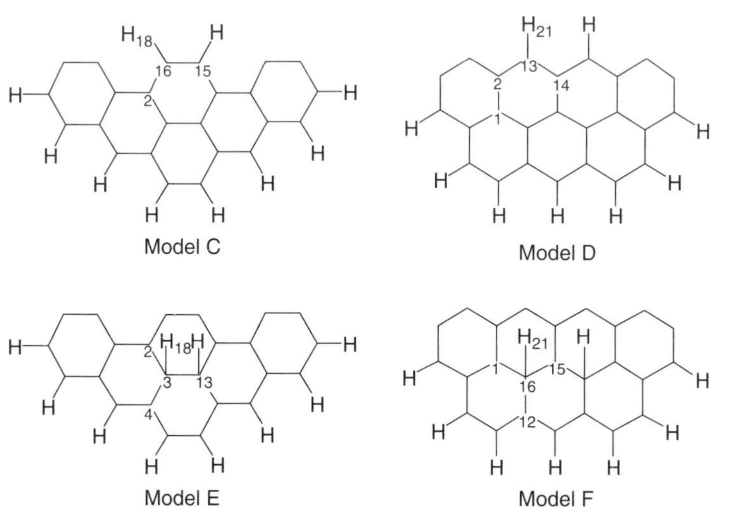

Figure 10.28. Models for H atoms chemisorbed on different faces of graphite: armchair (Model C), zigzag (Model D), and basal plane (Models E and F) (Yang and Yang, 2002 with permission).

Table 10.7. Bond energies (expressed by energies of adsorption) of H atoms on surface sites on graphite (see Figure 10 for models used) from ab initio molecular orbital calculations

Model	Bond Energy (kcal/Mol)
C (H on armchair edge)	−85.27
D (H on zigzag edge)	−90.24
E (H adjacent basal sites)	−27.04
F (H on alternating basal sites)	−46.47

Yang and Yang, 2002.

Model E, the energy drops from 46.47 kcal/mol (two H on alternate sites) to 27.04 kcal/mol (two H on adjacent sites).

Molecular orbital calculations have been published recently on the bonding of F atoms (Seifert et al., 2000; Kudin et al., 2001) and H atoms (Bauschlicher, 2000 and 2001; Froudakis, 2001) on SWNT. Both Bauschlicher and Froudakis studied the bonding of H atoms on the *exterior* wall of the SWNT (both consisting of 200 C atoms, but with different diameters). Bauschlicher calculated the bond energies for the tube with 1H and 2H and 24%, 50%, and 100% coverages (assuming 1H/1C). The average C–H bond energy for the first H was 21.6 kcal/mol, and 40.6 kcal/mol for the first two H atoms. The average bond energy for 50% coverage was 57.3 kcal/mol, decreasing to 38.6 kcal/mol for 100% coverage. Froudakis studied the bonding of 1H with the tube where the H atom approached the tube wall in two ways: direct approach to the top of a carbon atom, and approach along the centerline of a hexagon (Froudakis, 2001). The energy minima were, respectively, 21 kcal/mol and 56 kcal/mol. Froudakis also reported the C–C bond lengths in the nanotube after H bonding. With 16 H bonded to 64 C on the 200-atom tube, the C–C bond length increased from 143 to 159 pm.

The C–H bond energies on the tube are in general agreement with that on the basal plane of graphite. The energies on the basal plane are 46.47 kcal/mol for two alternating or separated H atoms, and 27.04 kcal/mol for two adjacent H atoms. A similar crowding effect was also seen on the curved tube, that is, 57.3 kcal/mol for 50% coverage and 38.6 kcal/mol for 100% coverage. The increase in the C–C bond lengths from 143 pm to 155–159 pm by adsorption of H was also seen on the basal plane of graphite, and are, in fact, in excellent agreement with the results of Yang and Yang (2002).

From the above comparison, it is unlikely that the adsorption of H atoms would differ significantly on the basal plane of graphite and on the exterior wall of the nanotube. Based on the results above, it is possible that the binding energies could be lowered further for adsorption inside the tube. Also, no calculations have been made on H atoms adsorbed between two layers of graphite at a *d*-spacing of 3.35 Å. This could be the situation for hydrogen storage on GNF and MWNTs.

Lee and Lee (2000) performed density-functional calculations for hydrogen chemisorption in SMNT, both inside and outside the tube. Their calculations were performed at zero Kelvin and showed two configurations for chemisorption. They predicted that 14 wt % could be adsorbed in (10,10) tubes.

Multi-wall Nanotubes. The platelet-structure of GNF is described in a preceding section. When the angle between the platelets and the fiber axis is small, there is little or no distinction between MWNTs and GNF. Both are grown catalytically, and the TEM images of these two types of materials can be essentially the same. In fact, it is known that the surfaces of the vapor-grown MWNTs are rarely perfect graphite planes and that their surfaces have functionalities (e.g., Kuznetsova et al., 2000).

It is these materials on which the widest range of hydrogen capacities have been reported. The pretreatment condition and the residual catalyst are clearly important in the hydrogen capacity, and yet these have not been clearly characterized in the previously published reports. Lueking and Yang (2002) have studied hydrogen storage in MWNTs grown from $Ni_{0.4}Mg_{0.6}O$ catalyst. The growth of MWNTs with graphitic layers parallel to the tube axis on this catalyst has been documented by Chen et al. (1997). The catalyst was removed from the MWNTs to various degrees by acid wash (using 6N HNO_3 solution). Hydrogen uptake was measured gravimetrically (by TGA) in H_2 flow at 1 atm pressure. The samples were pretreated in hydrogen at 700 °C for 1 h. The results are shown in Figure 10.29. The sample shown by curve (A) contained residual

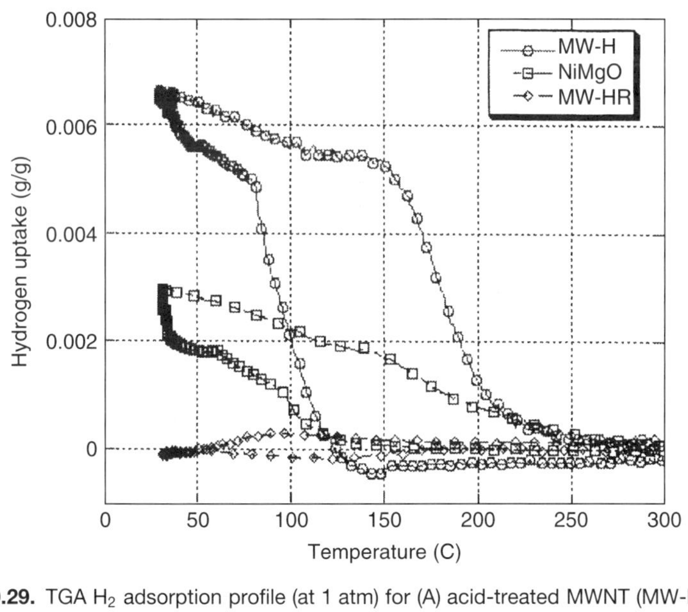

Figure 10.29. TGA H_2 adsorption profile (at 1 atm) for (A) acid-treated MWNT (MW-H), (B) the $Ni_{0.4}Mg_{0.6}O$ catalyst, and (C) the MWNT with extended acid treatment to remove the residual catalyst (MW-HR) (from Lueking and Yang, 2002, with permission).

catalyst, 17.4 wt % Ni and 1.9 wt % Mg. The BET surface area of this sample was 184 m^2/g. The pure catalyst after the pretreatment had a hydrogen capacity, due to MgO and its reduced forms. However, the hydrogen capacity of sample (A) was far greater than that due to the residual catalyst. The catalyst-free MWNTs (sample C) showed no hydrogen capacity. The hydrogen capacity of 0.65 wt % for sample A was at 1 atm, hence was significant. This result is a clear indication that a dynamic process was at work that involved hydrogen dissociation and spillover to the nanotubes. TPD runs at different heating rates were performed on sample A. From the hydrogen desorption peak temperatures, the activation energy for desorption was determined to be 23.3 kcal/mol. This value was close to the bond energy of 27 kcal/mol for chemisorption of H atoms on the basal plane of graphite, predicted from ab initio calculations (Yang and Yang, 2002).

To further elucidate the hydrogen uptake mechanism, TPD behaviors of the sorbed hydrogen from two samples were compared: the $Ni_{0.4}Mg_{0.6}O$ catalyst and the MWNT with residual catalyst (Lueking and Yang, 2003). The results showed that the MWNT had an enhanced desorption peak at 140 °C when compared with the initial $Ni_{0.4}Mg_{0.6}O$ catalyst, which had a dominant desorption peak at 250 °C. The dominant adsorption temperatures for these two samples were the same. These results suggest that the adsorption process for MWNT is the same as for the $Ni_{0.4}Mg_{0.6}O$ catalyst, whereas desorption from the nanotubes occurs directly from lower-energy carbon sites. Lueking and Yang (2003) also measured hydrogen uptake of the MWNTs with residual catalysts at high pressures by using the volumetric technique. The samples were pretreated in 1 atm H_2 at various temperatures (500–800 °C), transferred to the high-pressure apparatus, followed by heating at 500 °C in vacuo. The hydrogen uptake increased with the pretreatment temperature, being the highest at 800 °C. Considerable amounts of weight loss accompanied the pretreatments. Adsorption conditions were either at a constant temperature at 69 atm, or under a temperature-pressure (T-P) cycle. The T-P cycle consisted of a series of step changes where the temperature was lowered as the pressure was increased: 122 °C and 100 psia; 50 °C and 500 psia; 25 °C and 1000 psia. After adsorption, desorption measurements were also made. Higher hydrogen uptakes were obtained with the T-P cycle. The resulting uptake amounts are summarized in Figure 10.30.

As discussed by Lueking and Yang (2003), the pretreatment not only removed inactive materials but also activated the sample for hydrogen uptake. Other studies have explained high temperature activation to be due to removal of chemisorbed species (Park et al., 1999), destruction of surface functionalities that may block pores (Kuznetsova et al., 2000), or graphitization of the nanotubes (Li et al., 2001). However, these studies did not consider residual metal content, which has been shown to affect the reactivity and gasification of nanotubes (Chiang et al., 2001). The absolute hydrogen storage shown in Figure 10.30 is comparable with results reported by others on MWNTs. At ambient temperatures, the reported values range from 1.97% at 40 atm (Lee et al., 2002) to 4% at 100 atm (Li et al., 2001) to 6.3% at 148 atm (Hou et al., 2002). Although comparison of the results by Lueking and Yang (2003) to other hydrogen storage reports is not

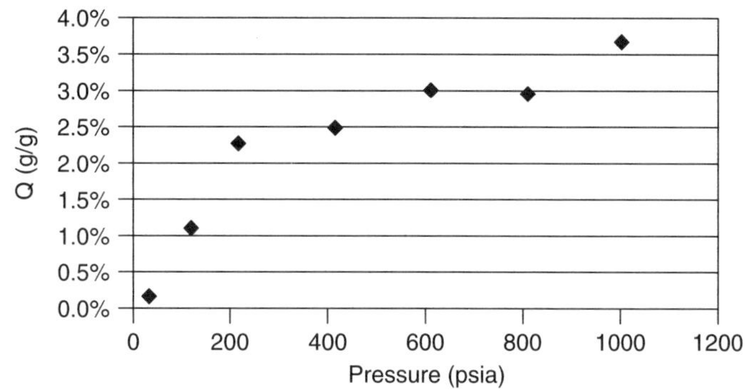

Figure 10.30. Adsorption isotherm of hydrogen on MWNT/NiMgO system after external pretreatment in 800 °C H_2 followed by 500 °C vacuum in the adsorption apparatus. The corresponding value of desorption showed a maximum value of 3.6% (from Lueking and Yang, 2003, with permission).

straightforward due to different pressures and pretreatments, it is interesting to speculate how residual metal content may have affected various hydrogen storage reports. For example, the SWNT sample that had an estimated 5–10% hydrogen storage value also had a ~20 wt % cobalt content (Dillon et al., 1997). The 4% hydrogen storage reported by Li et al. (2001) was for a sample with no attempt to remove the iron catalyst. The pretreatment used by Liu et al. — an acid wash followed by high temperature heat-treatment, which that led to a twofold increase in hydrogen storage (Liu et al., 1999) — is surprisingly similar to the pretreatment process that Chiang et al., showed to expose metal particles (Chiang et al., 2001).

10.4. METHANE STORAGE

Sorbent development for methane storage has been an active research area since the 1980's. The aim of the research is to develop a competitive storage system for natural gas for vehicular use, known as adsorbed natural gas (ANG).

The relative abundance and clean-burning characteristics of natural gas make it attractive as a vehicular fuel, especially for urban areas. Compared with gasoline, natural gas combustion produces less hydrocarbon emission as well as less emissions of sulfur and nitrogen oxides. The history of natural gas vehicles (NGV) dates back to the 1930's when Italy launched an NGV program (Cook et al., 1999). Today there are more than one million NGVs worldwide (mostly in Argentina, Italy, and countries of the former Soviet Union, where the relative price of gasoline/natural gas is high). These vehicles use compressed natural gas (CNG), at about 3000 psig (20 MPa or 200 bar) and ambient temperature. Liquefied natural gas (LNG) is usually stored at the boiling point of about 112 K (−161 °C) in a cryogenic tank at 1 atm pressure. LNG is not used for passenger vehicles due to safety hazards, leakage, and other problems, although over 1000 LNG fueled trucks and buses are in use worldwide (Cook et al., 1999). For CNG, about 230 unit volumes of natural gas at 1 bar are compressed to one unit volume of storage container, often designated as 230 V/V storage. For the same driving range, the size of the CNG vessel is at least three times the volume of a gasoline tank. Moreover, compression to 200 bar requires a four-stage compressor. For simplicity and reduced cost, a storage system using a single-stage compressor is attractive, which puts a pressure limit of about 5 MPa. This limit poses no problem for ANG since adsorption of methane on known sorbents has reached an isotherm plateau at this pressure range. A benchmark pressure of 3.5 MPa (35 bar) has been widely adopted for comparing different sorbents for ANG use.

A number of practical problems need to be considered in ANG technology: higher hydrocarbons and impurities in the natural gas (Mota, 1999), mass transfer rates, and heat effects. The first problem has been solved by the use of guard beds, which is well-known in the PSA technology. The guard bed traps these impurities during charging and releases them during discharge. The mass transfer and heat effects (heating during charging and cooling during discharge) have been studied and are well understood (Mota et al., 1997; Biloe et al., 2001; Mota and Rodrigo, 2002; Biloe et al., 2002). These problems have also been minimized with clever designs of monolithic sorbent and storage vessels (Cook et al., 1999). An excellent review of ANG has been given by Cook et al. (1999). Reviews on sorbents for ANG are also available (Mullhaupt et al., 1992; Menon and Komarneni, 1998).

For sorbent development, the U.S. Department of Energy set the target of 150 V/V deliverable capacity at 3.5 MPa and 25 °C (Wegrzyn et al., 1992). This deliverable amount is the total amount between 35 bar and 1 bar (isothermal), including the gas phase.

To date, the best sorbents are carbons. The theoretical limits for storage in carbons as well as the optimal pore dimension of the carbon have been studied extensively (Tan and Gubbins, 1990; Mastranga et al., 1992; Cracknell and Gubbins, 1992). Tan and Gubbins (1990) used GCMC and DFT to calculate methane adsorption in model porous carbons for a wide range of pore sizes, and determined that the optimal dimension for a slit pore is 11.4 Å. This is the center-to-center distance between the two graphite layers. Thus the free spacing is less than 11.4 Å, and is ~8 Å. Mastranga et al. (1992) reached the same conclusion as Tan and Gubbins. Using this slit width and assuming that each two slits are separated by a single layer of graphite, the theoretical limits (e.g., at 209 V/V) are substantially higher than the DOE target. Myers and Glandt (1993) concluded that the theoretical limit is 220 V/V.

The two key factors for methane storage are micropore volume and sorbent-packing density. Optimal storage will occur when the micropore volume is maximized. However, for fast mass-transfer rates, some mesopores and macropores are also needed as feeder pores. The micropore volumes of carbons are correlated well with the BET surface areas measured with N_2 at 77 K. In fact, a linear correlation was obtained for the amount of methane adsorbed at 3.5 MPa and 25 °C and the BET surface area based on the data on 35 commercial carbons (Mullhaupt et al., 1992).

A high packing density is required for high V/V storage. A clever approach for achieving high packing density is to form monoliths (such as discs) by using a polymeric binder (Bose et al., 1991). Typically, the sorbent powder is mixed with a binder and the mixture is moulded into discs under a high pressure. The monoliths are subsequently produced upon heat-treatment (e.g., 800 °C in N_2) (Chen and McEnaney, 1995). The percentages of binder shown in Table 10.8 are the optimal amounts for maximum V/V methane storage.

Table 10.8 summarizes the methane storage capacities of all sorbents reported in the literature, including various forms of carbon, zeolite, polymeric resin, and

Table 10.8. Methane storage capacities at 3.4 MPa (or 500 psia) and 25 °C, expressed by adsorbed amount, q (mmol/g) and V/V (in CH_4 at STP per volume, including gas phase)

Sorbent	Note	Bet	q(mmol/g)	V/V	Ref. (W/et al.)
AX-21 Type					
Amoco GX-32		2500	10.25	98	Barton (1983)
Anderson AX-21 monolith	37% PVDC binder	1860	8.75	163	Quinn (1987)
Anderson AX-21	No binder	2800	10.8	109	Bose (1991)
Anderson AX-21 monolith	2% PVA binder		9.25	186	Bose (1991)
Anderson AX-21 monolith	31% phenolic binder	2054		152	X. Chen (1995)
Kansai Maxsorb	Powder	3272	12.5	110	Manzi (1997)
Kansai Maxsorb monolith	Compacted disks	2043	8.2	155	Manzi (1997)
KOH-activated anthracite		2758	~13	155	Lozano-Castello (2002)
Activated carbon fibers (ACF)	H_2O-activated rayon fibers	1539	15.69	179	Lewis (1993)
ACF (KF-1500)	Cellulose base	1500	5.2 (at 313 K)		Jiang (1994)
ACF (pitch-based)	CO_2 activated	2400	~9.2	163	Alcaniz-Monge (1997)
Calgon BPL GAC		1030	4.68	~80	Barton (1983)
Norit GAC (R3)		1270	5.72		Quinn (1994)
CaX Zeolite			5.12	~100	Zhang (1991)
Polymer resin (polystyrene)	Dow XV43546	1600	4.18		Cook (1999)
MCM-41 (SiO_2)	~16 Å pores	1070	4.06		Ioneva (1995)

BET surface area is given in m^2/g.

silica (MCM-41). These represent the highest reported storage capacities from each category of sorbents. The V/V capacity is expressed in volume of methane stored (cc STP, including gas phase) per volume of sorbent. This is not the deliverable storage. The adsorbed amount at 1 atm needs to be subtracted to get the delivered amount. Empirically, the delivered capacity is about 15% lower than this amount. The isotherms of methane on selected carbons are shown in Figure 10.31.

The most promising sorbents are the "super" activated carbon (AX-21 type) and activated carbon fibers (ACF). Both have rather uniform pore sizes in the 10–20 Å micropore range. As discussed in Chapter 5 (5.1), the "super" activated carbons are produced by activation with molten KOH, invented by Wennerberg and O'Grady (1978). The process is remarkably simple and reproducible, and it works for a variety of precursors (cokes, coals, nut shells, and wood). Typically, KOH and coke are mixed at a ratio of about 3/1 KOH/coke, heated to around 700–800 °C (the melting point of KOH is 360 °C) in an inert atmosphere (or in a closed system) for about 2 hs. A small amount of water is used for pasting. After such an activation process, the carbon is washed to recover potassium. A large microporosity is formed during the activation, with "cage-like" pores,

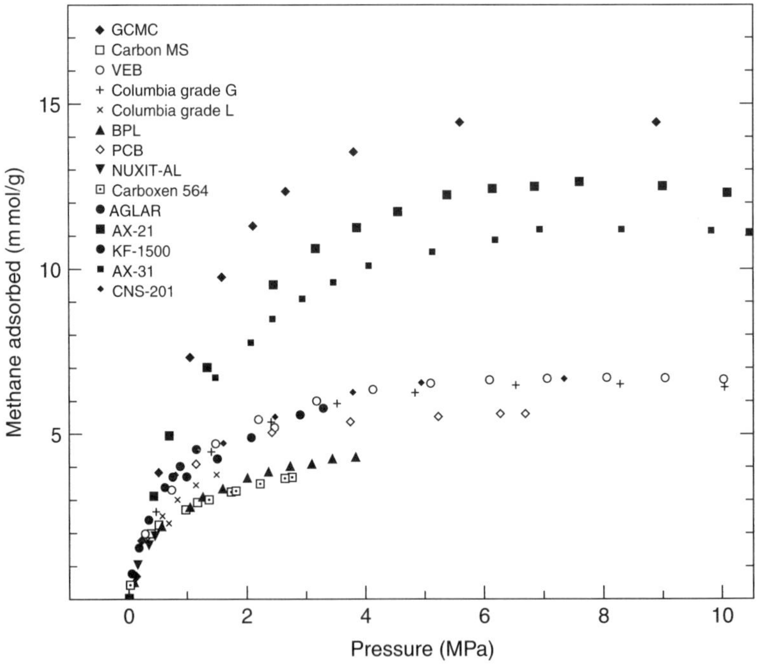

Figure 10.31. Adsorption isotherms at 25 °C for various carbons (Mastranga et al., 1992, with permission). The carbons include molten KOH activated carbons (AX-21 and AX-31), activated carbon fibers (KF-1500), carbon molecular sieves (Carbon MS), and conventional activated carbons (all others). GCMC: simulation for an "ideal" carbon with slit pores at 11.4 Å width formed by single graphite sheets.

mainly <2 nm. The BET surface area is typically 2800 m^2/g, reaching as high as 4057 m^2/g, depending on the activation condition (Wennerberg and O'Grady, 1978). The mechanism of activation is not understood. This carbon was first marketed by Anderson Development Company as Anderson AX-21, and is now made by Kansai Coke & Chemicals Company under license from Amoco and is marketed as Maxsorb. The original process was aimed at petroleum coke as the precursor. Kansai has developed carbons by using coconut shell (Otowa, 1991). Similar products using cellulose-based precursors were developed by Westvaco for ANG (Baker, 1997).

As shown in Table 10.8, the V/V storages in the AX-21 type carbons are very high. Good methane storage capacities were also reported on KOH-activated anthracite (Lozano-Castello et al., 2002, who also optimized the activation conditions) and KOH-activated bituminous coal (Sun et al., 1996). Monolithic sorbents of AX-21 prepared with polymer binders showed extremely high V/V storage capacities (Bose et al., 1991; Chen and McEnaney, 1995, see Table 10.8). Manzi et al. (1997) measured V/V storages for Kansai Maxsorb in three forms: powder, pellets, and disks. The pellets and disks were prepared by compaction alone. The bulk density increased with compaction: 0.28 g/cm^3 (powder), 0.32 g/cm^3 (pellets), and 0.56 g/cm^3 (disks). The V/V storage capacities were, respectively, 110, 100, and 155. For the pellets, the increase in packing density by compaction was not large enough to offset the decrease in surface area. Comparing these results, it is clear that it is more effective to use binder (then compaction alone) for forming monoliths. Based on the data shown in Table 10.8, the DOE target of 150 V/V delivered methane is well within reach or has already been reached with the AX-21 type and ACF sorbents.

Effects of chemical modification of carbons on their methane storage capacities have been reported. Kaneko and Murata (1997) modified AX-21 by MgO impregnation and showed a substantial increase in methane adsorption at 10 MPa and 303 K. However, as pointed out by Cook et al. (1999), on a V/V basis there would be no advantage as the micropore volume is significantly reduced. Surface treatment by oxidation and reduction of activated carbon (BPL) showed essential no effects on methane adsorption (Barton et al., 1991).

Zeolites have higher packing densities than carbons. The interactions of methane with zeolites are stronger than that with carbon. For example, the isosteric heat of adsorption on CaX is 24.6 kJ/mol (Zhang et al., 1991), compared with 13.3 kJ/mol for 11.4 Å slit pores of graphite (Tan and Gubbins, 1990). However, the isotherms of methane on zeolites level off quickly (Zhang et al., 1991), and hence do not yield high V/V capacities at 3.5 MPa. The strong adsorption of water in zeolites is a practical problem. The methane capacities on high-surface-area polymeric resins and silicas (such as MCM-41) are not high, as shown in Table 10.8.

The search for sorbents for methane storage remains to be an active research area worldwide. Methane storage could play a key role in fuel cell technology if an effective catalyst is developed for direct methane fuel cell (i.e., methane is fed directly to the fuel cell).

New possibilities for developing better sorbents for methane storage are still available. For example, no reports have appeared on activation of carbon fibers by using molten KOH, which could yield the optimum micropore distribution and volume. Single-wall carbon nanotubes with the right sizes, particularly bundled in the aligned forms, could result in the ideal sorbent for methane storage.

10.5. OLEFIN/PARAFFIN SEPARATIONS

Cryogenic distillation has been used for over 70 years for the recovery of ethylene and propylene from olefin plants, refinery gas streams, and other sources (Keller et al., 1992). These separations are difficult to accomplish because of the close relative volatilities. The ethane/ethylene distillation is performed at about $-25\,^{\circ}$C and 320 psig in a column containing over 100 trays. Propane/propylene distillation is performed at about $-30\,^{\circ}$C and 30 psig. These are the most energy-intensive distillations in the chemical and petrochemical industry (Safarik and Eldridge, 1998).

As discussed in Chapter 10.1, about 20% of the oxygen and nitrogen is now produced from air by PSA technology. A single PSA unit is capable of producing more than 200 tons/day of oxygen, and the capacity is steadily increasing as the technology further improves. Distillation is necessary for larger scales. In the following discussion, a case will be made that the situation for olefin/paraffin separation should parallel that of air separation. That is, PSA is capable of small-scale recoveries of ethylene and propylene, while for large-scale production of ethylene and propylene in the olefin plants, distillation will clearly remain to be the process of choice.

A significant amount of the light olefins produced during the refining of crude oil is used as a fuel or is simply flared. These streams are generally small. In the production of polypropylene and polyethylene, a significant amount of propylene and ethylene are lost in the purge gas. These gas streams are generally available at super-atmospheric pressures. For example, the refinery gas streams are at 100–250 psia pressure range containing 10–35% ethane/ethylene, 2–10 propane/propylene, 8–20% H_2, 20–45% CH_4, and trace amounts of higher hydrocarbons. The modern gas-phase polymerization reactors for both ethylene and propylene are operated at 20 atm pressure and 85–100 °C, and the purge streams contain about 80–85% olefins and 10–15% paraffins (Rodriguez, 1999). These ranges of pressure, temperature, and gas composition are ideally suited for PSA using π-complexation sorbents. Moreover, hydrocarbon emissions are no longer allowed by federal regulations. Small-scale recovery of olefins is necessary for both economic and environmental reasons. In fact, PSA using 4A zeolite (under the name of Petrofin Process) was used for recovery of propylene from the purge gas streams of polypropylene reactors. As discussed below, 4A zeolite is not a good sorbent for this separation and hence this process has been discontinued.

10.5.1. Sorbents

Two types of sorbents have been examined for ethane/ethylene and propane/propylene separations: zeolites/molecular sieves and π-complexation sorbents. The former type may be used for kinetic separation, that is, based on the higher diffusivities of olefins over that of paraffins. The use of 4A zeolite (Ramachandran and Dao, 1994) is one such example. $AlPO_4$-14 molecular sieve (Padin et al., 2000) is another possibility. Syntheses and characterizations of the π-complexation sorbents for olefin/paraffin separations have been discussed in Chapter 8 (8.3 and 8.4). The aluminophosphates ($AlPO_4$s) and silicoaluminophosphates ($SAPO_4$s) are discussed in Chapter 7 (7.2.2).

A general comparison can be made of these two types of sorbents. The adsorption of olefins under practical conditions is limited by the pore volume of the sorbent, that is, adsorption at pressures above ambient or adsorption at near room temperature and atmospheric pressure. The pore volumes of zeolites and molecular sieves are substantially lower than that of the π-complexation sorbents, which are based on silica gel and activated alumina. Thus, for propylene, the limiting adsorbed amounts for zeolites and molecular sieves are approximately 2.1–2.4 mmol/g, whereas that for the π-complexation sorbents supported on silica gel is well over 5 mmol/g. As will be shown below, direct comparisons of the PSA separation performances with these two types of sorbents show indeed that the π-complexation sorbents are significantly better than zeolites and molecular sieves.

The isotherms of ethane/ethylene and propane/propylene on $AgNO_3/SiO_2$ are shown in Figures 8.10 and 8.11. The isotherms of the same hydrocarbons on $CuCl/\gamma\text{-}Al_2O_3$ are shown in Figures 10.32 and 10.33. Because the π-complexation bonds between olefins and Cu^+ are stronger than that of olefins and Ag^+ (Huang et al., 1999), the olefin/paraffin selectivities of $CuCl/\gamma\text{-}Al_2O_3$ are higher

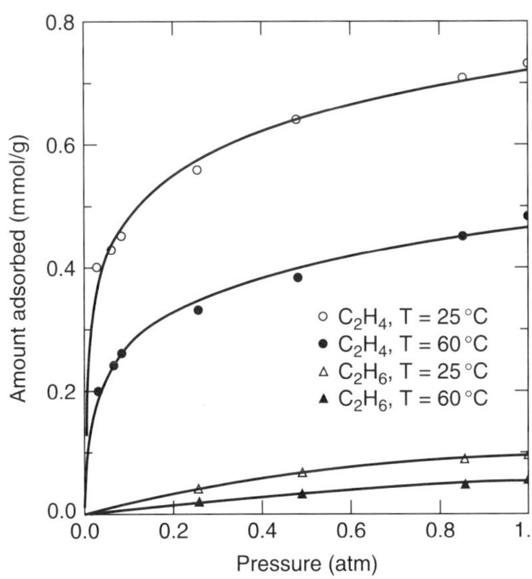

Figure 10.32. Isotherms of C_2H_4 and C_2H_6 on $CuCl/\gamma\text{-}Al_2O_3$ (Yang and Kikkinides, 1995, with permission).

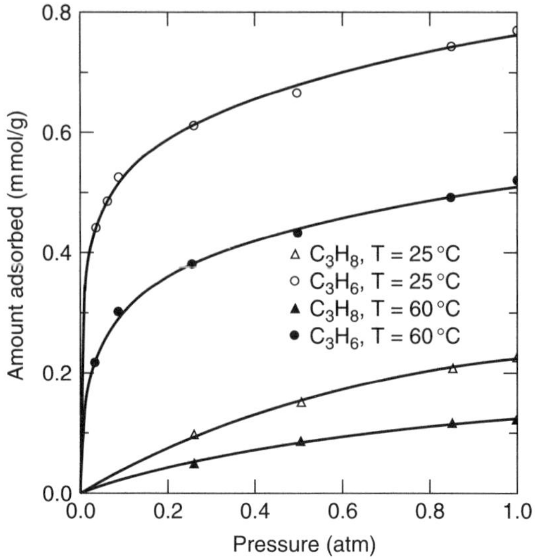

Figure 10.33. Isotherms of C_3H_6 and C_3H_8 on CuCl/γ-Al_2O_3 (Yang and Kikkinides, 1995, with permission).

than that of $AgNO_3/SiO_2$. However, the olefin isotherms on CuCl/γ-Al_2O_3 also exhibit greater curvatures due to the stronger bonds. The curvature is not desirable for PSA separation. This problem can be alleviated by raising the temperature of operation to, for example, 120 °C. Also, because the surface area of the γ-Al_2O_3 was low (340 m^2/g, compared with 670 m^2/g for the silica gel), the olefin capacities for CuCl/γ-Al_2O_3 were low. The olefin capacities can be increased substantially by using substrates with higher surface areas. The relatively low cost of CuCl makes supported CuCl a highly promising sorbent.

10.5.2. PSA Separations

PSA simulations have shown that the π-complexation sorbent, $AgNO_3/SiO_2$, outperforms 4A zeolite by a substantial margin (Rege et al., 1998). It was found that the zeolite with 5% Na^+-95% Li^+ had the optimal characteristics for the kinetic separation of a propane/propylene mixture (Padin et al., 2000). The C_3H_6 uptake rate for this NaLiA zeolite was four times faster than commercial 4A zeolite due to its optimized pore structure, and, moreover, its pure component equilibrium selectivity for C_3H_6 was marginally higher. However, this sorbent had limited capacity for C_3H_6 above 1 atm and its PSA performance was not as good as that of $AgNO_3/SiO_2$.

A number of molecular sieves with pore sizes near 3.8 Å (i.e., the kinetic diameter of propane) were examined for their kinetic separation properties for propane/propylene (Padin et al., 2000). One of the promising molecular sieves was $AlPO_4$-14 (see Chapter 7.2.2, Figure 7.8). However, based on the isotherms at ambient temperature, $AlPO_4$-14 did not appear to be a suitable sorbent for propane/propylene separation. Padin et al. (2000) suggested that at higher temperatures (e.g., 120 °C) the propylene isotherm became linear while maintaining adequate capacity such that $AlPO_4$-14 was an effective sorbent for the separation.

A direct comparison of PSA separation using the two types of sorbents ($AgNO_3/SiO_2$ vs. $AlPO_4$-14) was made by Rege and Yang (2002). A four-step PSA cycle was used in the simulations. The steps involved in each cycle were as follows: (1) pressurization with the feed gas mixture; (2) high-pressure adsorption with feed gas (i.e., feed step); (3) high-pressure co-current purge with part of the compressed C_3H_6-rich product obtained in step 4; (4) countercurrent blowdown to a low pressure. All the steps were of equal time duration.

Figure 10.34 shows the schematic of the PSA cycle used. As can be seen, a portion of the high-purity C_3H_6 from the countercurrent blowdown step was compressed to the feed pressure and used for rinsing the bed co-currently in step (3). It has been shown earlier that purging with the strongly adsorbed component results in a significant increase in the purity of that component in the product stream (Tamura, 1974; Cen and Yang, 1986; Yang, 1987). In the present work, the product of the high-pressure rinse step is recycled and mixed volumetrically with the feed gas supplied to step (2). The concentration of the feed gas to step (2) was averaged volumetrically while its temperature was kept the same as the feed gas temperature. The model and the numerical technique used in this study are described in detail in previous literature (Rege et al., 1998). The three important PSA performance parameters are product purity, recovery,

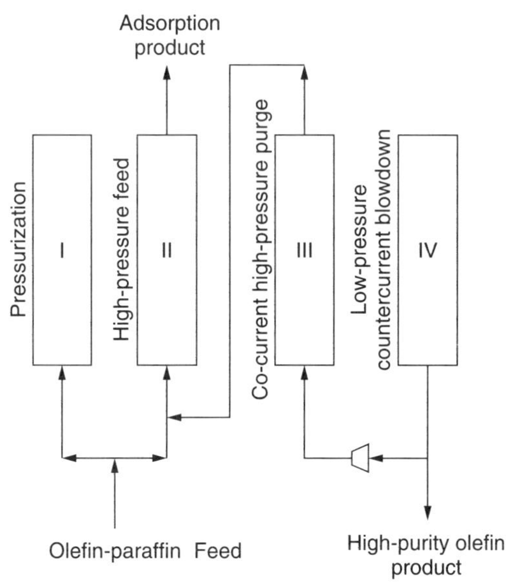

Figure 10.34. Schematic of the four-step PSA cycle used for C_3H_6/C_3H_8 separation (Rege and Yang, 2002, with permission).

and throughput. A fair comparison of sorbent performance can be made by using similar cycle conditions for each of the sorbent systems and by comparing one of the performance parameters with the other two parameters being kept nearly the same. It should be noted that the term "product" refers to the C_3H_6-rich effluent obtained in the desorption step (4). The various process variables were defined as follows:

$$\text{Product recovery} = \frac{\begin{pmatrix}\text{C}_3\text{H}_6 \text{ from} \\ \text{step 1}\end{pmatrix} - \begin{pmatrix}\text{C}_3\text{H}_6 \text{ used for purging} \\ \text{in step 3}\end{pmatrix}}{(\text{C}_3\text{H}_6 \text{ fed in step 1 and step 2})} \tag{10.4}$$

$$\text{Purge/feed ratio (P/F)} = \frac{(\text{C}_3\text{H}_6 \text{ used to purge in step 3})}{(\text{C}_3\text{H}_6 \text{ fed in step 1 and step 2})} \tag{10.5}$$

Another important parameter used to gauge the adsorbent's productivity is the product throughput:

$$\text{Product throughput} = \frac{\text{Amount (kg) of C}_3\text{H}_6 \text{ product per hour}}{\text{Amount (kg) of adsorbent}} \tag{10.6}$$

The equilibrium adsorption isotherms of propane and propylene on $AgNO_3/SiO_2$ at up to 8 atm are shown in Figure 10.35. The high-pressure isotherms for $AlPO_4$-14 are shown in Figure 10.36. From a comparison of Figures 10.35 and 10.36, it can be seen that adsorption capacity of the $AgNO_3/SiO_2$ for the olefin is higher than that of $AlPO_4$-14 due to their different pore volumes. The diffusion time

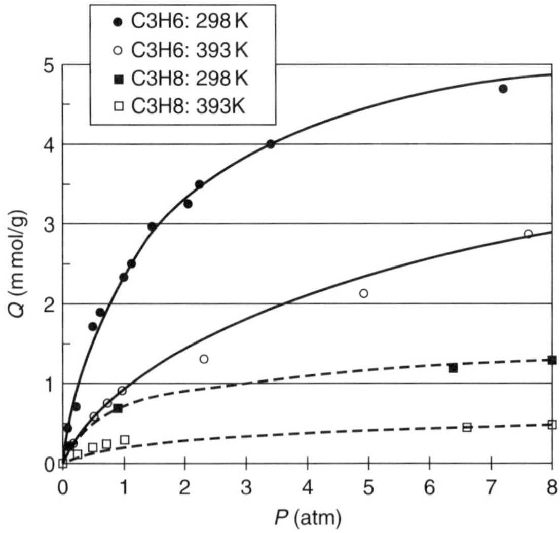

Figure 10.35. Adsorption isotherms of C_3H_6 and C_3H_8 on $AgNO_3/SiO_2$ sorbent at 298 K and 393 K. Lines: isotherm fitting (Rege and Yang, 2002, with permission).

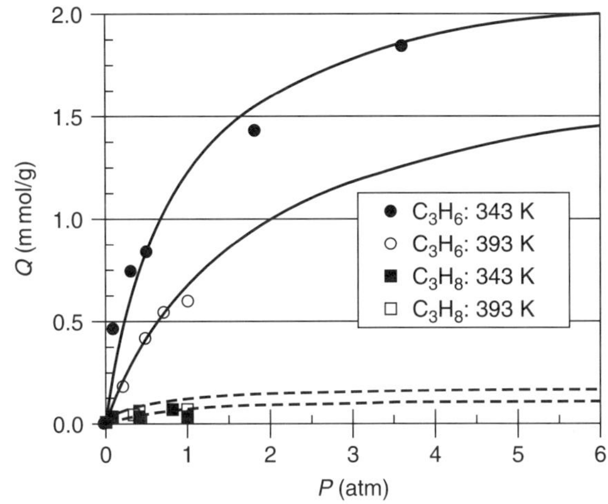

Figure 10.36. Adsorption isotherms of C_3H_6 and C_3H_8 on $AlPO_4$-14 sorbent at 343 K and 393 K (Rege and Yang, 2002, with permission).

Table 10.9. Operating parameters and performance results for separation of a mixture composed of 85% C_3H_6 and 15% C_3H_8 by PSA using $AgNO_3/SiO_2$ and $AlPO_4$-14 sorbents

Run No.	P_L (atm)	U_F (m/s)	U_P (m/s)	C_3H_6 Product Purity (%)	C_3H_6 Product Recovery (%)	C_3H_6 Product Throughput (kg Product/h/kg of Sorbent) × 10^3
$AgNO_3/SiO_2$						
1	1.0	0.40	0.30	99.33	42.99	3.56
2	1.0	0.10	0.30	99.14	59.60	3.45
3	0.2	0.40	0.30	99.17	54.51	7.95
4	0.2	0.01	0.35	99.24	75.11	6.54
$AlPO_4$-14						
5	1.0	0.50	0.25	99.34	37.69	3.77
6	1.0	0.05	0.30	99.27	65.16	2.00
7	0.2	0.25	0.20	99.29	56.88	6.14
8	0.2	0.10	0.20	99.18	72.33	6.07

Feed pressure $P_H = 7.0$ atm., step time = 60 s, feed T = 120 °C, P_L = des. press., U_F = feed velo., U_P = purge velo. (Rege and Yang, 2002, with permission).

constants (D/R^2) for C_3H_6 and C_3H_8 at 120 °C for both sorbents were $>10^{-3}$ 1/s, and the diffusion rates did not play an important role in the separation.

The results of the PSA simulations for the four-step cycle described above for the $AlPO_4$-14 and $AgNO_3/SiO_2$ sorbents with a feed composed of 85% C_3H_6 and 15% C_3H_8 at a temperature of 120 °C are summarized in Table 10.9. The simulation results for an olefin-leaner feed composed of 50% C_3H_6 and 50% C_3H_8 at the same temperature are shown in Table 10.10. The first two runs under each sorbent heading show the PSA performance at a desorption pressure of 1.0 atm. The next two runs were conducted at a lower desorption pressure of 0.2 atm.

A comparison of runs 2 and 6 in Table 10.9 for a feed of 85% C_3H_6 and 15% C_3H_8 shows that for a nearly similar product purity (99.2%) and product recovery (60–65%), the product throughput of the $AgNO_3/SiO_2$ sorbent exceeded that of $AlPO_4$-14 by over 70%. Similarly, for the lower desorption pressure of 0.2 atm, an inspection of runs 4 and 8 shows that at a product purity of about 99.2%, both the product recovery as well as the product throughput for the $AgNO_3/SiO_2$ sorbent are higher than those shown by the $AlPO_4$-14 sorbent. Thus clearly, $AgNO_3/SiO_2$ is a better sorbent for a hydrocarbon feed, which is richer in olefin compared with paraffin. This fact holds true irrespective of the operating pressure range of the PSA process.

The results of the PSA performance for a more olefin-lean feed of 50% C_3H_6 and 50% C_3H_8 can be obtained from Table 10.10. The comparison for this feed was less straightforward than that in Table 10.9. In general, it was difficult to exceed 99% purity in the case of $AlPO_4$-14 at reasonable product recovery and throughput. From runs 9 and 14, it can be seen that for a product purity in the range of 98–99%, the product recovery and product throughput were nearly the same (46% and 2×10^{-3} kg product/kg sorbent/h, respectively). However for

Table 10.10. Operating parameters and performance results for separation of a mixture composed of 50% C_3H_6 and 50% C_3H_8 by PSA using $AgNO_3/SiO_2$ and $AlPO_4$-14 sorbents

Run No.	P_L (atm)	U_F (m/s)	U_P (m/s)	C_3H_6 Product Purity (%)	C_3H_6 Product Recovery (%)	C_3H_6 Product Throughput (kg Product/h/kg of Sorbent) $\times 10^3$
$AgNO_3/SiO_2$						
9	1.0	0.40	0.35	99.32	45.92	2.09
10	1.0	0.10	0.35	99.46	58.71	0.63
11	0.2	0.40	0.40	99.42	59.18	5.16
12	0.2	0.05	0.45	98.52	71.65	2.61
$AlPO_4$-14						
13	1.0	0.25	0.20	98.41	47.03	1.91
14	1.0	0.20	0.20	98.29	51.35	1.88
15	0.2	0.30	0.25	98.83	57.53	4.22
16	0.2	0.25	0.23	98.65	63.91	4.81

Feed pressure $P_H = 7.0$ atm., step time = 60 s, feed T = 120 °C, P_L = des. press., U_F = feed velo., U_P = purge velo. (Rege and Yang, 2002, with permission).

the lower desorption pressure (0.2 atm.), runs 11 and 15 clearly demonstrate the superiority of $AgNO_3/SiO_2$ sorbent. All the three parameters, namely the product purity, recovery, and throughput for $AgNO_3/SiO_2$ exceeded that for $AlPO_4$-14.

The feed pressure was taken arbitrarily at 7 atm in the above simulation. The available pressures in the refinery gas streams and the purge streams from polyolefin reactors are higher than 7 atm. At higher feed pressures, the PSA performance with $AgNO_3/SiO_2$ will continue to improve because more olefin will be adsorbed, whereas that with $AlPO_4$-14 will not improve because its capacity is already saturated.

Due to the relatively high heats of adsorption, multiplicity of cyclic steady states can occur easily with this system. One unstable and two stable steady states were observed within certain regions for identical cycle conditions depending on the initial temperature or sorbate concentration in the PSA bed at startup. The exact initial bed temperature and concentration at which there was a switch in the stable steady states was identified (Rege and Yang, 2002). An example is shown in Figure 10.37. This figure shows the product purity for two initial bed temperatures: (1) 120 °C, as shown by the circles joined by a dotted line, and (2) 125 °C, denoted by diamonds connected by a solid line. Thus it was seen that product purity suddenly rose from 92 to 99.4% when the feed velocity was increased beyond 0.305 m/s for the bed that was started at 120 °C. However for an initial bed temperature of 125 °C, the jump in product purity occurred at an earlier feed velocity of 0.275 m/s. For the intermediate values of feed velocity between 0.275–0.305 m/s, two different cyclic steady states with distinctly different product purities and recoveries were obtained. Further details of the multiplicity are given in Rege and Yang (2002).

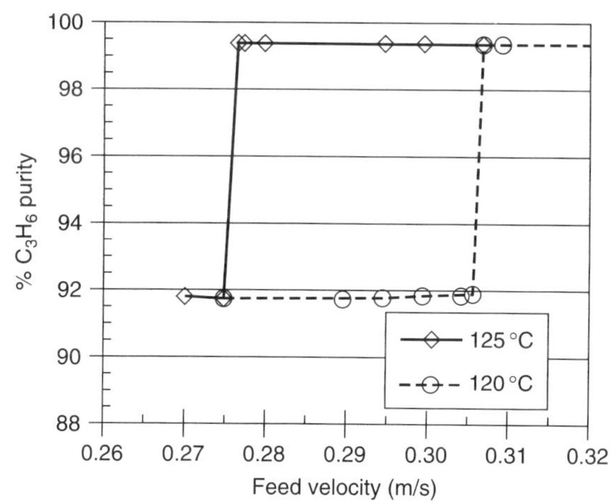

Figure 10.37. Variation of the C_3H_6 product purity as a function of feed velocity obtained at the cyclic steady state of the PSA process by using the $AlPO_4$-14 sorbent starting with two different initial temperatures in the bed. ($P_H = 1.0$ atm, $P_L = 0.1$ atm, step time = 60 s, $u_P = 0.30$ m/s, feed T = 120 °C, initial bed concentration = 5% C_3H_6, 95% C_3H_8) (Rege and Yang, 2002, with permission).

10.5.3. Other Sorbents

5A and 13X zeolites have also been considered for olefin/paraffin separations based on stronger adsorption of olefins. However, additional and substantial operations for the PSA are needed (e.g., Javelin and Fair, 1993). Rodriguez and co-workers have made interesting studied on the adsorption properties of these sorbents (Da Silva and Rodriguez, 1998; Grande et al., 2002). A vacuum swing cycle was studied for propane/propylene separation by using 13X (Da Silva and Rodriguez, 2001). Compared with the π-complexation sorbents, however, the separation results with π-complexation sorbents are significantly better.

Cho and co-workers (Cho et al., 2001; Choudary et al., 2002) have recently developed $AgNO_3$ supported on acid-treated clays, named Olesorb-1. Subatmospheric isotherms have been reported. The diffusion time constants for the C_2 and C_3 olefins and paraffins at ambient temperature are around 10^{-3} 1/s. From the isotherms and the diffusion time constants, the sorbent is suitable for olefin/paraffin separation. The isotherms are shown in Figure 10.38 and Figure 10.39. The isotherms of $AgNO_3/SiO_2$ are compared in these figures. For C_2 separation, the two sorbents appear to be comparable, although the isotherm of C_2H_4 on $AgNO_3/SiO_2$ is more linear, which helps separation. For C_3 separation, the isotherm of C_3H_6 on $AgNO_3/SiO_2$ is substantially higher as well as being more linear, although $AgNO_3/SiO_2$ also adsorbs more C_3H_8. No data on the Olesorb-1 were reported for higher pressures. It appears that the olefin capacity of Olesorb is also limited by its pore volume, since the isotherms seem to be leveling off at a low pressure. The Olesorb-1 has been evaluated in an experimental four-bed PSA system, using a VSA cycle similar to the one described above, except the feed pressure was at ~1.2 atm and the VSA was run at ambient temperature. Using a feed containing 83.1% ethylene (16.8% ethane and traces of other gases), Choudary et al. (2002) obtained a sorbent productivity of 0.04 kg C_2H_4/h/kg sorbent, at 99.8% ethylene product purity and 85% recovery.

From these results, it is seen that olefins can be recovered at high purities and reasonably high recoveries. The sorbent productivities are higher than that obtained in air separation by PSA/VSA. As mentioned, for air separation, a single PSA/VSA can produce over 200 tons/day of oxygen. Thus, it is feasible to recover olefins at that scale with a single PSA unit.

10.6. NITROGEN/METHANE SEPARATION

The separation of nitrogen from methane is becoming increasingly important for natural gas recovery and enhanced oil recovery. Natural gases that contain significant amounts of nitrogen need to be upgraded in order to meet the pipeline quality for minimum heating value specifications, typically >90% methane. This is the case with a large amount of natural gas reserves as well as aging natural gas wells. This also applies to enhanced oil recovery where nitrogen is injected into the reservoir. Nitrogen injection increases the level of nitrogen contamination in the gas fraction recovered from the reservoir, that is, petroleum gases,

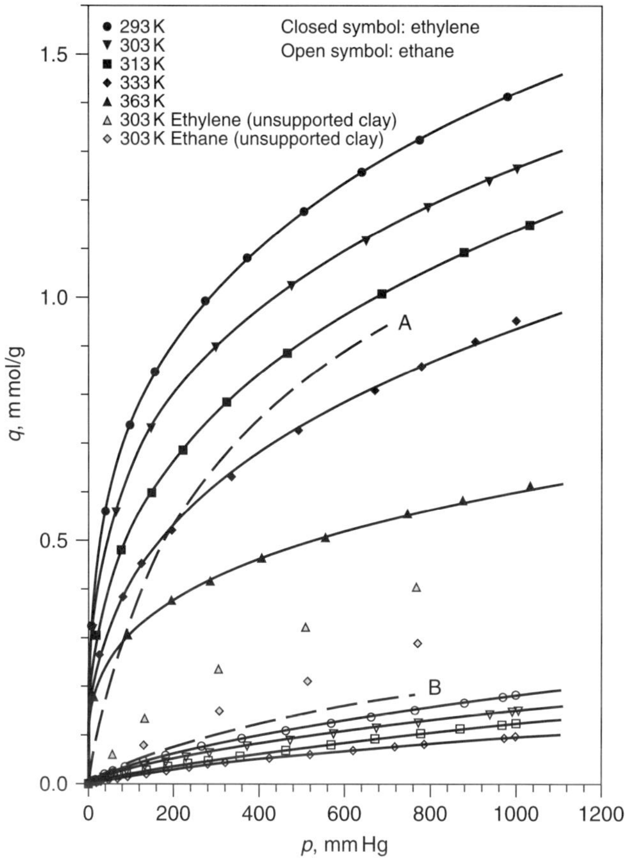

Figure 10.38. Isotherms of C_2H_4 and C_2H_6 on Olesorb-1 ($AgNO_3$/acid-treated clay) at various temperatures (solid lines) and that on $AgNO_3/SiO_2$ (dashed lines, A: C_2H_4 at 343 K, and B: C_2H_6 at 343 K). From Choudary et al., 2002 and Padin et al., 2000.

above their naturally occurring nitrogen concentration. Another application for this separation is the recovery of methane form coal mines when nitrogen concentration is also high. This separation is accomplished by cryogenic distillation. The more desirable separation technique is obviously adsorption, particularly pressure swing adsorption, because a high feed pressure is already available.

Despite the advantages of using adsorption for methane upgrading, this separation has been found particularly difficult because of the lack of a satisfactory sorbent. Such a sorbent needs to have a high nitrogen/methane selectivity. Methane has a higher polarizability (26×10^{-25} cm^{-3} for methane vs. 17.6×10^{-25} cm^{-3} for nitrogen). Both are nonpolar although nitrogen has a quadrupole. The equilibrium selectivity favors methane over nitrogen on all known sorbents, such as activated carbon, large-pore zeolites and molecular sieves, silica gel, and activated alumina. Therefore, the search for a sorbent has been directed toward kinetic separation, because there is a small but workable

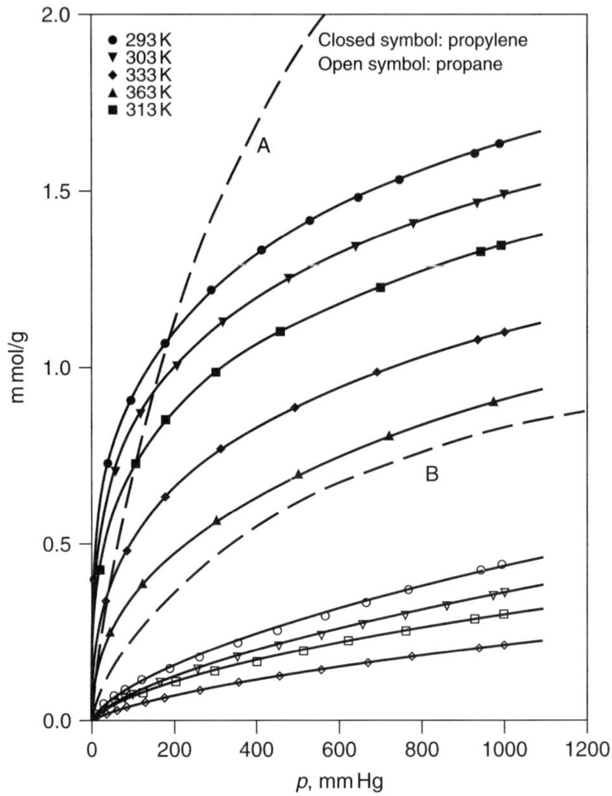

Figure 10.39. Isotherms of C_3H_6 and C_3H_8 on Olesorb-1 ($AgNO_3$/acid-treated clay) at various temperatures (solid lines) and that on $AgNO_3/SiO_2$ (dashed lines, A: C_3H_6 at 298 K, and B: C_3H_8 at 298 K). From Choudary et al., 2002; Rege and Yang, 2002.

difference in the kinetic diameters of these two molecules (3.8 Å for methane and 3.64 Å for nitrogen).

4A zeolite and carbon molecular sieves (CMS) have been examined for N_2/CH_4 separation. A process using 4A zeolite for this separation was developed by Habgood (1958), but this process was limited to low temperatures (−79 to 0 °C) and a high-methane feed content (≥90%). Ackley and Yang (1990) have demonstrated the use of carbon molecular sieve (CMS) for separation of N_2/CH_4 mixtures in pressure swing adsorption (PSA) processes but have also shown that the potential for CMS to achieve the desired pipeline quality (90% methane) is doubtful. The only two promising sorbents are clinoptilolites and titanosilicates, as discussed below.

10.6.1. Clinoptilolites

Clinoptilolite is a member of the heulandite group, and is the most abundant of the natural zeolites. Using Na^+ as the only charge-compensating cation, the formula for the ideal clinoptilolite is

$$Na_6Al_6Si_{30}O_{72} \cdot 24H_2O$$

The unit cell is monoclinic and is usually characterized on the basis of 72 O atoms and 24 water molecules, with Na^+, K^+, Ca^{2+}, and Mg^{2+} as the most common charge-balancing cations. The channel structure and the cation sites of clinoptilolite are discussed in Chapter 7 (7.4.1). It has a 2-dimensional channel structure that is formed by 8-oxygen rings and 10-oxygen rings. The transport channels have been discussed in detail by Ackley and Yang (1991b). The selectivity and rate of uptake of gases are strongly influenced by the type, number, and location of the cations in these channels.

Frankiewicz and Donnelly (1983) showed the promise of a calcium-exchanged clinoptilolite for N_2/CH_4 separation by a PSA process, but the product was below pipeline quality. This work stimulated a good deal of interest in further investigation. Two Japanese patent applications (61-255,994, in 1986, and 62-132,542 in 1987, see Chao, 1990) disclosed the use of natural clinoptilolite and Ca-exchanged forms for nitrogen removal from methane. Chao (1990) suggested the use of Mg-exchanged clinoptilolite for N_2/CH_4 separation.

The equilibrium isotherms of N_2 on various ion-exchanged clinoptilolite were reported by Ackley and Yang (1991a), and are shown in Figure 10.40. The clinoptilolite has been modified by ion exchange to fully exchanged forms of the monovalent cations K^+, Na^+, and H^+ and highly exchanged forms of the

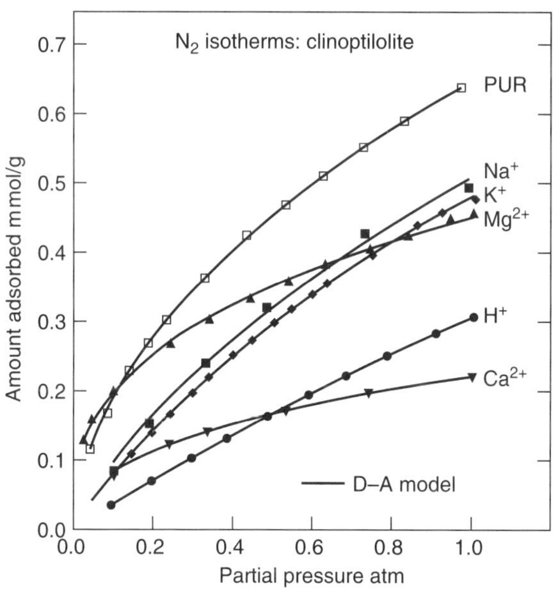

Figure 10.40. Isotherms of N_2 on ion-exchanged clinoptilolites at 300 K (Ackley and Yang, 1991a, with permission). PUR: purified form.

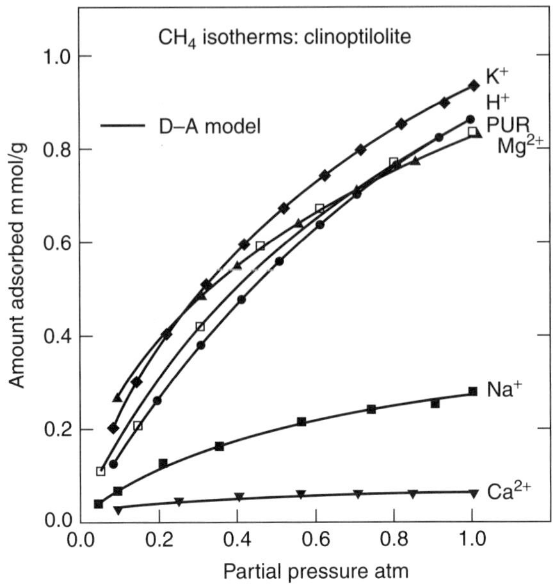

Figure 10.41. Isotherms of CH_4 on ion-exchanged clinoptilolites at 300 K (Ackley and Yang, 1991a, with permission). PUR: purified form.

bivalent cations Ca^{2+} (89%) and Mg^{2+} (72%). The detailed chemical analyses of these clinoptilolites were given elsewhere (Ackley and Yang, 1991a). The corresponding equilibrium isotherms of CH_4 are given in Figure 10.41. The isosteric heats of adsorption and the D–A isotherm fitting parameters are also given in Ackley and Yang (1991a).

The pore volumes were estimated from the D–A model parameters for both CH_4 and N_2 and they ranged approximately from 0.12 to 0.35 cm^3/g. From n-hexane adsorption, the pore volume was calculated as 0.138 cm^3/g (Ackley and Yang, 1991a). Barrer and Makki (1964) estimated a pore volume of 0.135 cm^3/g by measuring the amount of water vapor adsorbed on Na^+-exchanged clinoptilolite at a relative pressure ratio of $P/P_0 = 0.5$. Assuming 24 water molecules/unit cell, Barrer (1982) gave the porosity of Na^+ clinoptilolite as 0.34 cm^3/(cm^3 of crystal). This porosity is equivalent to a theoretical pore volume (based upon the assumed water content) of 0.186 cm^3/g. The pore volumes estimated from the N_2 data are higher and more variable than those determined from CH_4 isotherms. Breck (1974) has indicated that nitrogen always projects higher pore volumes in zeolites than other adsorbates of similar molecular size.

The selectivity ratio, defined here as the amount of CH_4 adsorbed per amount of N_2 adsorbed, is shown as a function of pressure in Figure 10.42 for the ion-exchanged clinoptilolites. The selectivity ratio varies for this class of adsorbents by more than one order of magnitude and includes a reversal of preference from CH_4 (H^+, K^+, Mg^+, PUR) to N_2 (Na^+, Ca^{2+}). Selectivity favoring N_2 has been induced by molecular sieving of CH_4. This range of selectivities is quite

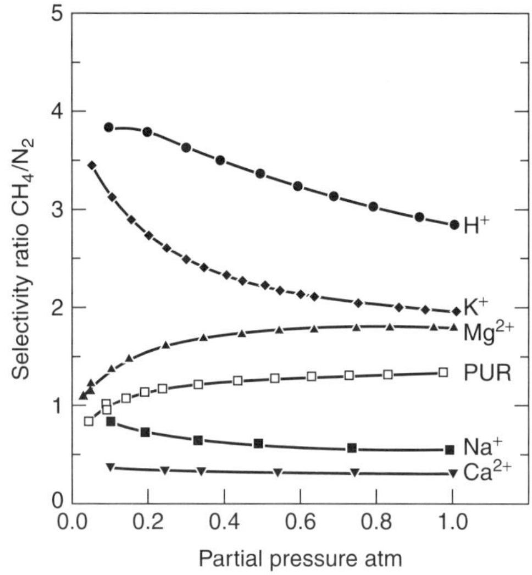

Figure 10.42. Pure-component equilibrium selectivity ratio of CH_4/N_2 adsorption on ion-exchanged clinoptilolites at 300 K (Ackley and Yang, 1991a, with permission). PUR: purified form.

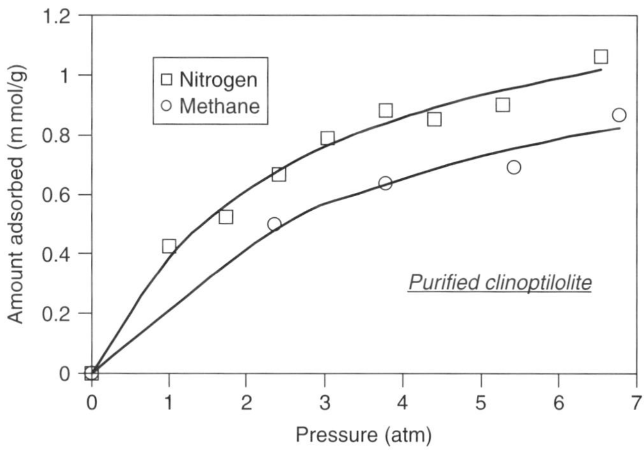

Figure 10.43. High-pressure nitrogen and methane isotherms on purified clinoptilolite at 22 °C (Jayaraman et al., 2002).

remarkable given the limited differences in properties of the two gases, which can be exploited for separation.

For PSA simulation, high-pressure isotherm data are needed. The high-pressure N_2/CH_4 isotherms for purified clinoptilolite and Mg-exchanged clinoptilolite are shown in Figures 10.43 and 10.44, respectively. An equilibration time of 2 h was used for each data point of the CH_4 isotherms. The N_2/CH_4 selectivity reflects molecular sieving.

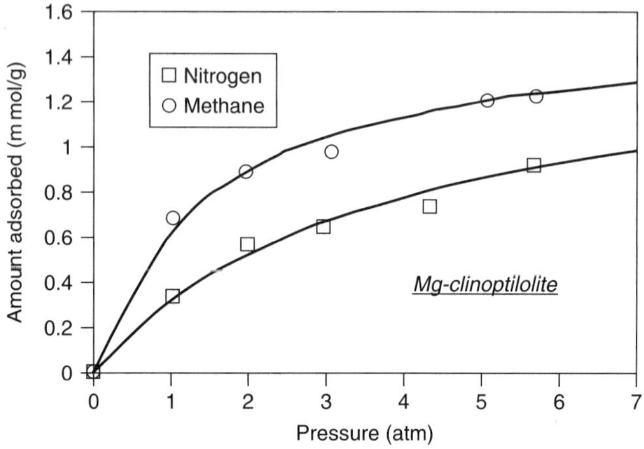

Figure 10.44. High-pressure nitrogen and methane isotherms on Mg-clinoptilolite at 22 °C (Jayaraman et al., 2002).

The high-pressure N_2 isotherms showed that the N_2 capacity is still increasing noticeably with pressure at 7 atm. This trend reflects the large pore volumes of clinoptilolites. Using a conservative value of 0.138 cm^3/g for the pore volume, the N_2 capacity would be 4 mmol/g (assuming pore filling with liquid nitrogen at a density of 0.808 g/cm^3).

The N_2/CH_4 uptake rates in the same ion-exchanged and purified forms of clinoptilolite have been measured and analyzed by Ackley and Yang (1991b). Diffusion takes place in two types of channels. Thus, the overall uptake rates can be analyzed with two different models, resulting in two sets of diffusion time constants. A better fit was obtained with the two-channel model. However, for PSA simulation, results from the two-channel model would make the simulation overly complicated. Hence an overall diffusion time constant will be used for each diffusion system. The overall uptake rates at 300 K are shown in Figures 10.45 and 10.46. The diffusion time constants obtained from the two-channel model are available in Ackley and Yang (1991b) and are not given here. Since low concentrations of N_2 and CH_4 (at 1 atm) were used, the diffusivities so obtained were at low loadings. The concentration dependences of the diffusivities (which were not strong) were also reported (Ackley and Yang, 1991b).

The true kinetic selectivity or initial separation factor can be determined as the ratio of the N_2/CH_4 mole uptake rates, as shown in Figure 10.47 for the various clinoptilolites at 300 K. The results in Figure 10.47 were obtained from the initial uptakes on a completely desorbed sample. The gas-phase concentrations of N_2 and CH_4 for the adsorption step are indicated in the figure. The time period for an adsorption step in a PSA cycle is typically in the range of 30–300 s, and a N_2/CH_4 selectivity >1 is necessary for the separation. The selectivities in Figure 10.47 approach the equilibrium selectivities at sufficiently long times. From the results, the Na^+, Ca^{2+}, and Mg^{2+} forms and the purified form are the best candidates for kinetic separation of N_2/CH_4.

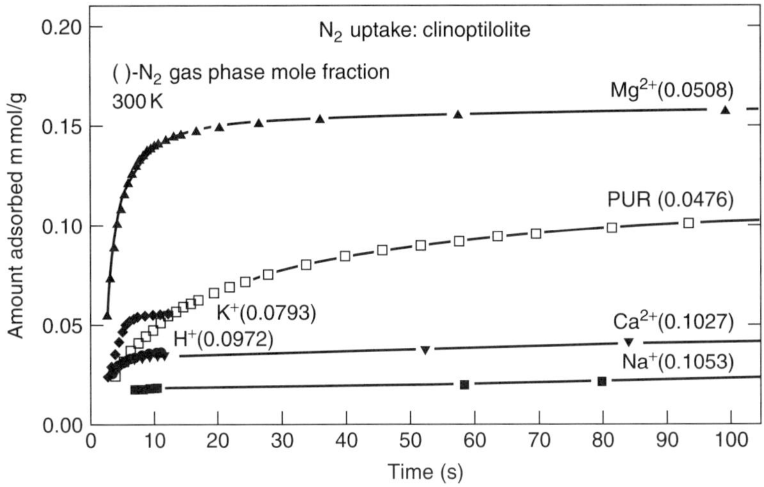

Figure 10.45. Uptake rates of N_2 by purified (PUR) and ion-exchanged clinoptilolites at 300 K and at N_2 partial pressures indicated in parentheses (in atm) (Ackley, 1991, with permission).

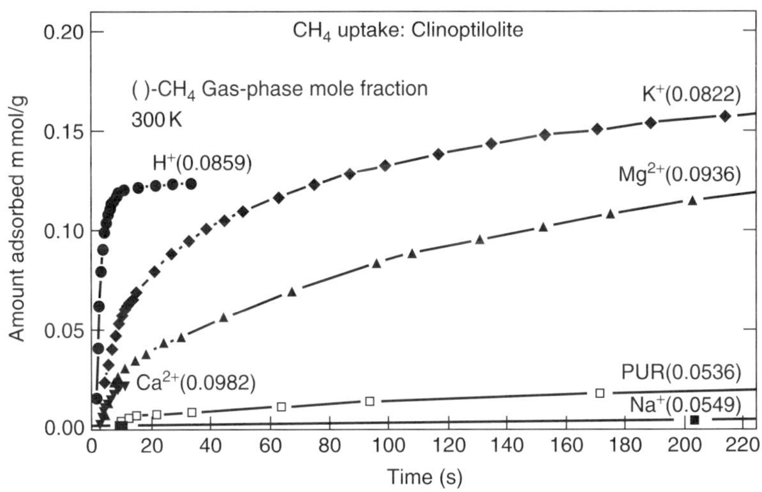

Figure 10.46. Uptake rates of CH_4 by purified (PUR) and ion-exchanged clinoptilolites at 300 K and at CH_4 partial pressures indicated in parentheses (in atm) (Ackley, 1991, with permission).

10.6.2. ETS-4

The syntheses of titanosilicate (TS) molecular sieves have been discussed in Chapter 7 (7.2.2). A particular form, named ETS-4, was synthesized by Kuznicki (1990) without the use of a template (7.2.2). The structure of this form is similar to that of the natural mineral zorite. Other similar titanosilicates have also been successfully synthesized (Chapman and Roe, 1990; Liu et al., 2000). The

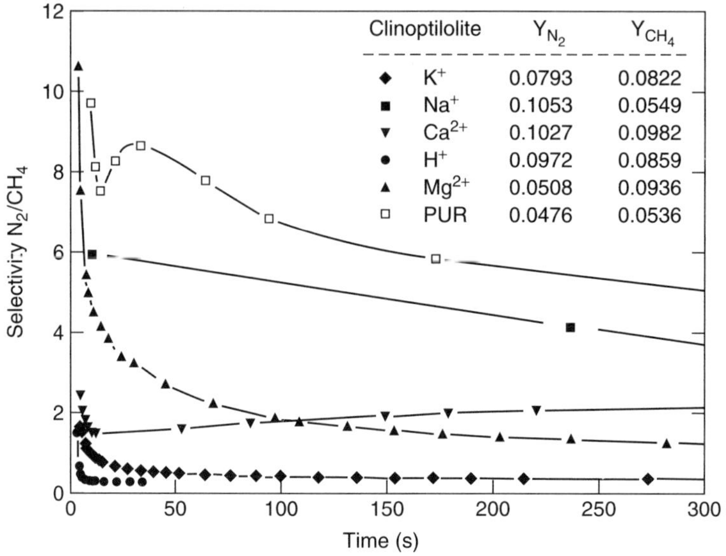

Figure 10.47. Kinetic selectivity expressed as ratio of pure-component mole uptake rates on clinoptilolites at 300 K and partial pressures (in atm) given as Y_{N_2} and Y_{CH_4} (Ackley and Yang, 1991b, with permission).

structure of the hydrated form has been determined (Philippou and Anderson, 1996; Cruciani et al., 1998; Braunbarth et al., 2000).

ETS-4 has a mixed octahedral/tetrahedral framework, and the structure is highly faulted in two directions (Braunbarth et al., 2000). With the faulting, the structure is accessible only through an open channel formed by 8-oxygen rings (of about 4 Å dimension) in the *b* direction. Upon heating, the structure of the hydrated form of Na-ETS-4 starts to collapse below 200 °C due to the loss of structural water chain present along the channel system. The ion-exchanged forms, exchanged with Sr^{2+} or Ba^{2+}, have a higher thermal stability, and the collapse temperature is extended to higher temperatures. For example, the structure of Sr-ETS-4 collapses at about 350 °C.

Through a series of studies by Kuznicki et al. (see references by Kuznicki et al., 2000; Braunbarth et al., 2000 and Kuznicki et al., 2001), the Ba-ETS-4 and Sr-ETS-4 have the best potential for tailoring as molecular sieves. The structure of the hydrated Sr-ETS-4 has been characterized by using neutron diffraction (Braunbarth et al., 2000). The ideal formula for Sr-ETS-4 is $NaSr_4Si_{12}Ti_5O_{38}(OH)\cdot 12H_2O$. Interestingly, upon heating, there is a gradual reduction of the 8-membrane ring pore opening with increasing temperature (Kuznicki et al., 2001). Unfortunately, the pore volume (as measured by the water sorption capacity) also decreases. Figure 10.48 shows the "equilibrium" sorption capacities of Sr-ETS-4 after heat-treatment at various temperatures. The Sr-ETS-4 heat treated at 300–315 °C is being used for N_2/CH_4 separation by PSA (Engelhard, 2001).

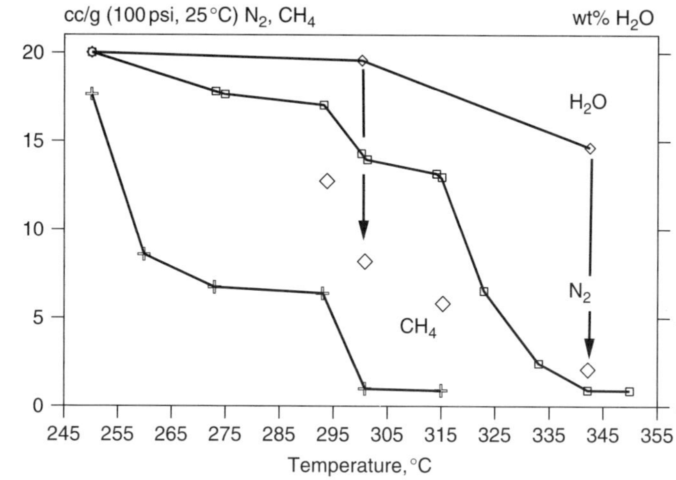

Figure 10.48. Equilibrium adsorption at 25 °C of N_2 and CH_4 at 100 psia and water at $P/P_0 = 0.84$ on Sr-ETS-4 pretreated at various temperatures (Kuznicki et al., 2000). The water capacities were later corrected in Kuznicki et al. (2001), shown by arrows and diamonds.

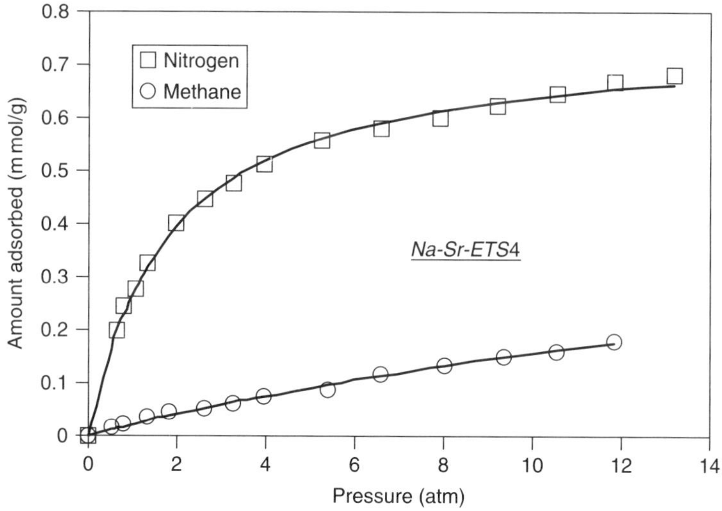

Figure 10.49. Nitrogen and methane isotherms on Sr-ETS-4 (heat-treated at 315 °C) at 22 °C (Jayaraman et al., 2002).

The N_2/CH_4 isotherms on Sr-ETS-4 that have been heat-treated at 315 °C are shown in Figure 10.49. The results are similar to those reported by Kuznicki et al. (2001). Comparing the N_2 isotherms of Sr-ETS-4 and clinoptilolites, the capacity of the latter is substantially larger. For example, at 5 atm pressure, the nitrogen capacity of the purified clinoptilolite is nearly twice that of Sr-ETS-4.

This comparison is a direct reflection of the relative pore volumes. The tendency of the N_2 isotherm on Sr-ETS-4 to form a plauteau at above 5 atm is also undesirable for PSA separation, particularly when a high feed pressure of natural gas is available (which is usually the case).

10.6.3. PSA Simulation: Comparison of Sorbents

From the N_2/CH_4 isotherms and diffusivity data, the most promising sorbents appeared to be Sr-ETS-4, Mg-clinoptilolite, and purified clinoptilolite. Thus, these three sorbents were compared for N_2/CH_4 separation (Jayaraman et al., 2002) by using a proven numerical PSA model (Rege et al., 1998). The overall diffusion time constants (D/R^2) at 295 K for N_2 were (in 1/s) 1.1×10^{-3} (purified clinoptilolite); 1.8×10^{-2} (Mg-clinoptilolite); 3.1×10^{-3} (Sr-ETS-4). The corresponding values for CH_4 were 2.0×10^{-5} (purified clinoptilolite); 6.0×10^{-5} (Mg-clinoptilolite); 1.2×10^{-3} (Sr-ETS-4) (Jayaraman et al., 2002). Heats of adsorption and other input data are available elsewhere (Jayaraman et al., 2002). The standard five-step PSA cycle was employed, which consisted of: pressurization, high-pressure feed, co-current depressurization (or blowdown), counter-current blowdown and evacuation, and low-pressure purge. The cycle conditions were optimized for each run. The results are summarized in Table 10.11.

With a feed mixture containing 85% CH_4 and 15% N_2, the results show that over 90% methane product purity at high recoveries and high throughputs are possible with both purified clinoptilolite and Sr-ETS-4. From the comparison, it is seen that for a feed pressure of 7 atm, the PSA results are comparable for purified clinoptilolite and Sr-ETS-4. For Mg-clinoptilolite, both methane purity and recovery were low. The sorbent productivities for both purified clinoptilolite and Sr-ETS-4 were higher than those obtained for air separation with zeolites. At feed pressures higher than 7 atm, the separation will further improve for clinoptilolites because the N_2 adsorption amount will continue to rise with pressure, although the improvement will be less for Sr-ETS-4 as the N_2 capacity is already nearly saturated at 7 atm. PSA upgrading of natural gas by Sr-ETS-4 has already been commercialized (Engelhard, 2001).

10.7. DESULFURIZATION OF TRANSPORTATION FUELS

Due to worldwide environmental mandates, refiners are facing the challenge of producing increasingly cleaner fuels (Avidan et al., 2001). The primary focus of the new regulations is the reduction of sulfur in gasoline and diesel. Other rules stipulate reduced levels of aromatics, especially benzene, benzene derivatives, olefins, and oxygenates. Sorbents for removal of aromatics will be discussed in Section 10.8 of this chapter.

In 1998, the European Union first mandated new sulfur specifications for drastically reduced levels that started to be phased in from the year 2000 (Knudsen et al., 1999). Similar regulations were legislated in the United States and elsewhere soon after. The EPA Tier II regulations require reductions of sulfur in

Table 10.11. Comparison of various sorbents for N_2/CH_4 separation using the standard five-step PSA cycle

Run No.	Co-Current Blowdown Pressure (bar)	Feed Velocity (m/s)	CH_4 Product Purity (%)—Average of Co-Current Blowdown and Ads Step	CH_4 Product Recovery (%)	CH_4 Product Throughput (kg Product/h/kg of Sorbent)
ETS-4					
1	1	0.05	97.255	53.192	0.081
2	1	0.12	95.244	71.281	0.172
3	1	0.20	92.821	80.392	0.276
4	2	0.20	92.732	79.327	0.272
Purified clinoptilolite					
5	1	0.05	94.486	80.653	0.123
6	1	0.12	92.050	87.812	0.212
7	1	0.20	90.148	91.634	0.315
8	2	0.05	95.382	73.131	0.112
Mg-clinoptilolite					
9	1	0.05	89.057	63.151	0.114
10	2	0.05	90.630	55.165	0.100
11	4	0.05	93.273	41.648	0.075

Feed = 85% CH_4 + 15% N_2 at 298 K and 7 atm. Desorption pressure = 0.4 atm. Column size: 0.2 m diameter and 2.0 m length. Cycle step times: $t_{press} = 30$ s; $t_{ads} = 60$ s; $t_{co\ bdn} = 10$ s; $t_{cn\ bdn} = 30$ s; $t_{des} = 60$ s (Jayaraman et al., 2002).

diesel from the current average of 500 ppmw to 15 ppmw by June 2006, and that in gasoline from 350 ppmw to 30 ppmw by January 2005 (Avidan et al., 2001). The detailed sulfur standards are complicated because different standards are applied to individual refinery, corporate pool, and per gallon basis (Avidan et al., 2001).

Removal of sulfur-containing compounds is an important operation in petroleum refining, and is achieved by catalytic processes operated at elevated temperatures (~300 °C) and pressures (20–100 atm H_2) using Co-Mo/Al_2O_3 or Ni-Mo/Al_2O_3 catalyst (Gates et al., 1979). The hydrodesulfurization (HDS) process is highly efficient in removing thiols, sulfides, and disulfides, but less effective for thiophenes and thiophene derivatives. Thus, the sulfur compounds that remain in the transportation fuels are mainly thiophene, benzothiophene, dibenzothiophene, and their alkylated derivatives. The least reactive derivatives are the dibenzothiophenes with methyl groups at the 4- and 6-positions, that is, positions adjacent to *S*. 4-methyldibenzothiophene and 4,6-dimethyldibenzothiophene are the refractory sulfur compounds that make deep desulfurization by HDS extremely difficult. Accompanying deep desulfurization is the saturation of olefinic compounds resulting in octane loss of about 10 numbers (Avidan and Cullen, 2001).

Another need for deep desulfurization is for potential application in fuel cell. Gasoline is the ideal fuel for fuel cell because of its high-energy density, ready availability, safety, and ease in storage. However, to avoid poisoning of the catalyst for the water–gas shift reaction and that in the electrode of the fuel cell, the sulfur concentration should be preferably below 0.1 ppmw. To reduce the sulfur content of diesel from 500 ppmw to this level, an estimate showed that the HDS reactor size needed to be increased by a factor of 7 (Whitehurst et al., 1998). Another estimate showed that in order to reduce the sulfur level in diesel from 300 to less than 10 ppm, the HDS reactor volume needed to be increased by a factor of about 15 at 600 psi, or by a factor of 5 at 1,000 psi (Parkinson, 2001; Avidan and Cullen, 2001).

Faced with the severely high costs of environmental compliance, a number of new technologies have been contemplated for post-treating of the FCC naphtha (Avidan et al., 2001). One commercialized technology, named *S Zorb*, has been announced by Phillips Petroleum Company, and it is claimed to remove the refractory sulfur species such as 4,6-dimethyldibenzothiophene (brochure of Phillips). No detailed information is available on the "sorbent." The sorbent is actually a highly sulfur-poisonable catalyst, used in a fluidized-bed reactor, for a hydrogenation reaction. The reaction is performed at 650–775 °F and 100–300 psig in H_2 for gasoline, or 700–800 °F and 275–500 psig in H_2 for diesel. The reaction reported for benzothiophene is: benzothiophene + $H_2 \rightarrow$ S(ads.) + ethyl benzene. The S atom is deposited on the catalyst/sorbent. The catalyst is regenerated with air (forming SO_2), followed by reduction with H_2. The reduced catalyst is recycled to the reactor.

Another new technology using ultrasound was disclosed by Avidan and Cullen (2001), Yen et al. (2002) and Mei et al. (2003). This is a two-step process, called *SulphCo* process, named after the company that developed it. In the first step, thiophenic compounds are oxidized in an ultrasonic reactor to form sulfoxides (with 1 oxygen attached to the sulfur atom) and sulfones (with 2 oxygen atoms attached to sulfur). The thiophenic compounds are oxidized in gas-bubble cavities generated by ultrasound. The sulfoxides and sulfones are subsequently removed by solvent extraction. Following Collins et al. (1997), H_2O_2 (oxidant) and a heteropolyanion (catalyst) were used for oxidation. Still another process was disclosed by Research Triangle Institute (Chemical Engineering Progress, 2001). A sorbent is reacted with the sulfur compounds in the naphtha from FCC in a "transport reactor" designed to minimize the reactor volume. The sorbent is regenerated with air to form SO_2. Both reactions (sorption and regeneration) are conducted at elevated temperatures.

The new challenge is to use adsorption to selectively remove these sulfur compounds from transportation fuels (gasoline, diesel, and jet fuels). Since adsorption would be accomplished at ambient temperature and pressure, success in this development would lead to a major advance in petroleum refining. However, success would depend on the development of a highly selective sorbent with a high sulfur capacity, because the commercial sorbents are not adequate for this application.

10.7.1. Fuel and Sulfur Compositions

The compositions of transportation fuels vary widely depending on the crude oils used, the refining process, the product demand, and the product specifications. The approximate compositions of gasoline, diesel, and jet fuel are given in Table 10.12. Branched and *n*-alkanes are the main ingredients of these fuels, typically 70–80%. The major alkane is *n*-hexane and the main branched alkanes are C_5 and C_6 compounds. The aromatics are mainly benzene, toluene, xylenes, and alkyl benzenes, totaling about 20–30%.

The sulfur compounds in transportation fuels can be analyzed with X-ray fluorescence spectroscopy or by gas chromatography equipped with a capillary column plus a flame photometric detector. The remaining sulfur compounds after hydrodesulfurization (HDS) are mainly thiophene (T), benzothiophene (BT), dibenzothiophene (DBT) and their alkylated derivatives. Figure 10.50 shows the substitution positions of DBT. The alkylated derivatives with alkyl groups at the 4- and 6-positions are most difficult to remove and are referred to as refractory species. Ma et al. (2001) showed GC-FPD chromatograms of a sample each of gasoline, diesel, and jet fuel, reproduced in Figure 10.51. The FPD detects only sulfur compounds. The dominant sulfur compounds in the gasoline were (in decreasing order) 3-MT, BT, T, 2-MT, and 2,4-DMT. Those in the diesel were 4-MDBT, 4,6-DMDBT, 2,4,6-TMDBT, 3,6-DMDBT, DBT, 2,3,7-TMBT, 2,3,5-TMBT, 2,3-DMBT, and others. The sulfur compounds in the jet fuel were

Table 10.12. Typical compositions of transportation fuels (vol %)

	Gasoline[a]	Diesel[b]	Jet Fuel[c]
Boiling range (°C)	40–204	232–350	330–510
Aromatics	30.5	17	18
Olefins	1.8	5	2
n-alkanes	17.3	—	—
Branched alkanes	32	—	—
Cycloalkanes	5	—	—
Saturates	—	78	—
Paraffins	—	—	60
Naphthenes	—	—	20

[a] Sciences International, Inc., "Toxicological Profile for Gasoline," Report to Department of Health and Human Services, June, 1995.
[b] Ma et al., 1994.
[c] Ma et al., 2002.

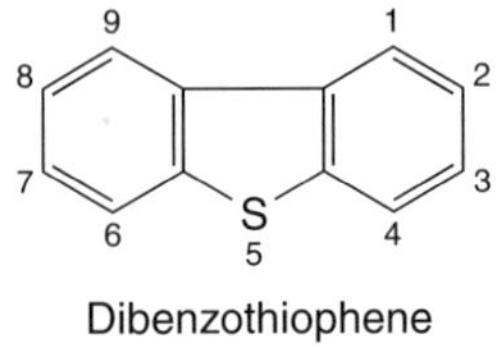

Figure 10.50. Dibenzothiophene (DBT). Alkylated DBT with alkyl (e.g., methyl) groups at the 4- and 6-positions, that is, 4-MDBT and 4,6-DMDBT, are the refractory sulfur compounds that are most difficult to remove by HDS or sorbents aiming at bonding with S due to steric hindrance.

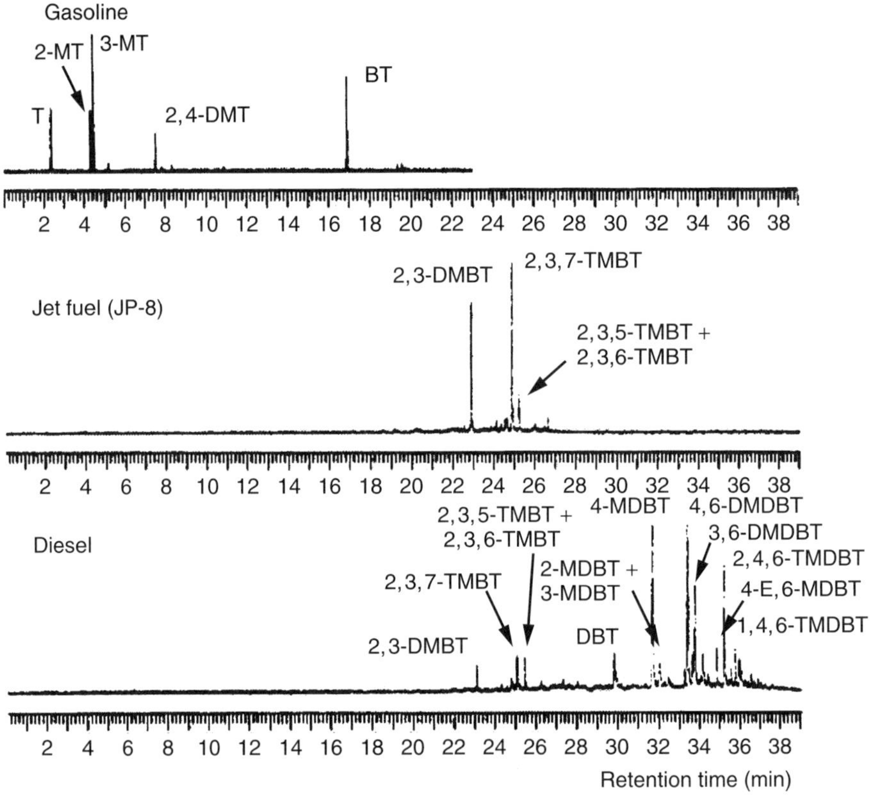

Figure 10.51. GC-FPD (flame photometric detector — for sulfur only) chromatograms for three transportation fuels (Ma et al., 2001, with permission).

2,3,7-TMBT, 2,3-DMBT, and the minor species 2,3,5-TMBT and 2,3,6-TMBT. The more detailed analyses of a diesel fuel by Ma et al. (1994) showed 61 sulfur compounds. These were, again, almost exclusively alkyl benzothiophenes and alkyl dibenzothiophenes. An example for the sulfur composition of gasoline is given in Table 10.13.

Table 10.13. Example of heteroatom contents in the FCC gasoline

Heteroatoms	Content, ppmw
Nitrogen	16.0
Oxygen	14.0
Mercaptan sulfur	24.2
Sulfide sulfur	7.3
Thiophene sulfur	61.9
C_1 thiophene sulfur	115.0
C_2 thiophene sulfur	130.6
C_3 thiophene sulfur	90.9
C_4 thiophene sulfur	88.0
Benzothiophene and dibenzothiophene sulfur	238.1
Total	786.0

Irvine, 1998.

From these analyses, the sorbent to be developed must have the highest affinities for the thiophenic compounds, medium affinities for the aromatics, and lowest possible affinities for the alkanes and branched alkanes.

10.7.2. Sorbents Studied or Used

The use of sorbents for sulfur removal dates back to the use of bauxite (i.e., the ore for aluminum smelting) to adsorb mercaptans from various petroleum fractions (Purdy, 1958). Red mud (the iron-rich waste from the Bayer process for alumina extraction from bauxite) has also been used as a sorbent for sulfur removal from petroleum oils.

The recent search for sorbents for liquid fuel desulfurization has taken two paths, both based on trial-and-error. One approach has been simply testing the commercial sorbents. The other approach has an origin in HDS catalysis. This approach is aimed at sorbents that are good catalysts for HDS, with the hope that such sorbents would form a bond with the sulfur atom of the thiophenic compounds. Forming such a bond may require above-ambient temperatures. The commercialized process discussed above, *S Zorb*, is based on an interesting hybrid of catalyst and sorbent. The use of π-complexation is an entirely different approach and will be discussed separately.

The commercial sorbents (activated carbon, activated alumina, and zeolites) have all been studied for sulfur removal. The zeolites included 5A; 13X (Salem, 1994; Salem and Hamid, 1997); various ZSM's, including ZSM-5 and silicalite (Weitkamp et al., 1991); and ion-exchanged zeolites (Michlmayr, 1980; Vansant et al., 1988). The ion exchange was intended for the exchanged cation to form a bond with the sulfur atom in thiophene. Although some of these sorbents showed selectivities toward thiophene, the thiophene capacities were fairly low in all cases. A summary of the thiophene capacities is given in Table 10.14. Even with the low thiophene capacities, an adsorption process, named Irvad Process, has already been commercialized by using activated alumina as the sorbent (Irvine, 1998; Ondrey, 1999).

The sorbent used in the Irvad process is Alcoa Selexsorb (CD grade), which is tailored for polar compounds. The process is essentially a TSA process. Adsorption is performed in a countercurrent moving bed with a slurry of fine-sized alumina. Desorption is performed with hydrogen at various temperatures up to 520 °F. Hydrogen is used for its high thermal conductivity and heat capacity, as well as its ready availability in the refineries. Gasoline products with sulfur as low as 0.5 ppmw were claimed (Irvine, 1998).

Ag-exchanged faujasite was claimed for thiophene removal (Michlmayr, 1980). Curiously, the preferred temperature for adsorption was 200–350 °C, and the sorbent was not dehydrated by heat-treatment prior to adsorption. Consequently, the sulfur capacities were very low, at 0.07–0.15 wt % for Ag-Y. The highest sulfur capacity (of 0.2 wt %) was obtained with the lowest Ag content, using USY zeolite. This sorbent was apparently intended for bonding the sulfur atom with Ag. It was clearly not intended for π-complexation (Michlmayr, 1980).

Vansant et al. (1988) investigated Cu(II)Y for thiophene removal. Curiously also, the Cu^{2+} exchanged Y zeolite was purposely heat-treated in *air* (to 200–550 °C) in order to maintain the Cu^{2+} in the divalent state, rather than treating in an inert or reducing atmosphere. Again, adsorption by π-complexation was clearly not intended. As a result, low thiophene capacities (the highest was 1.6 wt %) were obtained.

Ma et al. (2001, 2002a) described a sorbent for removal of thiophenic compounds from a jet fuel. Their sulfur breakthrough result is shown in Figure 10.52. The jet fuel contained 490 ppm sulfur. The sulfur capacity was given as 0.015 g sulfur per ml of sorbent. No details were given on the sorbent except that it was a transition metal compound supported on silica gel at 5 wt % loading (Ma et al., 2001) or simply given as a transition metal (Ma et al., 2002). It was obviously intended for the transition metal to form a bond with the sulfur atom of the thiophenic compound. The sulfur capacity of this sorbent was not high. Moreover, from Figure 10.52, it is seen that the breakthrough had already begun at the second sampling data point (~13 min), which indicates that the sorbent lacked selectivity. Ma et al. (2002b) subsequently reported the breakthrough curve of a model diesel on the same sorbent, showing a breakthrough capacity of only 1 cc/g.

10.7.3. π-Complexation Sorbents

A systematic approach has been taken by Yang and co-workers in the search for a thiophene selective sorbent, leading to the π-complexation sorbents (Yang et al., 2001; Yang et al., 2002; Takahashi et al., 2002; Hernandez-Maldonado and Yang, 2003). Effective π-complexation sorbents for sulfur removal (Yang et al., 2002) include Cu(I)Y, AgY, $CuCl/\gamma\text{-}Al_2O_3$, $AgNO_3/SiO_2$, and others. The preparation and characterization of these sorbents are described in Chapter 8.

Table 10.14. Thiophene capacities for various sorbents

	Capacity (wt %)	Adsorption Conditions/Notes	Reference
Activated carbon	0.8	2×10^{-5} atm, 120 °C	Takahashi et al., 2002
Act. Alumina Alcoa Selexsorb	nil	2×10^{-5} atm, 120 °C	Takahashi et al., 2002
NaY	~0.3	2×10^{-5} atm, 120 °C	Takahashi et al., 2002
USY	0	2×10^{-5} atm, 120 °C	Takahashi et al., 2002
AgY (not dehydrated)	0.07–0.15	100 ppmw thiophene added to gasoline ($S = ?$) (liq., 22 °C)	Michlmayr, 1980
AgUSY (not dehydrated)	0.2	100 ppmw thiophene added to gasoline ($S = ?$) (liq., 22 °C)	Michlmayr, 1980
Cu(II)Y	1.6	4,320 ppm thiophene in benzene liq., 25 °C	Vansant et al., 1988
ZSM-5	1.5–1.7	1,800 ppm (gas, 50 °C)	Weitkamp et al., 1991
Silicalite	2.2	1,800 ppm (gas, 50 °C)	Weitkamp et al., 1991
Transition metal X	~1 wt % (0.015g/cc)	490 ppmw thiophenics in jet fuel	Ma et al., 2002a
AgY (vapor)	12.6	2×10^{-5} atm, 120 °C	Takahashi et al., 2002
AgY (liquid)	7.5	2,000 ppmw thiophene/ *n*-octane	Yang et al., 2002 Patent pending; Hernandez-Maldonado and Yang, 2002
Cu(I)Y (liquid)	21.42	2,000 ppmw thiophene/ *n*-octane	Yang et al., 2002, Patent pending; Hernandez-Maldonado and Yang, 2002
Cu(I)Y (liquid)	10.75	500 ppmw thiophene/ *n*-octane	ditto

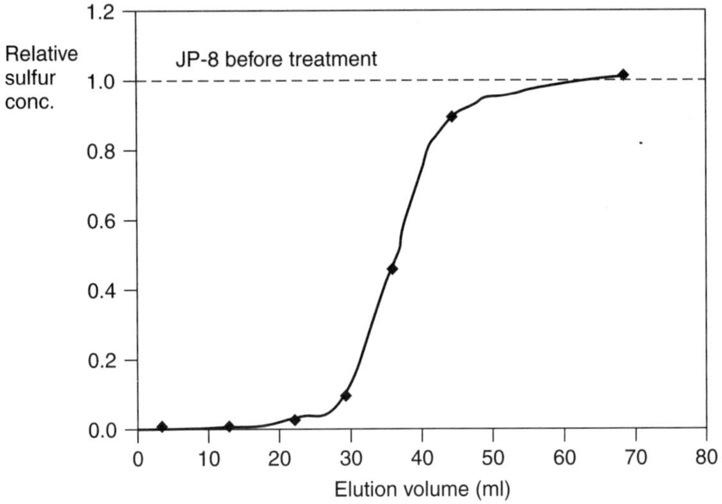

Figure 10.52. Breakthrough curve of thiophenic sulfur compounds from a jet fuel from a column containing 3.2 ml of a "transition metal" sorbent at ambient temperature (Ma et al., 2002a, with permission).

Table 10.15. Properties for evaluating van der Waals and electrostatic interactions

	Benzene	Thiophene
Polarizability, cm^3	10.3×10^{-24}	9.7×10^{-24}
Magnetic susceptibility, cm^3/molecule	9.1×10^{-29}	9.5×10^{-29}
Dipole moment, debye	0	0.55
Quadrupole moment, 10^{-26} esu cm^2		
Q_{xx}	2.8 ± 1.6	1.7 ± 1.6
Q_{yy}	2.8 ± 1.6	6.6 ± 1.5
Q_{zz}	-5.6 ± 2.8	-8.3 ± 2.2

Quadrupole moments taken from Sutter and Flygare, 1969.

Thiophene vs. Benzene. The sorbent needs to have stronger interactions with thiophene than benzene. The basic properties for van der Waals and electrostatic interactions for these two molecules are compared in Table 10.15. Benzene has a slightly higher polarizability, whereas Thiophene has a slightly higher magnetic susceptibility. Thus, for a given nonpolar sorbent, benzene may be slightly more strongly adsorbed due to slightly higher van der Waals interactions. Thiophene is slightly more polar than benzene. Hence, on polar surfaces thiophene would be adsorbed more strongly by a small margin. However, adsorption by the van der Waals and electrostatic interactions alone would lead to heats of adsorption below 10 kcal/mol, which are not high enough for efficient purification. Thus, we resort to weak chemical bonds such as π-complexation.

Table 10.16. Summary of energies of adsorption (*E*) for thiophene and benzene in kcal/mol, calculated from molecular orbital theory

Sorbent	E_{ads}(Thiophene)	E_{ads}(Benzene)
Cu^+Z^-	21.4	20.5
Ag^+Z^-	20.0	19.1

(Z^- denotes zeolite anion using the cluster model shown in Figure 8.3; From Yang et al., 2002.)

Ab initio molecular orbital calculations were performed for the bonding between benzene or thiophene with Cu^+ and Ag^+ exchanged zeolites (Yang et al., 2002; Takahashi et al., 2002). The results on energy calculations are summarized in Table 10.16. As described in Chapter 8, the energy predictions from *ab initio* methods (e.g., at high levels of basis sets of Gaussian) are fairly accurate. The calculations showed that the π-complexation bonds for thiophene are stronger than that with benzene, and that Cu(I)Y adsorbs more strongly than AgY does; that is, the relative strengths of the π-complexation bonds follow:

For the same sorbate: $Cu^+ > Ag^+$

For the same sorbent: thiophene > benzene

Furthermore, the predicted bond strengths in the neighborhood of 20 kcal/mol are well-suited for purification.

Vapor-phase adsorption isotherms of benzene and thiophene on various sorbents including CuY and AgY were measured first in order to assess their suitabilities for sulfur removal, as well as for the removal of aromatics. Direct correlations have yet to be established between the vapor-phase isotherms and the liquid-phase isotherms. However, some efforts have already been made in this direction by using the potential theory approach where log (C_s/C) replaces log (P_s/P) in the potential energy expression (C_s is the saturated concentration in solution and P_s is the saturated vapor pressure, see Manes, 1998). A sorbent capable of adsorbing thiophene at very low partial pressures in the gas phase should also be able to do so in the liquid phase.

NaY, AgY, and Cu(I)Y. Figure 10.53 shows the equilibrium isotherms of benzene and thiophene on NaY at 120 and 180 °C. More benzene was adsorbed on NaY than thiophene at pressures $<10^{-2}$ atm. From Table 10.15, benzene has a higher polarizability. This result indicates that the van der Waals interactions and the interaction between the field (of Na^+) and induced dipole dominated the adsorption; both interactions depend on the polarizability (see Chapter 2). At higher pressures over 1×10^{-2} atm, pore filling occurred and the benzene adsorption amount became smaller because benzene has a larger molar volume than that of thiophene. Na-Y does not have selectivity for thiophene;

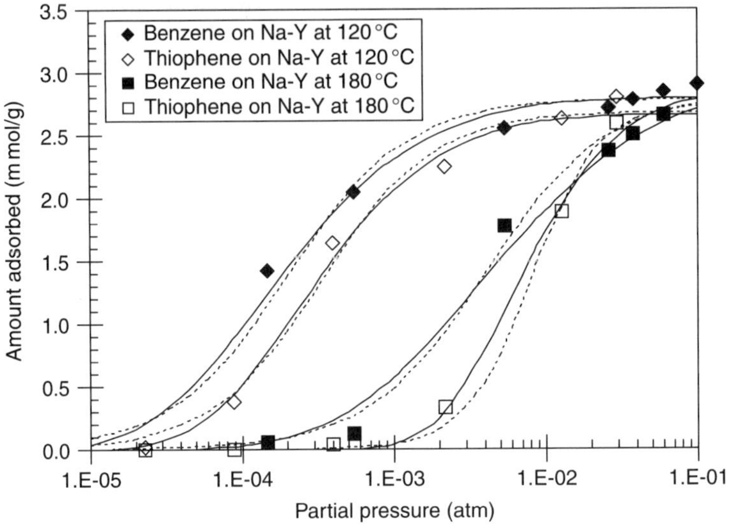

Figure 10.53. Pure-component equilibrium isotherms of benzene and thiophene on Na-Y (Si/Al = 2.43) at 120 and 180 °C. Curves are fitted with Dubinin–Astakhov (solid line) and Langmuir–Freundlich (dotted line) isotherms (Takahashi et al., 2002; Yang et al., 2002).

however, it adsorbs both benzene and thiophene quite strongly, as evidenced by the still measurable amounts adsorbed at partial pressures as low as 10^{-4} to 10^{-3} atm.

Figures 10.54 and 10.55 show the isotherms of benzene and thiophene on AgY and CuY, respectively. Compared with NaY, these sorbents adsorbed significantly more thiophene/benzene at pressures $<10^{-3}$ atm, and nearly the same amounts at higher partial pressures. This result was a clear indication of π-complexation with Ag^+ and Cu^+; since Na^+ could not form π-complexation bonds. However, the difference of thiophene/benzene adsorption amounts did not reflect the relative strengths of π-complexation between Cu^+ and Ag^+ because the Cu^+ exchange was not complete (Takahashi et al., 2002). The neutron activation analyses of the sorbent samples showed that the Ag^+ exchange was 100% but the Cu^+ exchange was only 46%. According to EPR analysis (Takahashi et al, 2001) only a half of the Cu^{2+} was auto-reduced to Cu^+ after heat-treatment at 450 °C for 1 h in He. On a per-cation basis, the π-complexation with Cu^+ was stronger than that with Ag^+. This result was in agreement with the molecular orbital prediction. To understand the relative strengths of π-complexation between Ag^+ and Cu^+, the thiophene adsorption amounts at 2×10^{-5} atm were normalized based on Ag^+ and Cu^+ contents, and the results are shown in Table 10.17. It is seen that Cu^+ could adsorb more thiophene molecules per cation. In fact, 0.92 thiophene molecule per Cu^+ was obtained at 2×10^{-5} atm at 120 °C. This amount was due to Cu^+, since the amount adsorbed by NaY at the same pressure was negligible. At the same pressure, only 0.42 thiophene/Ag^+ was obtained. This result indicated strong π-complexation bonds between both Cu^+ and Ag^+, and that the bond

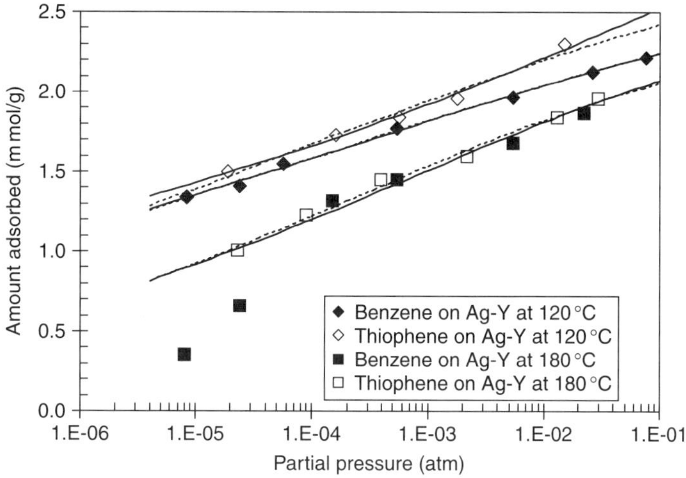

Figure 10.54. Pure-component isotherms of benzene and thiophene on Ag-Y (Si/Al = 2.43) at 120 and 180 °C. Curves are fitted with Dubinin–Astakhov (solid line) and Langmuir–Freundlich (dotted line) isotherms (Takahashi et al., 2002; Yang et al., 2002).

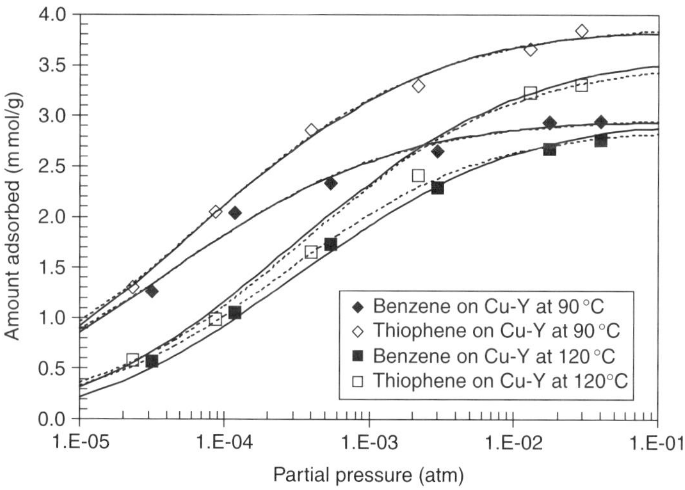

Figure 10.55. Pure-component isotherms of benzene and thiophene on Cu(I)Y (Si/Al = 2.43) at 90 and 120 °C. Curves are fitted with Dubinin–Astakhov (solid line) and Langmuir–Freundlich (dotted line) isotherms (Takahashi et al., 2002; Yang et al., 2002).

with Cu^+ was stronger. The results are in full agreement with predictions from molecular orbital theory.

The molar density of liquid thiophene is 83 cc/mol at 120 °C. 1.5 mmol thiophene adsorbed per gram of AgY at 2.3×10^{-5} atm corresponded to 0.125 cc liquid thiophene per gram of AgY. The thiophene liquid volume of 0.125 cc

Table 10.17. Normalized adsorption of thiophene on AgY and Cu(I)Y at 2 × 10^{-5} atm and 120 °C

	AgY	Cu(I)Y
Cation content (wt %)	38.2	8.29
Thiophene adsorbed (molecules/cation)	0.42	0.92

Ion exchange for Cu^{2+} was only 46% in this sample.

was nearly one-half of the pore volume of AgY, suggesting that one-half of the pore was filled with thiophene. As a result, no increase in thiophene adsorption was obtained by using Ag-exchanged type X zeolite (with Si/Al = 1.25, i.e., doubling the cation content) (Takahashi et al., 2002).

Na-ZSM-5, Activated Carbon, Activated Alumina, and Comparisons. As mentioned, Weitkamp et al. (1991) used Na-ZSM-5 for thiophene/benzene purification and showed that Na-ZSM-5 could remove thiophene from benzene in fixed bed breakthrough experiments. King et al. (2000) studied the breakthrough behaviors of thiophene, methyl, and dimethyl thiophenes in solutions of toluene and *p*-xylene from a fixed bed of H-ZSM-5, and showed that a sulfur-free effuent could be obtained. Figure 10.56 shows isotherms on Na-ZSM-5. Although thiophene and benzene isotherms turned out to be virtually the same for Na-ZSM-5, it is important to note that thiophene isotherms were fairly flat over the pressure range for purification applications. The small difference in adsorbed amounts

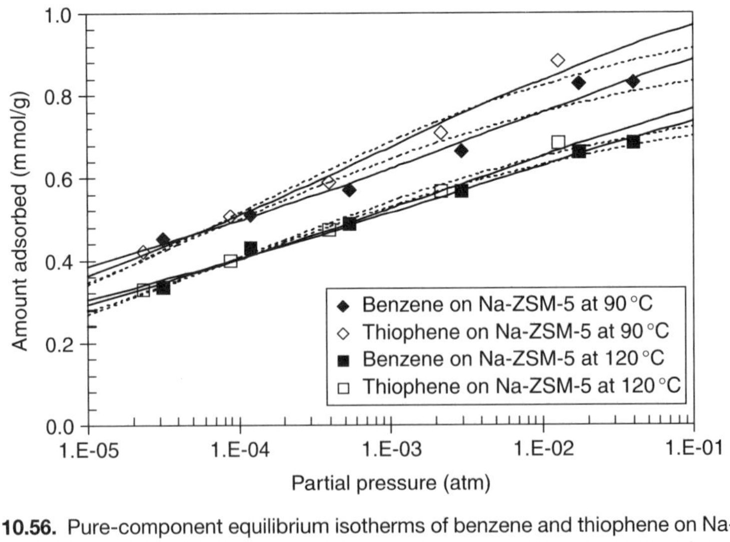

Figure 10.56. Pure-component equilibrium isotherms of benzene and thiophene on Na-ZSM-5 (Si/Al = 10) at 90 and 120 °C. Curves are fitted with Dubinin–Astakhov (solid line) and Langmuir–Freundlich (dotted line) isotherms (Takahashi et al., 2002; Yang et al., 2002).

at 3×10^{-5} atm and 2×10^{-2} atm was preferable for the purification of benzene by removal of thiophene. A disadvantage of Na-ZSM-5 was that the small adsorption capacity was limited by the small pore volume of ZSM-5. Moreover, no studies have been reported on the adsorption of benzothiophene, dibenzothiophene, and their derivatives on ZSM-5. The pore dimensions of ZSM-5 are 5.2×5.6 Å. These sulfur compounds (with more than one ring) will likely be sterically hindered or excluded by ZSM-5.

The adsorption isotherms of thiophene and benzene on activated carbon (Type PCB from Calgon) and modified activated alumina (Selexsorb CDX from Alcoa) are shown in Figures 10.57 and 10.58. PCB activated carbon is designated for commercial use in liquid phase or vapor phase applications, such as recovery of alcohols, chlorinated hydrocarbons, hydrocarbons, and aromatics. Selexsorb CDX is specially formulated for adsorption of polar organic compounds, including sulfur-containing molecules (mercaptans, sulfides, disulfides, thiophenes), nitrogen-based molecules (nitriles, amines, pyridines), and oxygenated hydrocarbon molecules (alcohol, glycols, aldehydes, ketones, ethers, peroxides). More benzene than thiophene was adsorbed on both activated carbon and modified activated alumina at pressures below 1×10^{-2} atm. This is surprising because Selexsorb was thought to adsorb thiophene more selectively than benzene. This result indicates that polarizability plays a more important role (than polarities and magnetic susceptibility) in adsorption on these two sorbents. At pressures higher than 1×10^{-2} atm, pore filling dominated hence benzene adsorbed less than thiophene because of the larger molar volume of benzene than thiophene.

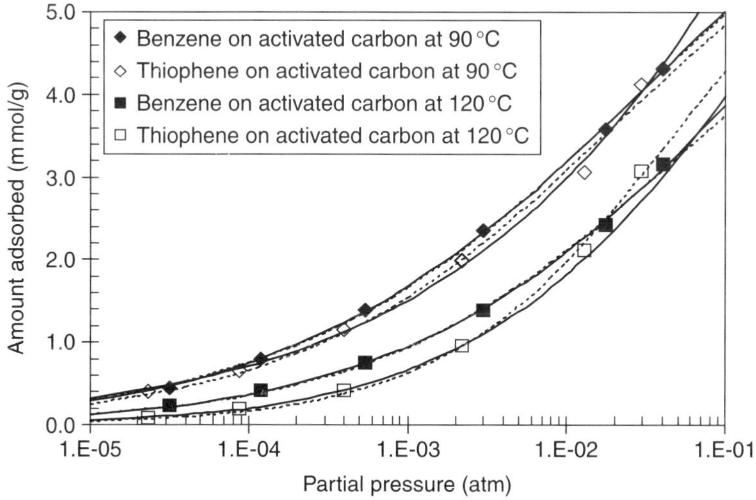

Figure 10.57. Pure-component equilibrium isotherms of benzene and thiophene on activated carbon (Type PCB) at 90 and 120 °C. Curves are fitted with Dubinin–Astakhov (solid line) and Langmuir–Freundlich (dotted line) isotherms (Takahashi et al., 2002; Yang et al., 2002).

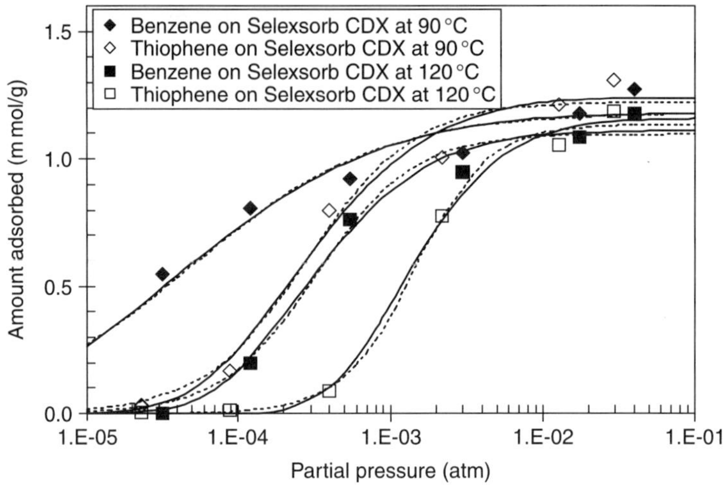

Figure 10.58. Pure-component equilibrium isotherms of benzene and thiophene on modified activated alumina (Selexsorb CDX) at 90 and 120 °C. Curves are fitted with Dubinin–Astakhov (solid line) and Langmuir–Freundlich (dotted line) isotherms (Takahashi et al., 2002; Yang et al., 2002).

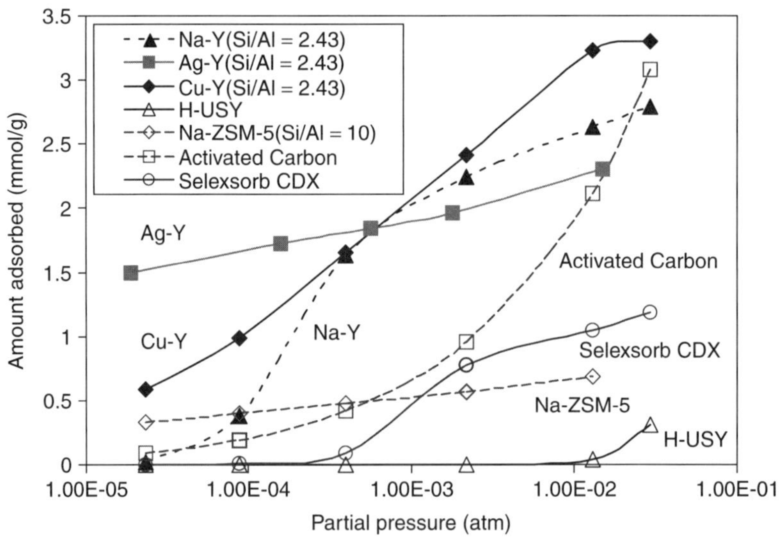

Figure 10.59. Comparison of equilibrium adsorption isotherms of thiophene at 120 °C (Takahashi et al., 2002; Yang et al., 2002).

The adsorption isotherms of thiophene on all sorbents are compared in Figure 10.59. It is clearly seen that AgY and CuY could adsorb significantly larger amounts of thiophene than all other sorbents particularly at very low vapor pressures.

Table 10.18. Approximate heats of adsorption in kcal/mol

	NaY	AgY	Cu(I)Y	Na-ZSM-5	Activated Carbon	Selexsorb CDX	H-USY
Thiophene	19	21.3	22.4	18	16	17	8
Benzene	18	20.1	21.8	18	16	18	8

From Takahashi et al., 2002.

Table 10.19. Diffusion time constants, D/R^2 (in 1/s), of thiophene and benzene at 120 °C in various sorbents

	Thiophene	Benzene
AgY	1.2×10^{-4}	4.2×10^{-4}
Cu(I)Y	1.4×10^{-4}	3.7×10^{-4}
NaY	2.8×10^{-4}	5.6×10^{-4}

The heats of adsorption are given in Table 10.18. The heats of adsorption on CuY and AgY are in good agreement with molecular orbital calculations. The values on H-USY are also included. This zeolite had a Si/Al ratio of 190, hence was nearly free from cation effects. The low heats of adsorption on this sorbent were due to only van der Waals and electrostatic interactions.

The diffusion time constants (D/R^2) are shown in Table 10.19. These values would not cause diffusion limitation for practical adsorber operations.

Liquid-Phase Breakthrough Curves. Liquid-phase breakthrough experiments were performed with a fixed-bed adsorber (Hernández-Maldonado and Yang, 2003a). Solutions of thiophene in *n*-octane were used as the feed. The zeolites were activated in situ prior to cooling and breakthrough experiments. For Cu(I)Y, high exchanges for Cu^{2+} were performed and auto-reduction of Cu^{2+} to Cu^{+} was also performed in situ. The breakthrough results are shown in Figures 10.60 and 10.61.

Cu(I)Y can be fully regenerated by first air oxidation (e.g., at 350 °C) followed by auto-reduction in an inert atmosphere (Hernández-Maldonado and Yang, 2003a).

As indicated from the vapor-phase isotherm results, NaY has similar total capacities for thiophene and benzene as CuY and AgY, because their pore volumes are similar. NaY, however, does not form a π-complexation bond with thiophene, and hence does not have selectivity for thiophene. The lack of selectivity for thiophene results in the premature breakthrough of thiophene, shown in Figure 10.60. Hence a high-purity *n*-octane product cannot be obtained with NaY as the sorbent. With AgY and CuY and feeds containing 2000 and 500 ppmw thiophene, on the contrary, the sulfur content in the effluent was below the

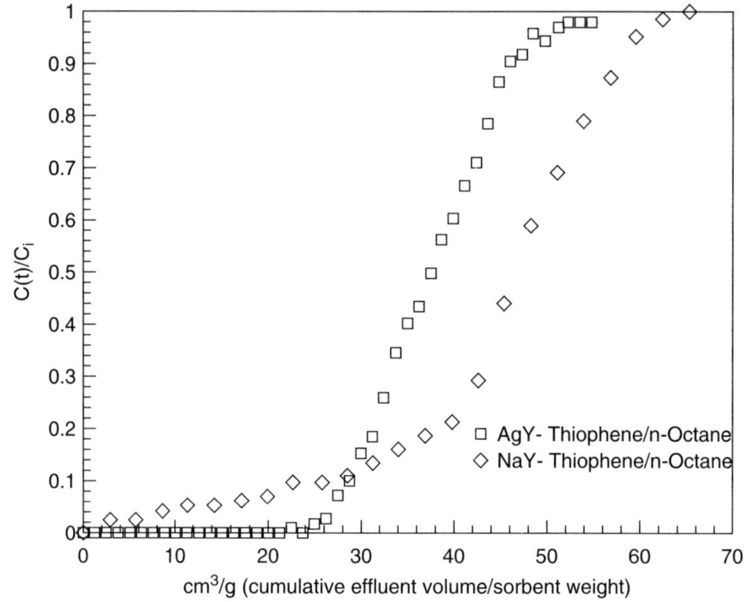

Figure 10.60. Breakthrough of thiophene in fixed-bed adsorber with NaY or AgY sorbent, with a liquid feed containing 2000 ppmw (C_i) of thiophene in *n*-octane, at 25 °C (Hernandez-Maldonado and Yang, 2003a).

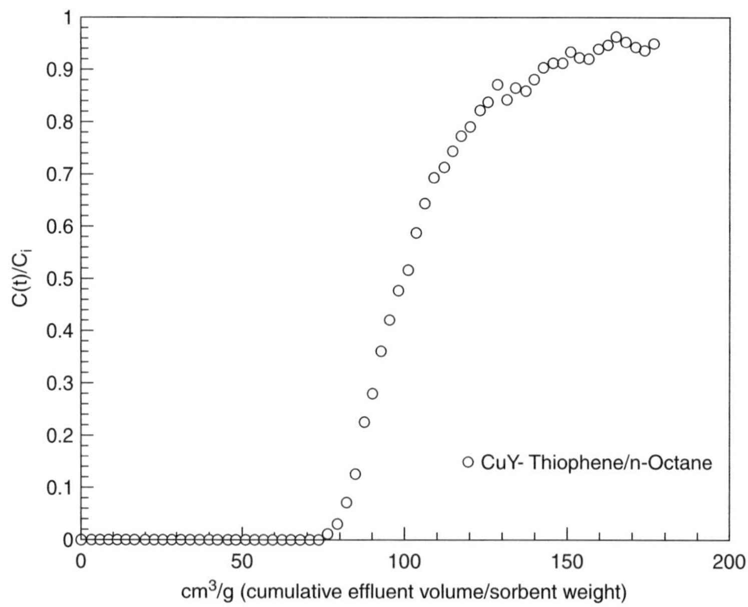

Figure 10.61. Breakthrough of thiophene in fixed-bed adsorber with Cu(I)Y sorbent, with a liquid feed containing 2000 ppmw (C_i) of thiophene in *n*-octane, at 25 °C (Hernandez-Maldonado and Yang, 2003a).

detection limit (~0.28 ppmw) for prolonged periods of time, and clear breakthroughs occurred only after these periods of time. The thiophene capacity for AgY was 7.5 wt % for a feed containing 2000 ppmw. This was higher than all other sorbents reported in the literature (Table 10.14). From the results shown in Figure 10.61, the thiophene capacity of Cu(I)Y was 21.42 wt % for a feed containing 2000 ppmw. The breakthrough curve for Cu(I)Y with a feed containing 500 ppmw thiophene is not shown. A "sulfur-free" effluent was also produced and the total thiophene capacity was 10.75 wt % (Hernández-Maldonado and Yang, 2003). Thus, Cu(I)Y is by far the best sorbent, in both sulfur selectivity and capacity, for sulfur removal from gasoline and diesel fuels (Yang et al., 2002 U.S. and foreign Patent application).

Hernández-Maldonado and Yang (2003b) also reported breakthrough curves of commercial gasoline and diesel on Cu(I)Y as well as Cu(I)Y with a layer of (15%) guard bed of activated carbon. The curves are similar to that in Figure 10.61. Thus, sulfur-free (<0.28 ppmw) diesel of 34.3 cm^3 and gasoline of 19.6 cm^3 can be produced per gram of total sorbent. The use of guard bed can enhance the sulfur capacity of Cu(I)Y. As expected, a lower content of aromatics (as in diesel) is favorable for desulfurization.

10.8. REMOVAL OF AROMATICS FROM FUELS

Purification of aliphatics by the removal of aromatics is important in the petrochemical industry as well as for pollution control. Current worldwide environmental mandates require reduction of aromatics, particularly benzene, in the transportation fuels. For example, the European Union standards are already in place that require the benzene concentration in gasoline to be lowered to 1 vol % by 2000 and to be lowered still by 2005, while the concentration of the polycyclic aromatics in diesel fuel needs to be lowered to 11 vol % by 2002 and to lower levels still by 2005 (Avidan et al., 2001).

In a typical benzene removal process, a combination of extraction and distillation is used (Jeanneret, 1997). Improvements by other processes have been considered, such as pervaporation (Hao et al., 1997), liquid membranes (Li, 1968; 1971), and adsorption by temperature swing adsorption (TSA) in the liquid phase (Matz and Knaebel, 1990). In the work of Matz and Knaebel, commercially available sorbents were used: silica gel, activated alumina, activated carbon, zeolite 13X, and polymeric resin (XAD-7). Among these sorbents, silica gel was considered the best due to its superior thermal-exchange capacity. However, the selectivities were low.

Takahashi and Yang (2002) studied adsorption of benzene and cyclohexane on various Y-zeolites. AgY showed superior benzene/cyclohexane selectivities to NaY and H-USY. Separation factors as high as 10^4 were obtained with AgY at low concentrations of benzene. The high selectivities were achieved by the strong interaction between benzene and AgY, while the interaction with cyclohexane was not influenced by cation exchange. Figure 10.62 shows the isotherms of benzene and cyclohexane on AgY at two temperatures. The isotherms on NaY are

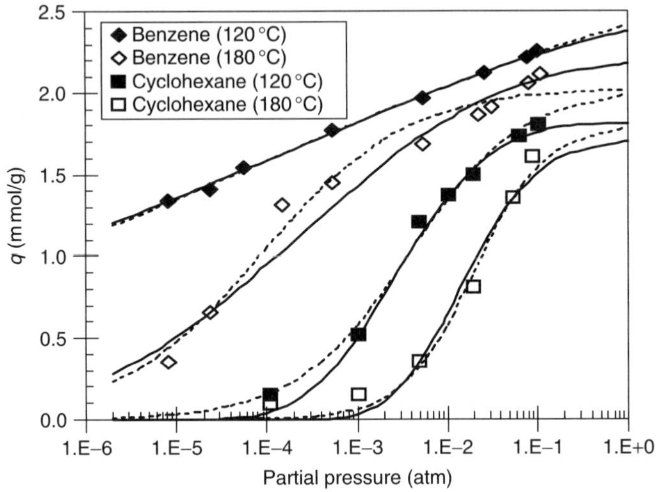

Figure 10.62. Pure-component equilibrium isotherms for benzene and cyclohexane on AgY at 120 and 180 °C. Curves are fitted with isotherm models (Takahashi and Yang, 2002, with permission).

Table 10.20. Separation factors (α) for benzene/cyclohexane at dilute benzene concentrations

Pressure (atm)		Temp. (° °C)	α (Benzene/Cyclo)	
Benzene	*Cyclohexane*		AgY	NaY
0.01	1	120	6	12
0.001	1	120	45	48
0.0001	1	120	335	191
0.00001	1	120	2462	767
0.01	1	180	19	5.4
0.001	1	180	57	4.2
0.0001	1	180	163	3.3
0.00001	1	180	470	2.6

Data taken from Takahashi and Yang, 2002.

given elsewhere (Takahashi and Yang, 2002). The separation factors for purification are shown in Table 10.20. The separation factors are expressed for dilute mixtures of benzene in cyclohexane. These separation factors were calculated from pure component isotherms by using a mixed-gas isotherm model. NaY showed good separation factors at 120 °C, but poor figures at 180 °C. AgY showed superior separation factors at both temperatures. Based on ab initio molecular orbital calculation results (see Chapter 8 and the preceding section, 10.7), still higher separation factors are expected from Cu(I)Y. Cu(I)Y remains to be studied for benzene removal.

10.9. NO_X REMOVAL

Adsorption has been playing an increasingly important role in environmental control. The sorbents being used in common industrial adsorption systems for the removals of SO_2 and volatile organic compounds (VOCs) are quite well-established. The VOC removal systems often use activated carbon, polymeric resins, and hydrophobic zeolites, for both gas and aqueous systems. Activated carbon (and alkalized forms) and hydrophobic zeolites are used for SO_2 removal. Lime injection is used for SO_2 removal from hot gases. For NO_x removal, on the contrary, no suitable sorbents have been established. For this reason, selective sorbents for NO_x remain an active research area, and will be discussed. The search for CO_2 sorbents is also of interest. The subject of CO_2 sorbents has been discussed recently in an excellent review by Yong et al. (2002) and will not be covered here.

Selective NO_x removal is an area where the boundary between sorbent and catalyst tends to disappear. Many different types of sorbents have been investigated for NO_x sorption for both cold (near ambient temperature) and hot (200–400 °C) applications. Transition metal oxides appear to be the best. In both temperature ranges, a significant amount of adsorption is achieved usually when assisted by catalytic oxidation of NO to NO_2 as an intermediate step. In air near ambient temperature, about 5% of the NO is in the form of NO_2. NO_2 adsorbs more easily than NO for both chemisorption and physical adsorption. The normal boiling point of NO is −152 °C and that of NO_2 is 21 °C. Thus, NO_2 adsorbs readily near ambient temperature in micropores and mesopores by pore filling. In chemisorption, NO_2 readily forms nitrite and nitrate on transition metal oxides. These chemisorbed species can be decomposed or desorbed only at elevated temperatures, e.g., 200–300 °C. Other surface species are also formed, such as NO^+ (nitrosyl). These species could be desorbed at near-ambient temperature. The complex chemistry of NO is due to the fact that there is one electron occupying the antibonding orbital of NO, which is empty in most other molecules.

Near-Ambient Temperature. There has been a long search for sorbents for NO_x at near-ambient temperature. Table 10.21 is a summary of the equilibrium capacities for these sorbents. The equilibrium capacity can be deceiving, because of the following problems. The NO_x capacities are decreased by the presence of water vapor. The most severe case is with the ion-exchanged zeolites, such as MFI (or ZSM-5); exposure to water vapor destroys the sorbent quickly and completely. SO_2 and CO_2 also have some effects on some sorbents, as to be discussed.

Chemisorption of NO on various metal oxides was studied before 1982, as shown in Table 10.21. The partial pressure of NO is not important for the NO capacity because the isotherms are usually very steep for chemisorption. The sorbents will be discussed following the sequence listed in Table 10.21.

Kaneko and co-workers (summarized by Kaneko, 1998) studied the adsorption of NO on activated carbon fibers (ACFs). The capacity of NO on ACF is not high. However, large capacities were obtained after iron oxide or FeOOH was dispersed on the ACF. The reason for the increase is not clear. An explanation

Table 10.21. Amounts of NO adsorbed (Q in mg/g) on various adsorbents at NO = 13 kPa (or 12,830 ppm) and near room temperature, unless noted otherwise

Adsorbent	Q, mg/g	Reference and Notes
SnO_2	5	Solimosi and Kiss, 1976
CeO_2	5	Niwa et al., 1982
NiO	1	Yao and Shelef, 1978
Co_3O_4	6	Yao and Shelef, 1978
CuO/g-Al_2O_3	36	Yao and Shelef, 1978
NiO/g-Al_2O_3	36	Yao and Shelef, 1978
Co_3O_4/g-Al_2O_3	7	Yao and Shelef, 1978
Fe_2O_3/g-Al_2O_3	45	Yao and Shelef, 1978
Fe_2O_3/SiO_2	5	Segawa et al., 1982
Fe-Y zeolite	18	Yuen et al., 1982
Fe_3O_4	22	Lund et al., 1979
Fe_3O_4	24	Otto and Shelef, 1970
a-Fe_2O_3	9	Otto and Shelef, 1970
Jaosites	3–10	Inouye et al., 1982
a-FeOOH	10–20	Kaneko and Inouye, 1987
b-, g-FeOOH	4–6	Hattori et al., 1979
s-FeOOH	13	Kaneko and Inouye, 1988
Fe_2O_3/ACF	120–260	Kaneko, 1998
Styrenic polymer resins	119	Kikkinides and Yang, 1991; Yang 1993; 0.1 atm each of NO, SO_2 and CO_2, at 26 °C
Cu-MFI	19.9	Zhang et al., 1993; 997 ppm NO, 273 K
Co-MFI	26.3	Zhang et al., 1993; 997 ppm NO, 273 K
$H_3PW_{12}O_{40} \cdot 6H_20$ (Heteropoly compounds)	76	Yang and Chen, 1994 and 1995 1000 ppm NO, with $H_2O/SO_2/O_2$, 50–200 °C
Fe-Mn oxides	16	Huang and Yang, 2001 200 ppm NO, 10% O_2, 25 °C
Fe-Mn-Ti oxides	45	Huang and Yang, 2001 200 ppm NO, 10% O_2, 25 °C
Fe-Mn-Zr oxides	44	Huang and Yang, 2001 200 ppm NO, 10% O_2, 25 °C
Cu-Mn oxides 1 wt % Ru doped	>2	Yamashita et al., 2002 10 ppm NO, 30% R.H. air, 35 °C

Data before 1988 taken from Kaneko and Inouye, 1988.

based on the magnetic property of NO has been proposed by Kaneko. NO is paramagnetic, and it can form dimmers, $(NO)_2$, only at a very low temperature. The dimerization of NO could be enhanced through magnetic interaction with magnetic iron oxide, such that NO loses its supercritical nature even at ambient temperature. Thus, by interaction with the dispersed iron oxide, NO is dimerized and subsequently adsorbed in the micropores of ACF.

Adsorption of NO in the presence of O_2 on carbon is not listed in Table 10.21. Mochida and co-workers (Mochida et al., 1994; 1997) reported catalytic oxidation of NO to NO_2 on activated carbon fibers at ambient temperature. With water, the

NO_2 can be removed as nitric acid. The catalytic oxidation activity for NO was also observed with activated carbon by Rubel and Stencel (1997). The mechanism of the catalytic oxidation reaction is not understood, but it is possibly related to the surface oxides on carbon.

Iwamoto and co-workers (Zhang et al., 1993) studied NO adsorption on various ion-exchanged zeolites. The best sorbent was Cu-MFI, Co-MFI, and Ag-MFI. They measured both "reversibly" (that could be desorbed at ambient temperature) and "irreversibly" adsorbed NO. The amounts given in Table 10.21 are for the irreversibly adsorbed. IR spectroscopy indicated that most of the reversibly adsorbed NO was NO^+ adsorbed on Cu^{2+}, and that the irreversibly adsorbed NO was in the forms of NO^+, nitrate ($NO_3{}^-$), nitrite ($NO_2{}^-$), and $NO_2{}^+$. Unfortunately, exposure to water vapor severely deactivated the zeolites.

$H_3PW_{12}O_{40}$ belongs to a large class of heteropoly acids and salts, which have been the subject of long-standing investigation (Pope, 1983). The crystal structure of the $PW_{12}O_{40}$ anion belongs to the Keggin structure of $XM_{12}O_{40}$, shown in Figure 10.63. In this structure, 12 MO_6 octahedra surround a central XO_4 tetrahedron, where M is usually W or Mo and X can be P, As, Si, Ge, B, etc. Although the structure of the heteropoly anion (e.g., the Keggin structure) is well-defined and stable, the structure by which the Keggin structures are linked together is less understood. However, a distinct X-ray diffraction pattern is obtained for $H_3PW_{12}O_{40} \cdot 6H_2O$. As shown in Figure 10.63, the Keggin polyanions are linked in a 3-dimensional network through $H^+(H_2O)_2$ bridges, and these linkages can be easily (i.e., at ambient temperature) replaced by polar molecules such as alcohols and amines (Misono, 1987). Generally, 2–6 alcohol or amine molecules per Keggin anion irreversibly replace the water linkages (of $6H_2O$). The Keggin structure is called the primary structure and the linked structure is called the secondary structure. Due to the flexible form of the secondary structure, Misono referred to the solid structure as a "pseudo-liquid phase." The importance of the "pseudo-liquid phase" in many reactions catalyzed by the heteropoly compounds has been demonstrated (Misono, 1987).

Yang and Chen observed that a large amount of NO could be absorbed by the heteropoly compounds (Yang and Chen, 1994; Yang and Chen, 1995; Chen and Yang, 1995). The water linkages in the secondary structure can be substituted readily by NO linkages at 50–230 °C at low NO concentrations (i.e., under flue gas conditions). The absorption product for $H_3PW_{12}O_{40} \cdot 6H_2O$ is $H_3PW_{12}O_{40} \cdot 3NO$. Moreover, a substantial fraction of the absorbed NO is decomposed into N_2 upon rapid heating of the NO-linked compound (Yang and Chen, 1994; 1995). The bond energy of the NO linkages is 25 kcal/mol. CO_2 does not adsorb/absorb in the heteropoly compounds. SO_2 has no effect on the absorption of NO. Because of this unusual property of the heteropoly compounds, a number of studies were undertaken after the work of Yang and Chen. Herring and McCormick (1998) determined the mechanism of the NO absorption. In the presence of O_2, the surface of $H_3PW_{12}O_{40}$ is involved in the catalytic oxidation of NO to NO_2 via an adsorbed $NO_x{}^{y-}$ intermediate. This species can be absorbed into the $H_3PW_{12}O_{40}$ as a form of NOH^+, displacing water, or can leave as NO_2 gas molecules.

(a) Primary structure ($PW_{12}O_{40}$, "Keggin" structure)

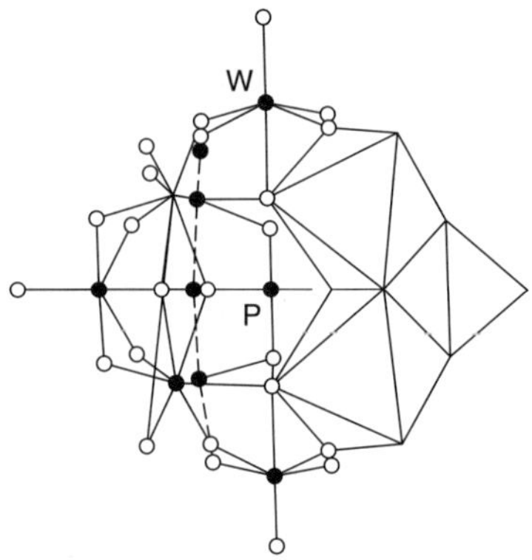

(b) Secondary structure ($H_3PW_{12}O_{40}{\cdot}6\ H_2O$)

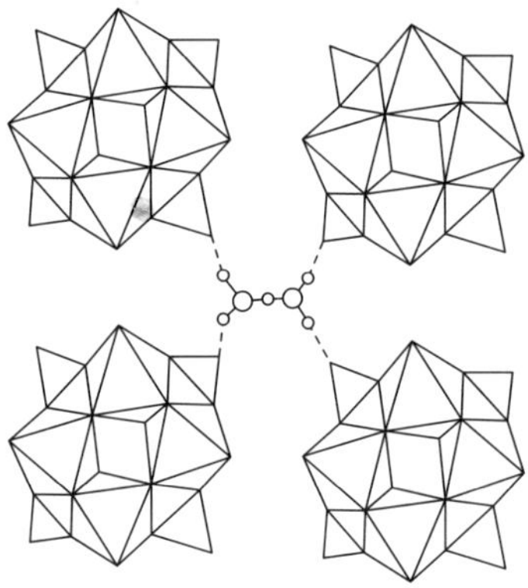

Figure 10.63. Heteropoly compounds. (a) Primary structure: Keggin structure. (b) Secondary structure: each Keggin anion is linked by $3H^+(H_2O)_2$ linkages to form $H_3PW_{12}O_{40} \cdot 6H_2O$.

They also found that not all structural water is displaced by NO, and water plays a role in holding the new structures together by hydrogen bonding. The mechanism of absorption and desorption was further investigated by Hodjati et al. (2001), for automotive applications under lean exhaust gas conditions. They found a capacity of 38 mg NO_x/g $H_3PW_{12}O_{40}$. The mechanism of absorption was reported as through substitution of linkage water molecules and formation of a $[H^+(NO_2{}^-, NO^+)]$ complex. Heteropoly compounds are promising sorbents for NO_x. Further work on their longevities under cyclic absorption/desorption conditions are needed.

In the search for a sorbent for selective SO_2 removal from combustion gases, it was found that styrenic polymeric sorbents had high selectivities for SO_2 over CO_2 and H_2O (Kikkinides and Yang, 1991; Yang 1993). It was also found that the polymeric sorbents had even higher selectivities toward NO_x than SO_2, and these sorbents were extremely hydrophobic. For example, over 100 mg/g of NO was chemisorbed at room temperature from a simulated flue gas. The chemisorbed NO forms monomers or dimers on the benzene rings of the surface of the polymer, possibly by π-complex bonding with the π-electrons. However, desorption of NO requires temperatures near the thermal stability temperatures of the styrenic polymers, that is near 200 °C. Polymeric sorbents with higher thermal stabilities such as the acrylics types need to be tested for this application.

The best sorbents for NO_x remain to be mixed metal oxides. Huang and Yang (2001) investigated a series of Fe-Mn based transition metal oxides for NO adsorption at 25 °C in the presence of O_2. Mixed metal-oxide sorbents are prepared from metal salts by standard coprecipitation method. The results of NO_x uptake on Fe-Mn based transition metal oxides at 25 °C are shown in Figure 10.64. For all sorbents, the initial rates are very high. This is related to their high activities in NO oxidation to NO_2 by oxygen. Among the mixed oxides, the uptake amount of NO_x decreased in the sequence of Fe-Mn-Ti, Fe-Mn-Zr > Fe-Mn-Ce, Fe-Mn-Ni > Fe-Mn-Co > Fe-Mn-Cu, Fe-Mn (Figure 10.64). Near 44–45 mg/g NO_x capacities were obtained on the Fe-Mn-Ti and Fe-Mn-Zr oxides, ~180% increase compared with Fe-Mn oxides. It is clear that TiO_2 and ZrO_2 are good storage components for NO_x. FTIR results showed that nitrates were formed on these oxides. Another reason for the enhancement is the increased surface areas of these oxides, from 54 m^2/g for Fe-Mn oxides to 183 m^2/g for these sorbents.

NO adsorption was also performed in a fixed-bed adsorber. The results are shown in Figure 10.65. After 500 ppm NO + 10% O_2 was passed over the sorbents, all NO was adsorbed and the NO_x concentrations in the effluents were zero during the first few hours. During this period, 100% NO removal was obtained.

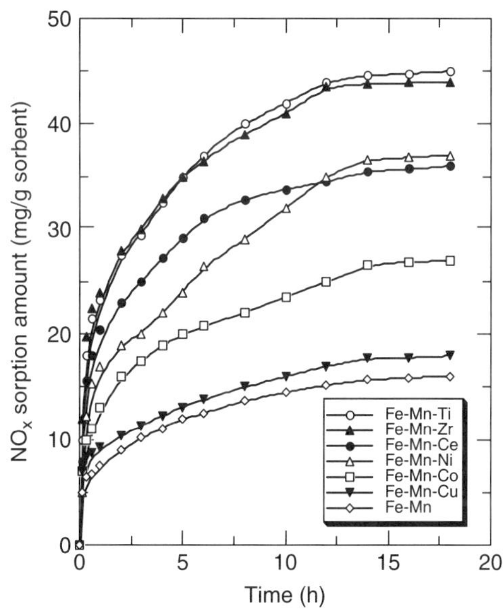

Figure 10.64. NO adsorption on the mixed, equimolar oxides at 25 °C. Reaction conditions: [NO] = 200 ppm, [O_2] = 10%, and balance He (Huang and Yang, 2001, with permission).

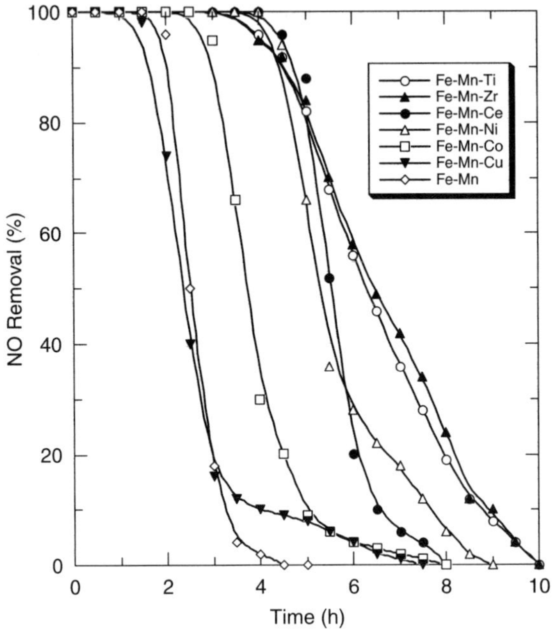

Figure 10.65. Fixed-bed adsorber breakthrough curve of NO in mixed, equimolar oxides at 25 °C. Feed conditions: [NO] = 500 ppm, $[O_2]$ = 10%, He = balance and GHSV = 6000 h^{-1} (Huang and Yang, 2001; with permission).

Then NO_x concentrations in the outlet gas increased gradually with time. The breakthrough time was 1.5, 1.5, 2.5, 4.0, 4.0, 3.5, and 3.5 hs, respectively, for Fe-Mn-Cu, Fe-Mn, Fe-Mn-Co, Fe-Mn-Ni, Fe-Mn-Ce, Fe-Mn-Zr and Fe-Mn-Ti (Figure 10.65). The total amounts of NO_x adsorbed from the breakthrough experiments were in good agreement with the equilibrium results measured gravimetrically. This agreement also indicates that the adsorbed amounts at 200 ppm (as in TGA) and 500 ppm (as in fixed bed) were the same.

The catalytic activities of these metal oxides for NO oxidation to NO_2 by O_2 at ambient temperature were also measured by Huang and Yang (2001). The sorbent capacities were directly dependent on the catalytic activities for NO oxidation. Many surface species were formed upon adsorption. Nitrate and nitrite were the main species, and they also desorbed at the highest temperatures. Desorption began at about 70 °C, and essentially all species were desorbed below 400 °C.

The effects of CO_2, SO_2, and H_2O on NO_x adsorption on these sorbents are shown in Figure 10.66. It is seen that CO_2 and SO_2 decreased NO_x adsorption slightly for the Fe-Mn-Ti oxide. However, when 2.5% H_2O was introduced into the reaction gas, the breakthrough time, NO_x capacity, NO conversion to NO_2 all decreased significantly. This indicates that H_2O greatly inhibited NO oxidation to NO_2, and consequently the NO_x adsorption capacity was also decreased. However, the inhibition was reversible. When the water supply was stopped, NO_x adsorption capacity was recovered (Huang and Yang, 2001).

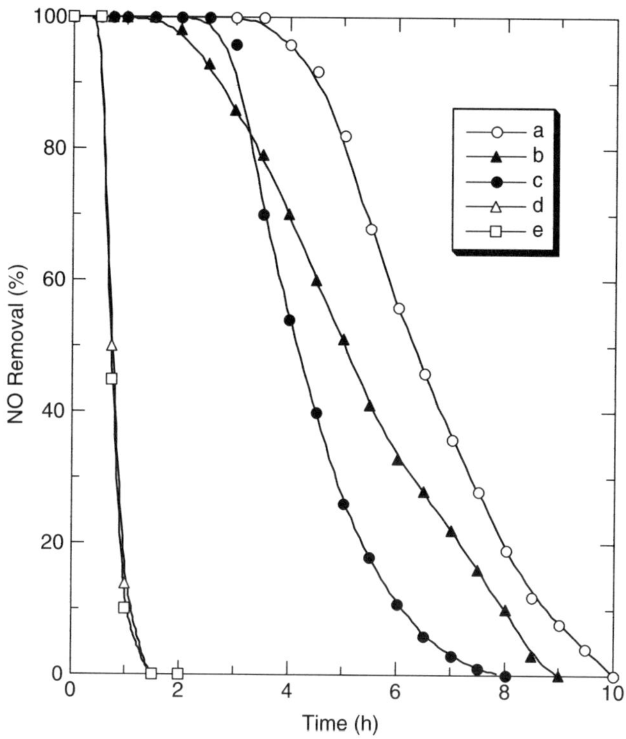

Figure 10.66. Effects of CO_2, SO_2 and H_2O on NO removal on Fe-Mn-Ti oxides at 25 °C. Feed conditions: GHSV = 6000 h^{-1}, (a) 500 ppm NO + 10% O_2, (b) 500 ppm NO + 10% O_2 + 10% CO_2, (c) 500 ppm NO + 10% O_2 + 200 ppm SO_2, (d) 500 ppm NO + 10% O_2 + 2.5% H_2O, (e) 500 ppm NO + 10% O_2 + 10% CO_2 + 200 ppm SO_2 + 2.5% H_2O (Huang and Yang, 2001, with permission).

An interesting application for the mixed oxide sorbents has been reported recently (Yamashita et al., 2002). The sorbent was a mixed Cu-MnO_x doped with 1 wt % Ru. It was reported that this sorbent had already been applied to the tunnel ventilation air cleaning system, and it had a high NO_x removal rate as well as longevity. Such sorbent was proposed to be used for NO_x removal from the underground highways in the Tokyo and Osaka areas and from tunnels. In their tests, air containing 10 ppm NO at various relative humidities was used, and a significant fraction of the NO was adsorbed. Some NO_2 was also formed. The mechanism of the reaction is the same as that on other transition metal oxides, that is through oxidation to NO_2. The sorbent was regenerated at 180–200 °C, where some ammonia was employed as the reducing gas.

A possible application for the mixed-oxide sorbents is for the removal of N_2O from ventilation air in hospitals, where N_2O is used in operating rooms as an anesthetic. This application would require further studies of adsorption of N_2O on these sorbents. Interestingly, N_2O has also been a problem in the cryogenic separation of air. Air contains 300 to 350 ppb N_2O, and it needs to be removed before the air is fed to the cryogenic separator. Essentially all commercial sorbents, including various zeolites (Golden et al., 2000; Miller et al., 2000; Centi et al., 2000) have been studied for N_2O removal from air, as reviewed by Ackley et al. (2002). From the comparison made by Ackley et al. (2002), clinoptilolite and chabazite appeared to be the best sorbents. For air prepurification, the sorbent for N_2O cannot be used alone, because other sorbents must be used to first remove CO_2, H_2O, C_2H_2, and other hydrocarbons. A transition metal sorbent would not have such limitation, that is it could selective adsorb N_2O.

Higher Temperatures. NO removal/reduction in power plants has been performed worldwide by the selective catalytic reduction (SCR) process, with ammonia injection and by using vanadia/TiO_2 (with W or Mo additive) as the catalyst. Over 80% NO is reduced to N_2 at an operating temperature near 350 °C. NO in automotive emissions is reduced by three-way catalytic converters. For the new generation of lean-burn engines (i.e., fuel-lean, for fuel economy), however, these three-way catalytic converters are not adequate. The most promising candidate (over the past 7 years) for lean-burn NO control has been the "NO trap." In this scheme, the engine is operated in lean-rich cycles (i.e., air rich–air lean cycles). The NO trap sorbent is added to the three-way catalyst. During the lean phase (or air-rich phase), the sorbent adsorbs/absorbs NO_x, and forms nitrates and nitrites. During the rich phase, these nitrates/nitrites decompose into N_2 (over the noble metal catalysts). The system is operated around 300 °C. BaO and SrO were the trap sorbents used in Toyota vehicles. These sorbents are deactivated by SO_2 (forming sulfate), as well as CO_2 (forming carbonate). The automotive NO trap has been an incentive for sorbent development for hot gases. Such sorbents would also be useful in many other possible applications.

Tabata et al. (1988) reported that NO and CO could be adsorbed rapidly on superconducting $YBa_2Cu_3O_7$. After pre-evacuation at 300 °C, the sample adsorbed approximately 2 mol/mol oxide for NO at the same temperature. The adsorbed NO molecules were almost completely desorbed when the temperature was increased to 400 °C. For these Y-containing oxides, Kishida et al. (1991) reported that the NO adsorption amount decreased according to the order: $YSr_2CO_3O_x > YBa_4Co_8O_x > YSr_2Mn_3O_x > YSr_2V_3O_x$. TPD and IR results showed that the adsorbed NO molecules were oxidized to NO_3^- by lattice oxygen. The adsorbed NO was desorbed as a mixture of NO/O_2. Arai et al. (1994) reported that Ba-Cu-O mixed oxides had a high adsorption capacity for NO/NO_2 at 200 °C. This uptake was accelerated by the presence of oxygen. XRD results indicated the formation of $Ba(NO_3)_2$/CuO. In the presence of O_2, a large amount of NO_x was liberated from the sample at temperatures above 500 °C. However, the NO adsorption capacity of this sorbent vanished completely after exposure to 8% CO_2 because of the formation of surface $BaCO_3$. Since the sorbents containing Ba are easily deactivated by CO_2, Eguchi et al. (1996) developed sorbents that did not contain alkaline earth metals. A series of mixed-oxide sorbents containing Mn and/or Zr were investigated. The mixed Mn-Zr oxide (at 1 : 1 mole ratio) was the best.

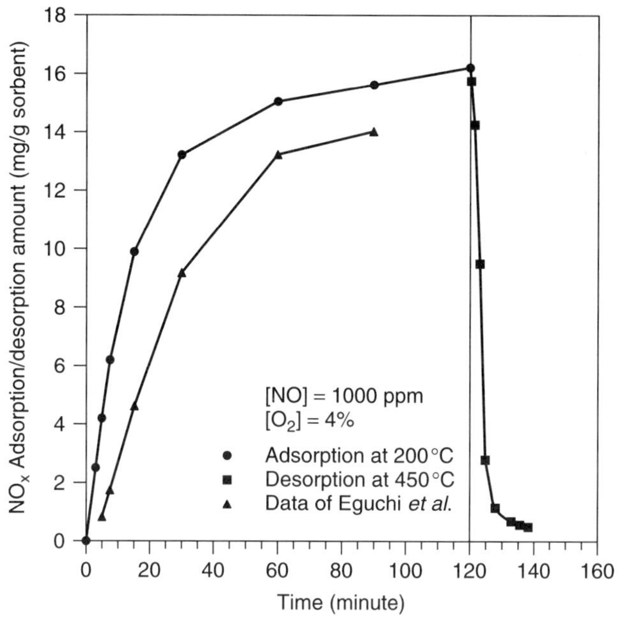

Figure 10.67. Adsorption and desorption (in the same gas flow) on Ce-CuO/TiO_2 (2% Ce, 5% CuO by wt) compared with the MnO_x/ZrO_2 (1 : 1 mole ratio) sorbent of Eguchi et al. (1996) under the same conditions (Yang and Krist, 2000).

CuO/TiO_2 has been shown to be a superior sorbent for selective, reversible adsorption of NO from hot combustion gases (Li et al., 1997; Yang and Krist, 2000). High NO_x adsorption capacities at 200 and 300 °C were obtained with a 5% CuO/TiO_2 sorbent, and the NO_x desorbed rapidly at 450 °C. Doping with 2% Ce on the CuO/TiO_2 sorbent further increased both uptake rates (50% increase in initial rate) and NO_x capacity (by 30%). In a direct comparison with the most promising sorbent reported by Eguchi et al. (1996), MnO_x/ZrO_2 (1 : 1 mole ratio), the Ce-CuO/TiO_2 sorbent showed both higher uptake rates (by 100% in initial rate) and higher NO_x capacity (by 15%). The results are shown in Figure 10.67. The effects of CO_2, H_2O, and SO_2 on NO_x sorption on the Ce- CuO/TiO_2 sorbent were studied at 200 °C. CO_2 slightly decreased the initial uptake rate but increased the NO_x capacity. H_2O coadsorbed with NO_x on different sites, both reversibly (i.e., desorbed at 450 °C). SO_2 irreversibly adsorbed (likely to sulfate the surface of TiO_2) and decreased the NO_x capacity by approximately 20%. The BET surface area of the TiO_2 support was 50 m^2/g. Further studies with TiO_2 of higher surface areas (such as xerogel) should lead to still better sorbents.

REFERENCES

Ackley, M. W. (1991) Separation of Nitrogen and Methane by Adsorption, Ph.D. Dissertation, SUNY at Buffalo.

Ackley, M. W., Rege, S. U., and Saxena, H. (2002) *Micropor. Mesoporr. Mat.* in press.

Ackley, M. W. and Yang, R. T. (1990) *AIChE J.* 36, 1229.

Ackley, M. W. and Yang, R. T. (1991a) *Ind. Eng. Chem. Res.* 30, 2523.

Ackley, M. W. and Yang, R. T. (1991b) *AIChE J.* 37, 1645.

Adduci, A. J. (1975) *Chemtech.* 6, 575.

Ahn, C. C., Ye, Y., Ratnakumar, B. V., Witham, C., Bowman, R. C., Jr., and Fultz, B. (1998) *Appl. Phys. Lett.* 73, 3378.

Akuzawa, K., Amari, Y., and Nakajima, T. (1990) *J. Mater. Res.* 5, 2849.

Alcaniz-Monge, J., de la Casa-Lillo, M. A., Cazorla-Amoros, D., and Linares-Solano, A. (1997) *Carbon* 35, 291.

Alefeld, G. and Volkl, J., eds. (1978) *Hydrogen in Metals II*. Springer-Verlag, Berlin, Germany.

Arai, H. and Machida, M. (1994) *Catal. Today* 22, 239.

Avidan, A., Klein, B., and Ragsdale, R. (2001) *Hydrocarbon Process*. 80 (2), 1.

Avidan, A. and Cullen, M. (2001) National Petrochem. & Refiners Assoc. Annual Meeting, March, Washington, D. C., paper AM-01-55.

Bailes, R. H. and Calvin, M. (1947) *J. Am. Chem. Soc.* 69, 1886.

Baker, F. S. U.S. Patent 5,626,637 (1997).

Baker, M. D., Ozin, G. A., and Godber, J. (1985) *J. Phys. Chem.* 89, 305.

Barrer, R. M. (1937) *Proc. Roy. Soc. A* 161, 476.

Barrer, R. M. (1938) *Proc. Roy. Soc. A* 167, 393.

Barrer, R. M. (1982) *Hydrothermal Chemistry of Zeolites*. Academic Press, London, UK.

Barrer, R. M. and Makki, M. B. (1964) *Can. J. Chem.* 42, 1481.

Barton, S. S., Dacey, J. R., and Quinn, D. F. (1983) *Fundamentals of Adsorption*. (A. L. Myers and G. Belfort, eds.). Engineering Foundation, New York, NY, p. 65.

Barton, S. S., Evans, M. J. B., and MacDonald, J. A. F. (1991) *Carbon* 29, 1099.

Bauschlicher, C. W. Jr. (2000) *Chem. Phys. Lett.* 322, 237.

Bauschlicher, C. W. Jr. (2001) *Nano Lett.* 1, 223.

Benard, P. and Chahine, R. (2001) *Langmuir* 17, 1950.

Biloe, S., Goetz, V., and Guillot, A. (2002) *Carbon* 40, 1295.

Biloe, S., Goetz, V., and Mauran, S. (2001) *AIChE J.* 47, 2819.

Bogdanovic, B. and Schwickardi, M. (1997) *J. Alloys Compounds* 253, 1.

Boscola, E. J. (1974) *J. Aircraft* 11, 444.

Bose, T., Chahine, R., and St. Arnaud, J. M. U.S. Patent 4,999,330 (1991).

Braunbarth, C., Hillhouse, H. W., Nair, S., Tsapatsis, M., Burton, A., Lobo, R. F., Jacubinas, R. M., and Kuznicki, S. M. (2000) *Chem. Mater.* 12, 1857.

Breck, D. W. (1974) *Zeolite Molecular Sieves*. Wiley, New York, NY.

Brown, C. M., Yildirim, T., Neuman, D. A., Heben, M. J., Gennett, T., Dillon, A. C., Alleman, J. L., and Fischer, J. E. (2000) *Chem. Phys. Lett.* 329, 311.

Browning, D. J., Gerrard, M. L., Lakeman, J. B., Mellor, I. M., Mortimer, R. J., and Turpin, M. C. (2002) *Nano Lett.* 2, 201.

Butter, S. A. and Kuznicki, S. M. U.S. Patent 4,606,899 (1986).

Calvin, M., Bailes, R. H., and Wilmarth, W. K. (1946) *J. Am. Chem. Soc.* 68, 2254.

Calvin, M. and Martell, A. E. (1952) *Chemistry of the Metal Chelate Compounds*. Prentice-Hall, New York, NY.

Cen, P. L. and Yang, R. T. (1986) *Separ. Sci. Tech.* 21, 845.

Centi, G., Generali, P., dall'Olio, L., Parathoner, S., and Rak, Z. (2000) *Ind. Eng. Chem. Res.* 39, 131.

Chahine, R. and Bose, T. K. (1994) *Inet. J. Hydrogen Energy* 19, 161.

Chambers, A., Park, C., Baker, R. T. K., and Rodriguez, N. M. (1998) *J. Phys. Chem. B* 102, 4253.

Chao, C. C. U.S. Patent 4,859,217 (1989).

Chao, C. C. U.S. Patent 4,964,889 (1990).

Chao, C. C., Sherman, J. D., Mullhaupt, J. T., and Bolinger, C. M. U.S. Patent 5,174,979 (1992).

Chapman, D. M. and Roe, A. L. (1990) *Zeolites* 10, 730.

Chemical Engineering Progress, September, (2001) p. 18.

Chen, D. and Martell, A. E. (1987) *Inorg. Chem.* 26, 1026.

Chen, D., Martell, A. E., and Sun, Y. (1989) *Inorg. Chem.* 28, 2647.

Chen, J. P. and Yang, R. T. (1989) *Surf. Sci.* 216, 481.

Chen, N. and Yang, R. T. (1995) *J. Catal.* 157, 76.

Chen, N. and Yang, R. T. (1996) *Ind. Eng. Chem. Res.* 35, 4020.

Chen, P., Zhang, H.-B., Lin, G.-D., Hong, Q., and Tsai, K. R. (1997) *Carbon* 35, 1495.

Chen, P., Wu, X., Lin, J., and Tan, K. L. (1999) *Science* 285, 91.

Chen, S. G., Yang, R. T., Kapteijn, F., and Moulijn, J. A. (1993) *Ind. Eng. Chem. Res.* 32, 2835.

Chen, X. and McEnaney, B. (1995) Carbon 1995 Extended Abstracts, 22nd Biennial Carbon Conference, San Diego, p. 504.

Chen, Y. D., Yang, R. T., and Uawithya, P. (1994) *AIChE J.* 40, 577.

Cheng, H.-M., Yang, Q.-H., and Liu, C. (2001) *Carbon* 39, 1447.

Chiang, I. W., Brinson, B. E., Smalley, R. E., Margrave, J. L., and Hauge, R. H. (2001) *J. Phys. Chem. B* 105, 1157.

Chlendi, M. and Tondeur, D. (1995) *Gas. Sep. Purif.* **9**, 231.

Cho, S. H., Han, S. S., Kim, J. N., Choudary, N. V., Kumar, P., and Bhat, T. S. G. U.S. Patent 6,315,816 (2001).

Choudary, N. V., Kumar, P., Bhat, T. S. G., Cho, S. H., Han, S. S., and Kim, J. N. (2002) *Ind. Eng. Chem. Res.* 41, 2728.

Coe, C. G. (1995) *Access in Nanoporous Materials*. (T. J. Pinnavaia and M. F. Thorpe, eds.). Plenum Press, New York, NY, p. 213.

Coe, C. G., Kirner, J. F., Pierantozzi, R., and White, T. R. U.S. Patent 5,152,813 (1992).

Coe, C. G. and Kuznicki, S. M. U.S. Patent 4,481,018 (1984).

Coe, C. G., MacDougall, J. E., and Weigel, S. J. U.S. Patent 5,354,360 (1994).

Collins, F. M., Lucy, A. R. and Sharp, C. (1977) *J. Mol. Catal.* A117, 397.

Collins, J. J. U.S. Patent 4,026,680 (1977).

Cook, T. L., Komodromos, C., Quinn, D. F., and Ragan, S. (1999) Adsorbent storage for natural gas vehicles. In *Carbon Materials for Advanced Technologies*. (T. D. Burchell, ed.). Pergamon Press, New York, NY, Chapter 9.

Cracknell, R. F. and Gubbins, K. E. (1992) *J. Mol. Liquids* 54, 261.

Cruciani, G., De Luca, P., Nastro, A., and Pattison, P. (1998) *Micropor. Mesopor. Mater.* 21, 143.

Da Silva, F. A. and Rodrigues, A. E. (1999) *Ind. Eng. Chem. Res.* 38, 2051.

Da Silva, F. A. and Rodrigues, A. E. (2001) *AIChE J.* 47, 341.

Darkrim, F. and Levesque, D. (1998) *J. Chem. Phys.* 109, 4981.

Darkrim, F. and Levesque, D. (2000) *J. Phys. Chem.* B 104, 6773.

Darkrim, F., Malbrunot, P., Levesque, D. (1998) *Hydrogen Energy Progress XII*. (J. C. Bolcich and T. N. Veziroglu, eds.). International Association for Hydrogen Energy, Corel Gables, FL, p. 985.

Darkrim, F., Malbrunot, P., and Tartaglia, G. P. (2002) *Int. J. Hydrogen Energy* 27, 193.

Department of Energy, A Multiyear Plan for the Hydrogen R&D Program, U.S. Department of Energy (Office of Power Delivery, Office of Power Technologies, Energy Efficiency and Renewable Energy), August, 1999.

Dillon, A. C., Gennett, T., Alleman, J. L., Jones, K. M., Parilla, P. A., and Heben, M. J. Proc. 1999 U.S. DOE Hydrogen Program Review, NREL/CP-507-26938 (1999).

Dillon, A. C., Gennett, T., Alleman, J. L., Jones, K. M., Parilla, P. A., and Heben, M. J. Proceedings 2000 U.S. DOE Hydrogen Program Review, NREL/CP-507-28890 (2000).

Dillon, A. C. and Heben, M. J. (2001) *Appl. Phys. A*, 72, 133.

Dillon, A. C., Jones, K. M., Bekkedahl, T. A., Kiang, C. H., Bethune, D. S., and Heben, M. J. (1997) *Nature* 386, 377.

Ding, R. G., Lu, G. Q., Yan, Z. F., and Wilson, M. A. (2001) *J. Nanosci. Nanotech.* 1, 7.

Doong, S. J. and Yang, R. T. (1986) *AIChE J.* 32, 397.

Doong, S. J. and Yang, R. T. (1987) *React. Polym.* 6, 7.

Drago, R. S., Bresinska, I., George, J. E., Balkus, K., and Taylor, R. J. (1988) *J. Amer. Chem. Soc.* 110, 304.

Dresselhaus, M. S., Williams, K. A., and Eklund, P. C. (1999) *MRS Bull.* 24, 45.

Drain, L. E. (1953) *Trans. Faraday Soc.* 49, 650.

Dzugan, S. J. and Busch, D. H. (1990) *Inorg. Chem.* 29, 2528.

Eguchi, K., Watabe, M., Ogata, S., and Arai, H. (1996) *J. Catal.* 158, 420.

Engelhard Corporation, Molecular Gate Technology for Nitrogen Rejection, brochure (2001).

Feuerstein, M. and Lobo, R. F. (1998) *Chem. Mater.* 10, 2197.

Fitch, F. R., Bulow, M., and Ojo, A. F. U.S. Patent 5,464,467 (1995).

Frankiewicz, T. C. and Donnelly, R. G. (1993) Methane/nitrogen gas separation over the zeolite clinoptilolite by the selective adsorption of nitrogen. In *Industrial Gas Separations*. (T. E. Whyte, Jr., et al., eds.). American Chemical Society, Washington, DC.

Frey, D. D. (1983) *Separ. Sci. Tech.* 17, 1485.

Froudakis, G. E. (2001) *Nano Lett.* 1, 179.

Gates, B. C., Katzer, J. R., and Schuit, G. C. A. (1979) *Chemistry of Catalytic Processes*. McGraw-Hill, New York, NY.

Gellens, L. R., Mortier, W. J., and Uytterhoeven, J. B. (1981a) *Zeolites* 1, 11.

Gellens, L. R., Mortier, W. J., Schoonheydt, A., and Uytterhoeven, J. B. (1981b) *J. Phys. Chem.* 85, 2783.

Gellens, L. R., Smith, J. V., and Pluth, J. J. (1983) *J. Am. Chem. Soc.* 105, 51.

Gleiter, H. (2000) *Acta Mater.* 48, 1.

Golden, T. C., Taylor, F. W., Johnson, L. M., Malik, L. H., and Raiswell, C. J. U.S. Patent 6,106,593 (2000).

Gordon, P. A. and Saeger, R. B. (1999) *Ind. Eng. Chem. Res.* 38, 4647.

Gordon, R. D. and Cussler, E. L. (1999a) *Langmuir* 15, 3950.

Gordon, R. D. and Cussler, E. L. (1999b) *AIChE J.* 45, 2313.

Grande, C. A., Gigola, C., and Rodrigues, A. E. (2002) *Ind. Eng. Chem. Res.* 41, 85.

Guerin de Montgareuil, P. and Domine, D. U.S. Patent 3,155,468 (1964).

Gupta, B. K. and Srivastava, O. N. (2001) *Int. J. Hydrogen Energy* 26, 857.

Habgood, H. W. U.S. Patent 2,843,219 (1958).

Hao, J., Tanaka, T., Kita, H. and Okamoto, K. (1997) *J. Membrane Sci.* 132, 97.

Habgood, H. W. (1964) *Canadian J. Chem.* 42, 2340.

Hattori, T., Kaneko, K., Ishikawa, T., and Inouye, K. (1979) *Nippon Kagaku Kaishi* 423.

Herden, H., Einicke, W. D., Schollner, R., and Dyer, A. (1981) *J. Inorg. Nucl. Chem.* 43, 2533.

Herden, H., Einicke, W. D., Schollner, R., Mortier, W. J., Gellens, L. R., and Uytterhoeven, J. B. (1982) *Zeolites* 2, 131.

Hernández-Maldonado, A. J. and Yang, R. T. (2003a) *Ind. Eng. Chem. Res.* 42, 123.

Hernández-Maldonado, A. J. and Yang, R. T. (2003b) *Ind. Eng. Chem. Res.* in press.

Herring, A. M. and McCormick, R. L. (1998) *J. Phys. Chem. B* 102, 3175.

Herron, N. (1986) *Inorg. Chem.* 25, 4714.

Hirscher, M., Mecher, M., Haluska, M., Dettlaff-Weglikowska, U., Quintel, A., Duesberg, G. S., Choi, Y.-M., Downes, P., Hulman, M., Roth, S., Stepanek, I., and Bernier, P. (2001) *Appl. Phys. A* 72, 129.

Hodjati, S., Petit, C., Pitchon, V., and Kiennemann, A. (2001) *J. Catal.* 197, 324.

Hou, P.-X., Yang, Q.-H., Bai, S., Xu, S.-T., Liu, M., and Cheng, H.-M. (2002) *J. Phys. Chem. B* 106, 963.

Howe, R. F. and Lunsford, J. H. (1975a) *J. Am. Chem. Soc.* 97, 18.

Howe, R. F. and Lunsford, J. H. (1975b) *J. Phys. Chem.* 79, 1836.

Huang, H. Y., Padin, J., and Yang, R. T. (1999) *Ind. Eng. Chem. Res.* 38, 2720.

Huang, H. Y. and Yang, R. T. (2001) *Langmuir* 17, 4997.

Huang, Y. (1974) *J. Catalysis* 32, 482.

Hutson, N. D. (2000) Ph.D. Dissertation, University of Michigan.

Hutson, N. D. and Yang, R. T. (2000a) *AIChE J.* 46, 2305.

Hutson, N. D. and Yang, R. T. (2000b) *Ind. Eng. Chem. Res.* 39, 2252.

Hutson, N. D., Rege, S. U., and Yang, R. T. (1999) *AIChE J.* 45, 724.

Hutson, N. D., Reisner, B. A., Yang, R. T., and Toby, B. H. (2000a) *Chem. Mater.* 12, 3020.

Hutson, N. D., Zajic, S. C., and Yang, R. T. (2000b) *Ind. Eng. Chem. Res.* 39, 1775.

Hynek, S., Fuller, W., and Bentley, J. (1997) *Int. J. Hydrogen Energy* 22, 601.

Ichimura, K. and Sano, M. (1992) *J. Vac. Sci. Tech.* A10, 543.

Imamura, S. and Lunsford, J. H. (1985) *Langmuir* 1, 326.

Inouye, K., Nagumo, I., Kaneko, K., and Ishikawa, T. (1982) *Z. Phys. Chem.* 131, 199.

Ioneva, M. A., Newman, G. K., and Harwell, J. M. (1995) *AIChE Symp. Ser.* 91 (No. 309), 40.

Irvine, R. L. U.S. Patent 5,730,860 (1998).

Ishikawa, Y., Austin, L. G., Brown, D. E., Walker, P. L., Jr., (1975) *Chemistry and Physics of Carbon*. (P. L. Walker, Jr., ed.). Dekker, New York, NY.

Jacobs, P. A., Uytterhoeven, J. B., and Beyer, H. K. (1979) *J. Chem. Soc. Faraday Trans. 1* 75, 56.

Jarvelin, H. and Fair, J. R. (1993) *Ind. Eng. Chem. Res.* 32, 2201.

Jayaraman, A., Yang, R. T., Chinn, D., Mohr, D., and Munson, C. L. Unpublished results (2002).

Jeanneret, J. J. (1997) *Handbook of Petroleum Refining Process*. (R. A. Meyers, ed.). McGraw-Hill, New York, NY, p. 2.13.

Jee, J.-G., Kim, M.-B., and Lee, C.-H. (2001) *Ind. Eng. Chem. Res.* 40, 868.

Jiang, S., Zollweg, J. A., and Gubbins, K. E. (1994) *J. Phys. Chem.* 98, 5709.

Jones, R. D., Summerville, D. A., and Basolo, F. (1979) *Chem. Rev.* 79, 139.

Kaneko, K. (1998) *Stud. Surf. Catal.* Vol. 120, 635.

Kaneko, K. and Inouye, K. (1987) *J. Chem. Tech. Biotech.* 37, 11.

Kaneko, K. and Inouye, K. (1988) *Adsorp. Sci. Technol.* 5, 239.

Kaneko, K. and Murata, K. (1997) *Adsorption* 3, 197.

Keller, G. E., Marcinkowsky, A. E., Verma, S. K., and Williamson, K. D. (1992) Olefin recovery and purification via silver complexation. In *Separation and Purification Technology*. (N. N. Li and J. M. Calo, eds.). Marcel Dekker, New York, NY.

Kikkinides, E. S. and Yang, R. T. (1991) *Ind. Eng. Chem. Res.* 30, 1981.

Kim, Y. and Seff., K. (1978a) *J. Am. Chem. Soc.* 100, 6989.

Kim, Y. and Seff, K. (1978b) *J. Am. Chem. Soc.* 100, 175.

King, D. L., Faz, C., and Flynn, T. (2000) SAE Paper 2000-01-0002, Society of Automotive Engineers, Detroit, Michigan, p. 9.

Kington, G. L. and Macleod, A. C. (1959) *Trans. Faraday Soc.* 55, 1799.

Kishida, M., Tachi, T., Kato, A., Yamashita, H., and Miyadera, H. (1991) *Shokubai* 33, 163.

Klein, G. and Vermeulen, T. (1975) *AIChE Symp. Ser.* 71(152), 69.

Knaebel, K. S. and Kandybin, A. U.S. Patent 5,226,933 (1993).

Knudsen, K. G., Cooper, B. H., and Topsoe, H. (1999) *Appl. Catal. A* 189, 205.

Kudin, K. N., Bettinger, H. F., and Scuseria, G. E. (2001) *Phys. Rev. B* 63, 45413.

Kuhl, G. H. (1987) *Zeolites* 7, 451.

Kuhl, G. H. and Sherry, H. S. GB 1,580,928 (U.K. Patent) (1980).

Kuznetsova, A., Yates, J. T., Jr., Liu, J., and Smalley, R. E. (2000) *J. Chem. Phys.* 112, 9590.

Kuznicki, S. M. U.S. Patent 4,938,939 (1990).

Kuznicki, S. M., Bell, V. A., Nair, S., Hillhouse, H. W., Jacubinas, R. M., Braunbarth, C. M., Toby, B. H., and Tsapatsis, M. (2001) *Nature* 412, 720.

Kuznicki, S. M., Bell, V. A., Petrovic, I., and Desai, B. T. U.S. Patent 6,068,682 (2000).

Lamari, M., Aoufi, A., and Malbrunot, P. (2000) *AIChE J.* 46, 632.

Leavitt, F. W. U.S. Patent 5,074.892 (1992).

Lee, C.-H., Yang, J., and Ahn, H. (1999) *AIChE J.* 45, 535.

Lee, H., Kang, Y.-S., Kim, S.-H., and Lee, J.-Y. (2002) *Appl. Phys. Lett.* 80, 577.

Lee, S. M., and Lee, Y. H., (2000) *Appl. Phys. Lett.* 76, 2877.

Lewis, I. C., Greinke, R. A., and Strong, S. L. (1993) Carbon 1993 Extended Abstracts, 21st Biennial Carbon Conference, Buffalo, p. 490.

Li, G. Q. and Govind, R. (1994) *Ind. Eng. Chem. Res.* 33, 755.

Li, N. N. U.S. Patent 3,410,794 (1968).

Li, N. N. (1971) *Ind. Eng. Chem. Proc. Des. Dev.* 10, 215.

Li, W. B., Yang, R. T., Krist, K., and Regalbuto, J. R. (1997) *Energy & Fuels* 11, 428.

Li, X., Zhu, H., Ci, L., Xu, C., Mao, Z., Wei, B., Liang, J., and Wu, D. (2001) *Carbon* 39, 2077.

Libowitz, G. G., Hayes, H. F., and Gibbs, T. R. P., Jr. (1958) *J. Phys. Chem.* 76, 62.

Liu, C., Fan, Y. Y., Liu, M., Cong, H. T., Cheng, H. M., and Dresselhaus, M. S. (1999) *Science* 286, 1127.

Liu, Y. L., Du, H. B., Xiao, F. S., Zhu, G. S., and Pang, W. Q. (2000) *Chem. Mater.* 12, 665.

Lozano-Castello, D., Cazorla-Amoros, D., Linares-Solano, A., and Quinn, D. F. (2002) *Carbon* 40, 989.

Lueking, A. and Yang, R. T. (2002) *J. Catal.* 206, 165.

Lueking, A. and Yang, R. T. (2003) *AIChE. J.* in press.

Lund, C. R. F., Schorfheide, J. J., and Dumesic, J. A. (1979) *J. Catal.* 57, 105.

Lunsford, J. H. (1975) *Catal. Rev.-Sci. Eng.* 12(2), 137.

Ma, X., Sakanishi, K., and Mochida, I. (1994) *Ind. Eng. Chem. Res.* 33, 218.

Ma., X., Sprague, M., Sun, L., and Song, C. (2002a) *ACS Petroleum Chemical Division, Preprints* 47(1), 48.

Ma, X., Sun, L. and Song, C. (2002b) *Catal. Today* 77, 107.

Ma, X., Sun, L., Yin, Z., and Song, C. (2001) *ACS Fuel Chemistry Div. Preprints* 46(1), 648.

Machida, M., Ogata, S., Yasuoka, K., Eguchi, K., and Arai, H. In Proc. 10th Int. Congress on Catalysis, L. Guczi, F. Solymosi, P. Tétényi, eds., Akadémiai Kiadó, Budapest, (1992), 2645.

Markarshin, L. L., Andreev, D. V. and Parmon, V. N. (1997) *Chem. Phys. Lett.* 266, 173.

Manes, M. (1998) *Encyclopedia of Environmental Analysis and Remediation*. (R. A. Meyers, ed.). Wiley, New York, NY, pp. 26–68.

Manzi, S., Valladares, D., Marchese, J., and Zgrablich, G. (1997) *Ads. Sci. Tech.* 15, 301.

Mastranga, K. R., Myers, A. L., and Glandt, E. D. (1992) *Chem. Eng. Sci.* 47, 1569.

Matz, M. J. and Knaebel, K. S. (1990) *Separ. Sci. Tech.* 25, 961.

McKee, D. W. U.S. Patents 3,140,932 and 3,140,933 (1964).

Meregalli, V. and Perrinello, M. (2001) *Appl. Phys.* A72, 143.

Menon, P. G. (1968) *Chem. Rev.* 68, 277.

Menon, V. C. and Komarneni, S. (1998) *J. Porous Mater.* 5, 43.

Mei, H., Mei, B. W. and Yen, T. F. (2003) *Fuel*, 82, 405.

Michlmayr, M. J. U.S. Patent 4,188,285 (1980).

Miller, C., Moreau, S., Kraus, G., and Gabillard, J.-P. EP 0995477 (2000).

Miller, G. W. (1987) *AIChE Symp. Ser.* Vol. 83 (No. 259), 28.

Miller, G. W., Knaebel, K. S., and Ikels, K. G. (1987) *AIChE J.* 33, 194.

Milton, R. M. U.S. Patent 2,882,243 (1959).

Misono, M. (1987) *Catal. Rev. Eci. Eng.* 29, 269.

Niwa, M., Furukawa, Y., and Murakami, Y. (1982) *J. Coll. Interf. Sci.* 86, 260.

Mochida, I., Kawabuchi, Y., Kawano, S., Matsumura, Y., and Yoshikawa, M. (1977) *Fuel* 76, 543.

Mochida, I., Kisamori, S., Hironaka, M., Kawano, S., Matsumura, Y., and Yoshikawa, M. (1994) *Energy Fuels* 8, 1341.

Mota, J. P. B. (1999) *AIChE J.* 45, 986.

Mota, J. P. B. and Rodrigo, A. J. S. (2002) *Fundamentals of Adsorption*. (K. Kaneko et al., ed.). IK International, Chiba City, Japan, p. 125.

Mota, J. P. B., Rodrigues, A. E., Saatdjian, E., and Tondeur, D. (1997) *Adsorption* 3, 117.

Mullhaupt, J. T., BeVier, W. E., McMahon, K. C., Van Slooten, R. A., Lewis, I. C., Grienke, R. A., Strong, S. L., Ball, D. R., and Steele, W. E. Carbon '92, Proceedings Carbon Conference in Essen (1992), p. 367.

Myers, A. L. and Glandt, E. D. (1993) *Adsorp. News, Quantachrome Corp*. 4(1), 3.

Niederhoffer, E. C., Timmons, J. H., and Martell, A. E. (1994) *Chem. Rev.* 84, 137.

Niwa, M., Furukawa, Y. and Murakami, Y. (1982) *J. Coll. Interf. Sci.* 86, 260.

Notaro, F., Ackley, M. W., and Smolarek, J. (1999) *Chem. Eng.* April Issue, 104.

Notaro, F., Schaub, H. R., and Kingsley, J. P. Proceedings 2nd Joint China/USA Chemical Engineering Conference, Beijing, China, AIChE, New York (1997), p. 558.

Ondrey, G. (1999) *Chem. Eng.* 106(4), 25.

Otowa, Y. U.S. Patent 5,064,805 (1991).

Otto, R. and Shelef, M. (1970) *J. Catal.* 18, 184.

Ozeki, S., Tazaki, T., Matsubara, Y., Miyamoto, J., and Sato, H. (2000) *Adsorption Science and Technology*. (D. D. Do, Ed.). World Scientific Publishers, River Edge, NJ, p. 492.

Ozin, G. A., Baker, M. D., and Godber, J. (1984) *J. Phys. Chem.* 88, 4902.

Padin, J., Rege, S. U., Yang, R. T., and Cheng, L. S. (2000) *Chem. Eng. Sci.* 55, 4525.

Park, C., Anderson, P. E., Chambers, A., Tan, C. D., Hidalgo, R., and Rodriguez, N. M. (1999) *J. Phys. Chem. B* 103, 10572.

Park, J.-H., Kim, J.-N., Cho, S.-H., Kim, J.-D., and Yang, R. T. (1998) *Chem. Eng. Sci.* 53, 3951.

Parkinson, G. (2001) Chem. Eng., February, 37.

Philippou, A. and Anderson, M. W. (1996) *Zeolites* 16, 98.

Pigorini, G. and LeVan, M. D. (1997) *Ind. Eng. Chem. Res.* 36, 2296.

Pinkerton, F. E., Wicke, B. G., Olk, C. H., Tibbetts, G. G., Meisner, G. P., Meyer, M. S., and Herbst, J. F. (2000) *J. Phys. Chem. B* 104, 9460.

Poirier, E., Chahine, R., and Bose, T. K. (2001) *Int. J. Hydrogen Energy* 26, 831.

Pope, M. T. (1983) *Heteropoly and Isopoly Oxometalates*. Springer-Verlag, New York, NY.

Purdy, G. A. (1958) *Petroleum Refining—Treating*. In *Petroleum*. McGraw-Hill, New York, NY, p. 216.

Quinn, D. F. and MacDonald, J. A. (1987) Natural Gas Adsorbents, Report to Ministry of Transportation, Ontario, Canada.

Quinn, D. F., MacDonald, J. A., and Sosin, K. (1994) *ACS Division Fuel Chemical Preprints*, 39, 451.

Ramachandran, R. and Dao, L. H. U.S. Patent 5,365,011 (1994).

Ramprasad, D., Pez, G. P., Toby, B. H., Markley, T. J., and Pearlstein, R. M. (1995) *J. Am. Chem. Soc.* 117, 10694.

Rao, A. M., Eklund, P. C., Bandow, S., Thess, A., and Smalley, R. E. (1997) *Nature* 388, 257.

Redmond, J. P. and Walker, P. L., Jr., (1960) *J. Phys. Chem.* 64, 1093.

Rege, S. U., Padin, J., and Yang, R. T. (1998) *AIChE J.* 44, 799.

Rege, S. U. and Yang, R. T. (1997) *Ind. Eng. Chem. Res.* 36, 5358.

Rege, S. U. and Yang, R. T. (2002) *Chem. Eng. Sci.* 57, 1139.

Reilly, J. J. and Wiswall, R. H. (1968) *Inorg. Chem.* 7, 2254.

Riggleman, R. A., Wang, J., Ebner, A. D., Ritter, J. A. and Zidan, R. (2002), presented at American Institute of Chemical Engineers Annual Meeting, November 4, 2002, Indianapolis, IN, paper 349d.

Rodriguez, F. (1999) *Principles of Polymer Systems*, 4th Ed. Taylor & Francis, London, UK.

Rodriguez, N. M. and Baker, R. T. K. U.S. Patent 5,653,951 (1997).

Rodriguez, N. M., Chambers, A., and Baker, R. T. K. (1995) *Langmuir* 11, 3862.

Rubel, A. M. and Stencel, J. M. (1997) *Fuel* 76, 521.

Rzepka, M., Lamp, P., and de la Casa-Lillo, M. A. (1998) *J. Phys. Chem.* 102, 10 894.

Safarik, D. J. and Eldridge, R. B. (1998) *Ind. Eng. Chem. Res.* 37, 2571.

Salem, A. S. H. (1994) *Ind. Eng. Chem. Res.* 33, 336.

Salem, A. S. H. and Hamid, H. S. (1997) *Chem. Eng. Technol.* 20, 342.

Sandrock, G., Gross, K. and Thomas, G. (2002) *J. Alloys Comp.* 339, 299.

Sastri, M. V. C., Viswanathan, B., and Srinivasa Murthy, S. (1998) *Metal Hydrides: Fundamentals and Applications*. Narosa Publishing House, New Delhi, and Springer-Verlag, Berlin, Germany.

Schlapbach, L., ed., (1988) *Hydrogen in Intermetallic Compounds I*. Springer-Verlag, Berlin.

Schlapbach, L. and Zuttel, A. (2001) *Nature* 414, 353.

Schollner, R., Broddack, R., and Jusek, M. (1981) *Z. Phys. Chemie, Leipzig* 262, 362.

Schwartz, J. A. Proceedings 1993 DOE/NREL Hydrogen Program Review (1993), p. 89.

Segawa, K., Chen, Y., Kubsh, J. E., Delgass, W. N., Dumesic, J. A., and Hall, W. K. (1982) *J. Catal.* 76, 112.

Seifert, G., Kohler, T., and Frauenheim, T. (2000) *Appl. Phys. Lett.* 77, 1313.

Sherman, A. and Eyring, H. (1932) *J. Am. Chem. Soc.* 54, 2661.

Simonyan, V. V., Diep, P., and Johnson, J. K. (1999) *J. Chem. Phys.* 111, 9778.

Simonyan, V. V. and Johnson, J. K. (2002) *J. Alloys Compounds* 330–332, 659.

Sircar, S., Conrad, R. R. and Ambs, W. J. (1985) U.S. Patent 4,557,736.

Skarstrom, C. W. U.S. Patent 2, 944,627 (1960).

Solimosi, F. and Kiss, J. (1976) *J. Catal.* 41, 202.

Sosin, K. A. and Quinn, J. A. (1995) *J. Porous Mater.* 1, 1.

Stan, G. and Cole, M. W. (1998) *J. Low Temp. Phys.* 110, 539.

Strobel, R., Jorissen, L., Schliermann, T., Trapp, V., Schutz, W., Bohmhammel, K., Wolf, G., and Garche, J. (1999) *J. Power Source* 84, 221.

Sun, J., Rood, M. J., Rostam-Abadi, M., and Lizzo, A. A. (1996) *Gas Sep. Purif.* 10, 91.

Sun, T. and Seff, K. (1994) *Chem. Rev.* 94, 857.

Sutter, D. H. and Flygare, W. H. (1969) *J. Am. Chem. Soc.* 91, 4063.

Tabata, K., Fukuda, H., Kohiki, S., and Misono, M. (1988) *Chem. Lett.* 799.

Talu., O., Zhang, S. Y., and Hayhurst, D. T. (1993) *J. Phys. Chem.* 97, 12894.

Takahashi, A. and Yang, R. T. (2002) *AIChE J.* 48, 1457.

Takahashi, A., Yang, R. T., Munson, C. L., and Chinn, D. (2001) *Langmuir* 17, 8405.

Takahashi, A., Yang, F. H., and Yang, R. T. (2002) *Ind. Eng. Chem. Res.* 41, 2487.

Tamura, T. U.S. Patent 3,797,201 (1974).

Tan, Z. and Gubbins, K. E. (1990) *J. Phys. Chem.* 94, 6061.

Taylor, R. J., Drago, R. S., and George, J. E. (1989) *Inorg. Chem.* 111, 6610.

Taylor, R. J., Drago, R. S., and Hage, J. P. (1992) *Inorg. Chem.* 31, 253.

Tibbetts, G. G., Meisner, G. P., and Olk, C. H. (2001) *Carbon* 39, 2291.

Tremblay, G., Vastola, F. J., and Walker, P. L., Jr. (1978) *Carbon* 16, 35.

U.S. Department of Energy, Proceedings of the 1998 U.S. DOE Hydrogen Program Review, August, 1998. NREL/CP-570-2315, Vol. 1.

Vansant, E. F., Enody, E. M., and Mingels, W. (1988) *European Patent Appl.* 275, 855.

Vaska, L. (1976) *Acct. Chem. Res.* 9, 175.

Wanatabe, K., Soma, M., Ohishi, T., and Tamaru, K. (1971) *Nature* 233, 160.

Wang, J. H. (1970) *Acct. Chem. Res.* 3, 90.

Wang, Q. and Johnson, J. K. (1999a) *J. Phys. Chem. B* 103, 4809.

Wang, Q. and Johnson, J. K. (1999b) *J. Chem. Phys. B* 110, 577.

Wankat, P. C. and Tondeur, D. (1985) *AIChE Symp. Ser.* 81(242), 74.

Watson, C. F., Whitley, R. D., and Meyer, M. I. U.S. Patent 5,529,610 (1996).

Wegrzyn, J., Wiesmann, H., and Lee, T. SAE Proceedings 1992, Automotive Technology, Dearborn, Michigan, 1992.

Weitkamp, J., Schwark, M., and Ernst, S. (1991) *J. Chem. Soc. Chem. Commun.* 1133.

Wennerberg, A. N. and O'Grady, T. M. U.S. Patent 4,082,694 (1978).

Whitehurst, D. D., Isoda, T., and Mochida, I. (1998) *Adv. Catal.* 42, 345.

Williams, K. A. and Eklund, P. C. (2000) *Chem. Phys. Lett.* 320, 352.

Yamashita, T., Nishiyama, T., Inoue, T., Takahashi, M., and Horii, Y. (2002) *Fundamentals of Adsorption 7*. (K. Kaneko et al., ed.). International Adsorption Society, IK International, Chiba City, Japan, p. 271.

Yang, F. H. and Yang, R. T. (2002) *Carbon* 40, 437.

Yang, R. T. (1987) *Gas separation by adsorption processes*. Butterworth, Boston, MA; (1997) Imperial College Press/World Scientific Publishers, River Edge, New Jersey.

Yang, R. T. U.S. Patent 5,248,321 (1993).

Yang, R. T. and Chen, N. U.S. Patent, 5,456,892 (1995).

Yang, R. T. and Chen, N. (1994) *Ind. Eng. Chem. Res.* 33, 825.

Yang, R. T., Chen, Y. D., Peck, J. D., and Chen, N. (1996) *Ind. Eng. Chem. Res.* 35, 3093.

Yang, R. T. and Doong, S. J. (1985) *AIChE J.* 31, 1829.

Yang, R. T. and Kikkinides, E. S. (1995) *AIChE J.* 41, 509.

Yang, R. T. and Krist, K. U.S. Patent 6,033,461 (2000).

Yang, R. T., Padin, J., and Rege, S. U. U.S. Patent 6,423,881 (2002) (Filed Oct. 1998).

Yang, R. T., Takahashi, A., and Yang, F. H. (2001) *Ind. Eng. Chem. Res.* 40, 6236.

Yang, R. T., Takahashi, A., Yang, F. H. and Hernández-Maldonado, A. J. Selective Sorbents for Desulfurization of Liquid Fuels, U.S. and foreign Patent Application (2002).

Yao, H. C. and Shelef, M. (1978) *The Catalytic Chemistry of Nitrogen Oxides*. (R. L. Klimisch and J. G. Larson, eds.). Plenum, New York, NY, p. 45.

Ye, Y., Ahn, C. C., Witham, C., Fultz, B., Liu, J., Rinzler, A. G., Colbert, D., Smith, K. A., and Smalley, R. E. (1999) *Appl. Phys. Lett.* 74, 2307.

Yen, T. F., Mei, H. and Lu, S. H. U.S. Patent 6,402,939 (2002).

Yong, Z., Mata, V., and Rodrigues, A. E. (2002) *Separ. Purif. Technol.* 26, 195.

Yoshida, A., Hirano, S., Harada, A., and Nakano, M. (2001) *Macrop. Mesop. Mater.* 46, 203.

Yuen, S., Chen, Y., Kubsh, J. E., Dumesic, J. A., Topsøe, N., and Topsøe, H. (1982) *J. Phys. Chem.* 86, 3022.

Zaluska, A., Zaluski, L., and Strom-Olsen, J. O. (2001) *Appl. Phys.* A72, 157.

Zhang, S. Y., Talu, O., and Hayhurst, D. T. (1991) *J. Phys. Chem.* 95, 1722.

Zhang, W. X., Yahiro, H., Mizuno, N., Izumi, J., and Iwamoto, M. (1993) *Langmuir* 9, 2337.

Zhou, L. and Zhou, Y. P. (1998) *Chem. Eng. Sci.* 53, 2531.

Zhu, H., Cao, A., Li, X., Xu, A., Mao, Z., Ruan, D., Liang, J., and Wu, D. (2001) *Appl. Surface Sci.* 178, 50.

Zhu, H., Chen, A., Mao, Z., Xu, A., Xiao, X., Wei, B., Liang, J., and Wu, D. (2000) *J. Mater. Sci. Lett.* 19, 1237.

Zidan, R. A., Takara, S., Hee, A. G., and Jensen, C. M. (1999) *J. Alloys Compounds* 285, 119.

Züttel, A. et al. (2002) *J. Alloys Compounds* 330–332, 676.

AUTHOR INDEX

Adsorbents: Fundamentals and Applications, Edited By Ralph T. Yang
ISBN 0-471-29741-0

SUBJECT INDEX

Adsorbents: Fundamentals and Applications, Edited By Ralph T. Yang
ISBN 0-471-29741-0